STEFANSSON, DR. ANDERSON AND THE CANADIAN ARCTIC EXPEDITION, 1913–1918

Figure 1 Vilhjalmur Stefansson and Dr. Rudolph M. Anderson on the wharf at Esquimalt, B.C., June 1913.
Photo: Library and Archives Canada, C-004805

STEFANSSON, DR. ANDERSON AND THE CANADIAN ARCTIC EXPEDITION, 1913–1918

A Story of Exploration, Science and Sovereignty

Stuart E. Jenness

MERCURY SERIES
HISTORY PAPER 56
CANADIAN MUSEUM OF CIVILIZATION

Published by the

Canadian Museum of Civilization Corporation
(CMCC)
100 Laurier Street
Gatineau, Quebec
K1A 0M8

Coordinator, Publishing: Ginette Côté
Mercury Series Design: Hangar 13
Production artist: Interscript
Printer: St. Joseph Print Group

Front cover photograph: The Canadian Arctic scientists at Nome, Alaska, July 1913.
Photo: © Canadian Museum of Civilization, 27790.

Back cover photographs
Top: Departure of *Mary Sachs* for Banks Island
Photo: © Canadian Museum of Civilization, photo by Kenneth Gordon Chipman, 43249
Bottom: The author and William McKinlay in Glasgow.
Photo: Courtesy of S.E. Jenness, photo by Jean M. Jenness.

Library and Archives Canada Cataloguing in Publication

Jenness, Stuart E. (Stuart Edward), 1925–

Stefansson, Dr. Anderson and the Canadian Arctic Expedition, 1913–1918: a story of exploration, science and sovereignty / Stuart E. Jenness.

(Mercury series)

(History paper; 56)

Includes bibliographical references and index.
ISBN 978-0-660-19971-9
Cat. no.: NM24-28/2011E

1. Stefansson, Vilhjalmur, 1879–1962. 2. Anderson, Rudolph Martin, 1876–1961. 3. Explorers—Arctic regions—Biography. 4. Canadian Arctic Expedition (1913–1918). 5. Scientific expeditions—Arctic regions—History. 6. Arctic regions—Discovery and exploration—Canadian. I. Canadian Museum of Civilization II. Title. III. Series: Mercury series IV. Series: History paper (Canadian Museum of Civilization) 56

FC3962.9 C35 J45 2011   910.92   C2011-980033-0

Object of the Mercury Series

This series is designed to permit the rapid dissemination of information pertaining to the disciplines in which the Canadian Museum of Civilization Corporation is active. Considered an important reference by the scientific community, the Mercury Series comprises over 400 specialized publications on Canada's history and prehistory. Due to its specialized audience, the series consists largely of monographs published in the language of the author.

But de la collection Mercure

La collection Mercure vise à diffuser rapidement le résultat de travaux dans les disciplines qui relèvent des sphères d'activités du Musée canadien des civilisations. Considérée comme un apport important dans la communauté scientifique, la collection Mercure présente plus de 400 publications spécialisées portant sur l'héritage canadien préhistorique et historique. Comme la collection s'adresse à un public spécialisé, celle-ci est constituée essentiellement de monographies publiées dans la langue des auteurs.

How to Obtain Mercury Series Titles

Email: publications@civilization.ca
Web: cyberboutique.civilization.ca
Telephone: (819) 776-8387 or toll-free, in
            North America only, 1 800 555-5621
Mail: Mail Order Services
            Canadian Museum of Civilization
            100 Laurier Street
            Gatineau, Quebec K1A 0M8

Comment se procurer les titres parus dans la collection Mercure

Courriel: publications@civilisations.ca
Web: cyberboutique.civilisations.ca
Téléphone: (819) 776-8387 ou sans frais, en
            Amérique du Nord seulement,
            1 800 555-5621
Poste: Service des commandes postales
            Musée canadien des civilisations
            100, rue Laurier
            Gatineau (Québec) K1A 0M8

Canada

To **Vilhjalmur Stefansson**, who had the vision for the Expedition, discovered several large Arctic islands, survived unbelievable journeys over frozen seas and wrote a highly popular account of the Canadian Arctic Expedition, 1913–1918;

To **Captain Robert Bartlett**, master of the Expedition's flagship, *Karluk*, whose heroic efforts resulted in the rescue of most of the survivors after the sinking of that ship;

To **Dr. Rudolph M. Anderson**, who led scientists of the Southern Party to the successful completion of their various field projects, guided the many resulting scientific reports through publication and left a wealth of documents about the Expedition;

To **William L. McKinlay**, the lone scientist to survive the sinking of the *Karluk* and subsequent Wrangel Island experiences, who encouraged me to write this account of the Canadian Arctic Expedition to "set the record straight"; and

To my father, **Diamond Jenness**, friend and admirer of the Copper Inuit, without whom I could not have written the following account of the Expedition that ended "the Heroic Age of Discovery."

ABSTRACT

In June 1913, the Dominion of Canada sent its first large exploration and scientific expedition into the Arctic, led by the ethnologist/explorer Vilhjalmur Stefansson. For the next three years, the Canadian Arctic Expedition, 1913–1916, was to carry out exploration and scientific work in the Beaufort Sea, planting the Union Jack on any new land it discovered as an indication of Canadian sovereignty. Additionally, it was to undertake methodical scientific studies on the coastal mainland between Cape Parry and Bathurst Inlet.

The Arctic ice quickly trapped the Expedition's flagship *Karluk*, carrying it northwestward for four months before crushing and sinking it. All twenty-five persons on board and a cat safely got off the ill-fated vessel, but eight men subsequently perished attempting to reach Siberia. The remaining seventeen reached Wrangel Island safely, but three of them perished before rescuers arrived, following a heroic journey by the *Karluk's* Captain Bob Bartlett to Siberia and Nome to alert authorities of the survivors' location.

Stefansson had left the *Karluk* to hunt caribou before it was carried westward in the ice. Deprived of ship, men and supplies for his Northern (Exploration) Party, Stefansson sledded eastward across the northern Alaskan coast, obtaining replacement personnel and supplies. He and his men then discovered all but one of the remaining large Arctic islands north of the North American continent, and carried out much of the exploration work he originally intended to do. In doing so, however, Stefansson ignored government instructions to return south in 1916 and remained in the North for another two years.

Meanwhile, zoologist Dr. Rudolph M. Anderson, in charge of the Expedition's Southern (Scientific) Party, with several scientists from the Geological Survey of Canada, successfully completed the methodical coastal mapping of much of the central Arctic mainland, its copper mineral potential, Copper Inuit studies, and biological research they had been assigned. In spite of many adversities, they returned south as expected in 1916. Over the next ten years Dr. Anderson was heavily involved in coordinating and editing more than sixty-four scientific government reports in thirteen volumes, which resulted almost exclusively from the work of the Southern Party scientists.

The name of the Expedition was changed in 1919, over Dr. Anderson's protest, to the "Canadian Arctic Expedition, 1913–1918," to include Stefansson's extended stay in the North.

This is the first comprehensive and authoritative account of the entire Canadian Arctic Expedition and its somewhat turbulent aftermath. It emphasizes for the first time the valuable contributions made by the scientists of the Southern Party and their dedicated leader, Dr. Anderson, to the history and knowledge of the central Canadian Arctic.

RÉSUMÉ

En juin 1913, le Dominion du Canada envoya sa première grande expédition scientifique en Arctique, expédition dirigée par l'ethnologue et explorateur Vilhjalmur Stefansson. Ainsi, de 1913 à 1916, l'Expédition canadienne dans l'Arctique avait pour but d'explorer la mer de Beaufort, d'y effectuer des recherches et de planter l'*Union Jack* sur tout nouveau territoire découvert afin d'affirmer la souveraineté du pays. Elle devait en outre mener une étude rigoureuse des côtes entre le cap Parry et Bathurst Inlet.

Bientôt, la glace emprisonna le vaisseau amiral de l'Expédition, le *Karluk*, et le fit dériver vers le nord-ouest pendant quatre mois avant de le broyer et de le faire couler. Les membres de l'équipe, 25 personnes accompagnées d'un chat, purent quitter en toute sécurité l'infortuné navire, mais huit hommes périrent par la suite en tentant de parvenir en Sibérie. Les 17 autres atteignirent sans encombre l'île Wrangel. Le courageux capitaine du *Karluk*, Bob Bartlett, entreprit un périlleux voyage en Sibérie et à Nome, en Alaska, pour alerter les autorités et leur indiquer l'endroit où se trouvaient les survivants. Malgré ses démarches, trois autres rescapés moururent.

Avant que le bateau ne soit emporté vers l'ouest par les glaces, Stefansson avait quitté le *Karluk* pour chasser le caribou. Privé du navire et des provisions embarquées en vue de l'Expédition nordique et coupé de son équipe, Stefansson se dirigea en traîneau vers l'est, le long de la côte septentrionale de l'Alaska, où il engagea du personnel et se procura des vivres. Avec ses nouveaux équipiers, il découvrit toutes les grandes îles arctiques encore inconnues situées au nord du continent américain, sauf une, et poursuivit une grande partie de l'exploration prévue à l'origine. Ce faisant, toutefois, Stefansson ne se conforma pas aux directives du gouvernement qui lui avait enjoint, en 1916, de retourner au sud. Il passa donc deux années de plus dans le Nord.

Pendant ce temps, le zoologiste Rudolph M. Anderson, qui dirigeait l'équipe du Sud regroupant plusieurs scientifiques de la Commission géologique du Canada, dressa avec succès la cartographie détaillée de la plupart des côtes de la partie continentale de l'Arctique et évalua leur potentiel minier, notamment en ce qui touche le minerai de cuivre. En outre, il réalisa des études sur les Inuits du Cuivre, ainsi que les recherches biologiques qu'on lui avait confiées. En dépit de nombreux revers, l'équipe retourna au sud comme prévu en 1916. Au cours des dix années suivantes, Rudolph Anderson participa activement à l'élaboration et à la publication de plus de 64 rapports scientifiques gouvernementaux réunis en 13 volumes : ces documents décrivaient presque exclusivement le fruit du travail des chercheurs de l'équipe du Sud.

À la suite de protestations formulées par Anderson, le nom de l'Expédition fut changé en 1919 et devint «Expédition canadienne dans l'Arctique de 1913-1918», afin qu'il désigne également le séjour prolongé de Stefansson dans le Nord.

Le présent ouvrage constitue le premier compte rendu détaillé et rigoureux de toute l'Expédition canadienne dans l'Arctique et de ses suites quelque peu tumultueuses. On y souligne pour la première fois la contribution exceptionnelle des scientifiques de l'équipe du Sud et de leur chef dévoué, Rudolph Anderson, à l'histoire et à la connaissance du centre de l'Arctique canadien.

CONTENTS

Abstract..vii

Résumé..viii

List of Figures and Maps..xi

Preface..xiii

Acknowledgments..xxiii

PART 1. STEFANSSON'S NEW ARCTIC EXPEDITION

Chapter 1. Preparations...3

Chapter 2. Victoria and Nome..17

Chapter 3. "Goodbye, Stefansson"...33

PART 2. THE *KARLUK* SAGA

Chapter 4. "Adios, *Karluk*"...43

Chapter 5. Shipwreck Camp to Wrangel Island...55

Chapter 6. Captain Bartlett's Journey to Alaska...65

Chapter 7. Survival on Wrangel Island...77

Chapter 8. Rescue..91

PART 3. STEFANSSON'S NORTHERN PARTY

Chapter 9. Revised Exploration Plans, 1913–1914 101

Chapter 10. The Search for "Crocker Land," 1914 117

Chapter 11. Banks Island, 1914–1915 127

Chapter 12. First Discovery of Land, 1915..143

Chapter 13. Enter the *Polar Bear*, 1915..157

Chapter 14. More Discoveries of Land, 1916...165

Chapter 15. Stefansson's Final Explorations, 1917 189

Chapter 16. Stefansson's Skirmish with Death, 1917–1918 207

Chapter 17. Storkerson's Drift on the Arctic Ice, 1918 211

PART 4. DR. ANDERSON'S SOUTHERN PARTY

Chapter 18. Northern Alaska, 1913 219

Chapter 19. Northern Alaska to Bernard Harbour, 1914 225

Chapter 20. Bernard Harbour, 1915 253

Chapter 21. Farewell to the Arctic, 1916 285

PART 5. AFTERMATH

Chapter 22. Publications and Other Post-Expedition Matters 307

Chapter 23. The Stefansson–Dr. Anderson Feud 339

Appendix 1. Personnel on the Canadian Arctic Expedition, 1913–1918 355

Appendix 2. Geographical Names and the Canadian Arctic Expedition, 1913–1918 361

Appendix 3. The Collections of the Canadian Arctic Expedition 365

Appendix 4. Reports of the Canadian Arctic Expedition, 1913–1918 369

References 375

Index 389

LIST OF FIGURES AND MAPS

Figure 1.	Stefansson and Dr. Anderson	Frontispiece
Figure 2.	The author and William McKinlay in Glasgow	xv
Figure 3.	The *Karluk* at Esquimalt	17
Figure 4.	Captain Robert Bartlett	19
Figure 5.	The scientists at Victoria	21
Figure 6.	Stefansson and his men on the *Karluk*	25
Figure 7.	The scientists at Nome	29
Figure 8.	The Iñupiaq Kuraluk and his family	36
Figure 9.	Stefansson leaving the *Karluk*	39
Figure 10.	Looking for the *Karluk* from Spy Island	40
Figure 11.	Camp at Cape Waring, Wrangel Island	82
Figure 12.	McKinlay helping Kuraluk construct a kayak	84
Figure 13.	Raising the flag at Rodger's Harbor	86
Figure 14.	The cat mascot on Wrangel Island	87
Figure 15.	Rescue party and survivors at Cape Waring	95
Figure 16.	The *King and Winge* with Wrangel Island survivors	96
Figure 17.	Stefansson reading at Collinson Point	114
Figure 18.	Departure from Collinson Point on Stefansson's first ice trip	115
Figure 19.	Pannigabluk and her son at Martin Point	119
Figure 20.	Johansen taking the seawater temperature north of Martin Point	121
Figure 21.	Stefansson at Cape Kellett	133
Figure 22.	Storkerson and Natkusiak preparing a sled boat, M'Clure Strait	146
Figure 23.	The *Polar Bear* camp, Prince of Wales Strait	162
Figure 24.	Storkerson family near Cape Armstrong	163
Figure 25.	Some of Northern Party men at Cape Kellett	166
Figure 26.	Guninana and Uttaktuak near Cape Armstrong	169
Figure 27.	Snowhouses and cliffs, Liddon Gulf, Melville Island	170
Figure 28.	Stefansson's cairn on Meighen Island	179
Figure 29.	Inuit women and children at Walker Bay	202
Figure 30.	*Polar Bear* camp on Barter Island, 1918	203
Figure 31.	CAE house at Collinson Point	221
Figure 32.	Christmas dinner at Collinson Point	237
Figure 33.	Three CAE schooners at Herschel Island	237
Figure 34.	CAE personnel at Herschel Island	239
Figure 35.	Departure of *Mary Sachs* for Banks Island	241
Figure 36.	Newly built CAE house at Bernard Harbour	255
Figure 37.	Wilkins, Crawford and Billy Natkusiak near Cape Bexley	265
Figure 38.	Snowhouse and Copper Inuit sleds, Coronation Gulf	267
Figure 39.	Chipman and Patsy Klengenberg set forth to survey Bernard Harbour	270
Figure 40.	Dr. Anderson and Patsy Klengenberg preserving specimens	271
Figure 41.	Copper Inuit fishing camp near mouth of Tree River	271
Figure 42.	Inuk girl, Mingeouk, and large lake trout at Tree River	273
Figure 43.	Schooner *North Star* leaving lonely base camp at Bernard Harbour	275
Figure 44.	*North Star* unloading supplies at Cape Barrow	277

Figure 45. Cox taking latitude reading near Cape Barrow 277
Figure 46. Chipman's camp near Kater Point.. 278
Figure 47. Cox, O'Neill and Chipman at base of 1839 monument........................... 281
Figure 48. Dr. Anderson and Siberia Mike in canyon of Croker River 287
Figure 49. Young Copper Inuit girl wearing priest's black robe 293
Figure 50. Copper Inuit heading inland towards Great Bear Lake........................... 294
Figure 51. Copper Inuit spearing salmon in fishing creek, Bernard Harbour 295
Figure 52. Ikpukhuak and Higilak in dress attire, Bernard Harbour 296
Figure 53. Jennie Kannayuk and Kila Arnauyuk in dress attire, Bernard Harbour 298
Figure 54. Southern Party leaving Bernard Harbour in 1916................................. 299
Figure 55. Aerial view of Bernard Harbour 300
Figure 56. Klengenberg family at Baillie Islands 301
Figure 57. Siberia Mike, Ikey Bolt and Mungalina at Baillie Islands 301
Figure 58. Cox taking final solar readings at the Canada-Alaska boundary................ 302
Figure 59. CAE schooner *Alaska* on the beach, Nome 303
Figure 60. CAE men and friends on coastal steamer heading south 304
Figure 61. Copper Inuit, RNWMP officers and judicial officials 304
Figure 62. Victoria Memorial Museum, Ottawa 308
Figure 63. Four deputy ministers.. 331
Figure 64. Plaque commemorating the CAE men who died 335

Map 1. Route of the *Karluk*, 1913–1914... 27
Map 2. Locations of Nome, Teller and Port Clarence 34
Map 3. Wrangel Island and Herald Island.. 61
Map 4. Captain Bartlett's route from Wrangel Island to Alaska 66
Map 5. The Alaskan north coast.. 105
Map 6. Camden Bay 118
Map 7. Route of Stefansson's 1914 ice trip on the Beaufort Sea.......................... 144
Map 8. Route of Stefansson's 1915 ice trip on the Beaufort Sea.......................... 148
Map 9. Brock, Mackenzie King and Borden Islands 176
Map 10. Route of Stefansson's 1916 ice trip 182
Map 11. Route of Stefansson's 1917 ice trip 191
Map 12. Storkerson's ice drift north of Alaska 213
Map 13. The Firth River and Herschel Island, mapped by Cox and O'Neill.......... 227
Map 14. The Mackenzie River Delta, mapped by Chipman, Cox and O'Neill....... 230
Map 15. Coast from Baillie Islands to Bathurst Inlet....................................... 257
Map 16. Cox's traverse from Bernard Harbour to Rae River.............................. 260
Map 17. Jenness' route about southwestern Victoria Island 262
Map 18. Southern Party's route around Bathurst Inlet, summer 1915 279
Map 19. Southern Party's route around Bathurst Inlet, spring 1916 291
Map 20. Chipman and Cox's map of Bernard Harbour 327

PREFACE

With the stroke of a pen on February 22, 1913, the Dominion of Canada agreed to send its first major expedition—both exploratory and scientific—into the Western Arctic. Canada's northern mainland coast west of Hudson Bay was then occupied by only a few hundred Inuit and had been visited by few white men. The islands farther north were known largely from the observations of British naval personnel in the mid-1800s while they were searching for the lost Franklin Expedition. The islands were thought to be uninhabited or almost so. More was known about the Eastern Arctic than about the Western Arctic, because it was more readily accessible to Europeans and Americans, fishermen and explorers alike. Britain had turned over to the new Dominion of Canada in 1870 any claims it had over the vast ill-defined northern lands then known as Rupert's Land and the North-Western Territory, lands that had long been travelled by traders working for Britain's Hudson's Bay Company. Ten years later, Britain relegated to Canada all of its claims to the Arctic islands north of the Canadian mainland. However, these two significant events raised little interest in the rest of Canada at the time, for the citizens were far more concerned with the relatively narrow band of land they occupied along their border with the United States than with anything or anyone farther north.

The Expedition headed north in June (four months after the February agreement) with the name "The Canadian Arctic Expedition of 1913–1916." In 1919, however, its name was changed to "The Canadian Arctic Expedition of 1913–1918" to accommodate the extra years Vilhjalmur Stefansson's Northern Party had remained in the North. Almost from its first day it encountered major trouble, including the worst summer in many years for navigation in the Western Arctic. Severe ice conditions prevented the two parts of the Expedition—the Northern (Exploration) Party and the Southern (Scientific) Party—from reaching their areas of operation in 1913, and resulted in the loss of its flagship *Karluk*, many of the men on board, and essential equipment and supplies. The advent of the First World War in Europe in the summer of 1914 then quickly displaced news about the Expedition from the front pages of North American newspapers and rendered the continuation of the Expedition in the Arctic difficult for the Canadian government to justify.

Notwithstanding these initial problems, the leader of the Northern Party (and of the entire Expedition), Vilhjalmur Stefansson, found replacements for the men lost with the *Karluk*, and managed in the next four years to discover several far northern islands, to determine the location of the Continental Slope and oceanic depths in several places west of Banks and Prince Patrick Islands, and to gather valuable information about oceanic currents in the Western Arctic. Meanwhile, the leader of the Southern Party, Dr. Rudolph M. Anderson, and his five scientists completed within the three years originally assigned them almost all of the intended scientific field tasks—geographical, geological, biological and anthropological—and brought south large collections of specimens in the last three of those fields for the Victoria Memorial Museum in Ottawa and for an assortment of other scientific organizations. Study of the many hundreds of specimens subsequently led to the publication by the Canadian government of thirteen volumes of scholarly scientific reports on the results of the Expedition.

Stefansson published *The Friendly Arctic*, his lengthy and popular narrative of the Expedition, late in 1921, which quickly attracted wide attention and acclaim. During the nearly ninety years since then, his account has remained the only book totally devoted to the story of this Canadian Arctic Expedition, although many others have included brief accounts of it, and still others have dealt in depth with certain parts of it.[1] Is another book on this subject justified?

Why Another Book on the Canadian Arctic Expedition?

Stefansson was a gifted writer and his many books and articles, published largely in the 1920s and 1930s, made interesting reading. Most of his audience at that time had little or no knowledge of the Arctic, however, and so found his book, *The Friendly Arctic,* highly informative as well as enjoyable. For the few who were familiar with the subject matter, such as Dr. R.M. Anderson, the leader of the Southern Party on the Expedition, R.A. Bartlett, the captain of the Expedition's ill-fated *Karluk,* and W. L. McKinlay, the only one of the five scientists on that ship who survived after it sank, there was much in the book that was totally false. Even the book's title gave a false impression. Thus it seems highly desirable to publish an account of the Expedition that is as historically accurate as the records permit.

Everyone connected with the Expedition is now dead, so it is no longer possible to obtain first-person accounts of Expedition events. However, there is today a considerable volume of documentary material relating to Expedition affairs that has become available since Stefansson wrote *The Friendly Arctic.* This includes interesting documents among the official files of G.J. Desbarats, the then Deputy Minister of the Naval Service, and Dr. R.W. Brock, the Deputy Minister of the Department of Mines as well as being the Director of the Geological Survey of Canada, and diaries and field notebooks of many members of the Expedition. There is also the correspondence of Dr. R.M. Anderson and of Mrs. R.M. Anderson pertaining to the Expedition, all of which is safely lodged in Library and Archives Canada in Ottawa. Less well known, additionally, is an extensive collection of documents related to the Canadian Arctic Expedition and the government publications stemming from it—letters, reports and even the odd sketch map, all carefully saved for many years by Dr. R.M. Anderson—which is archived today at the Canadian Museum of Nature's Gatineau (Aylmer) building in Quebec. Farther afield, new material about the Expedition has been added since 1921 to the Stefansson Collection in the Rauner Library, Dartmouth University in Hanover, New Hampshire. Captain R.A. Bartlett's papers were deposited at Bowdoin College Library, Brunswick, Maine. W.L. McKinlay's papers are archived in the National Library of Scotland, Edinburgh; and some of G.H. Wilkins' papers are now housed in the Byrd Polar Research Center, Ohio State University, Columbus, Ohio, in addition to those placed earlier in the Stefansson Collection at Dartmouth. The hypothetical benefits of time and quiet reflection have allowed perspective to do its work.

---

[1] See, for example, Bartlett and Hale (1916), LeBourdais (1963), McKinlay (1976), Diubaldo (1978), Hunt (1986), Levere (1993), Niven (2000) and Jenness (2004). Stefansson also published a series of popular articles about the Expedition in *Harper's Magazine* (Stefansson, 1919a–f) and *Maclean's Magazine* (Stefansson, 1919g–l), in addition to a 23-page summary in an English geographical journal (Stefansson, 1921a) prior to the appearance of his book *The Friendly Arctic.*

My reason for writing this new account of the Expedition is, in the words of W.L. McKinlay, "To set the record straight." That was why McKinlay published his account of the last voyage of the *Karluk* in 1976. He was a mathematician, I am a scientist. Facts are the building blocks to our thinking and writing, while seeking and interpreting that which is true forms our ultimate aim.

The story of the Expedition is important in the early history of Canada's northern development. Stefansson's long-surviving account is unreliable in many respects because of its errors of fact, exaggerations and fabrications. A more reliable account has long been overdue.

How and Why I Came to Write this Book

The idea for writing the present book may be traceable to the summer of 1975, when I decided to prepare a typed copy of my father's handwritten Arctic diary recording his experiences while he was a member of the Canadian Arctic Expedition 1913–1916. The following spring I was introduced to the lone scientist survivor of the ill-fated *Karluk*, William "Wee Mac" McKinlay,

Figure 2 The author and William McKinlay in the home of McKinlay's daughter Nancy Scott in Glasgow, examining an album of McKinlay's Canadian Arctic Expedition photographs in September 1980. The painting behind them was McKinlay's 90th birthday gift from friends in Ottawa, Canada. By Ottawa artist/scientist Maurice Haycock, the painting shows where McKinlay first met Haycock on Beechey Island, Nunavut, in 1975, the site where three members of Sir John Franklin's 1845 naval exploration party are buried.

Photo: Courtesy of S.E. Jenness, photo by Jean M. Jenness

when he visited Ottawa, and that introduction led to three subsequent delightful visits with him and his daughter at her home in Glasgow, in 1978, 1980 and 1983, the last one but a few weeks before his death. Our stimulating conversations during each of those visits greatly nourished my northern interests. My retirement in March 1985 then provided me with the time to commence my new career of Arctic research and writing. I chose to work first on editing my father's diary for publication.[2]

A highly fortuitous meeting in the summer of 1985 with Dr. David R. Gray, then the federal government's muskox expert with the National Museum of Natural Sciences[3] and now a science consultant to several Canadian government agencies, introduced me to the existence of several four-drawer file cabinets filled with little-known documents from and about the 1913–1918 Expedition. One cabinet contained documents pertaining exclusively to the Expedition; the others held documents having to do with the publication of the Expedition reports. All had been accumulated by Dr. R.M. Anderson. Many years later, long after Dr. Anderson's retirement in the 1940s, I understand that Dr. Gray rescued these files from disposal when the space they occupied was required for other purposes. With Dr. Gray's approval, I set about preparing an index of the Expedition documents that were in that one file cabinet, an activity that provided me with a basic knowledge of the Expedition and aided me in the preparation over the next two decades of several related publications, including my father's Arctic diary and an account of the Expedition's Australian photographer, G.H. Wilkins.[4]

One day in 2004, shortly after the publication of my Wilkins book, Mrs. Diana Rowley, wife of the archaeologist and Canadian government administrator, Graham Rowley, commented on it as we chatted after a meeting we both attended. Suddenly she remarked, "You really ought to write about the Arctic Expedition that your father was on." Her comment served as a challenge to me, and I commenced work on the manuscript within the next few months.

There are several reasons why I came to write this story.

(1) My father, Diamond Jenness, was the ethnologist in the Expedition's Southern Party, and made the greatest contribution to the published results of the CAE.

(2) In the late 1970s, with mother's permission, I deposited the three volumes of my father's handwritten Arctic diary in the National Archives in Ottawa (now Library and Archives Canada). That seemingly minor incident drew my attention to the fact that the diaries or field notes of quite a few of the other scientists on the Expedition were already held by the Archives, but several potentially significant ones were not included, and I decided to try to locate them. Over the next several months, in the belief that the documents I sought still existed somewhere, I made inquiry after inquiry. After a number of disappointments, I finally ascertained that Fritz Johansen's diary/field notes were housed in the Arktinsk Museum in Charlottenlund, Denmark; J.R. Cox's field notes were with his younger daughter Penelope (Penny) in Arizona; and J.J. O'Neill's field notes were in the hands of his son Melville in Ottawa. Cox and O'Neill evidently refrained

---

[2] My father's Arctic diary was published in 1991 as *Arctic Odyssey: The Diary of Diamond Jenness and the Canadian Arctic Expedition 1913–1916* (S.E. Jenness (Ed.), 1991).
[3] This is now the Canadian Museum of Nature.
[4] Jenness, 2004.

from keeping personal diaries while in the Arctic, but their field notes provide information on their activities as well as some personal observations. Cox's daughter graciously allowed me to bring her father's field notebooks back to Ottawa, where I prepared a typescript copy of their contents. Then I obtained her permission as well as that of her older sister Jane in Point Claire, Quebec, to turn the notebooks and typed copy over to the National Archives. I suggested to Melville O'Neill that he consider turning his father's notebooks and other Expedition documents over to the Archives in Ottawa, and this he subsequently did. I learned later that McKinlay had written two diaries, the condensed version of which he had deposited in the 1920s in the Public Archives in Ottawa, the fuller version ultimately being lodged in the National Library of Scotland in Edinburgh. (See Appendix 3 for further details on the Expedition diaries.)

My inquiries about Johansen's diary unexpectedly revealed the existence of twelve hitherto unknown articles written about the Expedition by him while in the Arctic and published in the Danish newspaper *Politiken*. From that publisher I obtained photocopies of the twelve articles about the same time that I obtained a photocopy of his diary from the Arktinsk Museum in Charlottenlund. A friend in Canada's Department of the Secretary of State then arranged to have both the diary and the articles translated into English and copies sent to me. I in turn photocopied the two documents and presented them to the National Archives of Canada. An interesting revelation from the translator was that Johansen's diary, skimpy as it is, was written in Old Danish. Presumably this was to foil Stefansson, whom Johansen knew could read modern Danish but not Old Danish.

(3) I have known or met during my lifetime seven (and perhaps eight) of the scientists who were on the Expedition: Stefansson, Dr. Anderson, K.G. Chipman, Cox, Johansen, McKinlay, Jenness (my father), and possibly Wilkins as well; also the sailor Karsten Andersen. Such a connection with the Expedition is likely to be unique today.

And (4), I have had the good fortune since 1985 of enjoying the stimulating friendship of Dr. David R. Gray, who has long been interested in the Canadian Arctic Expedition and Dr. Anderson's various important Arctic activities, and has done much to expand the knowledge and understanding of the Canadian Arctic. He also created the excellent Internet website about the Canadian Arctic Expedition—http://www.civilization.ca/cmc/exhibitions/hist/cae/splashe.shtml—a few years ago.

The Book's Organization

Upon undertaking to write about this Expedition, I realized that I was dealing with three distinct sub-stories, each sharing a common introduction and ending. This led to a five-part manuscript, each part dealing with a finite subject and following it more or less from start to finish. This format, unfortunately, produced a small amount of unavoidable chronological overlap. The book ends with four appendices.

Part 1 (Stefansson's New Arctic Expedition) tells of the planning and preparations for the Expedition prior to its departure for the North, fierce arguments arising between Stefansson and several of his scientists at Victoria and at Nome, and his sudden and often debated departure in September 1913 from the *Karluk* after it became trapped in the Arctic pack ice in August. At that point, what subsequently happened to the *Karluk* and the people on it became a story of its own. Information for this introductory part came largely from Dr. R.M. Anderson's correspondence and field notes, Stefansson's *The Friendly Arctic*, McKinlay's *Karluk* and diary, Jenness' *Arctic Odyssey*, K.G. Chipman's diary and B. Mamen's diary.

Part 2 (The *Karluk* Saga) relates the tragic story of the *Karluk* and many of the people on it from the time of Stefansson's departure in September 1913 until the dramatic rescue of the survivors from Wrangel Island a year later. Information was drawn from McKinlay's diary and book *Karluk*, Burt M. McConnell's diary, Jennifer Niven's *The Ice Master*, Captain Bartlett's *The Last Voyage of the Karluk*, and articles by Fred W. Maurer, McConnell and E.F. Chafe.

Part 3 (Stefansson's Northern Party) tells of Stefansson rebuilding his Northern Party after his departure from the *Karluk* in September 1913, his activities and accomplishments, and those of his revitalized exploration party until his departure from the Arctic in 1918. Included is an account of S. Storkerson's remarkable ice-drift experiment in 1918. In these activities Stefansson was greatly assisted by a number of Inuit hunters and seamstresses. Details were obtained from Stefansson's diary, correspondence and book, *The Friendly Arctic*, and from Harold Noice's *With Stefansson in the Arctic*, J.E. Ashlee's *An Arctic Epic of Family and Fortune*, and sundry correspondence.

Part 4 (Dr. Anderson's Southern Party) sets forth the multiple activities of Dr. Anderson and his Southern Party, primarily from the time they settled for the first winter at Collinson Point, on the North coast of Alaska, until they sailed west to Nome in the summer of 1916, their field investigations completed. Details for this part were drawn mainly from Dr. Anderson's excellent summary, *Canadian Arctic Expedition of 1913–1916: Report of the Southern Division* and his field notes, Chipman's diary, Jenness' diary, Wilkins' diary, the field notes of Cox, Johansen and O'Neill, and Dr. Anderson's correspondence with his wife and with Stefansson.

Part 5 (Aftermath) deals with the problems encountered during the publication of the scientific reports on the many aspects of the Expedition, the major role played by Dr. Anderson in getting all of the reports published, the costs and accomplishments of the Expedition, and the feud that arose between Dr. Anderson and Stefansson (and expanded to include many interested parties). Much of the information came from Dr. Anderson's letters and papers housed in the Archives of the Canadian Museum of Nature in Gatineau, Quebec.

The appendices list: (1) the persons employed on the Expedition; (2) geographical localities named after Expedition members; (3) the present whereabouts of the diaries, field reports and various collections (photographic, biological, ethnological, archaeological and mineralogical) brought back from the Arctic by the Expedition; and (4) the government reports stemming from the Expedition.

Six Remarkable Men

In resurrecting this story after nearly a hundred years, I seek to pay special tribute to six members of the Expedition who, despite isolation and severe hardships, dutifully carried out the many tasks assigned them with little fanfare but much success. To a man they were idealistic, dedicated and adventurous. I refer here to the six scientists—Dr. Anderson, Chipman, Cox, Jenness, Johansen and O'Neill—who comprised the Southern Party of the Expedition. Their largely unheralded activities in the Arctic, undertaken more for science and country than for fame or financial gain, provided Canada with a wealth of scientific information about its North.

Most deserving among these was the man in charge, Iowa-born, mammal-and-bird specialist, Dr. Rudolph Martin Anderson.[5] Charged with overseeing a multitude of different investigations, he succeeded in encouraging his group of scientists to complete almost all of their tasks, in spite of having only two of the three years they were originally allotted to carry them out. He then successfully oversaw them and all their assorted scientific collections safely south to the Victoria Memorial Museum in Ottawa. Later, in Ottawa, he worked diligently over the course of the next decade to ensure that all of the reports stemming from the field work and the collections were properly edited and published. Known more for his unhappy feud with Stefansson, it is high time to acknowledge his considerable contribution to the success of the Canadian Arctic Expedition.

McKinlay was supposed to have been a member of the Southern Party, but was on the *Karluk* when it started for Herschel Island and ended up instead on Wrangel Island, from where he was subsequently rescued. Stefansson is not included here because his contributions to the Expedition have been adequately publicized and honoured elsewhere, and his contribution to the government's published reports on the Expedition amounted to next to nothing.

Why the Title for Dr. Anderson?

Most of the professional people on the Expedition are referred to by just their surnames in the following pages, just as they generally referred to each other in their diaries or field notebooks. The one notable exception is the leader of the Southern Party, who was generally referred to as "Dr. Anderson" or "the Dr." It showed respect for him because he was several years older than most of the other scientists,[6] had a PhD degree (which was far less common in the early 1900s than today), had been a military officer in the Spanish-American War, and displayed the quiet dignified manner of an officer and leader. I have accordingly used the title "Dr. Anderson" for him throughout this book. The geologist J.J. O'Neill also had a PhD, but he was freshly out of college, eleven years younger than Dr. Anderson and did not have a leadership role, so was referred to simply as O'Neill or occasionally as Jack by the other men.

Changes in Terminology

Two changes in terminology since the Canadian Arctic Expedition went north have given me considerable difficulty during the preparation of this book: (1) the word "Eskimo" for the Aboriginal people encountered by the Expedition members along the Arctic coast, and (2) the words used for the various units of distance and temperature.

The word "Eskimo" (plural "Eskimo" or "Eskimos") was used until the early 1960s to identify the Aboriginal people bordering the Arctic Ocean. It is still used occasionally, especially in Alaska and the other forty-nine American states. From the late 1960s onward, however, improved levels of education among the North American Eskimos brought about calls for a

---

[5] Dr. Anderson became a British citizen and resident of Canada in 1920.

[6] The Scotsman James Murray was older than Dr. Anderson, but their paths crossed only for a few weeks in Victoria and Nome in 1913. Stefansson sometimes addressed him as Dr. Anderson in his letters, sometimes simply as Anderson, perhaps depending upon his mood at the time.

change in the name used for them by the general public. They disliked the term "Eskimo" because it was a foreign word, which many of them perceived to have a pejorative implication. It was widely believed that the word "Eskimo" came from an Algonquian word meaning "eaters of raw flesh," and was used by the Algonquian people to describe and insult their northern neighbours, with whom they shared no affection. The Eskimo, it might be noted, had their own insulting word for their unfriendly Indian neighbours.[7] The true origin of the word "Eskimo," however, remains somewhat uncertain. Anthropologists such as my father attached no such negative attitude to the word in the past and used it freely for the people from Alaska to Greenland.

Sometime shortly before 1970 Paul Lumsden, an employee of Canada's Department of Secretary of State in Ottawa, started to use the Eskimo's own word "Inuit"[8] in memos and reports circulating within that Department, initially at least adding the word "Eskimo" in parentheses afterwards for clarity.[9] In time the usage expanded to appear in the Canadian newspapers, magazines and books.

Then in 1977 at Barrow, Alaska, the attendees at the first Inuit Circumpolar Conference officially adopted "Inuit" as the replacement for the term "Eskimo."[10]

The simple substitution of "Inuit" for "Eskimo" does not always seem appropriate, however. For example, the people in the Mackenzie River Delta region prefer to be called by their own words, "Inuvialuk" (singular) or "Inuvialuit" (plural), and those in Northern Alaska prefer to be called "Iñupiaq" (singular) or "Iñupiat" (plural). This creates a bit of a terminology nightmare for persons writing about the Aboriginal people in the coastal regions from Northern Alaska to Greenland early in the twentieth century. Between 1914 and 1917, Stefansson gathered several such persons from Northern Alaska, Herschel Island and Baillie Islands for his revital- ized Northern Party, applying the term "Eskimo" to all of them, which was understood by all concerned at the time. A few of these, like Billy Natkusiak, came from known localities in

---

[7] Canadian Indians, so ably described by my father in his book *The Indians of Canada* (Jenness, 1932), now also prefer a new name and are today referred to as "First Nations" people.

[8] "Inuit" is the plural form of "Inuk," which means "person, or human being," both being the Inuktitut words the people use to describe themselves in their own language in preference to "Eskimo."

[9] I learned about the role of Paul Lumsden from Mary Carpenter Frost, whose father, the renowned hunter and trapper, Fred Carpenter, lived at Sachs Harbour, Banks Island, when she was born. Both her father and mother were Mackenzie River Inuit (Inuvialuit), but her father was part- white, with Huguenot ancestors from Ireland and France. Mary, the holder of MA degrees in both Journalism and Anthropology, told me in February 2009 that she did not regard the term "Eskimo" as pejorative, although many of her fellow Inuit do. In fact, she employs it as part of her username for Internet communications. Mary also told me that Paul Lumsden was formerly married to Mabel Pokiok, Canada's first Inuvialuk registered nurse and one of a well-known, very large family living in the Mackenzie River Delta region. Each email I have received from Mary Carpenter Frost in recent months ends with her father's amusing saying: "Always do something every day—even if it's nothing."

[10] Microsoft Encarta Online Encyclopedia 2008, "Inuit."

Northern Alaska and hence can be properly identified today as Iñupiat. A few others, such as Pannigabluk and Palaiyak, were known to have been from the Mackenzie River Delta, and could be identified as Inuvialuit. However, I found it almost impossible to determine the place of origin of many of the other Eskimos employed on the Canadian Arctic Expedition in order to distinguish the Iñupiat from the Inuvialuit. West-to-east migrations and intermarriages over many years further complicated the origin of such individuals. I recall that my father, in the 1960s, did not favour discarding the word "Eskimo" and replacing it with "Inuit," although I never asked him why. Perhaps it was because he knew the two terms were not truly interchangeable and that the adoption of the general term "Inuit" would create new problems. On the following pages I have used Inuvialuk and Inuvialuit, and Iñupiaq and Iñupiat respectively for those "Eskimos" whose places of origin are reasonably known, and the singular and plural terms "Inuk" and "Inuit" for those of uncertain or unknown geographic origin.

Anthropologist John L. Steckley recently discussed the problem of multiple terms in place of "Eskimo" under the clever but slightly rude heading "Who are you calling Inuit, White Man?" in his book *White Lies about the Inuit*.[11]

Units of measurement have been a different type of problem. Early in the 1970s, just about the same time "Inuit" was starting to replace "Eskimo," the Canadian government introduced the use of the International System of Units (SI), a basically metric system of measurements used in Europe, Asia and elsewhere throughout much of the world in place of the chiefly British units of measure.

All temperature readings during the Expedition were recorded in degrees Fahrenheit, as that was the normal form of public temperature readings then in Canada and the United States. Canada changed its system of temperature readings in the early 1970s when it adopted the SI system. Today daily temperatures are measured in Celsius degrees in Canada, while still being recorded in Fahrenheit in the United States.

To avoid introducing errors through conversion of original data to the current SI units, I have retained distances and elevations mentioned in this book in miles, yards and feet, and weights are given in pounds and tons, just as they were originally recorded. Similarly, temperature values are cited in Fahrenheit degrees. I have, however, converted the original oceanic-depth readings mentioned in the text from fathoms, in which they were recorded by Stefansson, McKinlay, Murray, Wilkins, and one or two others at the time, to feet, for easier comprehension. (One fathom is equivalent to six feet.) Metric equivalents for distances are not given in the text, but I have included metric-equivalence bars for distance measurements on the sketch maps scattered throughout the book.

The words "sled" and "sledge" (and their plural forms) were used by Stefansson and other members of the Expedition. "Sledges" were larger and heavier than "sleds," and were intended to carry heavier loads. The Expedition members were aware of the differences at the time, but the distinction has faded with the passage of time. I have, therefore, used the words "sled" or "sleds" in the following pages in the belief that these are the terms in more common use today.

---

[11] Steckley, 2008, p. 23.

ACKNOWLEDGMENTS

I am grateful to the French-Canadian ethnologist and folklorist C. Marius Barbeau for suggesting to V. Stefansson early in 1913 that he obtain the New Zealander Diamond Jenness as ethnologist for the Canadian Arctic Expedition, which was then in the planning stages. I am also naturally thankful that Jenness safely survived all the hardships he encountered in the Arctic during the next three years, and that after coming south to Ottawa in 1916 he liked not only his new boss at the Victoria Memorial Museum of Canada, Dr. Edward Sapir (after whom I received my middle name), but also the latter's secretary, Frances Eilleen Bleakney. Immediately following Jenness' safe return in 1919 from fighting in World War I, he married Miss Bleakney, and she in due course became my mother.

I extend my sincere thanks to my good friend, Dr. David R. Gray, zoologist and for many years Canada's muskoxen expert, for acquainting me with the fascinating story of the Canadian Arctic Expedition of 1913–1918 and the large collections of documents related to it so carefully saved by Dr. R. M. Anderson, now stored at the Canadian Museum of Nature in Gatineau, Quebec. We have both benefitted, I believe, from our exchanges of information over the years in regard to one or other aspect of that Expedition. David also graciously spent many hours carefully reading and supplying editorial suggestions to an earlier version of my manuscript, as well as to the present one, which greatly improved the quality of its content and for which I express my deep gratitude.

I am grateful for and received much inspiration from the opportunity offered me by the Polar Continental Shelf Project in the summer of 1989 to visit, with my wife Jean, several Arctic locations where Southern Party activities took place. These included: (1) the site of that party's base camp at Bernard Harbour, where my father and his colleagues lived for two years, a locality I found to be as bleak, windy and cold as one could envisage, yet with a strange indefinable beauty; (2) the mouth of a small fishing stream four miles southeast of the Southern Party's base camp, in which three rows of stones still mark where the companions of Ikpukhuak and Higilak or perhaps their predecessors, placed them before my father visited the site and photographed the stream; (3) a curious man-made stone monument near Cape Krusenstern at the western end of Coronation Gulf, about five feet in height, which appeared to be identical in appearance in 1989 when I photographed it as on the day in 1916 when my father photographed it and described it as a "stone house"; (4) Bloody Fall on the Coppermine River, infamous for its role in the historic clash in 1771 between Chipewyan Indians guiding Samuel Hearne and local Inuit who were peacefully asleep when attacked; (5) the town of Coppermine (now Kugluktuk) from which site (the town not then existing) the geographer Chipman in 1916 started overland south to Great Bear Lake on his way back to Ottawa; and (6) a temporary campsite on the bank of the Coppermine River where an Arctic wolf attacked my father (or vice versa according to some individuals) and savagely bit his right wrist in 1915.

My wife and I also enjoyed the opportunity of an exciting flight arranged by Dr. George Hobson, then Director of the Polar Continental Shelf Project, over that part of the Wollaston Peninsula on Victoria Island where my father wandered for several months in 1915 with his adopted Inuit parents, Ikpukhuak and Higilak, and their family. I had planned to have the small Cessna airplane land on one of the lakes where my father had camped with his Inuit family, in the hope of finding the exact location of their camp and perhaps some relics of their visit, but the lakes were still frozen, even though it was mid-July, and we were unable to land because our plane was equipped with pontoons. However, flying about 500 feet above the ground, we traced the probable route my father and his Inuit companions took west so many years ago, almost to Cape Bering, thereby obtaining a good understanding of the nature of the terrain over which they tramped in August and September 1915.

I wish to acknowledge the enjoyable and helpful discussions I have had over the years with various Arctic specialists, such as Dr. George Hobson, Dr. Denis St-Onge, (the late) Dr. Jette Ashlee, (the late) Dr. Graham Rowley and his wife Diana. All have influenced the input of this book in one way or another. I have also learned much from discussions with Mary Carpenter Frost, an Arctic-born, residential-school and southern-university educated daughter of mixed Inuit and Caucasian parentage, whose knowledge of her people is considerable and personal experiences amazing.

I very much appreciate the encouragement and assistance offered me along the way by archaeologists Dr. David Morrison and Dr. Ian Dyck, and library and archival personnel at the Canadian Museum of Civilization, the Canadian Museum of Nature, Library and Archives Canada (formerly the Public Archives of Canada and then National Archives of Canada), the Geological Survey of Canada, Rauner Special Library at Dartmouth College, Hanover, New Hampshire, and the Byrd Polar Research Center, Ohio State University Archives, Columbus, Ohio. I also would like to thank Cecelia McGuire, who demonstrated a remarkable talent for finding and correcting the many errors that somehow remained in my manuscript when I submitted it for publication. Special thanks go to Ginette Côté, one of the Canadian Museum of Civilization's Publishing Coordinators, who cheerfully and carefully guided this manuscript through the many time-consuming steps necessary to get it published.

And lastly, I am extremely grateful for the considerable encouragement and helpful editorial input I received during the preparation of this manuscript from both of my children, Dr. John Diamond Jenness and Mary G. Montgomery. Having their continuing affection and assistance over the years has meant more to me than words can express. All but two of the twenty maps on the following pages (Maps 6 and 20) were created by me. Sources of the photographs are included with their captions.

PART 1
STEFANSSON'S NEW ARCTIC EXPEDITION

PREPARATIONS

1

The year 1912 is remembered chiefly for the tragic sinking of the ocean liner RMS *Titanic*. That same year, however, a markedly different event took place at Seattle, Washington, which although considerably less dramatic, attracted a great deal of publicity at the time, and had scientific and political significance throughout much of North America.

Pre-Expedition Publicity

On September 8 of that year, a tall, well-tanned man in his early thirties strode down the gangplank of the newly arrived steamer from Nome, Alaska, bearing news of a little-known, isolated tribe of Inuit[1] he had visited on Victoria Island in the central Canadian Arctic. He was Vilhjalmur Stefansson, an American ethnologist and explorer with a distinct flair for self-publicity, who had spent the previous four years wandering about the Arctic coast of North America for the American Museum of Natural History, studying the Native people he encountered there and recording numerous observations on the terrain and off-shore ice.

Waiting on the Seattle pier was an experienced newspaper reporter from the *Seattle Daily Times*, whom Stefansson later described as "an authority on Alaska."[2] By his own assertion, John (Jack) J. Underwood had spent twenty months "… with an exploring expedition on the Mackenzie River, long before Stefansson was heard of,"[3] and had heard stories from the Inuit about the presence of white people far to the north. He had also spent several days with Roald Amundsen in 1906 when the latter was leaving the Arctic following his historical crossing of the Northwest Passage. On that occasion Underwood had asked Amundsen if he had seen "any trace of the lost followers of Leif Erikson, and he told me that he had not" although he had seen "some tribes that theretofore had never seen a white man," but "they did not even have a legend relating to Erikson and his followers."[4]

Underwood met Stefansson when the latter disembarked from the steamer and spent several hours interviewing him. During their meeting Stefansson spoke glowingly of the Aboriginal people he had encountered, many of whom had never before seen a white man. Among these people, he said, were some who had blond to reddish hair and beards, bluish eyes, and fair skin, features that were totally foreign to the Inuit along the Arctic coast. Being of Icelandic origin, Stefansson was familiar with the Norse sagas and the disappearance of Norse settlers from Greenland shortly before Christopher Columbus reached America in 1497.

---

[1] During Stefansson's lifetime they were known as Eskimos.

[2] Stefansson, 1964, p. 133.

[3] Underwood, 1921 (December 20, 1921).

[4] *Ibid.*

During their interview Stefansson said he believed some of the isolated people he had encountered on Victoria Island included their descendants. He then showed Underwood several photographs of individuals he claimed were examples of those long-lost Vikings. Underwood noticed the long-tailed caribou-skin coats on the men in the pictures and remarked that it seemed more likely these people were descendants of Inuit who had been in contact with survivors of British naval personnel of the lost Franklin Expedition, whose coats had similar long tails. Stefansson dismissed that suggestion, however, saying that his cephalic measurements of these people left no doubt that they were descendants of the lost Vikings.[5]

Underwood's article about Stefansson's discovery appeared in the *Seattle Daily Times* on the following day, September 9, under the headline "AMERICAN EXPLORER DISCOVERS LOST TRIBE OF WHITES, DESCENDANTS OF LEIF ERIKSSEN." Other newspapers around the country quickly picked up this story and another one that Underwood published in the *New York Sun*, coming out with such bold headlines as "Blond Eskimo Story confirmed" and "Natives of the Far North Had Caucasian Characteristics." The term "Blond Eskimos" attracted the public's attention and spread rapidly across the United States, arousing considerable interest and controversy, providing Stefansson with exactly the kind of publicity he wanted as he started his campaign to obtain money for a new expedition to the North. Almost overnight, he had become a celebrity. Yet, Underwood reported later, when Stefansson reached New York and was questioned by scientists about the validity of his startling claim, he promptly denied responsibility for making it, invariably replying "Why I never said any such thing! That statement came from an irresponsible newspaper man named Jack Underwood."[6]

Stefansson's partner in the just-completed expedition, Dr. R.M. Anderson, commented dryly years later that "The famous 'blonde Eskimo' hoax was not original with him [Stefansson]. Richardson started it, and old Captain Charles Klengenberg seems to have brought the idea to Stefansson …"[7] In the fall of 1912, however, Stefansson appeared to be turning the topic into a rewarding form of publicity.

Although mention of the term "Blond Eskimos" is still encountered occasionally, as in the recent books of Gisli Pálsson,[8] John Steckley[9] and, of course, the present book, Stefansson's hypothesis linking the Copper Inuit genetically with the vanished Norsemen of Greenland has now been disproved. The final nail in the coffin of his claimed linkage with Viking ancestry came with the DNA studies carried out in 2003 by anthropologists Gisli Pálsson and Agnar Helgason from the University of Iceland. Their research revealed no similarity between the DNAs of one hundred Copper Inuit elders in Cambridge Bay on Victoria Island, and those of Norse descendants from Iceland.[10]

---

[5] *Ibid.*

[6] *Ibid.*

[7] Anderson, 1926 (March 4, 1926).

[8] Pálsson, 2003, pp. 89–128

[9] Steckley, 2008, pp. 77–102.

[10] "DNA tests debunk Blond Inuit legend," Internet website seen on June 25, 2008: http://www.cbc.ca/news/story/2003/10/28/inuit_blond031028.html.

On November 1, 1912, the soft-spoken, mid-thirtyish, intensely dedicated American zoologist Dr. Rudolph M. Anderson reached San Francisco from Nome on the whaling ship *Belvedere*, bringing with him a great number of Arctic specimens but receiving no publicity whatsoever. He had formed the second half of "The Stefansson-Anderson Arctic Expedition 1908–1912." He returned south later than Stefansson because he had been left to wind up their northern business matters at Cape Bathurst and to ensure that their ethnological and zoological collections were safely shipped south from Alaska. Stefansson had left him to complete these thankless tasks while he (Stefansson) went on south to publicize his exciting discovery of the "Blond Eskimos."

Unlike Stefansson, Dr. Anderson made no effort to create any publicity in the local or national newspapers. Instead he proceeded directly to Iowa to see his parents and to marry his fiancée, Mae Belle Allstrand. He and his bride then proceeded east so that he could commence his studies of the numerous animal and bird specimens that he had collected in the Arctic and shipped to the American Museum of Natural History in New York.

That Museum had sponsored their four-year Arctic expedition, with minor additional financial support coming from the Geological Survey of Canada, the Royal Ontario Museum in Toronto and private individuals. As a result, the American Museum received most of the natural history, ethnological and archaeological specimens the two men collected, which it regarded as a very satisfactory return for its relatively small monetary investment. Duplicate ethnological material was forwarded to the Geological Survey for its newly constructed Victoria Memorial Museum in Ottawa, and duplicate archaeological material went to the museum in Toronto.

Dr. Anderson quickly settled in at the American Museum to study his specimens and prepare his reports, and both he and Stefansson subsequently submitted manuscripts to the American Museum of Natural History on their four years of work in the Arctic.

The New Expedition

Unknown to Dr. Anderson, however, Stefansson had already formed new and imaginative plans for a larger expedition to the Arctic. Stefansson has been called an "Arctic Dreamer,"[11] a very good description although the term "visionary" might be somewhat more suitable because of the practicality of many of his unusual ideas. "To do a unique thing in polar exploration" was one of two dreams he entertained on his return to the United States in 1912; the other was "to organize a comprehensive scientific expedition."[12] By December 1912 he was setting his plans in motion. Expressed simply, he contemplated an expedition with a dual purpose: to seek new lands and life in the Beaufort Sea, about which both British explorer Sir Clements Markham and Norwegian explorer Fridtjof Nansen had stated that its exploration was the most important remaining problem in north polar discovery;[13] and to study the "Blond Eskimos" Stefansson had seen in the Coronation Gulf region. The expedition party searching

---

[11] This was the title of an award-winning documentary film produced in 2003 by Peter Raymont of White Pine Pictures, Toronto, Ontario.

[12] Stefansson, 1921b, p. 286.

[13] *Ibid.*

for land in the Beaufort Sea was expected to depend largely on obtaining its food as it progressed; the party investigating the Coronation Gulf region, on the other hand, was expected merely to augment with local wildlife the provisions it had brought with it.[14] Stefansson would be in charge of both the exploration and the ethnological studies. They would almost certainly guarantee that he would receive headlines in major U.S. newspapers and elsewhere, and such publicity would help immeasurably to promote his anticipated future lectures, and the sales of books and articles he intended to write.

For this new expedition, Stefansson originally planned to employ six scientists and one ocean-going vessel, with provisions and equipment for three winters and four summers in the Arctic. For these he needed funds, and these funds he expected would soon be forthcoming from the American Museum of Natural History, in view of the fine ethnographic collections he had just brought back from the Arctic for that institution. His plans included the exploration of the islands and seas in the western Canadian Arctic, an activity he believed would attract additional financial backing from the National Geographic Society in Washington.

Evolving Plans

In the fall of 1912, Stefansson confidently approached Henry Fairfield Osborn, president of the American Museum of Natural History, for funds for his new expedition. His prime interest at this point was to undertake further ethnographic studies of the little-known Inuit he had seen around southwestern Victoria Island and Coronation Gulf. Osborn showed considerable interest in Stefansson's plans, but informed him that the American Museum was already committed to finance an expedition into the Arctic the following summer, which would be led by the explorer Donald B. MacMillan, and could not finance a second one at that time.

MacMillan had been a member of Robert E. Peary's second expedition to the eastern North American Arctic, 1905–1906, which failed to reach the North Pole in 1905, but got to the southwestern end of Ellesmere Island in the summer of 1906. Looking to the northwest of that island, Peary thought he saw a large mountainous land mass, to which he later gave the name "Crocker Land."[15] MacMillan did not see the mountainous terrain Peary claimed to have seen, although he was a member of Peary's party at the time. Now he was preparing to head north to explore Peary's "Crocker Land," if it really existed.

Disappointed by the American Museum's refusal to finance his trip, Stefansson sought funding from the National Geographic Society in New York. There he received a more favourable reception—indeed, a promise of $22,500 spread over four years, to undertake geographical work on Victoria Island, Banks Land (now Banks Island) and Prince Patrick Island in the western North American Arctic, with a goal of discovering and delineating new land in that still little-known part of the Arctic. Little was then known about the islands in the

---

[14] *Op. cit.*, pp. 287–288.

[15] Peary, 1907, p. 202. In *Nearest the Pole*, his account of his expedition, Peary did not mention the name "Crocker Land," merely referring to his supposed observation of it on p. 202 with the words "… my glasses revealed the faint white summits of a distant land …" However, the name "Crocker Land" is clearly printed on the two maps at the end of his book, positioned about 120 miles northwest of the southwest end of Ellesmere Island.

eastern North American Arctic other than what had been observed and mapped by British naval men in the 1840s and 1850s while they were searching for the missing Sir John Franklin Expedition. The National Geographic Society also wanted Stefansson to take soundings in the Beaufort Sea in order to determine the limit of the Continental Shelf.

The $22,500 in funds the Society offered, however, were inadequate to purchase a ship and to carry out the plans Stefansson had in mind, so that he had to seek additional funds from other sources. He turned again to the American Museum of Natural History. Meanwhile, Gilbert H. Grosvenor, the Director of the National Geographic Society, wrote Dr. Anderson urging him to join Stefansson's new expedition.[16] Stefansson may have suggested that Grosvenor do so.

When Osborn at the American Museum learned that the National Geographic Society intended to sponsor Stefansson, he reconsidered his initial refusal. He then told Stefansson that the American Museum would match the National Geographic Society's offer of $22,500, but attached several conditions to his offer: (1) the Museum would provide the funds only if Stefansson's former co-partner, Dr. Anderson, accompanied the expedition; (2) Stefansson was not to be in charge of the funds, for the Museum did not want him to have the same financial decision-making power that he had had on his previous expedition; and (3) the results of Stefansson's new expedition should be divided between the two organizations—the National Geographic Society would receive his geographical findings and reports, while the American Museum would receive his ethnological and zoological findings and collections.

Stefansson now approached Dr. Anderson, whom he found quite unenthusiastic about participating in the new northern expedition. There were several reasons why Dr. Anderson felt as he did. He had a large number of Arctic specimens he had already collected to study and write about; he wanted time to settle into married life; Gilbert Grosvenor had been trying to persuade him to give up his museum job and work for the National Geographic Society, writing the scientific results of the just-completed Stefansson-Anderson Expedition; and the American Museum of Natural History had already asked him to go as second-in-command on the MacMillan "Crocker Land" Expedition.[17] Knowing that the reality of his expedition depended upon the participation of Dr. Anderson, Stefansson desperately pressured him to join his expedition. He even pleaded with Mrs. Anderson to urge her husband to go North again with him, pointing out the wonderful opportunity the expedition offered to advance his scientific career. Dr. Anderson finally promised to accompany Stefansson's new expedition as its second-in-command, a decision he later regretted, writing some years later, "I had promised to my sorrow to go with him."[18]

Stefansson, the American Museum and the National Geographic Society reached agreement late in January 1913, with Stefansson recognized as leader of "The Stefansson-Anderson Expedition of 1913, 14, 15, and 16." To reach that stage, Stefansson had to agree to shift the

---

[16] Grosvenor, 1912 (December 17, 1912).

[17] Anderson, 1928b (December 13, 1928). "Stefansson actually wept tears over that," wrote Dr. Anderson wryly, knowing that if he went with MacMillan the American Museum would drop its financial backing of Stefansson.

[18] Anderson, 1920b (July 20, 1920).

principal objective of his expedition from the ethnographic study of the little-known Aboriginal people of Victoria Island and Coronation Gulf, which until then had been his principal objective, to geographical exploration in the Beaufort Sea and its immediately adjoining islands. Completion of the mapping of Victoria Island and Prince Patrick Island was set as the expedition's high-priority objective, along with the exploration of the Beaufort Sea for new land or, in the least, to determine the edge of the Continental Shelf.

The Revised Plans

Within short order Stefansson concluded that the $45,000 he had been promised collectively, by the American Museum of National History and the Geographical Society, plus an additional $5,000 offered by the Harvard Travelers Club in Boston, was still insufficient for his needs, and that he would need an additional $25,000. Somehow he found three friends in Philadelphia who were willing to supply him with that additional amount,[19] but instead he decided to seek his additional funds in Canada. Well-connected friends in Toronto from whom he sought assistance advised him to go to Ottawa and talk to both the Director of the Geological Survey of Canada, Dr. Reginald W. Brock, and to Canada's Prime Minister, Robert Borden (later Sir Robert Borden). Responding to this advice, Stefansson proceeded to Ottawa in the first week of February 1913. Dr. Brock, whose organization had contributed a small amount of money to Stefansson's previous expedition in exchange for a collection of ethnographical specimens, expressed interest in Stefansson's new plans, being fully aware that the northwestern North American Arctic was "practically the one remaining place in the world where great geographical discovery was possible."[20] Both Dr. Brock and Stefansson then went to the Prime Minister's office, where Stefansson presented his Arctic plans. While there, Dr. Brock took the opportunity to urge the expansion of the expedition by the addition of several scientists from his department in order to gather scientific information and make biological, anthropological and geological collections from the Arctic coast regions. Such an arrangement would almost guarantee valuable results for the government's financial outlay should the exploration part of Stefansson's expedition prove disappointing.[21]

Sovereignty over the Arctic islands and the possible discovery of new lands were of considerable interest to the Canadian Prime Minister in 1913 because of discoveries by Scandinavian and American explorers a few years earlier among the islands north of the Canadian mainland, islands which Canada assumed it possessed. He was also disturbed to learn

---

[19] Stefansson, 1921b, p. 286. One of the three "anonymous" donors was a man named Fay, after whom Stefansson later sought unsuccessfully to name one of the Arctic islands he had encountered; Stefansson, 1921 (June 1, 1921).

[20] Brock, 1913a (February 4, 1913). Dr. Brock's letter was to the Honourable W.J. Roche, who was Minister of Mines from only October 20, 1912 to February 9, 1913 (Johnson, 1968, p. 501.)

[21] I found it difficult to determine the factual story of Stefansson's efforts to raise funds for his new expedition. He offers somewhat different versions in his *Geographical Journal* article (1921a), *The Friendly Arctic* (1921b) and in his autobiography (1964). The account presented here follows those by historians Diubaldo (1978, pp. 58–63) and Levere (1993, pp. 390–393).

that two new American expeditions (MacMillan's and Stefansson's) were intending to enter the Arctic waters north of the Canadian mainland that very summer.

After Borden had discussed Stefansson's request for supplementary funding with his cabinet, he had the Minister of Mines telegraph Stefansson, "Government will pay total cost expedition on condition you become naturalized British subject before leaving and expedition [flies] British flag."[22] Other stipulations were that the National Geographic Society and the American Museum of Natural History were to relinquish their interests in the expedition, and that the British flag would be planted on any land Stefansson discovered, thereby asserting Canada's claim to it. Later the name of the expedition was changed from "The Second Stefansson-Anderson Expedition" to the "Canadian Arctic Expedition of Vilhjalmur Stefansson."[23] The latter name was, in turn, modified a little while later.

Stefansson accepted the offer of the Canadian government even before consulting with the American Museum of Natural History, the National Geographic Society or Dr. Anderson. The two institutions, when informed by the Canadian government of the change of plans,[24] readily agreed to relinquish their commitments. The National Geographic Society, however, took the occasion to stipulate that "If the Canadian Government cannot assist you this year, the Society reserves its claim upon your expedition, and is prepared to send you North in May or June next."[25]

The June 1914 deadline imposed by the National Geographic Society exerted considerable pressure upon the Canadian government to organize and equip its Expedition without further delay. There was scarcely four months to assemble it and have it head north. Dr. Anderson then had to be convinced that the change of sponsorship would not mean a change in the original plans that he and Stefansson had discussed or in Dr. Anderson's role in the management of part of the Expedition. He also had some concerns about his future at this time, for his position with the American Museum of Natural History lacked permanence.

Under Canadian sponsorship, Stefansson considered himself free of the exploration limitations concerning "Crocker Land" originally imposed upon him by the American Museum of Natural History. (The Museum wanted no rivalry for MacMillan's expedition in that search.) He therefore again took up his interest in searching for the suspected "Crocker Land" in the Beaufort Sea, determined to discover it, if it existed, before MacMillan. The latter, however, had a head start, for he was ready to leave for the Eastern Arctic that summer, while Stefansson was just beginning to organize his Expedition.

---

[22] Roche, 1913a (February 8, 1913). The British flag was stipulated because Canada was still a British Dominion in 1913, flew the British Union Jack as its official flag, and its residents were British citizens rather than Canadian citizens. Residents of Canada did not officially become Canadians citizens until 1947.

[23] Brock, 1913b (May 28, 1913).

[24] Roche, 1913b (February 10, 1913). Telegram to Henry Fairfield Osborn, President, American Museum of Natural History. New York.

[25] Grosvenor, 1913 (February 11, 1913).

The Canadian government's decision to add the scientific men from the Geological Survey created a division of Expedition duties and a duality of government departmental authority, which soon proved seriously confusing, aggravating and divisive. What Stefansson had envisioned as a single expedition with himself as the sole leader, now became an Expedition with two separate parties and purposes—a Northern Party, with exploration as its prime purpose and a Southern Party, with scientific activities as its main purpose. In addition, the Expedition would now have two leaders, each answerable to a different government department. However, Stefansson was put in command of the entire Expedition. Persons in the Northern Party, with the exception of the geologist George Malloch, who worked for the Geological Survey of Canada, and Stefansson, were paid and equipped by, and responsible to the Department of the Naval Service.

Stefansson, eager to ensure his own independence, convinced the Naval Service Department's Deputy Minister to let him serve without pay, an arrangement similar to that which he had made with the American Museum of Natural History for his 1908–1912 expedition. In exchange, he sought, and the Department of the Naval Service granted him, sole publication rights for the Expedition while it was in the Arctic, plus for a year thereafter. Members of the Southern Party, under the leadership of Dr. Anderson, were paid and equipped by, and responsible to the Geological Survey of Canada and its federal government department, the Department of Mines. Three other federal government departments—Marine and Fisheries, Interior, and Customs—also had very small roles in the activities of the newly expanded Expedition, their roles largely being assigned on a minor basis among the scientists of the Southern Party.

Stefansson's Northern Party would include several scientists as he had planned, most of whom he had to find and hire. The Northern Party's main purpose was formalized by three men: G.J. Desbarats, the Deputy Minister of the Naval Service, Dr. R.W. Brock, the Director of the Geological Survey of Canada, and Stefansson. It was to explore for new land in the Beaufort Sea, take possession of any land Stefansson found in the name of His Majesty King George V of Great Britain, complete the geographic mapping of Banks Island, Prince Patrick Island and Victoria Island, and pursue oceanographic, meteorological and geological studies north of the continental mainland. Little was known at that time about the lands that lay north of Canada's mainland, the map available to Stefansson at the time being chart No. 2118 produced in the 1850s by the British Admiralty. This chart had been updated slightly by the later discoveries of a few American and Scandinavian explorers, but it contained many inaccuracies, as later work revealed.

Dr. Anderson's Southern Party, whose instructions were formalized by Dr. Brock and one of his senior geologists, O.E. LeRoy, was expected to map the geology and geography of a 400-mile strip of the Arctic coast from Cape Parry, east of the Mackenzie River, to the Kent Peninsula, with special attention given to the copper deposits known to occur south of Coronation Gulf and on Victoria Island. It was also to study the mammals, birds, plants, insects, fish and other aquatic life, and the Inuit they encountered, making collections in all of these fields for the National Museum of Canada. Dr. Anderson was expected to see that the scientific work of his men was carried out regardless of what was going on with the Northern Party and to submit regular reports of progress to Ottawa, in addition to studying and collecting the bird and mammal species in the region for the museum.

Without frequent communication—no radio,[26] telegraph or telephone, and with mail received only two or three times per year at best—and with two powerful government departments in Ottawa not always communicating and cooperating amongst themselves—it is easy to see how confusion and disagreement between the leaders of the two Arctic teams could arise. In its haste, the Canadian government had created an administrative nightmare—a large expedition with, in the words of historian Levere, "… respective leaders of parties with different styles, different values and different priorities.…"[27] Unforeseen at the time were unpleasant consequences resulting from the differences between Stefansson and Dr. Anderson, as well as from the Canadian government's many hasty, ill-planned and uncoordinated decisions.

Stefansson's Trip to Europe

Late in February Stefansson suddenly informed Dr. Anderson that he was going to Europe until May, and sailed on March 1 for England. Canada's Prime Minister Borden had suggested Stefansson make the trip in order to see Donald Smith (Lord Strathcona), who was Canada's High Commissioner there.[28] One of the British Empire's wealthiest men, Lord Strathcona could probably aid him in obtaining certain oceanographic and ethnographic equipment not available in Canada and in seeking several specially qualified men for his Northern Party. In that regard, Stefansson was surprisingly successful. He obtained most of the scientific equipment he sought as well as a surgeon, a cinematographer and three scientists (a meteorologist/magnetician, an oceanographer and an ethnologist). His principal accomplishments, Dr. Anderson wrote somewhat caustically years later, were (1) attending the International Geographical Congress in Rome at the expense of the Canadian government,[29] where he publicized the Canadian government's intention of sending a large scientific expedition into the Western Arctic that very summer, and (2) selling "all the prospective results of the

---

[26] To give him due credit, Stefansson wrote one of Borden's cabinet ministers (the Honourable George Perley, then a Minister without Portfolio in the Canadian government, but a man whom Stefansson knew was interested in his Expedition) about his idea that the government establish wireless stations at select sites in the Canadian Arctic. Once a network of stations was established it would permit regular contact between wireless operators throughout the North. To start with, he suggested that a wireless station be established on his ship and another one at Herschel Island at a cost of $5,000. He claimed to have already a number of volunteers who were competent with wireless as well as an offer from a manufacturer to supply a free outfit. Then he added that he hoped Perley would discuss the idea with someone "who would not pigeonhole it." As further bait, he commented that the explorer Amundsen intended to take a wireless outfit on his next expedition and "… it will be undignified for a Government Expedition such as ours, to lag behind a private one such as Amundsen's." MacMillan's "Crocker Land Expedition" likewise intended to carry a wireless (Stefansson, 1913a [February 28, 1913]). It was an idea years ahead of its time, of course, but there was no follow-up.

[27] Levere, 1993, p. 400.

[28] Stefansson, 1964, p. 147.

[29] Anderson, 1919b, "The Canadian Arctic Expedition 1913–1918," p. 8. Dr. Anderson may have erred in his accusation that Stefansson spent government funds to attend the meeting in Rome, because Stefansson (1964, p. 149) later explained that Lord Strathcona "…either personally or acting for Canada, financed my trip to the International Geographical Congress in Rome …"

Expedition—newspaper, magazine and book rights—to the *London Chronicle* through the United Newspapers Ltd., and the pictures to be taken (still and cinematographic) by the Expedition members to the Gaumont Company Ltd."[30] As Stefansson remarked much later, being the only unpaid member of the Expedition (albeit at his own request) entitled him to seek compensation later through lecturing and writing, for which he needed exclusive publishing contracts.

Extra Responsibilities for Dr. Anderson

By going to Europe in March 1913, Stefansson left upon Dr. Anderson's shoulders the responsibilities of arranging and coordinating the acquisition of the Expedition's main ship, supplies and equipment, and an assortment of Expedition-related correspondence. Having these unexpected responsibilities thrust upon him without forewarning hardly pleased Dr. Anderson, for he was duty-bound at the time to complete his report on his previous four years of work for the American Museum of Natural History before he left that museum in May. The Geological Survey of Canada undertook to provide the variety of instruments needed by their own men, as well as notebooks and stationery supplies, while the Department of the Naval Service in Ottawa was responsible for supplying the provisions and clothing for all of the men. But questions invariably arose, and with Stefansson traipsing around Europe, those two agencies turned to Dr. Anderson for answers.

In the weeks following Stefansson's departure Dr. Anderson had to write letter after letter about the Expedition's flagship, supplies, equipment and personnel. His letters went to the manufacturers of equipment, clothing and canned foods, food suppliers, to persons in Alaska inquiring about sled dogs and sleds, and to several scientists concerning their employment and travel arrangements. Once news of the Expedition appeared in the American newspapers, many adventurous young men submitted their requests for employment. These, too, Dr. Anderson had to answer. He also wrote to several publishers requesting donations of Arctic and scientific books for the Expedition. Several of them responded favourably; the Macmillan Company, for example, sent three hundred books.[31] When these were received, they were divided, with one lot placed on the *Karluk* for the Northern Party, the other set aside to be placed on the *Alaska* at Nome for the Southern Party. Each party had, in addition, a complete set of the most recent Encyclopaedia Britannica.

Fortunately, Dr. Anderson did not have to arrange for the renovations required by the twenty-eight-year-old whaler *Karluk*, which Stefansson's San Francisco friend, Captain C. Theodore Pedersen, had duly selected and arranged to purchase for the Canadian government for $10,000 from the Stabens & Friedman Company in San Francisco. The Department of the Naval Service then arranged for Captain Pedersen to hire a temporary crew in San Francisco and take the *Karluk* from there to Esquimalt on Vancouver Island for renovations. It also provided an agent (George Phillips) in Esquimalt, British Columbia, who oversaw the receipt and storage of the masses of supplies that soon began to arrive for the Expedition from all parts of Canada, the United States and even Europe.

---

[30] *Idem.*

[31] Brett, 1913 (May 6, 1913).

The Problem with the Pemmican

Telegraphic cables flowed regularly back and forth across the Atlantic Ocean between Stefansson and Dr. Anderson during March and April. In one of his cables to Stefansson, Dr. Anderson stated that he wanted to have quality tests run on the purity of the Underwood brand of pemmican, which was produced in the U.S., because he had heard reports of glass or metal pieces being found in it. Pemmican was a canned-food concentrate then used by Arctic explorers and their dogs. The pemmican eaten by humans contained meat and suet, with added raisins and sugar and was packed in blue-labelled tins, whereas the dog-food variety contained only meat and suet and was packed in red-labelled tins for easy identification. Public demand was small, the product took time to manufacture, and pemmican manufacturers did not generally keep large supplies on hand. Stefansson had insisted that his Expedition be supplied with a large amount of this essential food product, and so became quite concerned that the tests Dr. Anderson advocated might prevent the supplies being available on time. Angrily he cabled back "DAMN PURITY TESTS ORDER PEMMICAN IMMEDIATELY WE HAVE NO ALTERNATIVE."[32] Despite such agitated instructions, the more cautious Dr. Anderson had the quality tests run, with satisfactory results, and the Expedition received its desired quantity on time. Many months later, however, Stefansson complained about the quality of some of the pemmican. McKinlay even suspected that the deaths of some of the survivors of the *Karluk* on Wrangel Island might have resulted from the questionable quality of the pemmican they ate. Following his return from the Arctic, however, Dr. Anderson reported in response to a question about the pemmican from the Deputy Minister of the Department of the Naval Service, "We had four different kinds of pemmican and never found a defective can of any kind." Some of his men found the Underwood-brand man-pemmican overly sweet and that company's dog-pemmican a trifle salty for human use, but that was all. The pemmican of preference to all of the men in his Southern Party was the brand made in Moss, Norway.[33]

Stefansson's Citizenship

Towards the end of March 1913, George Perley, the Minister without Portfolio with whom Stefansson had previously been in contact, raised the matter of Stefansson's nationality in a confidential letter to the Minister of Marine and Fisheries, J.D. Hazen.[34] He felt that Stefansson should take the oath immediately, but Stefansson had evidently replied that to do so would be unwise "...as the newspapers would probably comment on it a good deal and it might look as though he had been persuaded to do so because the Canadian Government had taken over the expedition."[35] Stefansson was then in London, England, so the oath-taking was arranged for a later date, before the Expedition headed north.

Stefansson returned to New York from Europe towards the end of April, and on April 30 made a hasty trip to Ottawa. The following day, the federal cabinet discussed his citizenship and he was advised "to make formal declaration of his desire and intention to be a Canadian

---

[32] Stefansson, 1913c (March 28, 1913).

[33] Anderson, 1916c (December 15, 1916).

[34] Hazen was also the Minister of the Department of the Naval Service.

[35] Perley, 1913 (March 21, 1913).

citizen." Stefansson argued that such action was unnecessary as he had long been a citizen of both countries, and then added that Prime Minister Borden had assured him some weeks earlier that in his opinion Stefansson's "American situation" did not invalidate his British citizenship.[36] Nonetheless, the Canadian government was not prepared to place an American citizen, not even one born in Canada as Stefansson had been, in command of a Canadian expedition into the Arctic. And so, on May 3, 1913, Stefansson quietly took and subscribed to the Oath of Allegiance to His Majesty King George the Fifth, the official document being signed by Rodolphe Boudreau, Clerk of the Privy Council.[37] Ironically, Stefansson resided in the United States for most of the rest of his life, rather confirming Dr. Anderson's comment some years later that Stefansson "...considered any means justifiable to accomplish any end he had in view."[38]

Last-Minute Activities

During his stay in Ottawa in the first few days in May, Stefansson worked on Expedition details and arrangements with G.J. Desbarats and Dr. R.W. Brock. He probably also spoke of the urgency of ensuring Dr. Anderson was part of his Expedition, because a few days later Dr. Anderson received an offer, which he accepted, of the position of mammalogist in the Geological Survey of Canada.[39] His appointment came with the prospects of continuing his field work in Arctic Canada after he returned from Stefansson's new Expedition.

Immediately after his visit to Ottawa early in May, Stefansson reportedly went to Boston for thyroid surgery, but was back in New York by the third week of the month.

"Stefansson took some time off in the spring of 1913, just before going north the last time, and said that he had to undergo a thyroid operation in Boston. He was not in his former good shape afterwards ..."[40]

Whether Stefansson had the surgery or any form of treatment on that occasion is not known.

Dr. Anderson dutifully wound up his work at the American Museum of Natural History and reached Ottawa on May 23, at which time he telegraphed Stefansson in New York that he had conferred with Dr. Brock and six members of the Southern Party, and planned to proceed to Vancouver on May 26. Stefansson hurried to Ottawa, met with Dr. Anderson and the scientists, and urged them to head west immediately in order to attend to matters at

---

[36] Anonymous, 1923c. Uncertainty on Stefansson's nationality continued to remain in the minds of Canadian government officials for many years. An argument about it arose in Parliament in 1925 between the former prime minister, Arthur Meighen, and the Minister of Mines Charles Stewart, the former insisting Stefansson was Canadian, the latter expressing some uncertainty on the matter. Another member (Mr. Forke) summed things up nicely by interjecting "Sometimes he is an American and sometimes he is a Canadian, just as it suits him or wherever he happens to be living" (Hansard, June 10, 1925, p. 4267).

[37] An authoritatively signed Naval Service Department document with that date is in RG42, Vol. 475, File 84-2-29, LAC.

[38] Anderson, 1921a (March 5, 1921).

[39] Foran, 1913 (May 3, 1913). Foran was Secretary of the Civil Service Commission of Canada.

[40] Anderson, 1921a (March 5, 1921).

dockside in Esquimalt, assuring them he would follow within a few days. In a letter to Dr. C. Camsell, Deputy Minister of Mines, eleven years later, Dr. Anderson wrote: "Mr. Stefansson was suspiciously anxious to get every member of the Expedition away from Ottawa as soon as possible in May 1913, being particularly urgent in my own case. I was finally induced to leave on May 27, 1913, in the assurance of Mr. Brock that the affairs of the members of the Geological Survey party would be looked after."[41]

Dr. Anderson was the last but one to leave Ottawa before Stefansson,[42] boarding the train for Vancouver at 1:20 a.m. on May 27 and reaching Victoria on May 31. He stayed the first night at the Empress Hotel but moved into the James Hotel the next morning where the other scientists were housed. Stefansson remained in Ottawa for a few more days, helping Desbarats and Dr. Brock finalize the government's formal written instructions for each of the scientists. This was when Desbarats, on Stefannson's advice, inserted several restricting sentences in each of the contracts. The new wording in Stefansson's contract, dated May 29, reads as follows:

> The members of the party will not engage in any private trading or make any private collections of specimens or of photographs. They will also refrain from giving out news of the Expedition until authorized. They will not be allowed to write paid magazine articles or any other popular description of the Expedition for one year after the return of the Expedition.[43]

Similar restrictive statements were included in the final contracts of each of the scientists. Stefansson took them west with him when he left Ottawa. In Victoria, after he distributed them, the added restrictive sentences caused the first of several heated arguments between Stefansson and the scientists, as Chapter 2 reveals.

---

[41] Anderson, 1924a (January 21, 1924).

[42] J.J. O'Neill left Ottawa on June 3, 1913 (Brock, 1913c).

[43] Copy of Instructions, Desbarats to V. Stefansson, May 29, 1913. MG30 B40, Vol. 1, File 13, LAC.

2

VICTORIA AND NOME

The *Karluk*

Shortly after his return from the Arctic in 1912, Stefansson had asked the experienced Arctic whaler, Captain C. Theodore Pedersen, with the assistance of three shipping inspectors, to examine all suitable ships then for sale on the Pacific coast and to advise him which one they considered best suited for ice work in the North. Pedersen subsequently selected the 28-year-old, 321-ton, 125-foot, wooden brigantine *Karluk*,[1] the largest and oldest of the three available vessels found to be even remotely suitable. Early in February 1913, after the Canadian government agreed to sponsor his Expedition, Stefansson recommended that the government purchase the *Karluk* from the Stabens & Friedman Company of San Francisco for $10,000. This the government proceeded to do, then Captain Pedersen gathered a temporary crew and sailed it from San Francisco to Canada's west-coast naval base at Esquimalt, near British Columbia's

Figure 3 The CAE flagship HMCS *Karluk* at Esquimalt, British Columbia, June 17, 1913.
Photo: Library and Archives Canada, photo C-032638

---

[1] Bockstoce, 1977, p. 97.

provincial capital, Victoria, on Vancouver Island. There the government engineer inspected it and quickly ascertained that the vessel needed to be overhauled from stem to stern. Repairs commenced immediately.

Karluk is an Aleutian word for fish. It was an appropriate name for the ship, for it had originally spent many summers successfully collecting salmon along the Alaskan coast for delivery to larger ships before it was put into the whaling service to operate near Herschel Island in 1892. It then made fourteen voyages to the Arctic as a steam-whaler, wintering there on five of them; then after the lucrative Arctic whaling industry collapsed, it had been put up for sale in 1911.[2]

Stefansson had wanted Captain Pedersen to be his ship's master on the three-year Expedition he was planning, but Pedersen suddenly quit early in May, shortly after taking the *Karluk* to Esquimalt. Evidently he had grown dissatisfied with trying to deal with Stefansson, for on May 5, he sent the following blunt telegram to Stefansson in New York:

> HAD TO ACCEPT FRISCO WHALING EXPEDITION CONSIDER YOU HAVE HAD SUFFICIENT TIME TO COME TO SOME UNDERSTANDING WITH ME JUDGING BY YOUR ACTIONS YOU MIGHT DITCH ME WHEN TOO LATE FOR OTHER OUTFIT YOUR WIRE FEBRUARY STATING CAN OBTAIN CANADIAN MASTER KARLUK THEREFORE WILL NOT PUT YOU OUT ANY PLEASE ARRANGE TO SETTLE UP WITH ME AT ONCE BEGINNING JANUARY EIGHTEENTH I EXPECT TWO HUNDRED PER MONTH FOR MY SERVICES EXPECT TO SAIL FOR FRISCO SATURDAY.[3]

It is obvious that Pedersen was annoyed with Stefansson when he departed, but the official explanation for his departure was that he was an American citizen and the Canadian government insisted that its flagship be captained by a British subject. That was also more or less how Stefansson explained the departure of Captain Pedersen.[4] Stefansson hastily consulted with Rear Admiral Robert Peary about a replacement and was most fortunate to be able to hire on short notice Captain Robert Bartlett of Brigus, Newfoundland, who had been Peary's ship's master on his famous 1905–1906 polar expedition and had the reputation as the world's best ice master.[5] Ironically, although by birth a British subject, Bartlett had become an American citizen.[6] That he was an able sea captain in the Eastern Arctic there was no question, but he had no sea-ice experience whatsoever in the Western Arctic, where conditions are markedly different.

The *Karluk* underwent renovations and vital modifications at the Canadian naval base at Esquimalt during April and May of 1913. Its hull and prow were strengthened, and more suitable quarters were created for the Expedition's scientists. At last, early in June, the vessel was declared ready to start loading for its northern voyage. However, the *Karluk's* new captain,

---

[2] *Ibid.*

[3] Pedersen, 1913 (May 5, 1913).

[4] Stefansson, 1921b, p. xi.

[5] *Op. cit.*, p. 27.

[6] *Op. cit.,* p. xi.

Bob Bartlett, remained skeptical of its condition in spite of official assurances, and telegraphed his views to the Department of the Naval Service in Ottawa that the ship was absolutely unsuitable for wintering in the Arctic ice. The *Karluk* had, in fact, wintered in the Western Arctic several times during its twenty-seven years of service there, but always in a sheltered location (mainly at Herschel Island near the Alaska-Canada boundary), never in the open sea, as Stefansson contemplated might happen on this Expedition. It subsequently turned out that even with its fortified prow, the *Karluk* could only attack very thin ice head on, lacked the body strength to withstand the pressure of ice broadside and its engine was much too underpowered to push the ship through loose ice.

The appearance of the *Karluk*—an old, square-sailed, steam brigantine, formerly used as a whaling ship—did little to comfort the Expedition members. As photographer George Wilkins later recalled, it stank of whale oil, and its living quarters were unpainted, crowded, smelly and swarming

Figure 4 Captain Robert A. Bartlett on the deck of the *Karluk* at Esquimalt, B.C., June 1913.
Photo: Library and Archives Canada, C-004511

with cockroaches.[7] Its engine was apparently no better, for its chief engineer, John Munro, "says this old coffee pot of an engine was never intended to run more than two days at a time."[8] The ship was certainly no prize, condemned by all.

By mid-June, the *Karluk* was reconditioned, loaded and ready to sail, its decks so thoroughly crowded with equipment and heavy bags of coal that there was scarcely room for its crew and passengers to move about.

As Stefansson was occupied with many other activities, he had left the hiring of the ship's crew to Captain Pedersen. The best that could be said of the resultant group of men was that it was a motley bunch. Fortunately, several of them proved to be good men, but two of them illegally smuggled liquor on board the *Karluk,* and the twenty-year-old Scottish cook was an admitted drug addict. A fireman named J. Ridley was discharged at the end of the first week in June before Stefansson reached Victoria, and a second fireman, T. Wiseman, was discharged

---

[7] Thomas, 1961, p. 65.
[8] Chipman, 1916 (July 2, 1913).

when the *Karluk* reached Nome, along with the first engineer, Thomas Anderson.[9] Reasons for their discharges are not known. Then shortly after taking charge of the *Karluk*, Captain Bartlett fired the first mate, James F. Allen, for incompetence. Dr. Anderson later stated that Allen "... became dissatisfied with the ship and other conditions and resigned."[10] Allen's date of discharge was June 13. This action created a new problem, however, for Allen had overseen the loading and stowing of the cargo, and took with him the knowledge of how and where the supplies and equipment were distributed on the *Karluk*.

The anthropologist Diamond Jenness was the first of the scientists to reach Victoria, arriving from New Zealand on April 30. While awaiting the arrival of the others, he buried himself in the local parliamentary library, learning all he could about the Arctic and its people. The other scientists arrived intermittently between May 23 and June 5. Last to arrive was Stefansson, who made his appearance on Saturday, June 7.

The Controversial Victoria Meeting

Shortly after his arrival at Victoria, Stefansson called a meeting of the Expedition's scientific members for the morning of June 8 at the James Bay Hotel, where most of them were staying. At the start of the meeting he explained that the Northern Party, under his command, would use the *Karluk* as its base for its exploration until new land was found or a suitable land base was established. The Southern Party would operate under Dr. Anderson's leadership from a shore base somewhere in the Coronation Gulf region and would have the use of a small schooner, *Alaska*, which Stefansson had arranged to purchase and take possession of when the Expedition reached Nome, Alaska. During this initial part of the meeting several of the scientists questioned Stefansson about the food and water supplies, equipment, clothing, and other details about the Expedition. The vagueness of his replies failed to satisfy their concerns and greatly agitated several of the men.

Stefansson then handed out new official instructions and contracts, which he had brought from Ottawa for each of them to sign. Prepared by G.J. Desbarats and Dr. R.W. Brock, with input by Stefansson and to a lesser extent Dr. Anderson, the instructions and contracts contained several new, restrictive and controversial statements that stirred the men's ire to a greater pitch. The feisty little Scotsman, James Murray, the Northern Party's oceanographer and a veteran of one Antarctica expedition with the explorer Ernest Shackleton, was especially outspoken, angrily telling Stefansson "... he was willing to exert himself to the utmost of his ability to collect material and data for the government, but that he had not contracted to sell himself 'body, mind and soul' for three years to Mr. Stefansson."[11]

---

[9] The names of these three men and their wages for employment with the *Karluk* in June 1913 appear on a "Monthly Pay-List and Board Bill" form of the Department of Marine and Fisheries Canada, a photocopy of which I obtained from the Stefansson Collection at Rauner Library, Dartmouth College, Hanover, N.H., in 2003.

[10] Anderson, 1919b, pp. 13–14.

[11] *Op. cit.,* p. 15.

Figure 5 The CAE scientists in front of the Empress Hotel, Victoria, B.C., June 1913. Front row (l. to r.): Wilkins (with camera), Chipman, Bartlett, Stefansson, Dr. Anderson, Murray; back row (l. to r.): McKinlay (seated), Beuchat, Mamen, Jenness, Malloch, Cox, O'Neill, Johansen.
Photo: Library and Archives Canada, photo by Curtis & Miller, Seattle, PA-074066

The details of the contract Jenness received that morning have survived[12] and are representative of those received by the other scientists. Each of the men residing in Ottawa or stopping off in Ottawa en route to Victoria (which excluded Jenness) had already signed a contract concerning Expedition conditions, but this new contract Stefansson had just handed out to them for their signatures contained new restrictions not even Dr. Anderson knew about. These were: (1) no personal trading for commercial purposes; (2) no public dissemination of news about the Expedition during its stay in the Arctic and for a year thereafter; (3) no magazine articles or popular books were to be written before one year after the return of the Expedition; and (4) all mail was to be sent to the Geological Survey of Canada for forwarding to the intended recipients.[13]

---

[12] Jenness, S.E. (Ed.), 1991, p. 728.
[13] *Ibid.*

The first restriction was understandable, for the scientists were in the employ of the Canadian government and could not use government time and means for personal gain. Restrictions (2) and (3) had been newly inserted to protect Stefansson's exclusive publishing contracts with news media in London, England, the *New York Times*, and *The Globe* in Toronto (the predecessor of the present *Globe and Mail*). It was through these contracts that he expected to receive a lot of the financial compensation for his leadership role in the Expedition but, of course, of this he made little or no mention. Restriction (4) was likewise new, and its hint of censorship infuriated the men.

Stefansson then dropped a verbal bombshell, by announcing:

> ... that although the orders did not so state, all private diaries and writings as well as scientific journals and notes made by any member of the Expedition must be turned over [to the government] upon the close of the Expedition (Stefansson himself to have access, of course, to these private diaries and notes to make such use of the material, literary or otherwise, as he wished). This was the day that Dr. Anderson and staff learned for the first time that Stefansson while in Europe the spring before had sold out the newspaper rights of both parties (quite contrary to his original agreement with Dr. Anderson) to the United Newspapers Limited and the Gaumont Company.[14]

The scientists had been instructed earlier to keep diaries of their daily activities, but these they regarded as their own personal property. As a result, they almost unanimously voiced their disapproval as soon as Stefansson finished speaking. Murray and Jenness objected strenuously on principle, indignant that Stefansson or anyone else would be reading their diaries following the completion of the Expedition. Stefansson refused to withdraw this controversial stipulation, however, even when Jenness informed him that he had a letter from his superior in Ottawa (the Chief Anthropologist Dr. Edward Sapir) assuring him that his diary would be his own private property. The subject remained both unsettled and a major bone of contention.

The feeling quickly spread among the scientists that what they had believed was going to be a great scientific expedition was fast becoming a newspaper and magazine exploiting scheme to enhance the reputation of Stefansson. None of the men objected to providing Stefansson with any scientific information he might want, or to him sending press releases to the newspapers from time to time, but they were troubled that the stories he was issuing to the press were already tinged with fiction. And they did object strenuously to Stefansson putting over an underhanded coup whereby he could exercise control over all the scientific and literary productivity of the Expedition, giving him a chance to skim over and plagiarize the best thoughts and results of a select body of scientific experts for three years for his own personal aggrandizement.

The scientists subsequently reacted in a variety of ways to Stefansson's instructions about their personal diaries. Most, like Jenness, chose to avoid discussing personal topics in the diaries they wrote, geographer J.R. Cox and geologists J.J. O'Neill and George Malloch did not keep formal diaries, and the geographer K.G. Chipman chose to ignore Stefansson's announcement

---

[14] Anderson, Mrs. R.M. (no date).

entirely. Instead he wrote what he liked, but inserted in the front of his diary the statement that as an employee of the Geological Survey of Canada he was … "not bound, unless it is of his own free will, by any expedition or any other regulations pertaining to his work."[15] The Danish biologist Fritz Johansen responded in an unusual fashion. Aware that Stefansson could only read New Danish, he deliberately wrote his personal comments over the next three years (when he bothered to write anything) in Old Danish in the skimpy notes in his field note-books.[16] Meteorologist W.L. McKinlay promptly decided to keep two diaries, one to turn in to the Canadian government, the other containing whatever personal views he decided to express. The photographer George Wilkins ignored Stefansson's instructions and wrote as he pleased in his diary, for he was not directly answerable to either the Canadian government or Stefansson. His penmanship, however, was almost illegible. Stefansson's secretary, Burt McConnell, included some interesting personal remarks in his diary, but he left the Expedition after only one year. Topographer Bjarne Mamen, one of the unfortunate scientists who lost their lives following the sinking of the *Karluk*, included only a small number of personal comments in his diary, which he wrote in Norwegian. Dr. Anderson did not keep a personal diary, but did write detailed field notes daily, among which are a few brief personal comments, and voiced many personal comments about the Expedition in letters to his wife. And Murray decided to express his private opinions and comments in letters to trusted friends in the "outside" world, as did McKinlay and others among the scientists. From a historical point of view, it is a great pity that this controversial issue arose when and in the manner it did, for it is probable that if Stefansson had not insisted upon it at the Victoria meeting, several of the scientists' diaries might have proven much more informative than they are.

Dr. Anderson,[17] Jenness, and one or two others became so incensed over the new contract restrictions, and other matters that arose at the Victoria meeting, that they contemplated resigning immediately from the Expedition. In fact, Dr. Anderson did offer his resignation to Stefansson and telegraphed Ottawa of his action. He retracted his offer, however, after Stefansson

---

[15] Chipman, 1916 (Inside front cover).

[16] Among Dr. Anderson's papers that I examined in the 1980s was a copy of a letter to V. Stefansson in Nome, Alaska, dated June 25, 1913 from Dr. R.W. Brock, the Director of the Geological Survey of Canada. In that letter Brock stated that he had a letter from Johansen expressing concern about publishing rights, evidently written after the controversial Victoria meeting. Brock responded that "… popular rights for a year after the return were yours (Stefansson's) … emphasizing the fact that the scientific results were the property of the Government, from whom permission to pub-lish would have to be obtained. He (Johansen) proposed keeping his note-books but promised to give the Government the scientific contents. I told him the Government must have the original notes … You had better speak to him as he seems to be on the wrong track." It is not known whether or not Stefansson spoke to Johansen, but Johansen kept his notebooks. (Brock, 1913d [June 25, 1913]).

[17] In November 2010, I unexpectedly received from Alan Purdy, husband of the late Isabel Anderson Purdy, a copy of a letter dated February 16, 1980, written to Isabel's sister Dorothy Smith (now deceased) by W.L. McKinlay in which he stated "… your dad (Dr. R.M. Anderson) was not pres-ent at any of the two conferences in Victoria and Nome …" The letter was my first indication that he was not present at the Victoria meeting.

agreed to have nothing to do with the direction of the Southern Party once the Expedition left Nome. The others also refrained from resigning after due reflection, probably out of a strong sense of duty and commitment to the Expedition and loyalty to their colleagues.

Curiously, Dr. Anderson signed a separate but equally severely binding document for Stefansson four days later, agreeing not to submit any reports, popular articles or books about the Expedition for publication until after the completion of the Expedition, and agreeing even then to submit his articles to the *Daily Chronicle* to market, subject to an agent's fee agreed upon by Stefansson.[18] As events of the next three years unfolded, however, Dr. Anderson was so busy he had little time or inclination to write any articles about the Southern Party, in spite of frequent urging by his wife in her letters to him. She apparently wanted him to publish popular articles and give popular lectures after his return from the Arctic, just like Stefansson, but Dr. Anderson, like the true scientist he was, did not yearn for popular acclaim by this means.

Thus, within hours of his appearance in Victoria, Stefansson succeeded in antagonizing almost all of the scientists he was depending upon to bring success to his Expedition. It was the first major friction to surface on the Expedition. Although this was Stefansson's third expedition, it was his first experience in charge of well-educated individuals and he had certainly gotten off to a bad start. The hostility of his scientific men towards him—rather than improving with time—with but few exceptions got worse, relieved only a little by the considerable distance that separated them and the almost total absence of communication between Northern and Southern Parties following the first year in the Arctic.

In an effort to prevent further interference, members of the Southern Party asked Stefansson before the Victoria meeting ended to clarify his role in the operations of their party, inasmuch as they were directly responsible to the Geological Survey of Canada. In response, "Mr. S. agreed to have nothing to do with direction of Southern Party after leaving Nome, stating that he only wished nominal control to protect his newspaper contracts."[19] They then insisted that he put in writing that the Northern and Southern Parties would operate separately after leaving Nome. Stefansson assured them that he would do so, but no written statement ever appeared.

Pre-Sailing Activities

The Expedition members kept busy in the days immediately before their departure, but also had time to enjoy several social events. These included an evening picnic on a nearby beach on June 14, a visit to the *Karluk* from the province's Premier Sir Richard McBride on June 15 and lunch on June 16 at the Empress Hotel, sponsored by the British Columbia government. On the evening before their departure, Wilkins and McConnell even enjoyed a noisy automobile ride with two young women from Victoria.

One of the last dockside activities during the loading of the *Karluk* was the secreting on board the ship of a stray young female black-and-white cat by a member of the crew so that

---

[18] Stefansson, 1913d.

[19] Anderson, 1919a, p. 1. "History of Operations of Canadian Arctic Expedition, 1913–16 ..." This is an annotated list of 30 chapters prepared in May 1919, outlined by Dr. Anderson for volume 1, part B, in the series of Reports of the Canadian Arctic Expedition 1913–1916.

Figure 6 Vilhjalmur Stefansson standing on coal bags to address CAE scientific staff on the deck of HMCS *Karluk* at Esquimalt, B.C., June 1913. (l. to r.): Johansen, Murray, Dr. Mackay (with cap), Malloch, Mamen, Jenness, McKinlay, Stefansson, O'Neill, Dr. Anderson.
Photo: Library and Archives Canada, photo by Curtis & Miller, Seattle, PA-203451

it could serve as both rat-catcher and ship's mascot. It was carefully looked after in the days that followed because of a general belief among the *Karluk's* crew that the cat's survival ensured their safety.[20]

A formal ceremony, involving British Columbia's Lieutenant Governor T. W. Paterson and Premier McBride, took place on the crowded deck of the *Karluk* on June 17. Later that day, while the *Karluk* was on a test run, the president of the Canadian Club of Victoria presented members of the Expedition with a "Canadian flag."[21] Shortly after 7 p.m. the fully loaded *Karluk*,[22] manned by eleven men (its captain, two engineers, two firemen or stokers, a steward, a messroom boy and four sailors), steamed around Esquimalt Harbour, receiving cheers from the sailors on board two British naval vessels in the harbour, HMS *Shearwater* and HMS *Algernine,* and one Canadian naval vessel, HMCS *Rainbow,* and flag-dipping salutes from various other boats before commencing its 3,000-mile journey north to Nome, Alaska.

---

[20] McKinlay to S. E. Jenness, oral communication, 1980.

[21] Chipman, 1916 (June 17, 1913). The "Canadian flag" was the red ensign.

[22] So fully loaded was the *Karluk* when it left Esquimalt that Stefansson reported "My own cabin was filled to the doors with trunks and personal baggage of the scientific staff and with many odds and ends and will not be used at all on the voyage to Nome." In addition, there were fifty tons of coal on the deck (Stefansson to Desbarats, 1913e).

Stefansson was on board the *Karluk* for promotional purposes during its departure, but once it was out of sight of the well-wishers at Esquimalt, he got into the launch that had accompanied it and with a final farewell wave returned to Victoria. He stayed behind with Dr. and Mrs. Anderson and Mr. and Mrs. Murray to wind up local arrangements and to obtain additional supplies for the Expedition from supply houses in Seattle.[23] When that had been accomplished a week later, the Andersons, Murrays and Stefansson (with his two female secretarial assistants) all proceeded from Seattle to Nome on the coastal steamer, SS *Victoria*, on July 1. George Phillips, the Naval Service Department's local stores clerk who had been overseeing the Expedition's Esquimalt preparations, accompanied them to assist with the legal transfer of the schooner *Alaska* and purchasing arrangements for additional supplies at Nome.

Northward Bound

Complaints about the *Karluk* surfaced soon after its departure. First, the chief engineer John Munro[24] complained about the wretched condition of its engine. Then the supply of fresh water proved inadequate and had to be rationed, to the disgruntlement of all. Geologist Malloch's assistant Mamen grumbled that several of the Expedition men were too lazy to assist in such regular ship's chores as disposing of the burnt ash from the boiler, citing especially McConnell, who "... can do nothing but typewriting and stenography ..." and H. Beuchat, who "can only sleep."[25] And Captain Bartlett later expressed to Mamen his lack of enthusiasm for the *Karluk's* "phenomenal" speed, cursing both the boat and the leader of the Expedition.[26] Then some time during the voyage the twenty-year-old Scottish fireman, T. Wiseman, went on a hunger strike, claiming he was too light for the job of stoking the ship's boiler with coal and would not do it. He was read the ship's law in the captain's cabin, threatened with the docking of two days' pay for each day he refused to work, told that he was liable to thirty days in jail at Nome, all to no avail.[27] As a consequence, he was dismissed after the *Karluk* reached Nome, and his job was taken over by Fred Maurer. There was no one available to replace Wiseman.

---

[23] On July 26 Stefansson acknowledged receipt of the $5,000 in cash he had requested (Stefansson to Chief Accountant, 1913f). Three days later he informed Dr. Anderson of his new plan, which was to retain for the Northern Party all of the provisions then on the *Karluk*, to ensure it had a five-year supply in case it was caught in the ice, and to purchase supplies immediately for the Southern Party at Seattle for shipment to Nome and transfer to the *Alaska*. Needless to say, Dr. Anderson was far from pleased, having spent much time in recent months selecting and testing special condensed foods he expected his Southern Party to share while in the North. (Anderson, 1916a [June 24, 1913]).

[24] Munro had been a junior officer in the British Navy on the British warship HMS *Rainbow* when the vessel reached Esquimalt in the summer of 1913. His term of enlistment expired while the *Rainbow* lay at anchor there. The commandant of the Esquimalt Navy Yard subsequently recommended him to Captain Bartlett for the job of engineer on the *Karluk* (Bartlett and Hale, 1916, p. 99). The *Rainbow* had been turned over to the Canadian government by the time the *Karluk* sailed north on June 17.

[25] Mamen, 1914 (August 10, 1913).

[26] *Ibid.*

[27] Chipman, 1916 (June 27, 1913).

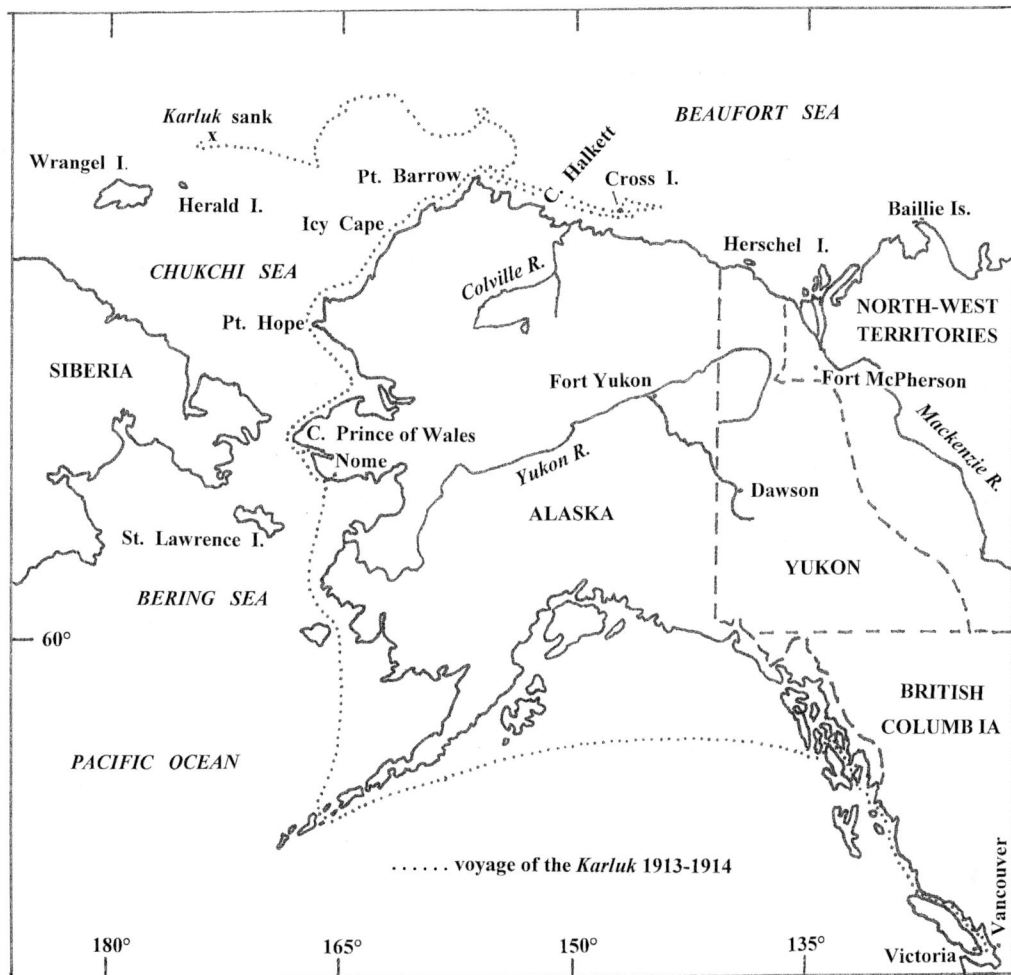

Map 1 The last voyage of the *Karluk*, 1913–1914. "X" marks the locality where it sank. "Shipwreck Camp" was nearby.

During the voyage north, someone (probably McConnell) recorded the ages, heights and weights of the scientists (and McConnell) on the *Karluk*.[28] From oldest to youngest these were:

Mackay, Surgeon	35	5′ 11″	178 lb
Malloch, Geologist	33	5′ 11″	168 lb
Beuchat, Anthropologist	31	5′ 9½″	154 lb
Johansen, Naturalist	31	5′ 10″	168 lb
Chipman, Topographer	28	5′ 9″	142 lb
Jenness, Anthropologist	27	5′ 6½″	128 lb
O'Neill, Geologist	26	5′ 9½″	152 lb

---

[28] A page with this information accompanies the daily entries known as the Karluk Chronicles. MG30 B40, Vol. 10, File 6, LAC.

Cox, Asst. Topographer	26	6' 1½"	160 lb
McKinlay, Magnetician	24	5' 6"	125 lb
Wilkins, Photographer	24	5' 10½"	165 lb
Mamen, Asst. Topographer	22	6' 1½"	165 lb
McConnell, Meteorologist	24	5' 10½"	164 lb
Stefansson, Commander	33		
Anderson, Head, Southern Party	37		
Murray, Oceanographer	?		

The last three men were not included on McConnell's list because they remained behind when the *Karluk* sailed, and travelled north later on the coastal steamer. Stefansson and Anderson were nearly six feet tall and sturdily built, and Murray was short and somewhat pudgy. Jenness and McKinlay were the shortest in height, light in weight and, according to the latter, were known initially as "the twins."[29]

Meanwhile, the *Karluk* worked its way slowly north, proceeding up the Pacific coast at a speed of about five knots per hour, stopping briefly to take on fresh water at Duncan, then Prince Rupert and then Wrangell. At this last settlement it dropped off the ship's pilot, Captain Josiah Gosse, and headed westward across the Gulf of Alaska for Unimak Pass in the Aleutian Islands. Boxing matches and the regular shifting of heavy bags of coal from the deck to the boiler room provided the scientists with breaks from the daily monotony of the long and tediously slow voyage.

On July 2 the *Karluk* passed safely through Unimak Pass and reached Nome shortly after midnight on July 9. Four hours later, the coastal steamer SS *Victoria* arrived with Stefansson, the Andersons, the Murrays and Phillips. It also brought two ladies—Gertrude Allen (Stefansson's secretary) and Bella Weitzner, an editorial assistant with the American Museum of Natural History in New York—who were helping Stefansson to complete his official report of his 1908–1912 expedition.[30] While en route north, Stefansson had also been dictating to Miss Weitzner chapters for his first popular book, *My Life with the Eskimo*, which set forth his experiences during the previous four-year expedition. It was published within a year and promptly became a source of friction between Dr. Anderson and Stefansson because of the latter's extensive inclusion of Dr. Anderson's text and photographs without sufficient credit.

Trouble at Nome

By the time the *Karluk* reached Nome, the morale of the men on board was in dire need of boosting. They were disheartened by the condition of the *Karluk*, the enormous array of supplies brought north for them from Seattle on the SS *Victoria*, which soon awaited them on the dock, and disillusionment with Stefansson's leadership. Several of the men wanted to return south, but were persuaded not to desert their companions. Dr. Anderson worked hard at the waterfront day after day, trying to locate and move the Southern Party's supplies off the *Karluk*. Because of the disorganized manner in which the ship's cargo was originally packed, however,

---

[29] McKinlay to S.E. Jenness, oral communication, 1980.
[30] Stefansson, 1919m.

Figure 7 The CAE scientists at Nome, Alaska, July 1913. Back row (l. to r.): Mamen, McConnell, Chipman, Wilkins, Malloch (with cap), Beuchat, O'Neill, Jenness, Cox, McKinlay; front row (l. to r.): Dr. Mackay, Captain Bartlett, Stefansson, Dr. Anderson, Murray and Johansen.
Photo: © Canadian Museum of Civilization, 27790

he was not completely successful, in spite of his determined efforts earlier at Esquimalt, and many of the supplies intended for the Southern Party remained unfound on the *Karluk* and were subsequently consumed or lost. And then there was a problem with Stefansson over the condensed rations—pemmican, dried foods and the like—for sled trips.

> The condensed rations, pemmican, chocolate, 'iron rations,' erbwurst and what-not for both parties were shipped to Nome on the *Karluk*, and the Southern Party's supplies were to be transferred to the *Alaska* at Nome. When we were at Nome, Stefansson began to worry about not having enough supplies on the ship and wanted to take all the pemmican (several tons) and the other condensed food on the *Karluk*, [claiming] that *my* party was going into a better game country, and could live on the country (not Stefansson's party to live on the country). I demurred, as we had work to do which required condensed food for sled trips, but finally I let him have all but about 1,000 pounds of pemmican, and I went into the well-supplied wholesale grocery stores at Nome and bought flour, rice and bacon enough for my party on the *Alaska*, also ordered some to be shipped in to Herschel Island on the *Belvedere*.[31]

---

[31] Anderson, 1921e (November 19, 1921).

Stefansson had arranged by correspondence, before heading north to Nome, for the purchase of the one-year-old, 64-foot, 50-ton gasoline schooner *Alaska* to transport the supplies and men of the Southern Party during the next three years. He was told that the schooner had been used in the coastal mail service, so assumed that it would be in fairly good condition. However, the *Alaska* had been ill-treated and proved to be in need of a new propeller and a great deal of engine repairs. In addition, it was soon found to be too small to carry all the supplies, equipment and men of the Southern Party, so Stefansson had to seek another vessel to transport the excess items to their destination.

On July 10, a few hours after the *Karluk's* arrival at Nome, O'Neill's camera and field glasses, and Chipman's camera disappeared from Johansen's lab on the ship. Much searching of the ship and then inquiries of every member of the Expedition and among the people in Nome finally revealed that a longshoreman of questionable reputation named "One-eyed Myers," who had been hired to help unload some of the cargo from the *Karluk*, had stolen the three items, and they were duly recovered. The *Karluk's* second mate, Charles Barker, had apparently told the thief that he could have them. Stefansson refrained from laying charges, however, because the Expedition would have been delayed beyond the navigation season.[32]

That same day Murray and Chipman, acting as spokesmen for the Northern and Southern Parties, respectively, persuaded Stefansson to attend a general meeting in the evening to discuss concerns some of the scientists had about the Expedition's plans or lack of them. They then in turn asked questions about the food supply for the Southern Party, Arctic clothing, sleeping bags, location of the base for the Southern Party, sled dogs, sledding provisions, lumber for their base camp, fuel, and several other basic items. In reply to Chipman's question about whether the twenty tons of provisions purchased in Seattle after the *Karluk* left Esquimalt were solid provisions suitable for sled work, or simply ordinary provisions, Stefansson "intimated that we had no right to ask such questions as he had thought over the question of provisions and we should have confidence in him." However, he made it clear that Dr. Anderson was responsible for anything wrong with it, for he was the one who had purchased it.[33]

Stefansson's responses to the various other questions asked by the men, and his increasing impatience that so many were asked, left them discernibly dissatisfied. The oceanographer Murray especially raged over Stefansson's vagueness about a base for the Northern Party and his assertion that lives were secondary to attainments of objects, and that Stefansson expected to "put the *Karluk* into the ice" for the winter[34] (just as the Norwegian explorer Nansen had done with his ship *Fram* earlier in the century). Murray was no novice, having previously been to the Antarctic with the British explorer Ernest Shackleton. He felt so certain that the *Karluk* would be crushed by the ice, with resulting loss of life, equipment and data (an unhappy premonition), that he spoke of resigning unless Stefansson provided better answers to their questions about his Northern Party. Instead, however, Murray wrote a strong note to Stefansson insisting that he needed to have a shore base for his oceanographic work or he would go no

---

[32] Chipman, 1916 (July 13, 1913).

[33] *Op. cit.*, July 10, 1913.

[34] *Op. cit.*, July 11, 1913.

farther.[35] Stefansson's response then was to try to persuade Johansen to take Murray's place as oceanographer on the Northern Party, but Johansen declined the offer. Faced with two irate Scots—the oceanographer Murray and the ship's surgeon, Dr. A.F. Mackay, who had also been with Shackleton—Stefansson agreed to place the newly acquired, 41-ton gasoline-powered schooner, *Mary Sachs*, at Murray's disposal for his oceanographic work; then the next day told Chipman the ship would be put to other use.[36]

The question of Dr. Anderson's authority over the activities of the Southern Party, originally aired in Victoria without satisfactory assurances, arose again at the Nome meeting. For a second time Stefansson assured the scientists that when the supplies had been properly divided at Herschel Island, the Northern and Southern Parties would separate and Dr. Anderson would be solely responsible for the Southern Party

Dr. Anderson was not at this meeting, being busy at the time trying to organize the supplies on shore and on the *Karluk*.[37] When he learned later of Stefansson's verbal assurance of separate operations for the two parties, Dr. Anderson again asked Stefansson to put his promise in writing, just as he had done in Victoria. Stefansson assured him that he would do so, but no document ever appeared. Stefansson later claimed that his document was lost with the *Karluk*.

Sizing up the discord among the scientists at Nome, George Phillips, the Naval Service Department's west-coast representative who had worked so diligently at Victoria and Esquimalt to get the Expedition under way and had come north to finish coordinating the supply arrangements at Nome, urged Stefansson to get rid of the Geological Survey scientists who were balking over his authority on the Expedition.[38] This Stefansson could not and would not do.

Stefansson later described the trouble at the Nome meeting as being primarily about dissatisfaction over the water tanks on the *Karluk*.[39] McKinlay, after reading Stefansson's explanation, commented, "That version makes amusing reading but the grossness of the misrepresentation is too obvious to be laboured."[40]

On July 13, three days after his somewhat unruly meeting with the scientists, Stefansson sent them and the *Karluk* north to Teller. This was "partly to get us out of the way and partly to avoid hotel expenses," quipped Jenness in a letter to his favourite New Zealand professor after the move.[41] At Teller the *Karluk* was to have its boilers blown down, and its water tanks repaired.[42] Meanwhile, Stefansson remained at Nome with Dr. Anderson, Murray and the photographer Wilkins. Dr. Anderson and Murray stayed in Nome because their wives were still with them, Wilkins because Stefansson wanted him to photograph Iñupiat scenes around

---

[35] Beuchat, 1913 (July 14, 1913).

[36] Chipman, 1913 (July 18, 1913).

[37] Anderson, 1922h, p. 7.

[38] Stefansson, 1921b, p. 115.

[39] *Op. cit.*, pp. 32–33. Stefansson further fictionalized his account of this meeting in his autobiography (Stefansson, 1964, pp. 154–155). See Chipman's diary for a much different version of the Nome meeting (Chipman, 1916 [July 10, 1913]).

[40] McKinlay to Anderson, 1922b.

[41] Jenness, 1914a.

[42] Anderson, 1916a (July 13, 1913).

Nome. Stefansson then decided, with little or no discussion with Dr. Anderson, that the Northern and Southern Parties would rendezvous at Herschel Island where the men would divide the supplies, store some if necessary and then the two parties would separate. He anticipated no further navigational problems.

"GOODBYE, STEFANSSON"   3

Teller

Following the departure of the *Karluk* for repairs at Teller on July 13,[1] Stefansson quickly set about acquiring a third ship to transport the vast amount of Expedition supplies and equipment still piled on the shore at Nome. Fortunately one was available, and on July 16 he bought for $5,000 the 30-ton, twin-propeller, gasoline schooner *Mary Sachs*, retaining its former owner, Peter Bernard, as its captain.[2]

On that same day Mrs. Anderson left Nome for Seattle on the steamer SS *Victoria*. She was returning to her family's home in Iowa to await the return of her husband in three years' time. They had been married for only nine months and she was pregnant.

On July 19 the *Alaska*, now well loaded with Expedition supplies, sailed for Teller in order to undergo a number of repairs to its engine and propeller. The *Mary Sachs*, likewise heavily loaded, followed four days later, taking with it the scientists Dr. Anderson, Chipman, Cox, Johansen, Stefansson's secretary McConnell and a "hitch-hiker," Ernest de Koven Leffingwell.[3] After sending his three ships north to Teller, Stefansson remained in Nome to wind up an assortment of unfinished business, including freighting arrangements for about seventy-five tons of Expedition supplies and equipment left at Nome after the departure of the fully laden

---

[1] Teller is a small Native settlement on a sandspit seventy-two miles northwest of Nome. A U.S. government reindeer station operated near it from 1892 to 1900. The town was established in 1900 and became a major regional trading centre with a population of about 5,000 after the discovery of placer gold fifteen miles to the south, which became the Bluestone Placer Mine. In 1913, the settlement showed some evidence of its former prosperity but was reduced to a small Eskimo village (Chipman, 1916 [July 24, 1913]). Teller Mission was built across the harbour from Teller in 1900 by the Norwegian Evangelical Lutheran Church and renamed Brevig Mission in 1903. The German *Graf Zeppelin* flew past Teller on its historic round-the-world flight in 1929, carrying as one of its passengers Sir Hubert Wilkins, the same man who was briefly at Teller in July 1913 with the Canadian Arctic Expedition, when he was known simply as George Wilkins.

[2] Writing from Teller, Alaska, Stefansson commented on July 26, 1913 that his new captain, Peter Bernard, a native of Prince Edward Island, signed his name sometimes Beneard, sometimes Bernard. Furthermore, "The correct spelling of the name seems to be "Beneard, but he is popularly known as Bernard." His nephew, Captain Joseph Bernard, whom the Southern Party encountered the following year, always spelled the family name, Bernard. (Stefansson to Chief Accountant, 1913f).

[3] The American geologist Leffingwell was returning to his field camp on Flaxman Island, on the north coast of Alaska, from a trip to his parents' home in the United States. He had been co-leader of the 1906–1907 Mikkelsen-Leffingwell Expedition, one on which Stefansson was hired as ethnologist, although the latter spent very little time with it.

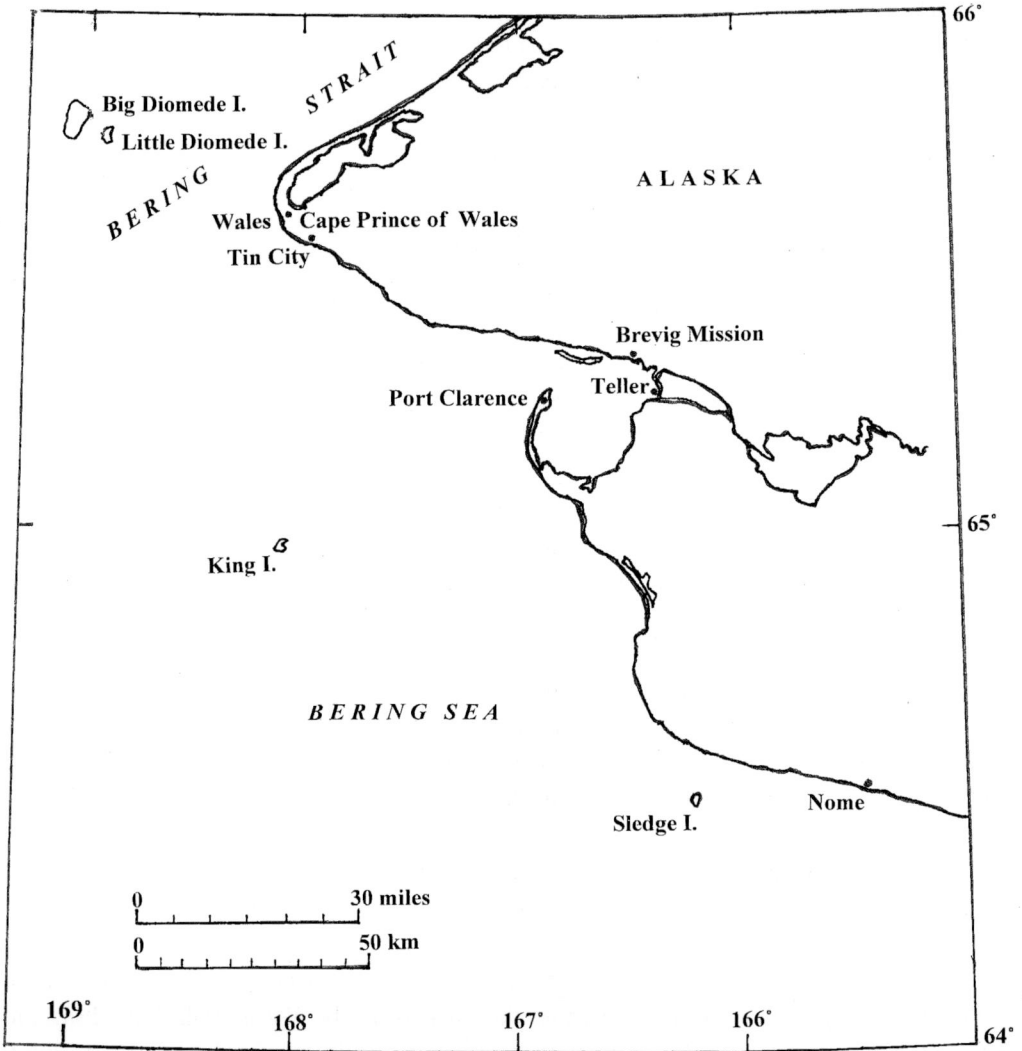

Map 2 Location of Nome and nearby settlements mentioned in Chapter 2.

Mary Sachs. These were later carried to Herschel Island on the *Belvedere.* The photographer Wilkins also remained in Nome to photograph a variety of local features and activities.

On July 25, Stefansson, his secretary Miss Gertrude Allen, Miss Bella Weitzner of the American Museum of Natural History in New York, Wilkins, and Mr. and Mrs. Murray proceeded to Teller from Nome on the coastal vessel SS *Corwin.* After the three men disembarked, the three ladies returned to Nome on the *Corwin,* from whence they found their way back to the United States.[4]

---

[4] Stefansson's two assistants were making proof corrections and final additions on his ethnological report about his 1908–1912 expedition, transcribing his popular book *My Life with the Eskimo,* and typing his correspondence. Mrs. Murray never saw her husband again after leaving him at Teller.

Continuing North

After assessing the situation at Teller and learning that the repair parts for the *Alaska* had not yet arrived, Stefansson decided to head northward with the *Karluk* and the *Mary Sachs*, and have Dr. Anderson follow later when the *Alaska* was repaired. He then asked the geographer Chipman to take command of the *Mary Sachs*, with "hitch-hiker" Leffingwell as its pilot, and the *Karluk* and the *Mary Sachs* sailed from Teller on July 27. The *Karluk* soon left the slower *Mary Sachs* far behind, separating the two ships, and they never came in contact again.

Stefansson had anticipated that the *Karluk* would reach Herschel Island well ahead of the *Mary Sachs* and the *Alaska*, and therefore arranged to have on board with him the two ethnologists, Beuchat and Jenness. It was his idea that they could initiate their studies of the Inuit living on that island while they awaited the arrival of the other two ships. The photographer Wilkins was also on the *Karluk* with all of his photographic equipment, as was the magnetician McKinlay, and the oceanographer Murray. Murray was to take charge of the *Mary Sachs* after the three ships reached Herschel Island, while the other four men—Beuchat, Jenness, McKinlay and Wilkins—were to join Dr. Anderson's Southern Party on the *Alaska*.

Unexpectedly, the summer of 1913 was a remarkably bad navigation season in the Western Arctic. Not a single ship managed to reach Herschel Island and return south that year, as they were normally able to do. Of this fact, however, neither Stefansson nor Captain Bartlett had any knowledge as they steamed north from Teller.

After passing Bering Strait, the *Karluk* tossed and rolled through a raging gale in the Chukchi Sea for a day, then dropped anchor briefly at Point Hope, a small Iñupiat settlement located where the spits of two counter-moving currents intersected like the peak of the letter "A." Here Stefansson persuaded two young Iñupiat men, Jimmy Asetsaq and Jerry Payuraq, to join the Expedition as hunters and dog drivers.

On August 1 the *Karluk* came upon the pack ice a number of miles south of Cape Smyth,[5] and two days later was icebound. The ship then drifted slowly northward within the ice. On August 4 Stefansson and the ship's surgeon, Dr. Mackay, drove a dog team across the intervening ice from ship to shore, then continued some twenty miles north along the shore to the little settlement of Cape Smyth. Some hours after their departure, the *Karluk* broke loose from the ice and continued northward, finally anchoring to the onshore ice a mile west of the settlement. While at Cape Smyth Stefansson hired three Iñupiat (a young widower and a married couple with their two small daughters),[6] purchased three *umiaks* (walrus-skinned open boats), two kayaks and some furs from which clothing could be made.

When Stefansson and Dr. Mackay returned to the *Karluk* they were accompanied by a white-bearded Englishman in his fifties named John Hadley, an old acquaintance of Stefansson with many years of experience living along the Arctic coast and elsewhere. Hadley's Iñupiaq wife had died recently, and he had asked Stefansson to take him to Banks Island where he

---

[5] Stefansson and others on the Expedition spelled the name of this small community "Smythe" at the time. The official spelling is "Smyth" as used here. Today the community is known as Barrow.

[6] As there was no room below deck for these newcomers, they established their living quarters on the deck near the bow of the *Karluk*.

Figure 8 The Iñupiaq Kuraluk and his family join the Expedition at Barrow, Alaska. (l. to r.): Mugpi, Keruk, Kuraluk and Helen, August 1913.

Photo: Library and Archives Canada, photo by W.L. McKinlay, C-070806

intended to establish a trading post. At Stefansson's suggestion, he moved into Stefansson's cabin on the *Karluk*.

Once the furs, *umiaks* and kayaks were loaded and Stefansson, Dr. Mackay and the Iñupiat were on board, the *Karluk* resumed its journey northward. Before long it again encountered the pack ice and for a second time became icebound, drifting slowly northwest beyond Point Barrow, the north-ernmost part of the North American conti-nent. On August 9, however, it succeeded in breaking free of the ice and renewed its voyage eastward past Point Barrow.

Two days later it anchored off Cross Island and a group of the scientists went ashore, where they observed evidence of recent Iñupiat presence and collected several kinds of plants, including the Arctic poppy.

Into the Ice

Then came a fateful decision. To the east and north of the *Karluk* lay the pack ice blocking any further progress. Stefansson told Captain Bartlett that the whaling ships customarily followed near-shore "leads" (bands of open water where long fractures in the ice had separated) when they steamed along Alaska's northern coast. Smaller vessels like the *Alaska* and the *Mary Sachs*, which drew less than ten feet of water, could follow a channel that extended for some sixty miles eastward along that coast to Flaxman Island on the shoreward side of a series of off-shore sandy islands. That channel, however, was too shallow in places for the larger, heavily laden *Karluk*, which was drawing sixteen-and-a-half feet of water. The *Karluk* had already run aground twice the previous day near Point Barrow, extracting itself with difficulty in both cases. Captain Bartlett reasoned, therefore, that he had to keep his ship well outside the chain of offshore islands, for the waters along the entire Alaskan north coast were shallow far out from shore. With its heavy load and seriously underpowered engine he did not dare risk a more severe grounding. The foggy and drizzly weather added to his concerns.

Captain Bartlett knew from his Eastern Arctic experience that one commonly encoun-tered open water by following leads in the pack ice to deeper water, so told Stefansson that he favoured following a good lead running northward, which he could see from the *Karluk's* masthead. Stefansson replied that he preferred to remain between the pack ice and the islands, but did not order Bartlett to do so. Instead, not wanting to argue with the fiery-tempered Newfoundlander, he retired to his cabin. Being the Expedition's commander, however, he expected the captain to follow his advice. Instead, sometime on the afternoon of August 12, Captain Bartlett headed the *Karluk* seaward along the lead until it suddenly ended. Within a very short time the *Karluk* became solidly icebound, from which it did not escape until five months later when it was crushed by the ice and sank northeast of Wrangel Island in the Chukchi Sea.

Nine years later, responding to an editorial in the *Ottawa Journal* about this incident, which had stated "The simple truth is that Stefansson put the 'Karluk' into the arctic ice …,"[7] Stefansson replied that it was true that in his book, *The Friendly Arctic,* he had "shouldered squarely the responsibility of the Karluk's going into the ice …"

However, he quickly added: "That is the only untruth I know of in the book," and went on to say, "The members of my expedition know privately, however, that the *Karluk* was taken into the ice when I was asleep, and that what I was guilty of was not having taken her into the ice but the lack of the necessary promptness in taking the command away from Captain Bartlett and turning the *Karluk* back towards the shore the moment I awoke."[8]

Stefansson presented a different explanation in a letter to Dr. Anderson's wife just five months after the incident:

> Our being adrift in the ice while the schooners are both safe near shore is due to my own weak-kneed yielding to Greenland ideas of ice navigation … I did it … on Aug. 12 when I allowed the *Karluk* to steam 20 miles off shore into the pack, when I knew from my own experience that the only safe way is to keep between the pack and the land, as every ship but ours did this year. A very amiable person I am, always yielding to everybody else's opinion—but a poor stick of a commander, don't you think.[9]

---

[7] Anonymous, 1923b (*The Ottawa Journal,* May 1, 1923).

[8] Stefansson, 1923b (May 5, 1923). The serious consequences of Captain Bartlett's decision were discussed for many years after the end of the Expedition and an "admiralty commission" report-edly found him guilty for putting the *Karluk* into the ice and for allowing eight men to leave the *Karluk* and seek to reach Siberia on their own (Niven, 2000, pp. 365–366). Arguments frequently centred on whether or not Stefansson was aware that Bartlett was heading seaward from Cross Island. Stefansson's May 5, 1923 letter to Dr. Camsell clearly states that he was asleep at the time. An earlier letter from Stefansson to Desbarats (1914a) had made the same claim. However, some-time in the 1980s, Canadian historian Dr. Gordon Smith told me that he had once had a discus-sion on that very subject with my father in 1960 or 1961 while they shared an office in the Department of Northern Affairs. Smith was researching Canadian Arctic sovereignty at the time, my father, Eskimo (Inuit) administration. According to Smith, my father said that he was in the cabin below Stefansson—McKinlay's diagram of the *Karluk* in Niven (2000, p. 395) does not indicate the presence of a deck or cabin below Stefansson's, so I suggest that my father might have been in the mess adjoining Stefansson's cabin—on that precise afternoon when Captain Bartlett headed the *Karluk* into the ice. In any case, he told Smith that Stefansson, far from sleeping, was pacing back and forth in his cabin, evidently fully awake and aware of what was happening.

[9] Stefansson, 1914b (January 19, 1914).

Icebound, the *Karluk* now drifted slowly eastward for several days, until it lay north of Camden Bay. Then it started a slow westward drift. On August 29 after several days of preparation, Stefansson sent off his two ethnologists, Beuchat and Jenness, together with the photographer Wilkins, the surgeon Dr. Mackay, the secretary McConnell, three Iñupiat men, and two heavily laden sleds and dogs in an attempt to get them to Flaxman Island, from whence they could proceed east to Herschel Island and the two ethnologists could commence their scientific studies. At Flaxman Island they expected to find the American geologist Leffingwell in his cabin and to ask him to inform the men on the *Alaska* and the *Mary Sachs* when he saw them about the ice-trapped status of the *Karluk*. If these two ships had already gone east, then the shore party was to hurry east to Herschel Island to await the arrival of the *Karluk*. At Herschel Island Jenness was to send word to the *New York Times* of the status of the Expedition. None of these plans came to fruition, however, for the shore party encountered nearly impassable ice and damaged its *umiak*. Two hours after their departure, Stefansson and Hadley set out after them with additional mail, and when Stefansson saw their damaged condition he ordered them to return to the ship.

Stefansson's Hunting Trip

While the days of being icebound grew into weeks, Stefansson grew increasingly restless, as did most of the men on the *Karluk*. Meanwhile, the *Karluk* continued its slow westward drift, its position on September 17 being several miles north of Thetis Island, which it had passed five weeks earlier on its way east. On that day, Stefansson sent Jenness, Dr. Mackay and the Norwegian topographer Mamen south to see if they could sight land, but they failed to do so and returned to the ship. Dr. Mackay and Mamen tried again the next day, but once more returned without success. Sometime on the 18th, someone—the fireman Maurer thought it was Jenness—suggested hunting caribou on the mainland, but Stefansson merely laughed, saying there were no caribou nearer than the mountains.[10]

The following evening, with the *Karluk* seemingly at a standstill in the ice some twenty miles north of Harrison Bay, Stefansson suddenly announced that he would take provisions adequate for twelve days and lead a hunting party in search of caribou on the mainland. His announcement came as a complete surprise to the men on the ship.

The following day Stefansson supervised the loading of two sleds. Captain Bartlett, watching the last-minute preparations from the deck of the *Karluk*, suddenly seized an enamel bowl and kicked it far out on the ice.[11] This spontaneous act was so out of character for him that it caused some of the others to suspect that it reflected his feelings over the departure of Stefansson. Perhaps he was thinking, "Goodbye, Stef!" Wilkins later told Dr. Anderson that the hunting party did not get a very enthusiastic farewell from the people lining the deck of the

---

[10]   Copy of an undated typed memorandum containing remarks made by F.W. Maurer to Mrs. R.M. Anderson in Kent, Ohio, in 1915, signed by Dr. Anderson (CMNAC/1996-077, Series A—R.M. Anderson Collection, Box 68, Folder 11, CMN).

[11]   Wilkins, 1916 (September 20, 1913).

Karluk. As the hunting party prepared to pull away from the ship, Wilkins heard the cook (Robert Templeman) say that he hoped they would not see them again.[12]

Stefansson's sudden decision to leave the ship may well have been, as he claimed, to hunt for fresh caribou as a welcome alternative to the seal meat, salt beef and salt pork his men and crew had been eating for days. However, few on board the *Karluk* believed this. Nor did Dr. Anderson when he learned about it several weeks later, for he knew how scarce the caribou were along the northern Alaskan coast. Perhaps Stefansson merely wanted to get away from the ship for a while to reduce his frustration and restlessness after being confined on it for five weeks.

The Stefansson party headed for the shore about 1:30 p.m. on September 19, immediately after dinner, with Stefansson breaking trail, followed by the eager sled dogs and their heavy loads, and Jenness, Wilkins, McConnell, and Jimmy and Jerry (the two young Point Hope Iñupiat). They were still in sight when darkness fell, causing McKinlay, standing on the deck of the *Karluk*, to wonder about Stefansson's wisdom in leaving so late in the day.[13] Two days later the hunting party camped on Spy Island, a small gravel island a few miles northeast of

Figure 9 Stefansson (white parka) leaving the *Karluk* with his hunting party, September 20, 1913.
Photo: Library and Archives Canada, photo by W.L. McKinlay, PA-203460

---

[12] Copy of undated typed memorandum re. comments made by F.W. Maurer, Wilkins and Jenness. CMNAC 1996/077, Series A—R.M. Anderson Collection, Box 68, Folder 11, CMN.

[13] McKinlay, 1914 (September 20, 1913).

Figure 10 Stefansson's hunting party looking for the *Karluk* from Spy Island, northern Alaska, September 26, 1913. (l. to r.): Iñupiat Jimmy and Jerry on tree trunk, Stefansson (white parka), Jenness and McConnell.
Photo: Courtesy of Stuart E. Jenness, photo by G.H. Wilkins

the mouth of the Colville River. That night the northeasterly wind, which had been blowing for several days, turned into a howling gale. After the gale finally passed Stefansson discovered that there was open water seaward of them and no sign of the *Karluk*. There was also open water between their island and the shore. They were stranded!

(The next chapter will continue the story of the *Karluk* and its personnel. See Chapter 9 for the continuation of the story of Stefansson and his hunting party.)

PART 2
THE *KARLUK* SAGA

"ADIOS, *KARLUK*"

4

The first big storm of the season, the ferocious September gale that enveloped Stefansson's hunting party when it reached Spy Island, also struck the *Karluk*. The next morning the men on the ship discovered that the large ice floe in which it was trapped had broken free from the shore ice and was drifting westward at the rate of thirty miles a day. An ever-widening body of water now separated the icebound ship from the shore and eliminated any chance of Stefansson's hunting party returning to it.

The *Karluk* had lost its commander. Stefansson was not on board, and Captain Bartlett was now responsible for the actions and safety of those still on the ship: twenty-two men, one woman, two children, twenty-two sled dogs and a black-and-white cat named Nigeraurak.[1] One of the captain's first orders to his men was to bring back on board all the provisions, equipment and dogs that had been put on the ice to provide more room for those on the ship. He also ordered everyone to stay on board to avoid accidents.

Strong winds, driving snow, and cloudy and foggy weather plagued the men on the *Karluk* for days, seldom permitting them to determine their ship's location. Geologist Malloch required sunshine at midday and the correct time in order to make the necessary sextant readings and calculate the ship's position accurately, and the sun was not favouring them. Oceanographer Murray's depth soundings indicated they were moving into deeper water, so they knew they were moving farther from shore. Captain Bartlett commented in his diary on the likelihood of the *Karluk* being crushed by the ice and on the survival chances of everyone on board, but otherwise kept his troubling thoughts to himself.

Everyone needed to keep busy. They also needed to be prepared for the cold and sunless days ahead. Accordingly, Captain Bartlett asked the Iñupiaq woman, Keruk,[2] to start sewing fur clothing for the men on the ship. She could not possibly sew for twenty-two men, however, so with her guidance the men were urged to start sewing their own clothing. The need was great, for they were all still wearing the clothes they had brought north, apart from the deerskin footwear she had already made for them.

Views on Stefansson's Departure

In spare moments the men on the *Karluk* wondered about Stefansson's hunting party and why he had insisted upon taking it ashore. As commonly happens under such circumstances, some

---

[1] The name "Nigeraurak," according to McKinlay (1976, p. 17), means "little black one." The cat was the ship's mascot.

[2] Captain Bartlett spelled her name Keruk, as did Stefansson, and described her as about twenty-eight (Bartlett and Hale, 1916, p. 19). McKinlay and Niven spelled it Kiruk, but mostly called her "Auntie."

of them came up with unflattering reasons for his departure. One such reason, expressed by McKinlay, was that Stefansson had deliberately abandoned the *Karluk* and its personnel in order to commence his planned exploration of the Beaufort Sea, beat his rival MacMillan to the discovery of "Crocker Land" and receive the acclaim he sought. Other persons along the north coast evidently entertained similar thoughts, once they learned that Stefansson had left the *Karluk*, for claims that he had deserted his ship and men appeared months later in some southern newspapers.

In her account of the *Karluk's* fate, published nearly ninety years later, author Jennifer Niven offered another explanation for Stefansson's supposed desertion. A few days before his departure in September Stefansson had been reading a book about the 1878–1881 expedition of George Washington De Long and the *Jeannette*. Like the *Karluk*, that vessel had been caught in the ice in the Chukchi Sea, although many miles farther west—just east of Wrangel Island actually—and drifted slowly westward for a year before being crushed by the ice far to the west of Wrangel Island with the loss of all hands. Murray and Dr. Mackay had been the first to discover the book shortly after Stefansson's departure and had come to their troubling suspicions about the similarity of the *Karluk's* route to that of the *Jeannette*. Subsequently they had discussed their ideas and speculations nightly with some other members on board. Mamen came upon the book that Stefansson had been reading early in October, and after reading parts of it suspected, as had McKinlay, Dr. Mackay, Murray, Malloch and Beuchat before him, that the *Karluk* was following a somewhat similar drift course and might very well experience a similar fate.[3] The conclusions they all reached cast a very dark and gloomy cloud over the ship. Perhaps the most dramatic response among them was that of the French anthropologist Beuchat, who remarked despairingly on at least one occasion that they were all lost. After Stefansson's departure, he spent much of his time huddled in his bunk trying to keep warm and bemoaning the probable fate of them all.

Many of the men on the *Karluk* thus came to the conclusion that Stefansson had deliberately deserted them and the *Karluk* in order to pursue his own interests. If Stefansson had done so, however, he gave absolutely no hint of it to Jenness, Wilkins or McConnell at any time during the next several weeks when they were together. Each one of them, including Stefansson, had left valuable possessions behind and fully expected to return to the *Karluk*. Jenness left his camera, his diary, personal clothing and some ethnographical equipment; Wilkins left most of his professional cameras and supplies, his clothing and his diary; McConnell left personal possessions, including his rifle; and Stefansson left a valuable rifle, camera, diary, 1908–1912 field notes, his partly created Eskimo-English dictionary, and hundreds of dollars of government paper money, silver and gold on the ship. We may never know the real reason or reasons why Stefansson left the *Karluk*, but from the evidence at hand it appears that he fully intended to return to it.

By October 3 the *Karluk* lay several miles north of Point Barrow, which the captain could just see on the horizon when he was in the ship's crow's-nest. Still captive in their giant ice floe, which measured several square miles in diameter, with open water beyond it in every direction, the *Karluk* was now moving in a direction that shifted periodically, sometimes a bit

---

[3] Niven, 2000, pp. 58–59.

easterly, sometimes more westerly, but chiefly northwesterly. Frequently their floe would encounter other ice masses with resounding crashing and tearing noises, which were frightening to hear both in daylight and in darkness.

Two Opposing Factions

Meanwhile, two factions soon developed among the scientific staff on the *Karluk*: the older men—Murray, Dr. Mackay and Beuchat—who had no confidence in Captain Bartlett and later favoured striking out for the Siberian coast, literally hundreds of miles to the south, and the younger men—McKinlay, Malloch and Mamen—who fully supported the captain. A temporary truce existed among them.

Murray and Dr. Mackay had both served with Ernest Shackleton on his Antarctica expedition of 1908–1909, which sailed from New Zealand and managed to reach within one hundred miles of the South Pole. Having acquired a sense of self-importance from their polar experiences on that expedition (although none of their experience was over ice), they seldom hesitated to suggest that both Bartlett and Stefansson were deficient in their knowledge and leadership abilities. Soon they hatched a plan to leave the ship and strike out for shore. For some unexplainable reason (except that he sided with them against Captain Bartlett), they included Beuchat in their plans. He was a gentlemanly and knowledgeable man in his thirties, but thoroughly impractical, lacking in field experience and physical fitness, and quite unsuited for the rigorous demands of Arctic life.

Night after night Captain Bartlett heard Dr. Mackay ranting against his leadership in the adjoining cabin, so was not surprised when the doctor came to him one day early in October with his plan to head for land. The captain responded, however, that he was not prepared to discuss the matter and brushed off Dr. Mackay's demand for a general meeting. Captain Bartlett would only say that he still hoped to break free of the ice floe in which the ship was trapped, and if that failed he would make plans for wintering with the ship in the ice. Meanwhile, he would endeavour to supply the needs of everyone on his ship. Ten days later Dr. Mackay again presented the captain with a letter requesting a general meeting, but the captain refused to accept it.

Fighting Boredom

To minimize boredom most of the men tried to involve themselves in useful activities or projects to while away the time. Murray dredged the sediments of the sea floor daily for samples of Arctic marine life, while Mamen, McKinlay and Malloch kept records on the air temperature, wind directions and other weather conditions, and regularly determined the ship's geographic location. Unhappily, none of this scientific information has survived.

The direction of the *Karluk's* drift showed striking similarities to the direction of drift of De Long's *Jeannette* in 1879–1881 and also that of Nansen's *Fram* in 1893–1896. Only the specially constructed *Fram* had survived the challenging Arctic forces. True, the *Jeannette* had survived for more than a year in the ice before being crushed, but it was more strongly built than the *Karluk*. From his reading of the De Long expedition, McKinlay suspected the *Karluk* would probably be crushed by the ice much sooner than the *Jeannette* had been, a correct intuition as it turned out, but kept his thoughts to himself.

The depth soundings that several of the men took regularly indicated early in October that the ship was in water more than a mile deep. By the end of October, however, they obtained shallower readings, and thereafter readings rarely exceeded sixty fathoms (360 feet). Following instructions from Captain Bartlett, crew members and the two Iñupiat men cut the ice around the *Karluk*, freeing it to rise if pressured by the ice, and banked the ship with snow blocks to insulate it in the hopes of keeping everyone warmer inside.

Winter Schedules, Meals and Personal Hygiene

At the end of October, with very few hours of daylight and colder weather, Captain Bartlett had the ship's boilers blown down, water drawn off and the engine dismantled by the ship's engineers and cleaned to ensure they were in top condition. All water was blown out of the pipes. He also cut the meals to two a day to conserve supplies and fuel. Breakfast was served at 9 a.m., dinner at 4:30 p.m., with tea at 1 p.m. and tea, coffee or hot chocolate before going to bed. Breakfast always started with oatmeal porridge with condensed milk, followed by eggs, ham, bacon, sausages, codfish and coffee. Dinner consisted of canned oysters or clams, shredded codfish, dessicated and frozen potatoes (thawed out for use), and a variety of vegetables. Fresh meat was served daily, normally seal meat, but polar-bear meat when it was available. The young cook, Templeman, produced a surprising variety of desserts: ice-cream of several flavours, sherbet, assorted pies, puddings, canned fruits and cakes.

The men were expected to be ready for work at 10 a.m. As the ship was well supplied with razors, soap and undergarments, Captain Bartlett established the rule that they were to shave at least three times a week and bathe at least once a week, making use of one or other of the several bathtubs on the ship.

Winter Darkness

On October 30 a northeasterly blizzard arose and raged for six days. The ice around the *Karluk* developed large cracks in all directions and opened up large watery gaps, but with the air temperature well below freezing these watery gaps generally froze over fairly quickly. They were nevertheless a constant concern. All persons were confined to the ship as it shook and groaned from the blizzard's onslaught. The floe in which they were trapped was at this time about half an acre in size and, fortunately, about thirty feet thick and thus able to withstand a considerable amount of bashing before it would break up.

After six days of raging winds, Captain Bartlett decided that the time had come to move vital supplies of emergency stores onto the ice in case it became necessary to abandon ship hastily. The possibility of ice damage to the ship or of fire was constantly on his mind. There were coal-burning stoves in several parts of the ship: in the engine room, in the galley where water was kept from freezing in a hundred-gallon tank, in the saloon (dining room), in the scientists' room and in the captain's room. There were also numerous lamps. The Iñupiat used blubber stoves of soapstone in their living facilities on the ship's deck, but these were a much lower fire risk. Both the scientists and the crew were regularly drilled in what to do in case of fire. Individual men were given twelve-hour shifts as watchmen to patrol various parts of the ship night and day.

The shifting of the supplies onto the ice lightened the *Karluk's* load considerably, allowing it to ride up two or three feet if it was squeezed laterally by the ice, and possibly sparing it from destruction if the ice started breaking up. Onto the ice went sacks of coal, Murray's drums of alcohol (he used the alcohol to preserve his oceanographic specimens), biscuit cases and casks of beef, most of which were then used to create walls for a house. Snow was banked against these walls to render them as snug as possible. Lumber off-loaded from the ship was used for the flooring of the house and for roof rafters; then spare sails were spread across the rafters to form the roof. A large snowhouse was erected a few weeks later, and the two buildings served temporarily to house the dogs injured through frequent fighting.

Next Captain Bartlett had Hadley and Second Mate Charles Barker construct several sleds. Having accompanied Peary during much of his run for the North Pole in 1909, the captain did not care for the Nome-type sleds Stefansson had purchased and had his men build them according to Peary's special design. The Peary-type sleds were able to turn around without unloading, which is a distinct advantage over the Nome-type sleds.[4]

While Hadley and Barker were thus employed, the captain sent the two Iñupiat men off to hunt seals. They were so successful that they soon supplied the camp with about fifty seals, all of which were carefully stored in a large natural ice-box the men had hewn out of a hummock near the ship.

Throughout the many weeks that the *Karluk* drifted, the oceanographer Murray continued to dredge regularly for marine life, examining the specimens he recovered in a small though cold laboratory constructed on the deck. Hour by hour he pored over his specimens, smoking continually as he worked, bottling the specimens carefully in alcohol when he had finished with them and keeping meticulous notes. Tragically, all of his specimens and notes were soon lost with the *Karluk*.

With the ever-dwindling daylight, the men spent more and more time confined to the *Karluk*. As a consequence they grew more easily aggravated by the ship's many deficiencies, which included the inadequate number of dishes, cups, chairs and stools to sit on, and lamps. As there were not enough chairs or stools, for example, McKinlay sat on a cannister of dynamite at mealtime! Captain Bartlett regularly sought activities to keep the men's minds occupied. One of his suggestions was a chess contest, to the winner of which he offered a box of fifty cigars. The cigars were originally intended as a gift for the police at Herschel Island, but since there was no longer any likelihood of seeing them, the captain felt free to put them to a different use, one he thought might boost the morale of his men. Many of the men sought to avoid boredom by reading some of the large number of books on the *Karluk*. Unfortunately the lack of comfortable chairs on the ship forced them to read in their bunks where lighting was generally poor. Some of them played bridge or chess in the dining room, smoked their pipes or cigarettes, and occasionally listened to records on the Victrola. Malloch took solar readings whenever the weather was clear in order to determine the ship's position, and Beuchat, McKinlay and Dr. Mackay received regular half-hour lessons in the Inuktitut language from the widower Kataktovik.

---

[4] Bartlett and Hale, 1916, pp. 46–47.

By mid-November the *Karluk* reached 73° N latitude, the farthest north it drifted. Thereafter the ice floe with the *Karluk* veered south to southwest, and their rate of drift, like their course, changed from time to time.

The sun remained below the horizon after November 11, providing more than two months of sunless darkness for all on the *Karluk*. As if that was not depressing enough, the *Karluk* then started to leak. By mid-December it was leaking so badly that it required more than an hour of manual pumping daily. Manual pumping was necessary because the engines were shut down for the winter. A violent storm then engulfed the ship for six days, with winds up to eighty miles per hour. It continued until the day before Christmas, greatly increasing the worries and stress among all of those on board. Captain Bartlett, normally self-controlled, became visibly anxious, frequently pacing about the deck. The three "conspirators" (Murray, Dr. Mackay and Beuchat) spoke more openly and gloomily about the fate of the *Jeannette* and its crew, and consequently were increasingly avoided by the other more cheerful individuals. Every time that Malloch determined the ship's location, Murray and Dr. Mackay would hurry to compare the course of the *Karluk* with that of the *Jeannette*.

Shortly before Christmas, McKinlay started visiting Captain Bartlett in his cabin, sometimes being joined by Mamen and Malloch. The visits commenced quite by accident one day when McKinlay was invited to the captain's cabin to pick up an item he had requested and noticed that the captain was reading a book about roses. McKinlay was proud of the roses he had grown in his own small garden in Scotland and mentioned this to the captain. The two men then enjoyed a most pleasant discussion on the subject. That led to the discovery of other topics of mutual interest, and further discussions, all of which helped relieve their minds briefly about the worries now troubling them so frequently. The captain would often speak of interesting or unusual events in his life and at other times of his fondness for the works of Shakespeare, Keats, Shelley and other authors—surprising literary interests for a man who seemed so rough and unpolished.

Christmas Festivities

McKinlay joined First Mate Sandy Anderson and the second engineer, R.J. Williamson, early on Christmas Day to decorate the dining area as cheerfully as they could, draping flags and ribbon obtained from Hadley around the room in an attempt to bolster everyone's spirits. After breakfast everyone participated in a series of challenging competitions on the ice, all carefully planned by McKinlay, Anderson and Williamson. The ice contests ended with a series of tugs of war, then all returned to the ship to warm up. Later they gathered for the special evening meal at 4:30 p.m. When all were seated, Captain Bartlett magically produced a bottle of whisky and passed it around, then proposed a toast. Calling upon all of the men to rise and hold their glasses high, the captain declared in a solemn voice, "To the loved ones at home." A brief silence followed while the men thought about their families and friends, then all sat down to what was truly an extraordinary meal, considering their circumstances.[5]

---

[5] McKinlay, 1914 (December 25, 1913). I do not know if Stefansson paid for the whisky himself or somehow charged it to the government, but such a gift to government employees, especially

The *Karluk*, being a government ship, was not supposed to carry any intoxicating liquor, but in fact did have on board a case of whisky, which Stefansson had intended to present to the Royal North-West Mounted Police at Herschel Island. That was the supply from which Captain Bartlett had produced his bottle.

The Christmas menu was typed by McKinlay, with each man receiving a copy as a souvenir. The menu was truly remarkable, considering the circumstances. It included pickles, oyster soup, frozen lobster, polar-bear steaks, ox-tongue, potatoes especially saved from the previous June for the occasion, peas, mince pies, plum pudding, nuts, tea and cake.

After all had finished eating, there were boxes to open, containing Christmas gifts provided the previous June by the ladies of Victoria, British Columbia. Among the boxes was one with shortbreads and candy, which were quickly shared. Music on the Victrola,[6] cigars, pipes and cigarettes topped off the day's celebrations.

A loud gunshot-like cracking sound on the morning after Christmas Day startled everyone on the ship. Their investigations revealed that the cause of the noise had been the formation of a major crack in the ice parallel and close to the ship. Its closeness brought home clearly how great a danger the ship was in, and preparations were immediately made to abandon it on short notice.

Land Sighted

On December 29 the sighting of land during a brief clearing of the dark horizon was a welcome event, although it was not possible at the time to determine whether it was Herald Island or Wrangel Island, the only two islands in the Chukchi Sea. The proximity of land increased the importance of being prepared for a sudden emergency, however, for it indicated increased ice pressures and thus greater risk of the loss of the *Karluk*, as some of the men knew. The men promptly renewed their efforts to sew fur clothing for themselves.

The prospect of being on land soon brought some comfort to the men. As they were uncertain of their location at the time, however, they thought that the silhouettes of the mountains they could see, with one distinctive high peak, suggested they were looking at Wrangel Island, for Herald Island was only four miles long. If it were Wrangel Island, then the high peak they saw would likely be Mount Berry, which, according to a map they had, rose 2,500 feet in the centre of the island and was its highest point. Meanwhile, the *Karluk* continued its drift, which for the moment was towards the land.

---

to the police, would certainly not have had government approval. Captain Bartlett proposed the toast, but was himself not a consumer of alcoholic beverages.

[6] The Expedition was presented with two Victrola gramophones prior to its departure, one by the Premier of British Columbia, Sir Richard McBride (Bartlett and Hale, 1916, p. 78), the other by the manufacturer. The gramophones were accompanied by both classical and popular 78-rpm recordings. The gramophone on the *Karluk* was the one presented by the premier. It was lost with the *Karluk*. The other one was evidently transferred to the *Alaska* at Nome and thereafter entertained the men of the Southern Party in northern Alaska and later at Bernard Harbour.

New Year's Day Activities

New Year's Eve was known as Hogmanay to the several Scotsmen on board the *Karluk*, a time when young Scottish people went about singing and seeking gifts to welcome in the New Year. Most of the men had retired to their bunks for the night, but at about 11:30 p.m. the few who were still up, led by the diminutive McKinlay, started up the gramophone, hoping with some singing and dancing to liven up the occasion. Another bottle of whisky mysteriously appeared, helping to persuade those who had long-since retired to their bunks to get up and join in. Someone struck the traditional "sixteen bells" at midnight and McKinlay paraded along the deck "raising the devil with the dinner gong."[7] Munro and Murray persuaded the captain to emerge briefly from his cabin and all joined in for a toast of hope. The captain was a tea-totaller, but went through the actions without drinking from his glass. The toast was followed by the singing of *Auld Lang Syne*, then the recitation by each Scotsman, in turn, in words largely incomprehensible to the others, of poems of Robert Burns. These activities probably raised their spirits briefly.

On New Year's Day, 1914, the men staged a rousing football game around midday, the five Scotsmen plus Malloch challenging the "All Nations" six. Captain Bartlett refereed as the dozen men shuffled about on the ice, barely visible in the winter darkness. They had created a ball of seal-gut, cut into sections, then sewn together, with surgeon's plaster covering the seams. This in turn was covered by a sealskin casing. The lone woman, Keruk, was attired in dress and bloomers and tended goal for the "All Nations" team, which won the match rather easily. There were a few minor bruises, but the only painful injury was to Captain Bartlett when the whistle he tried to blow froze to his lips.

"Abandon Ship"

The following morning a distinct humming sound wakened several of the men. Putting his ear to the side of the ship McKinlay heard a musical note, which later turned discordant, and then went silent. He concluded it was the ice crushing and forming pressure ridges a good distance away, so tried to put the matter out of his mind. The sounds recurred a short while later, but he could see nothing in the winter darkness that might explain them. The following day, the inclination of Murray's dredge line indicated the ship was again drifting westward after being more or less stationary for a day or two. The strange musical noise soon started to sound like a beating drum, then more like distant gunfire, which kept up all day. With their prospects looking ever more ominous, the captain urged everyone to renew their sewing efforts, for winter clothing was the one area in which they were the least prepared.

Sometime around 5 a.m. on January 10 several of the men were awakened by a harsh grating sound followed by a considerable shudder all through the ship. Inspection revealed a crack in the ice along the starboard side of the ship, with that side of the ship slowly rising. Then it was discovered that this apparent rising was because the ship was listing to port, soon nearly twenty-five degrees. Everyone remained tense and alert throughout the day.

---

[7] McKinlay, 1976, p. 62.

About 7:30 that evening, while the men were playing cards or chess, reading, or sewing in the dining area and the captain was standing near the engine-room door, there was a sudden splitting and crashing sound from below. This was immediately followed by the frightening sound of water rushing into the hold.

Captain Bartlett immediately investigated the damage and found that "… a point of ice on the port side had pierced the planking and timbers of the engine-room for ten feet or more, ripping off all the pump fixtures and putting the pump out of commission … the break was beyond repair. I went on deck again and gave the order, 'All hands abandon ship.' "[8]

The planning they had made for such an occasion proved invaluable. There was no confusion. The men quickly and systematically extinguished the fires in all of the stoves except the one in the galley, then put the emergency supplies overboard on the ice. The captain calmly asked Keruk and her two little girls to go to the house on the ice and start the fire in the stove to warm up the building. Then he instructed the cook Templeman to have coffee and food available for the men. By 10:45 p.m., the engine room was flooded to a depth of eleven feet. The removal of thousands of pounds of supplies from the ship together with the pressure of the ice buoyed up the ship for a while, preventing it from sinking. Captain Bartlett later wrote:

> … we could have saved everything on board but no attempt was made to save luxuries or souvenirs or personal belongings above the essentials, for it did not seem advisable to burden the sleds on our prospective journey over the ice with loads of material that would have to occupy space needed for indispensables.[9]

This decision, certainly the right one considering the haste required to get everyone safely off the *Karluk*, resulted in the loss of not only some personal items belonging to those who had just left the ship, but also of valuable documents, diary, money and other possessions belonging to Stefansson, Wilkins' considerable camera equipment, Jenness' diary, camera, books and clothing, and Murray's notebooks and unique collection of marine specimens.

All essential items were off the ship by 9:30 p.m. that evening. Someone then suddenly realized that their black-and-white cat, Nigeraurak, was nowhere to be found. Several men promptly searched the ship for her. She was well liked by all and considered a lucky mascot, so it was imperative that she be found. The poor creature had been so thoroughly frightened by the noises of the ice and ship, however, and then by the men's bustling activities around the deck, that she had hidden somewhere and could not be found. Fortunately she re-emerged early in the morning, and was promptly scooped up by Fred Maurer, the ship's fireman, who was especially fond of her. He then received the captain's permission to take her ashore.

When all essential supplies were on the ice Captain Bartlett instructed the men to move them by sled over to the house and the large snowhouse. The men finished moving them about 2:30 a.m., at which time the captain told them to go to bed. Dr. Mackay, Murray and Beuchat were already asleep by then. McKinlay had observed their uncooperative behaviour during the crisis and thought it inexcusable, commenting in his diary: "After hauling a sled

---

[8] Bartlett and Hale, 1916, p. 88.
[9] *Op. cit.*, p. 89.

with nothing but their own personal belongings—they would allow nothing else on it—they retired to the house & made no attempt to help us in the work but staked out their claim to about one third of the place."[10]

Captain Bartlett, meanwhile, remained on the ship awaiting its end, after seeing that all personnel and essential items were safely on the ice. Hadley and Munro stayed with him briefly and helped him move the Victrola into the galley where there was still a fire in the stove. The ship remained more or less static until early in the afternoon of January 11, being held up by the ice pressing on its sides, but water continued to pour in through its damaged port side. Captain Bartlett sat by the stove near the Victrola, solemnly playing records from the collection of some one hundred and fifty on the ship. As each record finished, he threw it in the stove.

Around 3:15 p.m. the ice opened up briefly and the ship started to settle. About 4 p.m. the captain noticed the water coming down the hatch, so put a recording of Chopin's *Funeral March* on the turntable, placed the needle on the record and dashed up the steps to the deck. Hastily he lowered the flag to half mast and then climbed onto the ship's rail. A moment later, with the mournful sounds of the music coming from the Victrola and the Canadian ensign flapping briskly from the main mast,[11] the *Karluk's* prow dipped under the water.

When the end became obvious, the survivors stood on the ice nearby to witness the last act of their marine home for the past seven months. The moment the ship headed below the surface of the water in her final plunge, the captain stepped off its rail and onto the ice, bared his head momentarily and said, "Adios, *Karluk!*"[12] A moment or two later she was gone to her cold grave. As the last of her disappeared beneath the icy water, "A feeling of intense loneliness came over us," reported E.F. Chafe, the youngest among the crew, "but we gave the old ship three hearty cheers …"[13]

Worth noting at such a challenging moment was the dedication of some of the scientists. Despite the Arctic darkness, a raging gale and an air temperature close to -20°F (-28°C), they thought to take a depth sounding for historical purposes just before the *Karluk* sank. They obtained a reading of thirty-eight fathoms (228 feet). The next morning, ice covered the place where she had sunk.

---

[10] McKinlay, 1914 (January 10, 1914). The quotation is from Niven (2000, p. 120), who gave its source as McKinlay's diary at the National Archives in Ottawa. However, neither copy of his diary that I examined at the National Archives included the quotation, nor did the copy in the Anderson Collection at the Canadian Museum of Nature. McKinlay wrote two versions of his diary, the "summarized" version being the one he sent to the Department of the Naval Service in Canada. The quotation in Niven's book probably came from the more detailed version in the National Library of Scotland (NLS/DEP 357).

[11] McKinlay, 1914b. The red ensign was an unofficial flag of the Dominion of Canada, authorized for use on its merchant ships in 1892. There were several variations of it, however, prior to 1922. Captain Bartlett (1914), in an equally early account, described the flag as the blue ensign.

[12] Bartlett and Hale, 1916, p. 91. Bartlett's version of what he said, which is presented here, seems more dramatic if less likely than Niven's version—"Good-bye, old girl" (2000, p. 124). An earlier account by Bartlett (1914) had picturesquely claimed "I stood on the ice, surrounded by the officer and crew of the Expedition, who lifted their hats, saying, one and all:'Adieus [*sic*], Karluk.'"

[13] Chafe, no date, but probably 1917, pp. 19–20. (Quoted in Niven, 2000, p. 123.)

By sheer coincidence, the location of the *Karluk's* sea grave, some sixty miles northeast of Herald Island in the Chukchi Sea, was not far from where the *Jeannette* and its crew had begun their fatal drift in the ice in 1878.

The loss of his ship weighed heavily on Captain Bartlett, but he was greatly relieved that all hands, essential supplies and equipment, and all of their dogs and the cat were safely off the ship. They were reasonably comfortable on their new floating quarters and had sufficient supplies to maintain them for a considerable time. With luck, he thought, they might all get back to their loved ones some day.[14]

---

[14] Bartlett and Hale, 1916, p. 92.

SHIPWRECK CAMP TO WRANGEL ISLAND

5

The disappearance of the *Karluk* beneath the Chukchi Sea left the survivors wondering about their future. Captain Bartlett's feelings must have been especially strong, for he had lost his ship, perhaps the worst thing that can happen to a sea captain. He must have taken some comfort, however, in knowing that the twenty-five persons, all of their dogs, and their cat mascot, for whose lives he was responsible, were in good health and they had ample supplies and equipment on the ice to carry on.[1]

Shipwreck Camp

The survivors named their emergency site on the ice floe "Shipwreck Camp," determining its geographical location on the sea ice as 73° N latitude, 178° W longitude.[2] Here all of them would remain for a few weeks until there was sufficient daylight to permit safe travelling. That delay gave them valuable time to sew more fur clothing, which they would certainly need in the days ahead. In addition to their clothing Captain Bartlett had the men sew tents and canvas covers for the sleds. Meanwhile he cut up his raccoon-skin coat, which supplied enough material to make several foot-bags—fur inside, Burberry cloth outside—to keep the men's feet warm at night. Almost everything had to be sewn by hand, for the reindeer skins and seal skins were too heavy to sew with the small hand-operated sewing machines, of which they had saved two before the ship disappeared. Fortunately, the chief engineer Munro proved handy with one of the sewing machines, no doubt learned during his days in the British Navy, and Keruk made excellent use of the other.

Daily routine in the ice camp remained as it had been on board the ship. Someone maintained the stoves in each of the two dwellings (the snowhouse and the wooden shack), meals were served at the same hours as before, and lamps were turned out at 10 p.m. Despite the daily chores, there was time for games (cards and chess), reading and occasional singing. Hadley would play his guitar and sing, sometimes so loudly that the others "could not hear ourselves think."[3] A number of books also had been moved to the ice camp from the ship before it sank. Captain Bartlett made sure that he had his own favourite book, the *Rubáiyát of Omar Khayyám*, which he had carried with him on many previous voyages and never tired of reading.

---

[1] McKinlay told me in 1980, when I visited him in Glasgow, that the *Karluk* survivors were careful to keep the cat alive throughout their winter ordeal, because they believed that as long as it was alive there was hope of them all being rescued.

[2] Chafe, E.F., 1918, p. 309.

[3] Mamen, 1914 (August 14, 1913).

Scouting Party

After the *Karluk* had disappeared, Captain Bartlett commenced planning how best to get his twenty-four companions to safety. His first plan entailed trying to lead the entire group to the Siberian coast by way of Wrangel Island, following that coast eastward to St. Lawrence Bay. If they encountered a settlement en route, he would try to arrange that the majority of his companions remain there while he and a small group continued on to seek their rescue.

Following discussions with Mamen and McKinlay, the captain modified his initial plan and decided to send off Mamen and the two Iñupiat men (Kataktovik and Kuraluk) to scout a route to Wrangel Island, leading at the same time a party of four men with three sleds, each loaded with 400 pounds of supplies and pulled by six dogs. Captain Bartlett selected First Mate Sandy Anderson to head the four men destined to create the camp on Wrangel Island, then picked Second Officer Charles Barker and two seamen, John Brady and Edmund Golightly, to complete the foursome. Golightly, for some unknown reason, had joined the crew under the alias of "Archie King." At the end of each day while en route the men were to build a snowhouse, which could serve as shelter for others who would follow. Upon reaching the island, they would unload most of the supplies and the four crew members would establish a camp and then scout their surroundings, gathering driftwood for fuel and searching for game. Meanwhile Mamen and the two Iñupiat men would return to Shipwreck Camp with all the dogs and two of the three sleds to help bring the rest of the people and many more supplies from Shipwreck Camp.

Once the route to Wrangel Island had been selected and a base camp established there, and the survivors of the *Karluk* were safely moved there, Captain Bartlett intended to strike out for Siberia to seek a ship to rescue his companions.

This scouting party of seven got under way on the morning of January 21 for Wrangel Island, which lay out of sight about seventy miles over the horizon to the southwest, according to Captain Bartlett.[4] During the captain's absence, after he set off for Siberia, the men on Wrangel Island were to move the rest of the supplies and equipment from Shipwreck Camp

---

[4] Captain Bartlett underestimated the distance to Wrangel Island, which was closer to eighty miles in a direct line. Known today as Ostrov Vrangelya, the island lies astride the 180th meridian on the western side of the Chukchi Sea, about ninety miles northeast of the Siberian coast at Cape Billings (Mya Billingsa). Lengthy off-shore bars flank the island's northern and southern coasts, and three parallel mountain ranges cover its southern half, with the highest peak (Gora Sovetskaya) reaching about 3,850 feet above sea level. Wrangel Island was discovered in 1867 by Thomas Long, an American whaling captain, and named after Russian Baron Ferdinand von Wrangel, who had attempted unsuccessfully to visit it between 1820 and 1824. It was first landed upon in 1881 and claimed for the U.S. by Captain Calvin L. Hooper of the U.S. Revenue Marine (later Coast Guard) steamer *Corwin*. Captain Hooper was exploring and searching for the De Long expedition and its ship, *Jeannette*. The only islands in the Chukchi Sea, Wrangel Island and Herald Island were part of the Beringia land mass that joined Asia and North America until about 13,000 years ago (Guthrie, 2004). Wrangel Island was not glaciated during the Quaternary Ice Age. Today it is known to have the highest density of polar-bear dens in the world, more than four hundred plant species, which is more than any other Arctic island, and to have served as a refuge for the last surviving mammoths in the world until 4,000 years ago, 6,000 years after they had become extinct on the Siberian mainland (Vartanyan et al., 1995). The island was unpopulated by humans when

to the shore camp. McKinlay sat up most of one night making two copies of their British Admiralty map of Wrangel Island, one for each of Anderson and Mamen, and Captain Bartlett prepared written instructions for both parties.

Both Iñupiat men had a rifle with cartridges and other equipment. For provisions the scouting party took cases of pemmican (for both men and dogs), hard biscuits, tea tablets, canned milk and sugar. Pemmican had been the staple food for Arctic explorers for many years. Each of their pemmican tins weighed six pounds, and there were eight tins in each case. Captain Bartlett had instructed them to flatten their used pemmican tins to serve as trail markers for those who returned or who followed later. Anderson's party also took a tent with fly, two primus stoves, a Winchester rifle and ammunition.

The people left behind in Shipwreck Camp continued with the various chores that had to be done to get ready for their move towards land once sufficient daylight had returned.

Reappearance of the Sun

Four days after the departure of the scouting party on January 21, the sun reappeared on the horizon at Shipwreck Camp for the first time in seventy-one days. The people there marked the event with a celebration dinner. Two days later the sun got above the horizon and cast sufficient light that Captain Bartlett recognized that the land they could see now in the distance did not resemble the description they had of Wrangel Island, hence must be the much smaller Herald Island. This latter island lay about forty miles east-northeast of Wrangel Island, and was described in Captain Bartlett's Coast Pilot as "a solid mass of granite about 900 feet high," and "almost inaccessible" (in July 2008, a Wikipedia Internet website described the island's maximum height above sea level as 372 metres (1,210 feet) and the rock type as granite gneiss). It was not exactly an attractive haven to head for.[5]

Departure of Dr. Mackay's Party

On January 31 Dr. Mackay and Murray, the two disgruntled Scottish Antarctica veterans, impatient to head for Wrangel Island and then possibly for Siberia, asked Captain Bartlett to provide them with a sled and supplies for four men for fifty days. The ethnologist Beuchat was going to accompany them to Wrangel Island, they said, and the sailor Morris had expressed his desire to accompany them as well. The captain urged the two Scotsmen to wait until there

---

Captain Bartlett's party reached its shores, but later had two small fishing villages on its south coast. In 2004 Wrangel Island was named the world's northernmost World Heritage Site.

5   The small, rocky and steep-cliffed Herald Island (Ostrov Gerald), about forty miles northeast of Wrangel Island, was first landed upon and named in 1849 by Captain Henry Kellett of the HMS *Herald*, a British naval vessel searching for the Sir John Franklin Expedition. The Captain thought he saw land far to the west and named it "Kellett Land," and it was later shown by that name on a British Admiralty Chart. This subsequently became known as Wrangel Island after Baron Ferdinand von Wrangel. Captain Kellett was the same man after whom Captain M'Clure in 1851 named a long curved sandspit in the southwest corner of Banks Island (Cape Kellett), near which George Wilkins later established his Canadian Arctic Expedition base in 1914 from which to search for Stefansson. In 1853 Captain Kellett repaid the compliment by rescuing Captain M'Clure and his men of the HMS *Investigator* from Mercy Bay on the north coast of Banks Island.

was a little more daylight and then to journey to Wrangel Island with all the rest of the people in the camp. They replied that they preferred to make the trip on their own.

Captain Bartlett reluctantly agreed to provide them with the supplies they requested but only if they all signed an agreement that released him from any responsibility if they should come to grief as a result of their actions. They willingly gave him the required document the following day, signed by all four men. The captain then assured them that they were welcome to rejoin the main party at any time if they ever wanted to and offered them a team of dogs to pull their sled.[6] The two Scotsmen stubbornly replied that they preferred to haul it themselves, as they had done in Antarctica on the Shackleton expedition. The four men got away about 9 a.m. on February 5, after McKinlay had checked the supplies issued to them and made a record of them.

Mamen, Kataktovik and Kuraluk arrived back at Shipwreck Camp on February 3. Mamen reported that Sandy Anderson and his party had managed to get within three miles of land after eleven days, but were then stopped by open water. Mamen's description of the land they could see from the place where their progress was stopped by the open water did not fit the known description of Wrangel Island and thus confirmed Bartlett's suspicion that it was Herald Island. The latter is small, about four square miles, hilly and rocky, while the former is much larger, about 2,937 square miles, with low-lying tundra on parts of its coastline, and mountains inland.

Ferrying the Supplies

After a brief respite, Mamen, Kataktovik and Kuraluk departed on February 7 with three sledloads and seventeen dogs to start moving the supplies as far along the prepared trail as they could, and also to link up with Sandy Anderson and his party and resupply them so they could continue to Wrangel Island. Crew members E.F. Chafe and Hugh Williams departed the same morning with a load, intending to ferry it to the fifth temporary camp established by Anderson. Mamen and Williams returned early in the afternoon, as both of them had had accidents; Mamen had dislocated his knee-cap and Williams had fallen through some young ice. When they headed back to Shipwreck Camp, Chafe joined Kataktovik and Kuraluk and continued on with their three loaded sleds. The three of them returned on February 16 with the news that they, too, had been stopped by open water within three miles of Herald Island. They could

---

6 In Jack Hadley's account of the ill-fated *Karluk* in Stefansson's *The Friendly Arctic*, he stated that Captain Bartlett refused to give dogs to Dr. Mackay's party after they asked for them (Stefansson, 1921b, p. 710). This completely contradicts Captain Bartlett's earlier account (Bartlett and Hale, 1916, p. 123), which stated that the Mackay party refused the dogs he offered them. McKinlay later revealed that Hadley kept no diary while on the *Karluk*, and the diary he kept later on Wrangel Island was merely a brief account of events without personal comments. Furthermore, Hadley had given McKinlay that diary. McKinlay also reported that Hadley shared Captain Bartlett's cabin on the *Karluk* and regarded him with much respect. It thus appears probable that the conflicting statement about the dogs that appears in *The Friendly Arctic* (p. 710) in the Hadley account of the fate of the *Karluk* was created by Stefannson, evidently to cast blame on Captain Bartlett (McKinlay, 1922b [March 30, 1922]). Indeed, McKinlay and Bartlett agreed that much of the supposed Hadley account of the *Karluk's* fate in Stefansson's book was created by Stefansson rather than by Hadley.

see it clearly, but had observed no sign of Anderson's party. Captain Bartlett hoped Anderson would have realized his mistake, turned west and headed for Wrangel Island, but for some unknown reason, Anderson did not do that. The skeletons of his four-man party were found on Herald Island by Captain Louis Lane in 1924.[7]

Dr. Mackay's Party in Trouble

While on their way back, about twenty miles from Herald Island, Chafe and his two Iñupiat companions came upon Dr. Mackay's foursome. Morris had cut his hand and had developed blood-poisoning. Beuchat was staggering along the trail alone and far behind the other three, with both feet and hands frozen, appearing close to death. Chafe urged them to return with him to Shipwreck Camp, but all four men refused, insisting that they intended to keep heading for Wrangel Island. They were never seen again.

More Loaded Sleds Head for Wrangel Island

On February 19 Captain Bartlett sent two more sledloads to Wrangel Island from Shipwreck Camp. Led by Hadley, Malloch, Williamson and George Breddy with the first, and Munro, Maurer, Williams and Chafe with the second, they were expected to draw supplies from the caches along the route in order to move as much of the supplies as they could to their destination. As only four dogs could be spared for each load, both sleds included man-harnesses so the men could help the dogs with the pulling. Hadley and Maurer had made a caribou-skin bag to carry Nigeraurak, the cat. She rode in the bag on one of the sleds or alternatively slung from Maurer's neck when they were on the trail. At night she was brought into their igloo, fed scraps of pemmican and slept in her footbag on top of one of the men's feet.

Captain Bartlett remained another five days at Shipwreck Camp with McKinlay, Mamen, Templeman, Kataktovik, and Kuraluk and his family, for he wanted to make an inventory of all the supplies they were leaving behind and to be sure that everything was in order before he closed the camp. He and his companions finally got away on the morning of February 24, with their three heavily loaded sleds, each carrying several hundred pounds of pemmican, biscuits, canned milk, tea and coal oil, supplies Bartlett considered sufficient for about sixty days. Kuraluk started off at 9 a.m., accompanied by his wife, Keruk, her eight-year-old daughter Helen, Templeman and Kuraluk's loaded sled, which was pulled by his five dogs. Keruk was carrying their young daughter, Mugpi, on her back.[8] The other two sleds got under way an hour later, led by Kataktovik and Mamen, the latter limping pronouncedly because his injured knee was causing him considerable pain. Kataktovik's sled was pulled by three dogs. Captain Bartlett and McKinlay brought up the rear, with their sled pulled by four dogs, the former guiding the dogs, the latter helping to pull their sled. Left behind at the camp were two surveying instruments (transits), about 3,000 pounds of pemmican, cases of biscuits, sacks of coal, gasoline, coal oil, and other odds and ends. These the captain fully intended to have picked up and brought to Wrangel Island later. The red ensign still flew over the camp as they headed in a southwesterly direction.

---

[7] Le Bourdais, 1931; Niven, 2000, pp. 368–370.
[8] She carried Mugpi all the way to Wrangel Island (Bartlett and Hale, 1916, p. 141).

Captain Bartlett and his party made twelve miles on the first day and eighteen more on the second day, while roughly following the route marked by the empty pemmican tins, igloos and caches. At the fourth cache the captain found a note from Munro stating that he had been held up there by open water and a heavy gale. Captain Bartlett hurried on, catching up to Kuraluk and his family, who were occupying the sixth igloo. The captain's party built a second igloo nearby. During the night a large crack in the ice developed through the middle of Kuraluk's igloo, forcing his family out lest the crack open while they slept and all tumbled into the icy water. The captain then let Keruk and her two children use his igloo, while he and his five men walked back and forth all night trying to keep warm in the –40° F temperature. They also kept a close eye on the ice, which continued to break apart around the camp and then close up again.

At daybreak the captain sent McKinlay and Kataktovik back to Shipwreck Camp with an empty sled and all twelve dogs to bring back thirty gallons of coal oil. They returned the next afternoon with the oil, as well as some alcohol, sealskins and tablets of tea.

Stopped by Pressure Ridges

On February 28 Captain Bartlett and his party caught up to the advance party of Munro and Hadley, who had been stopped by an elongated belt of pressure ridges twenty-five to one hundred feet high and perhaps three miles wide. These ridges blocked their way, with no visible way around them to left or to right as far as they could see. Storms had forced the mobile sea-ice against and over the land-fast ice, creating this wide band of extremely rough ice. Captain Bartlett had never encountered more impassible ice. To get around it was out of the question. The only solution was to cut their way through. Captain Bartlett therefore instructed his men to build igloos while he reconnoitred the formidable ice barrier to determine where best to start cutting a route through it.

Once he had reached his decision, Captain Bartlett organized his men to start hacking a four-foot-wide trail through the pressure ridges with their pick-axes, instructing them to make the trail smooth enough to avoid damaging their sleds as they passed through. It was a monumental undertaking. Bartlett wrote later, "Building a road across them was like making the Overland Trail through the Rockies."[9] Each day he sent his men ahead to continue the trail-cutting through the rough ice, while he and one or two others ferried sledloads of supplies forward to the next temporary camp from their temporary camp on the seaward side of the pressure ridges. Working strenuously each day in this manner, the men finally succeeded after four days in crossing the three miles of rough ice and emerging on the smoother ice on the landward side. There they camped and waited for McKinlay, Hadley and Chafe to return from Shipwreck Camp with two sledloads of pemmican, biscuits and tea.

When the McKinlay party arrived on March 6, they told a story of being suddenly confronted by three polar bears, one of which got between them and their rifles on one of the sleds. Fortunately their shouting frightened that bear, and they were able to get to their guns and shoot all three bears. McKinlay, Hadley and Mamen went back the next morning to collect the bear meat.

---

[9] *Op. cit.,* p. 154.

For the next week Captain Bartlett's men ferried sledloads of supplies forward by means of multiple short trips, approaching ever closer to their intended destination. Although the temperature remained far below zero throughout their entire journey, they were blessed with clear weather after leaving Shipwreck Camp.

Arrival at Wrangel Island

At last, early in the afternoon of March 12, Captain Bartlett and his companions reached Wrangel Island. Their exhausting eighty-mile trip from Shipwreck Camp had taken them two weeks. They landed on a long sandspit then known as Icy Spit, which jutted quite a distance seaward on the northeast side of the island. An area of tundra fringed the coast, with low mountains and valleys some distance inland and higher peaks still farther inland. To their considerable relief, they found a good supply of driftwood nearby along the shore.[10]

As soon as the *Karluk* survivors were safely on shore, they set about establishing their camp, building three igloos and gathering firewood. The Munro party of eight men occupied one

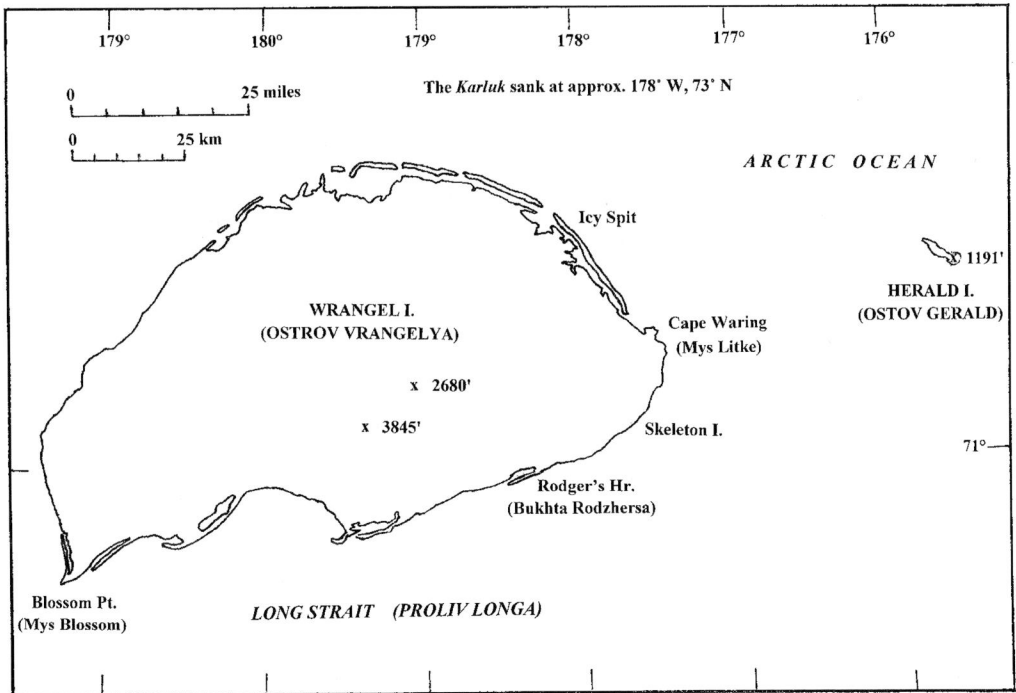

Map 3 Location of the four camps (Icy Spit, Cape Waring, Skeleton Island and Rodger's Harbor) occupied by the survivors of the *Karluk*, 1913–1914, Wrangel Island and Herald Island. Elevations are in feet. (Copied from part of Operational Navigation Chart ONC C-8 (Edition 2) 1973, published by Defence Mapping Agency Aerospace Center, St. Louis Air Force Station, Missouri 62118.)

---

[10] Wrangel Island's polar climate has kept the island devoid of trees for thousands of years, so this driftwood must represent trees from the interior of Siberia that were uprooted and floated down rivers to the Arctic Ocean and then to the island.

igloo, Kuraluk and his family the second, and Captain Bartlett, McKinlay, Mamen and Kataktovik the third. The captain then sent Kuraluk to scout the surroundings for signs of polar bears, caribou, reindeer or other forms of animal life. Kuraluk returned hours later having seen no animals and only the tracks of a few foxes and one bear. This did not bode well, for the survivors needed fresh meat to augment their meagre provisions. Furthermore, they had seen no seal holes in the ice closer than twenty-five miles from the island, and without seals, they could not expect to find polar bears.

Seven weeks had passed since the departure of Sandy Anderson's four-man party from Shipwreck Camp for Wrangel Island. Captain Bartlett had hoped to find signs of the men on Wrangel Island, but to his consternation and disappointment he saw none.

Once their camp was established, the first order of business was to get their boots and clothing dried out and repaired, and their injuries healed. Both men and dogs were greatly wearied from the strenuous trip. Once the clothing was taken care of and all had recovered, Captain Bartlett planned to send some of the men back to Shipwreck Camp to bring the supplies that had been left there. Meanwhile, he intended to head for Siberia and civilization to alert the world of the loss of his ship, the location of the survivors and to seek their rescue during the coming summer's short navigation period. He dared not delay his departure for long, for it was already mid-March and the breakup of the ice between Wrangel Island and the Siberian coast would soon open up leads he could not cross without a boat, and would increase the danger of his journey immeasurably.[11]

On March 17, Munro, Breddy and Williams left for Shipwreck Camp with one sled and sixteen dogs. They were to ferry a load of biscuits from the camp to the west side of the big pressure ridges, some twenty-five miles from Wrangel Island, then return to Shipwreck Camp for a load of pemmican. Then they were to bring all of these supplies from the pressure ridges to the new camp on Wrangel Island.

Before his departure from Icy Spit, Captain Bartlett handed McKinlay a letter of instructions to give to Munro when the latter returned from Shipwreck Camp, placing Munro in charge of the Wrangel Island camp during his (the captain's) absence. This was because Munro, as the *Karluk's* first engineer, was the next highest-ranking member of the crew to the absent First Mate Sandy Anderson. When he returned from collecting supplies from Shipwreck Camp, Munro was to make a trip to Herald Island to search for signs of the first mate and his party. On his way back from there he was to bring to their Wrangel Island camp the supplies that had been cached along Anderson's trail to Herald Island from Shipwreck Camp, as well as any additional supplies remaining in the latter camp. Munro was also to give each group at Icy Spit their proportional share of the supplies, do what he could to promote good feelings among the men, and assemble everyone farther south around the coast at Rodger's Harbor about the middle of July, where Captain Bartlett hoped to meet them with a rescue ship.

---

[11] The survivors had taken no form of water transport with them from Shipwreck Camp to Wrangel Island. Their whale boat (*umiak*) had gone down with the *Karluk*, and their two canoes had either served as fuel for their stoves at the camp or were left behind for later recovery owing to the limited capacity of their sleds and dogs.

Captain Bartlett then asked McKinlay to take inventory of the supplies here at their new camp on the spit and to apportion them among the party. He thought that the men should divide into three groups and live sufficiently far apart to provide large hunting areas for each. Each group was to be allotted rations for eighty days.

The captain had planned their survival activities well.

CAPTAIN BARTLETT'S JOURNEY TO ALASKA

6

Captain Bartlett and Kataktovik started south for Siberia on March 18, the day after the Munro party left for Shipwreck Camp. The entire responsibility for getting news quickly to Nome and Ottawa about the fate of the *Karluk* and the locations of the survivors now fell entirely upon the captain's shoulders, so that the survivors could be rescued during the brief summer navigation season only a few months hence. If he failed, all were doomed. Just before he departed, he spoke quietly to McKinlay, asking him to help Munro in any way he could and to do everything possible to maintain peace among the various survivors.

For his journey, Captain Bartlett took seven dogs, one sled, provisions for forty-eight days for Kataktovik and himself, and for thirty days for their dogs. He also carried outgoing mail from some of the men at Icy Spit. He intended to keep a sharp eye out for any signs of the missing parties of Dr. Mackay and Sandy Anderson while he travelled along the east coast of Wrangel Island to its south side. From there he intended to cross Long Strait to the Siberian coast, probably more than 110 miles of frozen ocean, then proceed east for another 600 miles to East Cape.[1] Once there he was confident that he would find some way to get to Nome, where he could telegraph the news to Ottawa and try to initiate rescue efforts.

Before he and Kataktovik had progressed a mile from Icy Spit they were enveloped in swirling snow by a strong northwest wind, which severely limited their visibility. Stubbornly they trekked onward, camping the first night near Skeleton Island, the second night at Rodger's Harbor (now known as Bukhta Rodzhersa). The captain looked for signs of the missing Anderson or Mackay parties as they travelled along the coast, but saw none. He had planned to cross Long Strait from Rodger's Harbor to Cape North, but the great ice ridges lying offshore persuaded him to change his plans. As a result, he and Kataktovik continued the next day to follow the shore, inside the ridges of upthrust ice, westward to Hunt Point. Their progress was slow, however, because they had to cut their way through rough heavy ice with their lone pick-axe. That night, as on subsequent nights, they built an igloo in which to sleep. The following day travelling conditions were much better and they reached Blossom Point[2] at the

---

[1] East Cape is the mountainous peninsula at the northeastern extremity of Asia, facing Bering Strait. Also known as Cape Dezhnev (Webster, 1980, p. 330) and Dezhneva Cape, it is the most easterly point of Asia (169° 40' W), and is immediately west of the International Date Line.

[2] The English names used by Captain Bartlett for geographic features around the coast of Wrangel Island are those shown on the chart accompanying the American Coastal Pilot book he carried. They were the names of officers on the U.S. revenue cutter *Corwin*, who had explored the island in 1881. Most of these names were subsequently changed to Russian names after the Russians formally claimed the island in 1924. Blossom Point (now Mys Blossom), the southernmost point on the island, was an extensive sandbar.

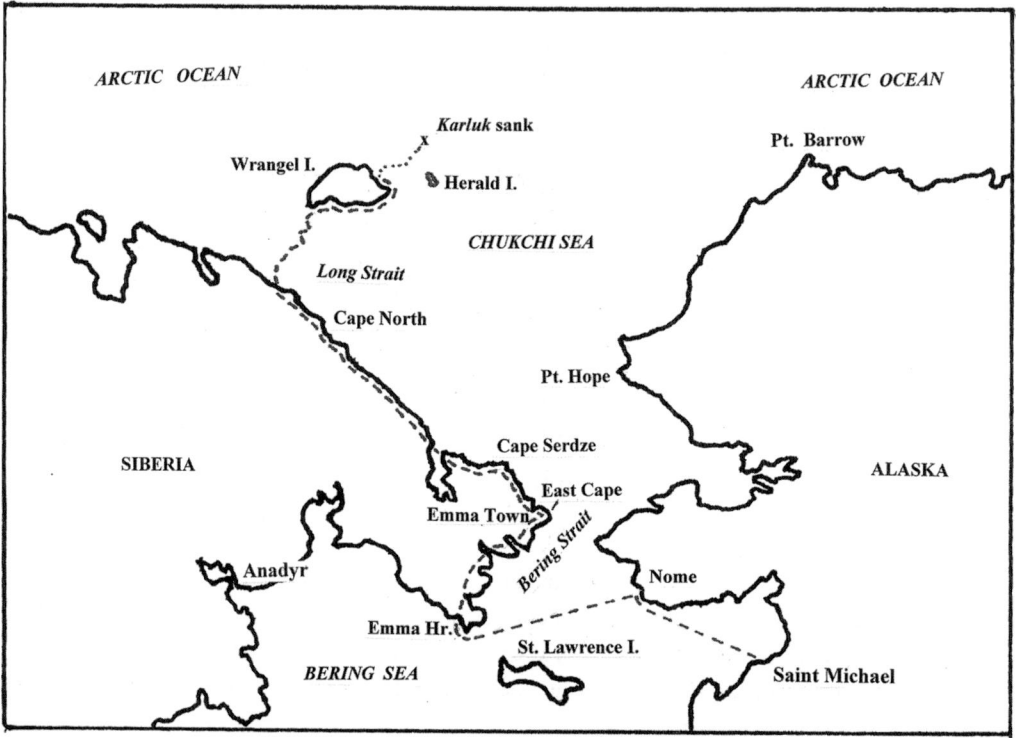

Map 4 The route taken by Captain Bartlett from "Shipwreck Camp" to Wrangel Island, Siberia and Alaska, 1914.

southwesternmost part of Wrangel Island in the evening. Their journey around the east coast of the island to this point, from which the captain now intended to cross to the Siberian coast, had taken them four days.

Crossing Long Strait

On March 23 they started over the ice in Long Strait (Proliv Longa), almost immediately encountering raftered ice[3] through which they had to cut a trail. Despite strenuous efforts they progressed only about five miles that day. The captain wore his snow goggles throughout the day and even during the night, but by evening one eye pained him a great deal. The following morning the two men got through the raftered ice, but immediately encountered open water and running ice that was kept in motion by a westerly gale. Their journey across the strait thereafter was constantly hampered by blowing snow, open leads and problems with their dogs, resulting in more delays and slow progress.

---

[3] "Raftered ice" is a term used by the Newfoundland sea captain Bartlett (Bartlett and Hale, 1916, p. 177). The *Dictionary of Newfoundland English* (Story *et al.*, 1982, p. 402) defines a "rafter" as "a large sheet of ice tilted or forced up by pressure of the sea."

The captain's American Coastal Pilot book provided a little information about the Siberian coastline and its people, as well as a highly generalized chart of the coastline towards which they were heading. He regularly looked at his compass to ensure they were going in the proper direction. Without both chart and compass, he and Kataktovik could easily have become lost and confused and would soon have perished.

Captain Bartlett's experiences as a seal hunter on the ice off Newfoundland proved extremely valuable in getting the two of them safely across many of the leads. They rafted across some of the leads on small blocks of ice, using their snowshoes as paddles. They bridged some narrow leads with their sled, and leaped across still narrower leads after first tossing their dogs over.

In a few places they encountered a lead covered with thin new ice that was much too risky to walk across, and here the captain used a novel lead-crossing procedure. First, they placed their tent poles on the thin ice in the lead and Kataktovik (a much lighter-weighing man than the captain) crawled to the opposite side between the poles. He had a rope tied around his waist to allow the captain to rescue him if he broke through the ice. When he reached the other side, Kataktovik pulled the partly loaded sled across and unloaded it. Then the captain pulled it back with a rope attached to the back of the sled and partly loaded it again. In this manner, they gradually moved all of their supplies and equipment across the thin ice. At that point, the captain pulled back the empty sled and lay flat upon it, and Kataktovik pulled it across the thin ice as quickly as he could. Then they reloaded the sled, harnessed their dogs to it and continued on their way.

They spotted seals in a few open leads and succeeded in shooting one on two different occasions. Kataktovik retrieved each of them with a *manak*, a device his people used that consisted of a fishing line and a wooden ball about the size of a baseball, with hooks projecting all around it. The ball was attached by a long fishing line to a wooden handle. Kataktovik whirled the ball and some of the fishing line around his head, like a cowboy with a lasso, then hurled it beyond the seal they had shot. By pulling in the fishing line he managed to hook the dead seal and draw it to the edge of the ice where they could retrieve it. After skinning the seals and taking whatever meat they could use, they deliberately left residual pieces of the dead seals on the ice, hoping to attract the sensitive nose of a polar bear. This ruse worked well, and from each abandoned seal a bear followed their trail to their next camp. Fortunately, in each case they saw and shot the bear before it could cause them any harm or damage. The addition of the bear meat to their supply of seal meat saved their pemmican for later days.

Each night they built an igloo and mended their fur clothing, which somehow always managed to get torn through contact with ragged pieces of ice.

March 30 was the first fine day since they had departed from Icy Spit thirteen days earlier. On that day Captain Bartlett and Kataktovik crossed lead after lead, one of them half a mile wide. Somehow they always managed to find some location along each lead where they could bridge it or ferry themselves across on one or more cakes of ice.

That evening the captain scanned the horizon to the southwest and thought he saw land, but Kataktovik was unconvinced when he looked through the binoculars. Early the next morning, however, the captain was certain he saw snow-covered land in the distance. Much cheered by the prospect that their long journey across the ice to Siberia was finally coming to an end, they pressed onward, crossing many more leads that day, and by the time they stopped for the night they could definitely see the land, though it was still perhaps forty miles away.

Travelling the Siberian Coast

On April 4, after another five days of cold and difficult travel, Captain Bartlett and Kataktovik finally stepped onto the Siberian coast, seventeen days and about two hundred miles after their departure from Icy Spit. The temperature, the captain guessed (for he had no thermometer with him), was about what it had been during most of their journey, that is, in the vicinity of -50° F. They were safely ashore now, but had only four of their original seven dogs to pull them hundreds of miles to the east on the second leg of their journey. One of their dogs had run away, one had died from exhaustion and the third one had been left behind when it refused to cross a lead.

The imperfections of his coastal chart and the lack of prominent land features prevented Captain Bartlett from identifying exactly where they landed, but he later decided it had been about sixty miles west of his intended original destination of Cape North (Mys Shmidta).[4] From the coast he could see that the tundra extended inland only a short distance, then the land rose sharply to heights exceeding three thousand feet. Narrow gravel islands extended eastward a short distance offshore and parallel to the coast.

Shortly after stepping ashore, he noticed the trail of a Native's sled heading eastward along the tundra. He and Kataktovik followed it, for it was heading in the direction they wished to travel. Furthermore, they were confident that the individual with the sled would select the easiest trail. Soon they were enveloped in a driving snowstorm and forced to construct an igloo for shelter.

Travelling proved easy the next day as they continued their journey eastward following the well-marked sled trail along the tundra. Early in the afternoon they spotted a group of people well ahead of them. At that point, the young Kataktovik showed a distinct unwillingness to continue leading the dogs. When pressed for an explanation, he said that he had been told by his parents that Siberian Natives killed Alaskan Iñupiat. Captain Bartlett assured him that the Chukchees were known to be kind and friendly people, but his companion remained unconvinced, so the captain went on ahead and Kataktovik followed behind the sled, reluctantly steering it forwards.

When they drew near the people, Captain Bartlett walked towards them, and according to his own account held out his hand, and said, "How do you do?"[5] His English words meant nothing to the local people, of course, but his visible friendliness was easily recognized. Soon

---

[4]   The captain's description would seem to place the landing site on the Siberian coast as being close to Laguna Kyanygtokynmankyn, from which offshore sandbars extend almost continuously across the mouths of coastal indentations southeastward to Mys Shmidta and beyond. I chanced upon the current name for Cape North in my father's copy of Stefansson's wartime book *Arctic Manual* (Stefansson, 1944, p. vii), where Stefansson called it Cape Schmidt. The U.S. Defence Mapping Agency Aerospace Center Map ONS - 8 (1974) shows it as Mys Shmidta.

[5]   Bartlett and Hale, 1916, p. 209.

he and Kataktovik found themselves inside the wooden house of a Chukchee family,[6] seated on a platform with uncooked reindeer food being offered them. The house, called an *aranga*, was constructed of driftwood with a rounded roof of saplings. A dozen or so other people sat nearby and chatted cheerfully with the two strangers. Their language was totally unfamiliar even to Kataktovik, which surprised him, but much to his relief, they showed no indication of wanting to kill him or the captain. The Natives all chatted back and forth as they feasted, first on reindeer meat, then on some tainted walrus meat, a small piece of which Captain Bartlett politely tasted, but Kataktovik ate with obvious pleasure. They were offered hot tea every few minutes.

The feast went on for hours. Afterwards the Chukchees smoked pipes with Russian tobacco. Most of them, Captain Bartlett observed, also coughed so regularly and violently that he concluded that tuberculosis was rampant among them. In response to their questions, he explained where he had come from and where he wished to go, using his coastal chart to illustrate his words. Whether his hosts understood a word he said he did not know. He guessed that they thought he was a trader, of which a number frequented their coast from time to time. Using his chart and sketching on paper, he determined that these people had to travel many days inland to where there were reindeer and trees.

After a day and a half with these new acquaintances, which gave the visitors time to rest and feed their four dogs and to dry and repair their clothing and sled, Captain Bartlett and Kataktovik renewed their long journey east on the morning of April 7, a clear but extremely cold day. A young Native accompanied them with his small sled and lone dog.

Two days of exceedingly cold travel, the coldest days Captain Bartlett had ever experienced, took them to Cape North, where eleven *arangas* fringed a small harbour. A few gold mines operated in the hills nearby. Several of the houses at the cape were occupied by reindeer men, the persons who take care of the herds of reindeer kept in the interior. These men occasionally came to the coast to trade reindeer meat for seal and walrus meat. The cape had been seen and first named in 1778 by England's famed Captain Cook. Here Captain Bartlett encountered two Russian men who seemed content to stand outdoors and converse in the cold. Presently, however, another Native invited them to his house, where many others were enjoying the indoor warmth and hot, strong tea. The Native host knew a few English words and managed to convey to the captain that he had visited the *Karluk* and other whaling ships at East Cape. He also seemed to understand the captain's explanation of the sinking of the *Karluk,* and how far he and Kataktovik had travelled since then.

After a time one of the two Russians who had remained outdoors conversing, a Mr. Caraieff, came into the house and invited the captain and Kataktovik to his house for the night. He also gave the captain a letter of introduction to a brother at East Cape, ensuring him a proper reception when he got there. Then they had a meal of Russian bread, salmon, tea and milk, and spent a comfortable night.

Captain Bartlett and Kataktovik left Cape North (Mys Shmidta) shortly before noon the next day, following a well-defined trail to the east. They built and slept in an igloo that night.

---

[6] Chukchee is the name for the Natives (and their language) in northeastern Siberia. Many of these Natives are reindeer herders. Their language differs greatly from that of the people of northern Alaska.

The next morning (April 10), shortly after they started, one of their dogs reached its physical limit and refused to go any farther. He was unharnessed and put on the sled, and they made little progress after that, finding shelter for the night in a Native's house. The captain had the good fortune to obtain a dog on loan from his new host. Several days later, after much hard travelling, he and Kataktovik reached a small settlement on Koliuchin Island (Ostrov Kolyuchin), a few miles outside the mouth of Koliuchin Bay (Kolyuchinskaya Guba). By then the captain was suffering greatly from snow-blindness and his five dogs were almost played out.

There they encountered two young men, one of whom knew a little English. Having heard of these two travellers who wished to get to East Cape, the man with the little knowledge of English asked Captain Bartlett how much he would pay to be taken there. With many miles still to travel, weary dogs and time running out to get rescue operations underway, Captain Bartlett offered US $40, which was almost all of the money he had.[7] This amount satisfied the young man and he agreed to the arrangement. Captain Bartlett then transferred his dogs and some of his goods to the two sleds owned by one of the two young men, abandoned his own badly worn sled, and on April 19 the four men started on their way. Before the day was out, however, the two men decided they would not continue to East Cape and dropped the captain, Kataktovik and their dogs at the house of a Native. This left Captain Bartlett with miles still to travel, five weary dogs, no sled and almost no money. However, in exchange for a snow-knife, pick-axe and two steel drills from his residual supplies, he arranged a ride with another Native to the house of a naturalized, thirty-eight-year-old American trader named Olsen on the west side of Koliuchin Bay. Olsen was an American trading agent for Olaf Swenson of Seattle, and there the two weary travellers were well received.

On to Emma Town

From Olsen's, the captain bartered his and Kataktovik's way to Cape Serdze (now Mys Netan), where they arranged to be taken to East Cape by a somewhat dare-devilish Native named Corrigan, reportedly the best-known hunter in Siberia, and the most prosperous Native Captain Bartlett had met. The distance to East Cape was about ninety miles, which Corrigan's sixteen dogs together with Captain Bartlett's five dogs covered in two days. Bartlett and Kataktovik mostly rode on the two sleds driven by Corrigan and a friend of his who accompanied them. Travelling on the sea ice just offshore, Corrigan enthusiastically entertained Captain Bartlett with tales of his hunting experiences as they went along until he finally realized that the captain had understood virtually nothing of what he was saying. Their trip actually ended in Emma Town,[8] a few miles southwest of East Cape, where lived Mr. Caraieff,

---

[7] He had obtained $45 in American money from Hadley before leaving Wrangel Island (Bartlett and Hale, 1916, p. 249). Whether this was truly Hadley's money or some of the large amount of government money Stefansson claimed he had left on the *Karluk*, remains an unanswered question. Hadley had occupied Stefansson's cabin on the latter's invitation when they left Cape Smyth.

[8] The settlement of Emma Town mentioned by Captain Bartlett is now known as Dezhnev, and is located on the southwest side of the East Cape peninsula. It should not be confused with Emma Harbor, which is in Providence Bay (Bukhta Provideniya), considerably farther south.

the brother of the man Captain Bartlett had stayed with at Cape North. Once they had arrived, the captain presented his letter of introduction to Mr. Caraieff and was promptly invited to stay as long as he liked.

It was now April 24 and thirty-nine days since Captain Bartlett and Kataktovik had left Icy Spit and the *Karluk* survivors. He and Kataktovik had successfully completed the second of their three-stage journey to Nome. The first stage had been to reach the Siberian coast, the second was their long journey along the Siberian coast. Still ahead was the challenge of getting to Nome, to notify the Canadian government of the plight of his men and to set in motion operations to rescue his companions on Wrangel Island.

Delayed by Illness

Captain Bartlett therefore spoke to Mr. Caraieff of his urgent need to reach Nome. The latter knew sufficient English that he was able to instruct the captain on the four courses of action available to him.

The captain's first option was to cross Bering Strait on the sea-ice via the two Diomede Islands to Cape Prince of Wales and then proceed southeasterly to Nome, but the season was now too late to sled across the strait safely, and he lacked both sled and healthy dogs. He would therefore have to wait until some time in May or early June for the strait to open up, at which time he could arrange to cross it in a whale boat. Once across, however, he would still be a considerable distance from Nome.

His second option was to wait until June when trading ships would likely appear at East Cape.

His third option was to try to go south to Anadyr, where there was a wireless station, and send his message from there. Mr. Caraieff felt, however, that the season was too far advanced for a safe journey to Anadyr,[9] because the ice in the rivers he would have to cross would already be breaking up. Furthermore, it was quite possible that the Anadyr wireless station was out of commission.

The fourth and last option required him to wait until the first week in June, when a Mr. Thompson of Emma Harbor would be taking his schooner to Nome and could take the captain with him.

Restless to keep going, Captain Bartlett decided to undertake the third option and head for the wireless station at Anadyr. With his host's assistance he found some Natives who would take him to Indian Point, from where he fully expected to arrange with other Natives to take him the rest of the way. That night, however, his legs and feet suddenly swelled up so much that he was unable to rise from his bed. Devoid of energy, his eyes bloodshot, he was so stiff he could scarcely move hand or foot. He had lost nearly forty pounds during his journey from Wrangel Island. Kataktovik was equally worn down. Both men were thus forced to remain and rest where they were.

A few days later the captain received a visitor, Baron Kleist, the Russian Supervisor of Northeastern Siberia, who lived in Emma Harbor some distance to the south. He had been

---

[9] Anadyr is about 275 miles west of Emma Harbor. Russian naval vessels visited Anadyr, which had a sheltered deep-water port.

on a tour of inspection of the Siberian coast and had heard about Captain Bartlett from the American trader, Olsen, at Koliuchin Bay. Olsen had thought Captain Bartlett was in his fifties on the basis of the latter's appearance when they met. After rest, proper food, washing and shaving, however, the captain now looked more his real age of late thirties, but he still was in no condition for further travel.

Then just when he appeared ready to get under way again, Captain Bartlett came down with an acute attack of tonsilitis and a high fever. These he treated with peroxide and alum and the infection slowly diminished. Nevertheless, he was still weak and not fully recovered by May 10, when the baron wished to start for his home in Emma Harbor, a community in Providence Bay (Bukhta Provideniya).

Captain Bartlett intended to continue on alone, so had to make arrangements for his young companion, Kataktovik. He soon managed to arrange for Kataktovik to wait at East Cape until the navigation season opened and then to proceed by ship across Bering Strait to Point Hope, where he wanted to go. The captain thanked the young Iñupiaq for his loyalty and trust, and said he would get him his pay after he (Bartlett) had gotten back to civilization. On parting he gave Kataktovik the rifle they had brought on their journey, as well as provisions and outfit to tide him over until he could return to Alaska. He also asked a local trader, an Australian named Charles Carpendale, to inform the Chukchees what a good man Kataktovik was, should he ever have occasion to return to the region, as this would ensure Kataktovik's future safety on that coast. The captain also left his dogs with his host, Mr. Caraieff, intending to collect them at a later date.

Overland to Emma Harbor

Captain Bartlett and the baron got underway for Emma Harbor late in the afternoon of May 10, under foggy conditions, but with the temperature above freezing. Two Natives drove the dogs that pulled the two men on their sleds, and they all stopped at Native houses for tea and for sleeping en route. There was sufficient daylight at this time of year that they could travel all night, so they continued on that first day until four in the morning. When they awoke it was snowing and too foggy to risk travelling, so they waited until early the following morning. A change of wind some hours later brought rain from the Bering Sea, then wet snow and a thick fog, but they somehow managed to reach a reindeer settlement on the north side of St. Lawrence Bay (Zalin Lavrentiya), where they were well fed by two young men who were in charge of the animals. These were the first reindeer the captain had seen since reaching Siberia. After crossing the ice on St. Lawrence Bay they proceeded overland to the mouth of Mechigmen Bay (Mechigmenskaya Guba), averaging almost five miles per hour. They then crossed that bay on the ice near its mouth and followed the coast for about twenty miles before heading overland to Cape Neegehan (Mys Mygchigen). On they went, in spite of the heavy fog, finally stopping to sleep at a place then called Mesigman. The captain was soaked by the time they arrived and, fearful of a return of his recent illness, removed his wet clothing, wrapped himself in a deerskin robe in the warm house and soon fell asleep. Renewing their journey that evening, they travelled over sea-ice for many hours, finally stopping and resting at a place the captain called Elewn. Late in the afternoon they renewed their journey, making use of the better ice conditions in the cooler hours overnight when the sun was lower in the sky. During one of several stops for tea the next day, some Natives told them that they had heard of a whaling ship at Indian

Point (Mys Chaplina) with a Captain Pedersen. This news was of much interest to Captain Bartlett because, of course, he knew of Captain Pedersen.[10] Later in the day they crossed a divide and early on the sixth morning, they reached Emma Harbor and the baron's home.

Two months had passed since the captain's departure from Icy Spit, and he now felt fully confident that in another two months he would be on a ship that was en route to rescue the survivors on Wrangel Island.

The baron had a fine, heavily timbered house, recently built, which was both warm and comfortable. The baroness was away visiting relatives in Russia for the winter, but the baron had an excellent chef. He also had a personal physician who promptly undertook to restore the captain to good health.

Shortly after arriving at Emma Harbor, Captain Bartlett learned that Captain Pedersen had recently been in the neighbourhood with his current ship, the *Herman*, so dispatched letters to people along the coast asking them to notify Captain Pedersen, if they saw him, of his wish to be picked up. His action brought quick results.

Heading for Nome

On the morning of May 19, just three days after his arrival at Emma Harbor, Captain Bartlett was delighted to see the *Herman* steaming into view. Captain Pedersen had heard of his need for transportation from some Natives farther along the coast, and hastened to Emma Harbor to get him. As soon as Captain Bartlett had taken leave of his kind host, he boarded the *Herman* and was warmly greeted by Captain Pedersen. The latter quickly provided Captain Bartlett with some American clothing to wear instead of his well-worn furs, and the *Herman* sailed for Nome, 240 miles across the Bering Sea.

Notifying Ottawa about the *Karluk*

When they reached Nome on May 24, shore-fast ice prevented the *Herman* from getting close to the shore and it was forced to anchor twelve miles off shore, waiting for a change of wind to drive the ice away. After putting up with Captain Bartlett's increasing impatience for two days, Captain Pedersen decided to head southeast along the coast to Saint Michael, where there was a wireless station operated by the Signal Corps of the United States Army. The *Herman* approached that settlement early on May 27, but thick fog prevented Captain Pedersen from getting his vessel close to the port for several hours. As soon as the *Herman* anchored, however, some crewmen rowed the two captains to the ice margin near the shore, and they walked the remaining distance to the wireless station, only to find it closed!

As they walked back to the town, the two captains met Hugh J. Lee, the local U.S. Marshall, who was known to Captain Bartlett. Lee arranged for Captain Bartlett's accommodation in town that night, and Captain Pedersen was then free to continue his summer's trading mission.[11]

---

[10] Captain C. Theodore Pedersen had arranged the purchase of the *Karluk* by the Canadian government just one year previously and had served as its captain from San Francisco to Esquimalt. He and Captain Bartlett had not met.

[11] The owners of the *Herman*, H. Liebes and Co., San Francisco, later charged the Canadian government $5,500 for transporting Bartlett from Emma Harbor to Saint Michael (Auditor General Report, 1920).

The next morning Captain Bartlett presented himself at the wireless office to send his message "collect" to the Department of the Naval Service in Ottawa. Unexpectedly the "by-the-book" army sergeant in charge of the office refused to send his message unless the captain paid for it. This Captain Bartlett could not do, as he had spent all of his cash on his journey to Alaska. Once again the U.S. Marshall came to his rescue, persuading the sergeant to bend the rules and send the wireless. Thus the following historical message finally went forth to Ottawa:

ST. MICHAEL'S, ALASKA[12]

May 29, 1914

Naval Service, Ottawa, Canada:

Karluk ice pressure sank January 11, sixty miles north Herald Island. Preparation made last fall leave ship therefore comfortable on ice. January twenty-first sent first and second mate two sailors with supporting party three months provisions Wrangell [*sic*] Island. Supporting party returned leaving them close Herald Island. They expected land island when ice moved in shore. February fifth Mackay, Murray, Beuchat, Sailor Morris left us using man power pull sleds. Sent again Herald Island three sleds, twenty dogs, pemmican, biscuit, oil. Open water prevented their landing. Saw no signs men, presumed they gone Wrangell [*sic*]. Returning left provisions along trail. Shortly after their return east gale sent us west. February twenty-fourth I left camp. March twelfth landed Munro, Williamson, Malloch, McKinlay, Mamen, Hadley, Chafe, Templeman, Maurer, Breddy, Williams, Eskimo family Wrangell [*sic*] eighty-six days' supplies each man.

March seventeenth Munro two men fourteen dogs left for supplies Shipwreck Camp. Plenty of driftwood game island. March eighteenth I left island

Eskimo landed Siberia fifty miles west Cape North. May twenty-first Captain Pedersen Whaler *Herman* called for me Emma Harbor going out of his way whaling to do so. Soundings meteorological observations dredging kept up continually. Successful. Twelve hundred fathoms animal life found bottom.

Need funds pay bills contracted Siberia and here. Wire Northern Commercial Company, San Francisco, five hundred dollars. Instruct them forward by wire St. Michael's.

BARTLETT, CAPTAIN, C.G.S.[13]

Rescue Arrangements

Captain Bartlett had now succeeded in alerting the Canadian government of the fate of the *Karluk* and its personnel, so it was up to the Department of the Naval Service in Ottawa to

---

[12] The name of the community is Saint Michael (Webster, 1980, p. 1048).

[13] Bartlett and Hale, 1916, pp. 283–284.

arrange for the rescue of the *Karluk's* survivors from Wrangel Island. Nevertheless, much still remained for the captain to do. The Naval Service Department wired back immediately asking for advice regarding arrangements for the rescue of the men and time of year when ice conditions would permit it. Captain Bartlett replied that July or early August were the best times to attempt the rescue, and suggested that the Canadian government ask the Russian government to have one or more of its ice-breakers—*Taimir, Vaigatch* or *Nadjeshny*—pick up the survivors from Wrangel Island, or the U.S. revenue cutter *Bear* if it could be accompanied by the Russian ice-breakers. There were no other suitable vessels in the region. The Russian ice-breakers were equipped with wireless, so they could be contacted at sea. He then added that he wanted to go with the rescue ship.

With the *Bear* to Wrangel Island

Captain Bartlett remained at Saint Michael receiving medical treatment for his legs and throat for the next two weeks, then late in June proceeded by motor boat to Nome. On July 13 he left Nome as a guest of Captain Claude Cochran on board the U.S. revenue cutter *Bear*. The *Bear* later received orders from Washington to go to Wrangel Island and to take Captain Bartlett. But before going to Wrangel Island, the revenue cutter had to visit a number of communities on both sides of Bering Strait, including Emma Town, where Captain Bartlett had left his dogs. When he reached Emma Town, however, on the *Bear*, the dogs were still not fully recovered so he agreed to leave them with the people who were caring for them. The *Bear* then stopped at Point Hope, where Captain Bartlett met his Iñupiaq companion, Kataktovik, and as promised paid him his wages for his work with the Expedition, together with a complete set of clothing supplied by the Canadian government. From Point Hope the *Bear* continued north, reaching Barrow on August 21, where Captain Bartlett unexpectedly encountered Stefansson's secretary, McConnell, who had left the Expedition in June and was on his way to Nome. McConnell quickly brought the captain up to date on news about Stefansson and the Southern Party. Upon learning that Stefansson had undertaken an exploratory sled journey in March with two companions over the ice north of Alaska in search of land, Captain Bartlett promptly " ... wanted to know 'What in Hell Stefansson meant by going off to his death ...' in that way."[14]

The *Bear* left Barrow on August 23, almost a month and a half after leaving Nome, and headed at last for Rodger's Harbor on Wrangel Island, where Captain Bartlett had instructed the survivors of the *Karluk* to gather in July. As the ship approached the island several days later, however, it encountered fog and much loose ice, requiring it to proceed with greater caution and further delays. Four days later Captain Cochran realized that his ship had insufficient fuel to complete its rescue mission and was forced to return to Nome for more coal. En route he made brief stops at Cape Serdze and East Cape, all of which delayed the rescue operations perilously longer. Captain Bartlett's concerns and frustrations increased with each unexpected delay.

At East Cape, Captain Bartlett learned that the Russian ice-breaker *Vaigatch* had been within ten miles of Wrangel Island on August 4 when it received a wireless message with news of the

---

[14] McConnell, 1914 (August 21, 1914).

start of the war in Europe and was ordered back immediately to Anadyr with the *Taimir*. How many more problems would they encounter, he wondered, to delay the rescue operations?

The *Bear* reached Nome on August 30. It had scarcely commenced loading the needed coal the next morning when a southwest wind sprang up, forcing it to put to sea until the winds abated or changed. It returned a few hours later, but another gale struck, forcing it to hurry back out to sea lest it be driven aground on the unsheltered coast. These unexpected and recurrent delays taxed Captain Bartlett's patience almost beyond endurance. Since the middle of March he had suffered undue hardships and illness in his efforts to get help for the men, lone woman and two children he had left on Wrangel Island. They were totally dependent on him to save their lives. Would it all have been in vain? More than two months had passed since he had reached Nome and here he was, still in Nome waiting to return to Wrangel Island with a rescue ship. Unless the *Bear*, or any ship, reached his friends on that island within the next two weeks, it would probably be too late, for the navigation season was fast drawing to a close for the year.

SURVIVAL ON WRANGEL ISLAND 7

Before leaving Wrangel Island for Siberia, Captain Bartlett had written instructions for Munro to take charge, for he was the *Karluk's* senior officer among the survivors on the island. He would have preferred to leave McKinlay—whom he held in high regard—in charge, but McKinlay was a member of Dr. Anderson's Southern Party and thus had no authority over Stefansson's Northern Party people from the *Karluk*.

As I mentioned at the end of Chapter 5, the captain asked Munro to lead a search party to Herald Island in hopes of finding the parties of First Mate Sandy Anderson and Dr. Mackay, then to return to Shipwreck Camp to retrieve as much of the remaining biscuits and pemmican as possible. He especially asked Munro to promote good feelings in camp and to assemble everyone in mid-July at Rodger's Harbor, where Bartlett would send the rescue ship. The captain asked McKinlay to make a full list of all of the supplies at Icy Spit and to ensure that each man received his share of the limited food supplies.

Friction among the Survivors

Shortly after the captain departed on March 18, quarrelling broke out among the survivors over the distribution of the biscuits and the use of fuel oil. With only 2,500 biscuits to last them all until they were rescued, the mathematician McKinlay quickly calculated that each person was entitled to 178 biscuits plus a fraction of one, but a few of the men argued for a different distribution. Complaints also arose that the Iñupiat were burning too much fuel oil to keep their fires going all day. It quickly became obvious that the sooner some of the men separated the better all would be.

Munro, Breddy and Williams arrived back at Icy Spit a few days after the departure of Captain Bartlett and Kataktovik. All had frozen faces and feet, and reported that their effort to return to Shipwreck Camp had been completely thwarted by impassable pressure ridges and a large body of open water beyond. This meant that their chances of obtaining the remaining food supplies from Shipwreck Camp were slight.

The stormy weather that commenced a few hours after the captain's departure continued for the next several days, making life generally miserable for the remaining survivors. Finally the weather cleared and, on March 23, Munro and McKinlay headed east for Herald Island, forty miles distant, with a fully loaded sled and five dogs to search for signs of the eight missing men. At about the same time, Mamen, Malloch, Templeman and Kuraluk started southeast around the coast for Rodger's Harbor to establish a second camp, taking with them a loaded sled with supplies and nine dogs. Mamen and Kuraluk were to return with the sled and dogs once they had established their camp. They ignored McKinlay's suggestion that they make their camp closer to the Icy Spit camp than Rodger's Harbor. Kuraluk returned on March 28 after quarrelling with Mamen; the latter arrived the following evening.

Munro and McKinlay soon encountered impassable pressure ridges that extended all the way from near Icy Spit to Herald Island, completely blocking their passage to that island. For three days they struggled against deep snow, bitterly cold and blowing winds with zero visibility, and stretches of glare ice where there was no snow to build igloos for shelter, all of which left both men and dogs nearing exhaustion. Taking advantage of a temporary clearing of the weather, they climbed one of the highest ice hummocks and scanned the rocky shore of Herald Island with their binoculars, but there was no sign of a camp or persons anywhere. Reluctantly, they turned back and commenced the return journey to Wrangel Island, soon forced to struggle wearily against another driving blizzard. One of their dogs gave out from exhaustion and had to be placed on the sled. Fortuitously, they came upon one of Mamen's dogs, which had broken free, and brought it back to the main camp. Both men were totally exhausted by the time they got back to Icy Spit about 11 p. m. on March 27 and only managed to drink some tea before going to bed. McKinlay had to remain in his sleeping bag until April 10, for his body ached from top to toe and his feet had swollen to twice their normal size.

On March 28 Munro and Kuraluk went to gather a polar bear and its two cubs that Kuraluk had shot the previous day on his way back from Mamen's party. Kuraluk shot another bear a few days later when it decided to investigate their camp and killed yet another with two cubs on April 2. Two days later he and Hadley unsuccessfully hunted six more bears, after which the survivors saw no more bears during the rest of their stay on Wrangel Island.

Disaster

Mamen returned to Icy Spit with the dogs on March 28, but departed again on April 1 for Skeleton Island, about two-thirds of the way around the east coast from Icy Spit to Rodger's Harbor, where he had left Malloch and Templeman, both of whom were ill.

A few hours after Mamen's departure, Munro, Williams and Chafe started out again to reach Shipwreck Camp to bring the rest of the supplies left there, taking two sleds and all twelve dogs. Nine days later, Munro and Williams arrived back at Icy Spit on foot without Chafe and minus both their sled and their dogs. They had managed to get through to the north side of the pressure ridge, but could not find the trail beyond that in spite of much searching for it. With their food supply running low and open water in the direction of Shipwreck Camp, they decided to turn back. Chafe led the way with his pick-axe, cutting a path where necessary. Munro and Williams followed with the two sleds. Suddenly there was a shout. The ice under Williams had given way and he had fallen into the icy water, with dogs and sled following him. Munro and Chafe promptly took hold of a pole on the remaining sled and held it for Williams to clutch. Then the ice under them gave way and they were both plunged into the water. Somehow they all managed to climb out of the water, but all were soaked and desperately cold. Next they rescued the dogs and the sleds, but a rifle and Chafe's camera and binoculars were gone, swallowed up by the frigid sea. Knowing they had to keep moving, they continued on, but a short while later, Chafe and his dogs and sled got safely across a narrow lead just as Williams again fell through the ice. Munro rescued Williams, but by then the lead had widened and Chafe and his dogs and sled were drifting away.[1]

---

[1] Niven, 2000, pp. 208–214.

After a very cold and troubling night, during which Munro and Williams forced themselves to stride up and down to keep warm, they struck off for Icy Spit, forced by their desperate conditions to abandon their sled and turn their dogs loose, hoping they would all stay together while they tried to work their way back through the pressure ridge. Both men had badly swollen hands and frozen feet when they finally reached Icy Spit.

The next day, using a small sled created by Kuraluk from a pair of skis, Munro went off alone to search for Chafe, but returned in the evening without finding him. Meanwhile, McKinlay was still confined to his tent, too ill to be of any assistance.

Chafe showed up unexpectedly at the Icy Spit camp in the morning of April 13, looking in an awful state, his fur pants ripped and filled with snow. He had managed to reach the pressure ridge and struggled desperately to work his way across it. By the time he was partway over the ridge, one foot and one hand were frostbitten and he was almost snow-blind, one of his dogs had died and another one had run away. In desperation he abandoned the sledload of provisions, took the leads of his remaining three dogs and stumbled onwards. The dogs somehow led him back to the base camp at Icy Spit, which he reached before a blizzard had attained its height. He rallied and told his story after consuming some hot tea.[2]

This latest attempt to reach Shipwreck Camp had proven to be a complete disaster. The men had lost two sleds, eleven dogs, two rifles, ammunition, food, tools, skins and some other equipment, all irreplaceable. Fortunately all three men had survived, although not without injuries. One of Williams' big toes became gangrenous and required amputation. Without anaesthetics, the engineer Williamson performed the surgery with a pair of tin shears on April 21. Several of the men held Williams down while Williamson carried out the gruesome task. McKinlay marvelled at Williams' courage, later writing "I have never known anyone who lived up so well to his nickname [Clam]; his lips remained tightly closed; there was never a murmur, only a slight twitching of the face muscles."[3]

Hadley and the Seals

McKinlay slowly regained his strength, forcing himself to walk about outside on good days. As the only scientist in the group at Icy Spit, he felt very much alone at times, largely because he had little in common to discuss with the others. He thought about joining Mamen and Malloch at Rodger's Harbor, because they were scientists and had some of his food rations and his stock of clothing, but he was still far too weak to attempt the sixty-mile journey.

When the supply of bear meat approached exhaustion, Hadley and Kuraluk went out on the ice on April 27 with the remaining three dogs and their own two rifles, leaving just one rifle in camp. By the first of May the supply of oil for cooking was also exhausted and the survivors at Icy Spit were totally dependent thereafter upon driftwood for heat and cooking. All cooking was now done outdoors.

On May 8, Kuraluk returned with a small seal and reported that Hadley had three more at the ridge thirty miles out on the ice. Kuraluk intended to take his family and some firewood, and return to where Hadley was located in the hopes of obtaining more seals. Several of the

---

[2] McKinlay 1914 (April 13, 1914).
[3] McKinlay, 1976, p. 105.

men in camp promptly assumed that Hadley intended to share the seals only with Kuraluk and his family, and became exceedingly angry, threatening violence against Hadley, but calmed down after eating their fill of the seal Kuraluk had just brought them. On May 11, Kuraluk started back to Hadley's camp, but lost his way.

When Kuraluk failed to get back to Hadley's camp, Hadley returned to the main camp but brought no seal meat, telling the others that there was only part of one left. Evidently he had eaten well in the meantime, which certainly did not please the hungry men in camp.

McKinlay's Journey to Rodger's Harbor

On May 12, Munro, McKinlay, Breddy and Maurer set out for the ridge where Hadley was seal-hunting and encountered a forlorn Kuraluk en route. Kuraluk developed a touch of snow-blindness and had lost his way. They all then returned to the base camp. By the time they got back, McKinlay was suffering from snow-blindness as well, which worsened, was extremely painful, and made further travelling for him out of the question until it cleared up.

At this time several of the men at Icy Spit needed medication for their various ailments, for which some medicine was in McKinlay's medical kit. However, Mamen's group had taken the kit to Skeleton Island to diminish the load for McKinlay. McKinlay had intended to join them there as soon as he was well enough, but as his eyes were still too painful for him to travel right away, Breddy volunteered to take his place and set off for Mamen's camp to get the medicine. At Skeleton Island Breddy found Mamen's abandoned camp and no sign of McKinlay's medicine kit. There was, however, a note from Mamen saying his group had moved twenty miles farther south to Rodger's Harbor. Instead of continuing on to that locality, Breddy returned to Icy Spit.

McKinlay's eyes were still painful when Breddy reappeared empty-handed. Nevertheless he felt that it was his duty to make the sixty-mile trip to Rodger's Harbor to get the medication for his invalided companions. And so, in spite of Captain Bartlett's instructions not to travel alone, McKinlay set off on foot for Mamen's camp on May 17. He was on foot, for there was neither sled nor dogs available for his use.

After walking all day and most of the night he reached Skeleton Island, having covered more than thirty miles. Crawling into Mamen's abandoned snowhouse, he found a few skins to cover himself and slept for several hours.

Then he struggled on for the additional twenty miles to Rodger's Harbor, one leg bleeding from chafing, the other cramping. He reached the camp about 9:30 that evening to be greeted with the distressing news that the geologist Malloch had died the previous night. The body still lay in the tent, surrounded by the foul odour of human decay. It had not been removed because gale winds and blowing snow forced Mamen and Templeman, both of them swollen and weak from illness, to remain inside. McKinlay did his best to console Mamen, who was terribly distressed by the loss of his friend. Then he helped Mamen to get Malloch's body outside, wrap it in the tent cover, and place logs round it. That arrangement would have to do for the time being, for the ground was too hard to try to dig any sort of grave and they lacked a shovel.

McKinlay then talked to the two men about returning at least to Skeleton Island to recover and preferably all the way to Icy Spit. Mamen was amenable, but Templeman was not up to undertaking the walk the next day. Then when Templeman thought he could go, Mamen proved to be too weak to walk. Convinced that their illness was caused by the brand of pemmican they had been eating, McKinlay set off alone back to Skeleton Island to fetch some of his rations which had been left there, taking with him only a tin of condensed milk, some tea and matches. After a few miles, he discovered that he had dropped his snow-goggles somewhere. Walking on the ice off shore, he was forced farther and farther seaward by intervening knee-deep water on the ice. Suddenly he realized he did not know where he was and fell into a state of panic. "It was the only time in all my experience … that I felt fear," he wrote later, "not fear of danger, but from the weight of my responsibility to Mamen …"[4]

Dutifully he stumbled on, with all sense of time and direction totally lost. As a result he missed Skeleton Island and arrived completely exhausted at the main camp at Icy Spit early in the morning of May 25 after covering sixty miles in thirty-eight hours. There were holes in the soles of his skin boots, his feet were raw and bleeding, and his stockings and legs were soaked well above his knees. He wakened Munro and gave him an initial report about the desperate conditions at Rodger's Harbor, then fell asleep, wet, miserable and totally exhausted.

With McKinlay no longer capable of travelling for a while, Munro and Maurer set off for Rodger's Harbor the next day to fetch Mamen and Templeman. They planned to assess the conditions while there, with the idea of possibly moving everyone from Icy Spit down there as Captain Bartlett's instructions had stipulated.

While he was recuperating, McKinlay pondered the cause of Malloch's death and the swelling and weakness of many of the men. Even the Iñupiaq woman Keruk, whom McKinlay preferred to call "Auntie," had been affected. He finally concluded that it was the brand of pemmican, the quality of which Dr. Anderson had questioned a year earlier. Stefansson later agreed with McKinlay.

The Move to Cape Waring

Early in June Kuraluk decided that he would move his family from Icy Spit south to Skeleton Island in the hope of finding more food and fuel. Hadley decided to accompany them. McKinlay agreed to go with them and bring the sled and three dogs back to the Icy Spit camp. This was their only sled and their only dogs. They made good time, but McKinlay soon lagged behind, his strength not having recovered fully from his recent exhausting journey from Rodger's Harbor. Near the cliffs at Cape Waring they met Munro and Maurer coming from Rodger's Harbor and bearing the disturbing news that Mamen had died on May 26 and Templeman was "out of his mind."[5] McKinlay was especially distressed by the news of Mamen's death, having so recently left Rodger's Harbor for Skeleton Island to fetch more palatable pemmican for Mamen but then having failed to return. He wrote "Although the news was not unexpected it still gave me a terrible shock. The last of my scientist colleagues had gone."[6]

---

[4] *Op. cit.*, p. 116.
[5] Niven, 2000, p. 266.
[6] McKinlay, 1976, p. 120.

Munro now suggested that the others move to a sandbar in a small bay just south of Cape Waring, where there were lots of driftwood and thousands of birds nesting on the nearby cliffs. He asked McKinlay, once the sled had been unloaded, to return with it and the dogs to Icy Spit and bring the sick men at Icy Spit south to the new site. After that he was to fetch the gear still at Skeleton Island. Munro and Maurer then continued on their way north to the Icy Spit camp to retrieve their things and move them to Rodger's Harbor. McKinlay, Hadley and the Iñupiat family proceeded to the suggested site near Cape Waring and made camp. Within a few hours Kuraluk had killed a small seal and a goose, and Hadley had obtained ten gulls from the cliffs nearby, all of which served to provide them with an adequate meal.

Shortly after supper on June 6, McKinlay started back to Icy Spit with the sled and dogs to bring the sick and injured from there to the new camp. Keruk gave him a cooked gull to eat on the trail and seven uncooked ones for the men at the main camp. He arrived there five hours later and was promptly asked if he had brought any food. The men were less than thrilled when he showed them the seven gulls, but did not refuse to eat them once they were cooked.

McKinlay and Munro then discussed how best to move the men and the Icy Cape camp equipment to Cape Waring. They finally agreed that McKinlay would use the sled and dogs to transport the camp gear halfway to the new site, unload it, and return to collect Williamson and Williams, who were too ill to walk. Meanwhile, Munro and the other three men— Breddy, Chafe and Maurer—would endeavour to walk to McKinlay's temporary half-way

Figure 11 Camp of some of the *Karluk* survivors on the beach at Cape Waring, Wrangel Island, summer 1914.
Photo: Library and Archives Canada, photo by W.L. McKinlay, C-071030

station, where they would set up the tent. McKinlay loaded the sled with the gear from the camp, and after breakfast attached the dogs and drove them roughly halfway to the new camp. There he hastily unloaded the gear and returned to Icy Spit to get Williamson and Williams. By mid-afternoon he had them both on his sled and was heading back to the half-way halting-place. On arriving there, he found to his disgust that the tent had not been erected, and Munro and his three companions were fast asleep. By then close to exhaustion, having been on the trail since leaving Cape Waring the evening before, McKinlay somehow managed to feed the dogs and brew some tea for himself and his two sick passengers before falling asleep himself.

The following morning Munro and McKinlay ferried Williams and some camp gear the rest of the way to the camp near Cape Waring, leaving Williamson and some gear for their second trip. Breddy, Chafe and Maurer meanwhile set off for the new campsite on foot. Munro argued with McKinlay almost the entire way about whether it was preferable to follow the coast or travel out on the ice. Their arguing troubled McKinlay, for he recalled Captain Bartlett's counsel to avoid causing ill feelings among the men. At the camp near Cape Waring, however, Keruk soon dispelled their ire with a good meal of seal meat. Breddy, Chafe and Maurer arrived on foot a few hours later.

After a brief rest, McKinlay returned to the half-way station to collect Williamson and the remaining gear. The three dogs by this time were worn out, as was he himself, but he urged them on, woke Williamson, loaded him and the gear on the sled, and started back to Cape Waring. Trouble arose almost immediately when the dogs crashed through some thin ice and the sled slipped partly into the icy water. While McKinlay struggled to recover both sled and dogs—not an easy task when one recalls he was only 5 ft. 6 in. tall, perhaps 125 pounds, and exhausted—the dogs managed to break free from the sled and ran off. McKinlay finally succeeded in getting the sled out of the water and then tried to pull it the rest of the way to Cape Waring with Williamson on it, but was too weak to do so. Williamson struggled off the sled and tried to walk independently, but collapsed. In desperation McKinlay wrapped him in two blankets, a travelling rug, and the tent cover, and then set off to walk the remaining ten miles to the new camp in hopes of retrieving the dogs. By the time he finally reached the camp, however, he was too exhausted and half-blind from the glare of the ice and snow to do anything further. Fortunately, Kuraluk had already collected the dogs and had gone to get Williamson and the sled. Kuraluk returned a few hours later with Williamson, by which time McKinlay had moved into the tent with Hadley and Kuraluk's family and was fast asleep. Someone, somehow during the foregoing series of movements from Icy Spit to Cape Waring, transported their mascot, the black-and-white cat, to her new home at Cape Waring. The incident went unrecorded.

With McKinlay unfit for further travel, both physically and because of his eyes, Munro mustered Maurer and headed south on June 10 to take the pemmican and other provisions cached weeks earlier at Skeleton Island to Templeman, who was then all alone at the Rodger's Harbor camp.

McKinlay's current attack of snow-blindness became so bad that from June 10 to June 12 he had to be led around the camp. Even the zinc sulphate with which Hadley bathed his eyes provided only temporary relief from the pain.

Figure 12 McKinlay (l.) helping Kuraluk (r.) construct a kayak frame from driftwood, Cape Waring, Wrangel Island, July 1914.

Photo: Library and Archives Canada, photo by W.L. McKinlay, C-071044

More Trouble among the Men

Rivalry over food soon resulted in festering resentments among the two groups at the Cape Waring camp. Initially the total bird catch was divided equally among the two tents. In the one tent, Keruk carefully doled out portions of meat and soup to stretch the meagre provisions for her family and for Hadley and McKinlay. In the other tent, however—Williamson, Chafe, Breddy and Williams—were less careful, rarely saved any food for their next meal and soon were apparently trying to cheat Hadley's group out of their share of the birds.

McKinlay's eyes improved to such a degree in the next four days that on June 13 he unwisely headed southwards from Cape Waring alone, unarmed and on foot for Skeleton Island in order to recover the useable gear left there. When he got there he discovered that Munro had taken most of it to Rodger's Harbor, but he brought back what remained, noting that much of his own bag, which had been left there previously, had been ransacked of its contents.

Tempers frequently flared up now among the men. An angry argument broke out after Munro and Maurer arrived back from Rodger's Harbor and demanded what the Cape Waring people considered more supplies and ammunition than was their rightful share. Charges and countercharges were made, and evidence surfaced of the unwillingness of some of the men to share food equally. Presently the Cape Waring people split into groups headed by Munro and Hadley respectively. Then Munro suddenly announced that he and Maurer were going back to Rodger's Harbor, where Templemen was alone, asking McKinlay to come there later with the sled and dogs, when ground conditions permitted, and bring the three of them back. McKinlay would have liked to have accompanied Munro and Maurer to Rodger's Harbor right away, in order to get away from the unpleasantness of Breddy, Chafe, Williams and Williamson, but thought it unwise to ask, and Munro and Maurer then headed south on foot.

Thereafter McKinlay and Hadley worked hard to provide food for their group. Being well aware their 140 rounds of ammunition were not likely to last them more than two months, they constructed a ladder with which to reach the nests of the seabirds on the lower part of the cliffs and collect their eggs. McKinlay had a nasty fall off the ladder one day while collecting eggs. After he had recovered, he decided he would carry out Munro's instructions by taking the sled and dogs and retrieving the gear that had been left at the Icy Spit camp. Alone and unarmed, he managed to make the round-trip safely, but was fortunate that he did not encounter a polar bear, and had a great deal of trouble coping with the water that lay on top of the rotting sea-ice over which he had to travel.

One day Kuraluk carved a bow and several arrows for the hunting of birds, and with these managed to secure fourteen birds in one day. Hadley continued hunting with his rifle, but wasted many bullets trying to shoot the birds. On June 23, however, he succeeded in killing a large seal, which was dutifully divided, with half being given to the four men in the second tent. Keruk carefully doled out portions from the second half of the seal in soup to the members of Hadley's group in her tent for the next five days, but the four crewmen in the other tent soon had consumed all of their share of the seal meat, in spite of remonstrances from Hadley to make it last.

Tragedy

On the morning of June 25 McKinlay was awakened by the sound of a shot, followed by a shout from Williamson in the adjoining tent, "Clam! Call Hadley! Breddy has shot himself!" McKinlay, Hadley and Kuraluk hurried to the other tent, where they found the *Karluk's* fireman, Breddy, lying with a Mauser revolver beside him, a bullet having passed through his right eye and out above his left ear.

Several men carried Breddy's body from the tent and laid it on the ground. Williamson then went through Breddy's belongings in the presence of the others, finding McKinlay's compass and several other items that had been taken from McKinlay's bag when it was at Skeleton Island.

Afterwards McKinlay and Kuraluk dug a make-shift grave on top of a small hill behind the camp, using an axe and a piece of board for tools, for they had no shovel. But by the time they got Breddy's body to the grave the body had swollen up and would not fit into the grave. McKinlay enlarged the grave the next day, encountering much difficulty digging in the permafrost. The men then succeeded in burying the body, covering it with driftwood, skins, soil and moss. Afterwards, Hadley removed the Mauser revolver and the Winchester rifle from the crewmen's tent to prevent any further "accidents."[7]

Thoughts of accident, deliberate suicide and even murder ran through McKinlay's mind, but he kept silent about his suspicions. Hadley was convinced that Williamson had shot his

---

[7] McKinlay, 1976, p. 135.

tent-mate, with whom he had recently argued, for the only other man in the tent when the shot was fired was Williams, who was too ill at the time to have undertaken such a deed. In the end no clear understanding of the man's death was ever reached.

Barely Surviving

Kuraluk shot three seals at the end of June, which provided the ten survivors and their three dogs at the Cape Waring camp for several days. After that for some reason he seemed indifferent to undertaking any further hunting. Rain and foggy weather followed. Large stretches of water developed between shore and the main ice, making the need for some form of watercraft essential. Kuraluk was urged to build a kayak, which the group needed to hunt seals now that there was an increasing amount of open water. Assisted by McKinlay and one or two others, Kuraluk soon created a frame, but then stopped work for a week for no apparent reason. Stirred into action again by a rainy spell, he made a double-bladed paddle and scraped the sealskins needed for the frame covers. Finally, on July 19, after his wife, Keruk, had skilfully sewn the sealskins onto the frame, Kuraluk launched the kayak.

The first walrus appeared near their camp shortly before the kayak was completed. Kuraluk was understandably reluctant to hunt an animal the size of a walrus alone in his small kayak, but quickly overcame his fear. On July 20 he killed a young walrus weighing nearly one thousand pounds. With much difficulty he manoeuvred the beast's ungainly carcass onto the

Figure 13 Templeman (l.) and Munro (r.) raise the Canadian red ensign to celebrate Canada's Dominion Day, Rodger's Harbor, Wrangel Island, July 1, 1914.
Photo: Library and Archives Canada, photo by F.W. Maurer, C-024948

grounded ice along the beach. It proved to be the only walrus he killed that summer, although they frequently saw or heard others, and on one occasion saw an entire herd drifting by on some floes. This lone walrus provided them with meals for several days, but after it was eaten they soon returned to their state of being constantly hungry.

From time to time McKinlay picked a small plant, which Hadley told him was called "scurvy grass,"[8] at Barrow to add some greenery to their diet, which undoubtedly was beneficial. His various activities about the camp, considerably greater than that of most of the others, left his clothing in very poor shape by mid-summer and he spent much time on foggy or rainy days trying to repair it. Parts of his caribou skin clothes were in such poor condition that they would no longer hold stitches, but there were no other skins with which to create new clothing. The skin clothing worn by Hadley and the Iñupiat family had been of much better quality originally, so the five of them remained the best dressed persons in the camp.

Slowly the days of summer passed without any sign of a rescue ship. What hopes some of the men continued to have for rescue diminished with the passing days. Their prospects looked increasingly grim, for food and fuel were rapidly approaching exhaustion and August was at hand. "If relief is coming" wrote McKinlay, "it should come then, for, after the end of August, says Hadley, the chances of relief, though not absolutely nil, are very, very small."[9] Like the others, he was thinking all too often about sighting a rescue ship and wondering if Captain Bartlett had succeeded in reaching Nome, knowing their lives depended upon his doing so. "This all depends on the captain; if he won through, we are safe; if he was lost, we must winter here."[10]

Then suddenly their luck improved. Kuraluk killed a bearded seal on each of the last two days of July and a third one on August 1. Hadley had also shot one, but his seal sank before Kuraluk could collect it

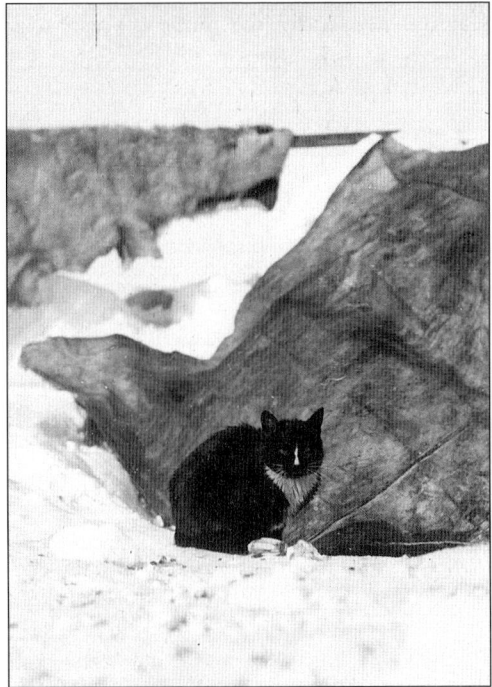

Figure 14 The cat mascot, Nigeraurak, resting in the sun at Icy Spit, Wrangel Island, May 1914.
Photo: Library and Archives Canada, photo by W.L. McKinlay, C-071039

---

[8] McKinlay, 1976, p. 141. As a boy in Scotland, McKinlay had called it "Sourocks." The term "scurvy grass" is applied to about thirty species of low leafy plants in the cabbage family, which have a tolerance of salt and hence are common in coastal regions. The leaves are rich in vitamin C and were eaten extensively by sailors in the past to avoid or cure scurvy.

[9] McKinlay, 1914 (July 31, 1914).

[10] *Op. cit.* (August 4, 1914).

with his kayak. They divided Kuraluk's three seals among the members of each tent. Hadley's group built a makeshift rack on which to dry some of their meat to ensure it lasted longer. Williamson's group, conversely, ate their supply without any attempt at conservation.

On August 18 Williamson unexpectedly announced that he was going to walk to Rodger's Harbor, more than thirty miles from Cape Waring around the coast to the southwest. The others were startled, for he had not walked more than a mile on any day since they landed on Wrangel Island in March. However, he returned after three days, claiming that all three men at Rodger's Harbor (Munro, Maurer and Templeman) were well, had obtained five seals and many duck eggs, but were now living on sealskin. He brought back a revolver and a number of cartridges, which needless to say did not sit well with Hadley, who minced no words in letting Williamson know that he wanted no more trouble with guns in Williamson's tent. The day after Williamson's return he asked for some of the dried meat saved by Hadley's group. Hadley agreed to give him some in exchange for tea tablets, of which Williamson's group had a good supply, for he had brought back 300 miles from Rodger's Harbor.

By late August, their daily menu was as follows: "Breakfast—cup of soup from the "starvation" tin; Lunch—cup of tea with a piece of walrus hide and raw blubber from [a seal] "poke"; Supper—Cooked roots with oil."[11] The scraps in their "starvation" tin were small pieces of meat they had set aside during more plentiful days, the "poke" was the sealskin that held the decaying seal fat, which was slowly turning to oil, and the roots were from an unidentified local plant that Keruk boiled, which was long and fibrous, and proved very chewy, tasted rather like the licorice stick McKinley knew as a boy and gave them frightful constipation.

The hunters Kuraluk and Hadley had absolutely no luck in shooting anything from August 1 to August 28. On this latter date they espied a large number of duck-like birds swimming in pools of water in ice that had been pushed close to the shore. McKinlay, starving and soon grateful to be one of the beneficiaries of this unexpected supply of food, called them *crowbills* in both his diary[12] and in his 1976 book (p.147), a name not found in standard North American bird books. Niven identified them as ducks.[13] "Crowbill" is, in fact, a name used locally in western Alaska for murres.[14]

Not daring to waste any of their sparse supply of ammunition to shoot the birds, Kuraluk gathered a large net that was among their supplies. With some ingenuity, several of the men stalked the birds and on a signal threw the net over them, successfully catching ninety in this way on that first day and 120 more the next day.

Their sudden good fortune suppressed briefly their worries that August was almost over and no rescue ships had appeared, but their elation did not last long. A few days later it snowed.

---

[11]   *Op. cit.* (August 25, 1914).

[12]   *Op. cit.* (August 28 and 29, 1914).

[13]   Niven, 2000, p. 328.

[14]   Moss and Bowers, 2007, p. 44. Murres are black-and-white seabirds that somewhat resemble ducks but have shorter necks, and are, in fact, more closely related to auks and puffins than to ducks. Thick-billed murres are known to nest now in the cliffs of Wrangel Island. McKinlay undoubtedly picked up the local name from Hadley, who had lived around Nome and Barrow for many years.

Farther south, at Rodger's Harbor, Munro, Maurer and Templeman were faring no better. By the end of August they were almost without food and starving. Munro realized they would have to swallow their pride and attempt the long walk to Cape Waring in the hope of obtaining food from the Hadley "crowd." None of them was in condition to attempt the trip and the blizzard-like weather was a huge deterrent. They had watched for a ship every day for nearly two months, always in vain, their hopes for rescue diminishing more each day. It appeared that only the Lord could save them from their present situation. "We are still hopeful trusting in the Lord for delivery," wrote Munro.[15]

Within days the ocean would freeze over again, as indeed it was already doing along the shore, and all of the survivors at the two camps would face the winter with almost no ammunition, no caribou skins to replenish their badly worn clothes, little driftwood and almost no provisions. And if Captain Bartlett had failed to reach the Siberian coast and to alert the Canadian government of their plight, no one would even know where they were. Their future looked hopeless.

---

[15] Munro, 1914 (August 27, 1914).

RESCUE

Dual Efforts to Send a Rescue Ship

Burt McConnell reached Nome from Barrow on the schooner *King and Winge* on August 30, shortly before the *Bear* returned for more coal and provisions after its aborted trip to Wrangel Island. On September 2, with the *Bear* still offshore waiting out stormy weather to complete its loading, McConnell concluded that its Captain Cochran was " … not going to risk his fat job and a $250,000 ship to rescue …" the *Karluk* survivors, because "Revenue Cutter Captains always steer clear of the ice …"[1] Furthermore, "The BEAR has no dogs, sleds, skin boats or Eskimos aboard, and their men are not equipped with clothing for a winter in the Arctic."[2] Recognizing that Captain Bartlett was a guest of Captain Cochran on the *Bear*, had to follow official procedures and could not very well advocate the *Corwin* and the *King and Winge* going to Wrangel, McConnell telegraphed Deputy Minister Desbarats in Ottawa asking permission to hire the SS *Corwin* to proceed to Wrangel Island to rescue the *Karluk* survivors. There was no reply.

On September 3, the first fine day in several, Captain Bartlett came ashore from the *Bear*[3] and lunched with Jafet Lindeberg, president of the local Pioneer Mining Company of Nome and the leading mining operator of western Alaska. The captain mentioned the plight of the *Karluk* survivors on Wrangel Island and his frustrations over the various delays in the *Bear's* schedule in its attempt to rescue them. Lindeberg responded by offering to outfit the SS *Corwin* with supplies and men, and to dispatch it to Wrangel Island to rescue the *Karluk's* survivors. And true to his word, Lindeberg did. The *Corwin* was the same old steam U.S. revenue cutter whose Captain Calvin L. Hooper and crew members had mapped Wrangel Island in 1881 and named various prominent features around it while searching for the missing Captain De Long expedition and his ship, *Jeannette*.

Later that same day, Captain Bartlett also spoke to the trader Olaf Swenson of the plight of the *Karluk* survivors on Wrangel Island. A well-known Scandinavian from Seattle, Swenson had charge of the trading schooner *King and Winge* and was about to leave Nome to hunt walruses off the Siberian coast. Swenson had brought McConnell to Nome on his return trip from Barrow. Referring to his visit with Swenson, Captain Bartlett wrote later: "I asked him, if he went anywhere near Wrangell [*sic*] Island, to call and see if the men had been taken off

---

[1] McConnell, 1914 (September 2, 1914).

[2] *Op. cit.* (September 4, 1914).

[3] McConnell made the curious remark in his diary that Captain Bartlett was heartily disliked at Nome, so did not spend much time ashore. McConnell, 1914 (September 3, 1914).

and he promised that he would do so."[4] Captain Bartlett's requests to Lindeberg and Swenson were not made lightly, for the voyage to Wrangel Island was six hundred miles each way, involved days of risky ocean travel and ice encounter, and was a costly voyage for whoever undertook it. Nonetheless, the lives of the *Karluk* survivors depended upon him to arrange their rescue that summer, while they were still alive. Weeks earlier he had asked the Canadian government to arrange to have the U.S. revenue cutter *Bear* hasten to Wrangel Island, but its many commitments that summer had to date prevented it from getting there. Now, Bartlett reasoned, with the Arctic navigation season fast coming to a close, the more ships that went to Wrangel Island the better the chances that one of them would manage to rescue the desperate survivors.

When McConnell did not receive a reply from Deputy Minister Desbarats in Ottawa the day after he sent his telegram concerning the SS *Corwin*, he decided that there was no time for any further delay; and on his own asked Olaf Swenson, whom he had known for several years, to undertake the rescue of the survivors with the *King and Winge* before he went walrus hunting. Swenson by then had talked with Captain Bartlett, and promptly agreed to go, invited McConnell to go along with him and headed for Rodger's Harbor that same day. The 110-foot *King and Winge* was a small but speedy, well-equipped schooner, with a captain (A.P. Jochimson) and an Iñupiat crew who were willing to go wherever owner Swenson wanted to go.

The *King and Winge* to the Rescue

The following afternoon, September 4, after crossing Bering Strait, the *King and Winge* schooner pulled in at East Cape and Swenson took on board fifteen *Yuit* men (Siberian Natives) and an *umiak*. Two days later the schooner encountered ice. Thereafter it had to navigate through it carefully, because it was not adequately strengthened to deal with ice. That evening, according to McConnell:

> The mountains of Wrangel Island came in sight … These mountains are not beautiful. Dull, pale-gray, forbidding, they could be dimly seen beyond the jagged white-and-blue ice-field … the Captain [Jochimsen] told me that we possibly were within 20 miles of the island, and that he would promise to put us ashore before breakfast.[5]

The *King and Winge* encountered more densely packed ice as it drew closer to the island and passed pressure ridges that were almost as high as its masts. A southwest wind rendered the ship's situation even more dangerous, for the farther north they proceeded the more ice they encountered. Captain Jochimsen halted operations for a few hours around midnight when visibility became too risky to continue, then took the helm again after 3 a.m.

After three hours of bumping, crashing and grinding against the densely-packed ice we drew within sight of the precipitous granite cliffs and sandy beach of Rodgers Harbour [*sic*] …

---

[4] Bartlett and Hale, 1916, p. 312.

[5] McConnell, 1914 (September 6, 1914).

within two miles of shore a tent was sighted by the lookout in the crow's nest … When within half a mile … Captain Jochimsen began blowing the ship's whistle at frequent intervals … no one appeared in answer to our blasts … we had expected to find 23 people at this place, yet the sight that greeted us was a four-man Burberry tent, a flag pole and a rough wooden cross … no sleds or dogs in sight.[6]

Rodger's Harbor

The schooner was within five hundred yards of the beach at Rodger's Harbor when a man (Munro) suddenly crawled out of the tent, rose slowly to his feet and gazed at the ship as if stunned.

He did not show any signs of joy. He did not wave his arms and shout for sheer happiness when he sighted the ship, as some of us had anticipated. He did not run up and down the beach to attract our attention. The poor creature simply rose and stood rigidly beside the tent, gazing at us as if dazed.[7]

Then he re-entered the tent and emerged a moment later with a British flag, which he raised to half-mast on the tent pole. On the other side of the tent stood a crudely erected wooden cross. Death had obviously pervaded the campsite. Some men on the schooner then lowered the *umiak* into the water, and Swenson, McConnell, two Los Angeles cameramen (Fred L. Granville and Charles Zalibra), and several *Yuit* crewmen climbed in and went ashore. Two more men emerged from the tent and stood silently watching the unfolding scene. Suddenly the first man picked up a rifle—it happened to be Stefansson's prize Mannlicher, which Stefansson had left on the *Karluk*—and slowly advanced towards the spot where the *umiak* intended to land, while appearing to load the gun. Swenson urged the paddlers to continue to the shore, at which point McConnell finally recognized the menacing gunman as Munro, the *Karluk's* chief engineer.

"He was a pitiable looking object. His shaggy, matted hair streamed down over his eyes in wild disorder. His grimy face was streaked and furrowed with lines and wrinkles. His clothes were in tatters, begrimed with seal-oil, blood and dirt … he had lost 30 or 40 pounds."[8]

The two emaciated and bearded men with Munro, although almost unrecognizable, were the fireman Maurer and the cook Templeman. They had only the one rifle among them and twelve cartridges, so would have had little means of obtaining subsistence after another week. They were alive, though just barely so, weak and half crazed, and clothed in the dirty, tattered furs they had lived and slept in for months. Their rescue had come just in time.

Munro soon informed his rescuers that the other nine *Karluk* survivors were camped near Cape Waring, just over thirty miles to the northeast. While McConnell wrote a message telling of the rescue, which he attached to the tent pole, the two photographers took both still and moving pictures of the desolate scene, and the three men gathered their possessions. Then all

---

[6] *Op. cit.* (September 7, 1914).
[7] McConnell, 1915a, p. 355.
[8] McConnell, 1914 (September 6, 1914).

climbed into the *umiak* and proceeded to the *King and Winge*, leaving the upright tent as a beacon.[9] Once they were safely on board, Swenson ordered Captain Jochimsen to head northeast for Cape Waring.

Cape Waring

Near Cape Waring, later that same September morning, Kuraluk was the first to notice the ship far in the distance. *"Umiakpik kunno!"* ("Maybe a ship!") he shouted to arouse the others in the camp.[10] It was indeed a ship! Some two miles off shore, a small schooner seemed to be steaming slowly northwest along the outer edge of the ice that enclosed Cape Waring. On board the *King and Winge* the crew watched for signs of a camp. From the schooner's masthead suddenly came a shout, announcing the sighting of two tents on the shore, and near them two dark figures moving up and down. Captain Jochimsen promptly ordered the *King and Winge* tied up to the ice, stopped and instructed some of his men to proceed to shore to bring the survivors to the ship. Swenson, the photographer Granville with his cinematographic equipment, a sailor, McConnell and the *Yuit* crew formed the rescue team on the ice, and within minutes they were slowly and carefully moving shoreward over the comparatively smooth ice.

On shore, McKinlay and the others were dazed by the realization that Captain Bartlett had truly succeeded in getting through to Nome and sending out a call for their rescue. "They had abandoned hope of rescue for that year, and had but forty cartridges left. Their flimsy tents were torn and full of holes, and their food supply was almost exhausted."[11] Their ordeal on Wrangel Island was finally at an end. They promptly sent Kuraluk hurrying out to greet the approaching crewmen, still far in the distance, then turned their attention to having something to eat.

Just that morning Keruk had seen a fifteen-inch tomcod in a crack in the ice and they had managed to "jig" the fish, using a bent pin attached to a sinew. The pin was held motionless until the fish swam over it, then with a quick tug on the sinew, the pin impaled the fish, and it was scooped out. The men were planning to have a good meal that morning before packing up to move to a more sheltered site, where they hoped to find enough driftwood to build a cabin for the oncoming winter. Instead they put pots of water on the fire and started cooking the fish and making tea.

The men from the ship finally reached the shore, greeted the nine survivors and told them that Captain Bartlett was on the *Bear*, which was also on its way to rescue them. With Granville turning the cranking handle on his cinematographic camera, the survivors obediently posed while being told that their three companions from Rodger's Harbor were already on the ship. Then Granville followed them about, filming them as they gathered their few possessions. That done, McKinlay and McConnell wrote notes that they posted conspicuously to let anyone else who might happen to come by know that the *Karluk* survivors had been rescued.

---

[9] McConnell, 1915a, p. 357.

[10] McKinlay, 1976, p. 149.

[11] McConnell, 1914 (September 7, 1914).

Figure 15 Rescue party from the *King and Winge* with some of the *Karluk* survivors by their tent at Cape Waring, Wrangel Island, September 7, 1914.
Photo: Library and Archives Canada, photo by W. McKinlay, C-071046

Abandoning their tents along with all unessential items, the nine survivors proceeded slowly across the two or three miles of ice to the ship. For dramatic effect, the cameraman insisted that each survivor, excluding the Iñupiat family, should be supported by two of the ship's company. The little girl Mugpi carried the cat, Nigeraurak, all the way to the ship. In true mascot fashion, it had been kept alive since their departure from Shipwreck Camp with scraps of food given it by the survivors, who fervently believed that as long as the cat survived, they too would survive.[12] Hadley's dog, Molly, trotted along near him, her three recently born puppies struggling to keep up with her.

The *Corwin* reached Cape Waring the day after the three men there had been rescued by the *King and Winge*. Its crew found the deserted camp and notes and, although they were

---

[12] The cat was subsequently shipped south from Nome in a specially made box to the home of the *Karluk*'s fireman Fred Maurer in New Philadelphia, Ohio, where it lived until 1926. Maurer took one of its first kittens to Ottawa as a gift for Mrs. R.M. Anderson in 1917 (*Ottawa Citizen*, 1930). Named Karluk by the Anderson family, this kitten in due course had kittens of its own, one of which found its way to the home of Miss Eilleen Bleakney, then secretary to the Chief Anthropologist of the same Victoria Memorial Museum where Dr. R. M. Anderson worked. Miss Bleakney later married the surviving anthropologist from the Canadian Arctic Expedition, Diamond Jenness, and subsequently became the mother of the author of this book.

disappointed over arriving too late, were grateful that the survivors had been rescued. The crew had been instructed to cache a load of supplies on the island for the survivors if they did not find them, but that was now unnecessary.

Once on board the *King and Winge,* the twelve Wrangel Island survivors were told that much of Europe was engulfed in a war with Germany and Austria. At the time of telling, however, such a topic seemed a bit too remote, and the men's thoughts were on more immediate matters, including decent food and replacement clothing. They ate, enjoyed hot baths, shaved, then got into clean clothing, drank coffee and relaxed on the deck. There they enjoyed their first smokes in eight months, ever since Captain Bartlett had ordered that all tobacco be discarded when the *Karluk* sank.

Heading for Nome

The *King and Winge* got underway early the next morning. Thin, fresh ice had formed overnight around the schooner, forcing it to twist and turn to reach open water. Then it headed east for Herald Island to see if there was any sign of the eight missing members of the Expedition parties of Dr. Mackay and First Mate Sandy Anderson. Solid ice soon blocked its way, however, preventing it from getting close enough to that island for its crew to get a good look.

Captain Jochimsen then turned the *King and Winge* southward and headed towards the smoke of another ship. Some while later, the schooner pulled alongside the revenue cutter *Bear,* one hundred miles from Wrangel Island and in loose ice, and the *Karluk* survivors spotted Captain Bartlett, the man to whom they owed their lives. According to McConnell,[13] Captain Cochran sent his Lieutenant Miller to the schooner with instructions to bring the rescued

Figure 16 The schooner *King and Winge* east of Wrangel Island, with rescued *Karluk* survivors on foredeck, September 8, 1914.

Photographer unknown

---

[13] McConnell, 1914 (September 8, 1914).

people aboard the *Bear if they wanted to come*, or if Swenson wanted *to get rid of them*. Captain Cochran apparently offered them the option to return to Nome on the *King and Winge*. One or two of the rescued men thereupon expressed their desire to remain on the schooner, but at that point Captain Bartlett boarded the schooner and announced that he would take charge of them all. He then thanked Olaf Swenson and Captain Jochimsen of the *King and Winge* for their trouble in rescuing the survivors and ordered his people to get on board the *Bear*, where they could be examined by the ship's doctor. The *King and Winge*, its rescue mission accomplished, then headed off to hunt walruses,[14] while the *Bear* steamed towards Herald Island in an attempt to look for the eight missing men of the Anderson and Dr. Mackay parties. They were, however, prevented by ice conditions from getting closer than ten miles to the island, saw no indication of human habitation and headed back to Nome.

The rescue of the Wrangel Island survivors was most timely. In view of the partly starved and physically weakened condition of the survivors, it is extremely doubtful if any of them, with the possible exception of the Iñupiat family and Hadley, could have survived much longer.

Subsequent Activities of Some of the Karluk Survivors

The *Bear* reached Nome on September 13. News of the safe arrival there of the *Karluk* survivors appeared the next day in the *New York Times*,[15] followed a day later by a lengthy account written by Burt McConnell about the rescue operations.[16]

Captain Bartlett and the *Karluk* survivors remained on the *Bear* and (except for the Iñupiat family, which remained in Alaska) sailed to Victoria, reaching it on October 25. There they were well feted, then scattered, not to meet again. Two of them, McKinlay and Chafe, subsequently joined army units in Scotland and Newfoundland, respectively, and fought in World War I in France.[17]

Fates of the Sandy Anderson and Dr. Mackay Parties

Nothing further was known of any of the eight missing members of the *Karluk* for the next ten years. Then in the summer of 1924, Captain Louis Lane, captain of the *Herman*,[18] and

---

[14] *Ibid*. The Hibbard-Stewart Co., Seattle, owners of the *King and Winge*, later billed the Canadian government $3,000 for rescuing the Wrangel Island survivors (Auditor General Report, 1920.)

[15] Anonymous, 1914.

[16] McConnell, 1914c.

[17] Maurer returned to Wrangel Island in 1921 with three male companions and a young Iñupiaq seamstress on a new expedition sponsored by Stefansson. Their mission was to settle the island and claim it for Great Britain. Bad ice conditions prevented a supply ship from reaching them in the summer of 1922, and during the following winter Maurer and two of the other three men perished trying to cross Long Strait to Siberia to get word of their activities and condition to Stefansson and the world. The fourth man died of scurvy at their camp on Wrangel Island in June. The lone survivor, the seamstress Ada Blackjack, was rescued near Rodger's Harbor in August 1923 (Niven, 2003).

[18] The *Herman* was the ship that rescued Captain Bartlett at Emma Harbor in May 1914. It keeps reappearing under different captains in this story about the Canadian Arctic Expedition. On board

some of his crew went ashore on Herald Island briefly while on a voyage to Wrangel Island. On a stony beach on the northwestern corner of that small and rocky island, Captain Lane and his men discovered the partial skeletons of four men, along with an assortment of artifacts, including a watch, a pocket compass, field glasses, snow glasses, hunting knives, pocket knives, tent poles, remnants of tent cloth and parts of a sled.[19] There was also a rusted rifle, with the letters "B" and "M" carved on its stock. These artifacts matched closely the items issued to the *Karluk's* First Mate Sandy Anderson and his three seamen companions—Edmund Golightly ("Archie King"), Charles Barker and John Brady[20]—whom Captain Bartlett had sent from Shipwreck Camp in February 1914 to establish a trail to Wrangel Island. The rifle was the personal rifle of Stefansson's secretary, Burt McConnell, who had taken a short-barrelled rifle the previous September when he left the *Karluk* with Stefansson's hunting party, leaving his own rifle on which his initials were carved back in his cabin.[21] No sign of the other four men from the *Karluk* was ever found, and it has long been assumed by all except author G.E. Crich that Dr. Mackay, Murray, Beuchat and the sailor Morris perished trying to reach land.[22]

Captain Lane's voyage to Wrangel Island was requested by Stefansson to remove one white man and twelve Iñupiat who had been left there in August 1923. Shore-bound ice had prevented Captain Lane from succeeding in his mission and after waiting three weeks for the shore conditions to improve, they had stopped at Herald Island on the return part of their voyage. The thirteen people they were intended to remove had been deposited on Wrangel Island by Harold Noice and the schooner *Donaldson* in the summer of 1923, acting on behalf of Stefansson. Noice's trip to Wrangel Island in 1923 was intended to leave the thirteen people to settle on the island and claim it for Canada, and to bring back four men and one Iñupiaq woman, Ada Blackjack, who had been left at the island by Stefansson's arrangements in 1921. Of the five, however, only Ada Blackjack was still alive. Thus ends the story of the ill-fated *Karluk* and its personnel. It is unlikely that the full story can ever be told. Too much of it remains forever on the sealed lips of the men who perished after the dramatic sinking of their ship.

---

when it visited Herald Island in 1924 was D.M. LeBourdais, a special correspondent for the North American Newspaper Alliance of New York, who later mentioned this discovery in a biography of Stefansson (LeBourdais, 1963, p. 171).

[19] Niven, 2000, p. 4–6.

[20] The artifacts from Herald Island were subsequently forwarded to the Canadian government in Ottawa, where they were carefully examined by Dr. Anderson and K.G. Chipman. In a letter replying to an inquiry from G.J. Desbarats, Dr. Anderson wrote "[I] am convinced that the remains found were those of First Mate Anderson's party instead of the scientists as reported in the press" (Anderson 1924c). His interpretations were agreed with by Chipman but were strongly disagreed with by Mrs. G.E. Crich (1990, p. 246), the niece of *Karluk* seaman Stanley Morris. The artifacts were, in due course, returned to Captain Lane's employer, the shipping company, Messrs. H. Liebes and Co., of San Francisco. Curiously, some of them (including snow goggles, a rusted pocket watch and a long snow knife) resurfaced in 1999 on the electronic Internet auction site, E-Bay, and were purchased by *The Ice Master* author Jennifer Niven (Niven, 2000, p. 370).

[21] McConnell, 1914 (October 26, 1913).

[22] Crich, 1990.

PART 3
STEFANSSON'S
NORTHERN PARTY

REVISED EXPLORATION PLANS, 1913–1914   9

The same violent storm that carried the *Karluk* westward a few hours after Stefansson and his small hunting party left it on September 20, 1913, also saw him and his companions stranded on Spy Island, several miles off the mouth of the Colville River. Spy Island was the westernmost of the Jones Islands, low-lying and composed of sandy gravel. Open water lay between them and the shore and now also between them and the *Karluk*. In addition, the *Karluk* had disappeared, evidently blown somewhere to the west, judging from the prevailing winds, and the hunters could no longer return to it. Gone with the ship were the personal belongings of Stefansson and his five companions, the men who formed his Northern Party and, for all practical purposes, Stefansson's dream of discovering an unknown continent north of Alaska. A major revision to his exploration plans was now necessary.

But first Stefansson had to get himself and his companions out of their immediate predicament. They had provisions for a week, so they promptly set to work to augment these, shooting a few migrating ducks and one seal during the next few days. Fortunately, driftwood was plentiful on their island, allowing them to cook on demand. Meanwhile, as they waited for the ice to thicken between their island and the coast, Stefansson instructed his companions on how to be comfortable and warm in their sleeping equipment, delegated cooking responsibilities among them on a rotating basis, discussed Iñupiat cultural matters and wrote notes.

A week passed before the ice between Spy Island and the coast was thick enough to support a man's weight. Stefansson and his companions then hurriedly crossed to the mainland and camped near a point just east of the mouth of the Colville River. For the next several days he left camp ostensibly to hunt caribou, but saw only one, which he told them was too far away to chase. Not once did he invite Jenness, Wilkins or McConnell to hunt with him, although all three had expected him to do so.

By October 2, their food supplies had become greatly depleted, Stefansson had seen little indication of caribou in the region and, in spite of frequent prolonged searching with field glasses, there was no sign of the *Karluk*. Stefansson therefore decided to lead his party 150 miles west to the small settlement of Cape Smyth, or Barrow as it is called now. It was his hope to get news of the *Karluk* from Iñupiat he met along the way or from someone at Barrow. He also anticipated getting news at Barrow about the Expedition's other two ships, the *Alaska* and the *Mary Sachs*, and about Dr. Anderson's Southern Party.

He and his men needed provisions and warmer clothing for the colder days ahead, which they could obtain at Barrow. He knew that there was a post office at Barrow, but no wireless, radio or cable service at that time. He also knew that the mail was scheduled to be taken south from Barrow before Christmas, curiously by reindeer-drawn sleds rather than the customary dog-team-and-sled method. Thus from Barrow, he could send word about his

Expedition and his separation from the *Karluk* to the government in Ottawa and to the three newspapers—*The New York Times, London Chronicle* and *Toronto Globe*—with which he had financial arrangements.

Heading West to Barrow

Having decided to head for Barrow, Stefansson instructed his men to move their camp west along the coast, while he hunted caribou for one more day. He asked Wilkins to return to Spy Island and leave a message on a tree stump telling of their plans to head west for Barrow, then rejoin the others en route. Two days later as they all trudged slowly across Harrison Bay for Cape Halkett, they sighted a driftwood house on the shore and headed towards it. As they approached, the Iñupiat family emerged from the house, first the man Aksiatak, then his two young daughters Kukpuk and Siniuna, and finally his wife, Otoyuk. They recognized Stefansson immediately, having met him the previous year, and came to shake his hand. Then they shook hands with the other members of his party, making them all welcome. During the conversation that followed they mentioned that they were living principally on fish they obtained from a large but shallow lake (Teshekpuk Lake) a few miles inland. Stefansson promptly arranged to obtain some fish from them to feed his men and dogs while en route to Barrow.

Stefansson and his party stayed with Aksiatak's family for two days. At some time during their visit he arranged for Jenness to return after a few days at Barrow to stay with them for the winter and study their language, culture and activities. Stefansson then took his departure and led his men to the next Iñupiat camp, which consisted of four houses near Cape Halkett. Here they were welcomed and stayed for a day to acquire a supply of whale meat that was stored in one of the houses. The meat belonged to Kuraluk, the Iñupiaq man who, with his family, was then on the *Karluk*.

Sojourn at Barrow

After enduring several more days of difficult travel, Stefansson's party finally reached Barrow on October 12. The local storekeeper, Charlie Brower, provided accommodation for Stefansson, Jenness, McConnell and Wilkins, while the two young Iñupiat, Jerry and Jimmy, slept elsewhere in the settlement. Brower had come north on a whaler more than fifty years before, stayed, married, sired a large family and flourished financially. He was widely spoken of throughout Alaska as "The King of the Arctic."

From Brower Stefansson learned that almost nothing was known of the *Karluk*—one Native had claimed he had sighted it far to the north several days before—but the *Alaska* and the *Mary Sachs*, with the scientific members of the Southern Party, were wintering in Camden Bay about three hundred miles to the east. He thereupon resolved to attend to his business affairs at Barrow and then head for Camden Bay.

For the next two weeks Stefansson dictated to his secretary McConnell a steady flow of letters to government officials in Ottawa, several newspapers, one or two scientific societies, and friends. He also arranged for local Iñupiat women to sew fur clothing, mitts and boots for his men, all of which took time.

The delay at Barrow permitted the healing of various afflictions from which Stefansson and most of his men were suffering when they reached the settlement—back injury

(Stefansson), fevers and chills (Jenness—ague attacks stemming from the malaria he had contracted the previous year in Papua-New Guinea), and foot injuries (McConnell and Wilkins). From Brower Stefansson obtained a pair of crutches, which enabled him to hobble about the settlement. Meanwhile, the ague attacks, which had progressively weakened Jenness since his departure from the *Karluk*, ceased soon after he started taking the quinine that he obtained from Brower. And McConnell and Wilkins slowly recovered from their foot injuries. Thereafter, while Stefansson and McConnell were occupied with the former's extensive correspondence, Jenness and Wilkins read old magazines and newspapers they found at Brower's, wrote letters, visited the local school and observed how the local residents went about their daily activities.

The Fishing Lake

Late in October Stefansson sent Wilkins and Jenness east with Jimmy Asetsaq to stockpile fish at Teshekpuk Lake, assuring them that he would be along in a week. At the time Jenness commented: "Stefansson is sending us on ahead to save the expense of our board at Cape Smythe, so he told us; it seems a very absurd reason, when he throws $1,000 away without second thought on things of little or no use."[1] More likely it was to get these two bright and inquisitive men out of the way while he was writing commercial articles about the Expedition and letters to the Deputy Minister of the Naval Service outlining the new plans he had to replace the Northern Party, which had disappeared with the *Karluk*.

Jimmy Asetsaq was one of the two Point Hope youths who had accompanied them from the *Karluk*.[2] The other youth, Jerry Payuraq, decided to stop working for the Expedition when they reached Barrow, so Stefansson arranged for another young Point Hope man, Angutisiak (Ikey Bolt), to accompany Wilkins and Jenness eastward in place of Payuraq.[3]

Jenness, Wilkins and their two Iñupiat assistants, Jimmy and Angutisiak, left Barrow on October 27, and had a miserable thirteen-day journey east because of the severe cold, winds and falling snow they encountered en route. Wilkins' beard, nose and forehead became coated with a mass of ice while they crossed Smith Bay during a driving blizzard, the lashes of one of his eyes were frozen shut and he could breathe only with difficulty because his mouth would only open a crack. Jenness' bearded face likewise was badly frostbitten. Both Wilkins and Jenness suffered greatly when their faces later thawed out. Somehow their two Iñupiat companions fared much better. After finally reaching the fishing lake (Teshekpuk Lake) on November 8 they established their camp and set a net beneath the ice on the lake, but caught barely enough fish each day to feed their dogs.

---

[1] Jenness, 1916 (October 27, 1913). Also in S.E. Jenness (Ed.), 1991, p. 34.

[2] Sometime shortly after leaving the *Karluk* with Stefansson and several others in September 1913, Jimmy composed a song about the *Karluk*, which he taught various persons along the north coast of Alaska that year. Jenness transcribed it at the time and later publicized it (Jenness, 1957, pp. 103–106).

[3] Angutisiak subsequently accompanied Stefansson east to Collinson Point where he joined the Southern Party. He remained with the Southern Party until it left the Arctic in 1916, and thereafter became an active trader around Coronation Gulf, by which time he was better known as Ikey Bolt.

Stefansson and McConnell remained at Barrow until November 7, the former dictating to the latter each day, first reports, then letters, and ultimately an article or two. Each night McConnell borrowed the missionary's typewriter and typed the dictated material until well after midnight, when it became too cold to continue, rising early in the morning to return the typewriter to the mission. Stefansson wanted all his letters and reports ready for the mail, which left November 1 for Nome. Meanwhile he arranged with local Iñupiat seamstresses to make winter skin clothing and footwear for himself, McConnell, Jenness and Wilkins, which delayed them longer than he anticipated. He also obtained supplies for his trip east and for Jenness for the winter, arranged to purchase several local sled dogs, and contracted for the construction of a sled to carry the extra supplies to the fishing lake.

Stefansson, McConnell and a teenaged boy, Alfred "Brick" Hopson, the son of Brower's cook, Fred Hopson and his Iñupiat wife, finally left Barrow on November 7. Two days later, McConnell informed Stefansson that they had forgotten a box with his papers and letters, and was sent back to Barrow with Brick to retrieve it, causing a delay of three days. They reached the fishing lake on November 20, two weeks later than Stefansson had promised Jenness and Wilkins. By then, Jenness, Wilkins and their two young Iñupiat companions had almost exhausted their food supply. Shortly after Stefansson's arrival, Jimmy suddenly announced that he was quitting the Expedition. He had been expected to obtain caribou meat as needed for Jenness during the winter, but with none apparently in the area he decided he would rather trap foxes for his own benefit.

Stefansson and McConnell stayed with Jenness only three nights. Before departing, however, Stefansson agreed to leave "Brick" Hopson to replace Jimmy and enough supplies for the winter for both Jenness and Hopson. The young Hopson, with his smattering of English, would serve as Jenness' interpreter, hunter and assistant. Stefansson also assured Jenness that someone from the Southern Party's camp at Camden Bay would come west in about two months to bring him east to Camden Bay. Stefansson and McConnell then headed east for Camden Bay, taking Wilkins and Angutisiak with them. As it was already November 24, the sun no longer rose above the horizon, and they travelled during the faint illumination offered by the few hours of twilight or by moonlight.

Jenness and Hopson remained with Aksiatak and his family until late in January, during which time Jenness compiled extensive notes on his host's daily life, language, folk-tales and string games. By late January, most of Jenness' food was gone. As his Iñupiat hosts had none to spare, he and "Brick" were forced to return to Barrow, where he awaited the arrival of someone from the Southern Party to take him east to Camden Bay.

On to Collinson Point

After leaving Jenness near the mouth of the Colville River, Stefansson and his three companions continued east. On December 15 they reached the Southern Party's base camp at Collinson Point, a prominent sandspit sheltering a small embayment, or lagoon, in Camden Bay. There he learned that Dr. Anderson had gone east to Herschel Island with the mail, leaving the geographer, Chipman, in charge of the camp and its men during his absence. Scarcely two days after arriving, Stefansson announced he was going east about ninety miles to visit the *Belvedere*, a large, three-masted, steam whaler and freighter that was caught in the ice off shore. It had on

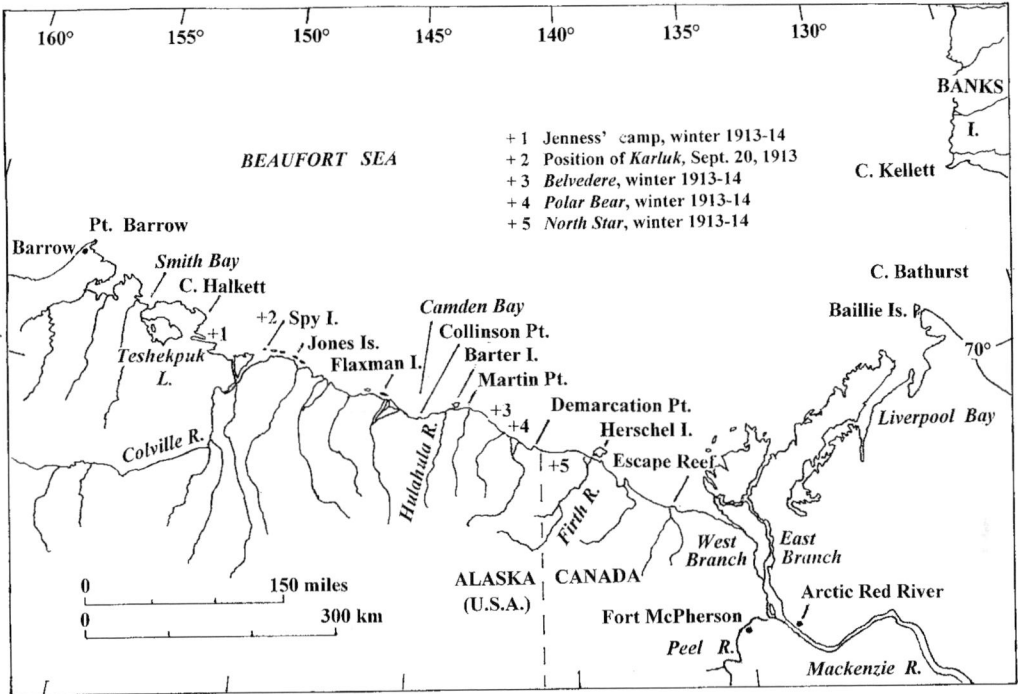

Map 5 The northern Alaskan coast showing locations mentioned. Jenness' camp (+1) was on the shore, and the schooner *North Star* (+5) was sheltered in Clarence Lagoon. The *Alaska* and the *Mary Sachs* were in Camden Bay.

board seventy-five tons of the Expedition's supplies and equipment,[4] which it was supposed to leave at Herschel Island, creating a depot there for Stefansson's Northern and Southern Parties. However, it had failed to reach that destination.

Stefansson's main reason for leaving Camden Bay, however, was to catch up with Dr. Anderson at the *Belvedere* before the latter left for Herschel Island, for Stefansson had developed new plans for increased field work by the Southern Party's scientists that spring and summer, and wanted to discuss them with Dr. Anderson. The latter could then modify the details of the spring and summer plans he had mentioned in his official letters to Ottawa before he turned his mail over to the Royal North-West Mounted Police at their station at Herschel Island. The police would take the mail south with them when they made their winter sled trip up the Mackenzie River to Fort McPherson shortly after Christmas. From Fort McPherson the mail would later be taken over the mountains to Fort Yukon, then down the Yukon River to Nome and thence south.

Stefansson and Dr. Anderson Argue

Stefansson left Collinson Point on December 17, accompanied by a sailor, Louis Olsen, from the *Alaska*, and Captain Peter Bernard of the *Mary Sachs*. The next evening he unexpectedly met Dr. Anderson at a cabin twenty-four miles to the east. Dr. Anderson was on his way

---

[4] Stefansson, 1913g (October 18, 1913).

back to Collinson Point. The small cabin was being used temporarily as a shelter by James Crawford, the engineer of the *Mary Sachs,* while he was trapping foxes. There Stefansson and Dr. Anderson discussed their mutual plans for the coming summer's work. As Stefansson later described their meeting, Dr. Anderson expressed his strong disapproval of Stefansson's plans for his new Northern Party, which would replace the men now presumed lost on the *Karluk,* because they entailed the use of some of the men and equipment from his Southern Party, which he could ill-afford to provide. He was also incensed by Stefansson's interference with the plans of the Southern Party, which Stefansson had assured the men at Victoria and at Nome he would not do. According to Stefansson, Dr. Anderson then stated that if Stefansson insisted on carrying out his new plans, he (Dr. Anderson) would resign as head of the Southern Party, although he was willing to continue as one of its scientific members. Stefansson naturally refused to accept Dr. Anderson's resignation and the latter quickly changed his mind when Stefansson said he would be forced to appoint Chipman as head of the Southern Party. This would have made Dr. Anderson, with his four years of Arctic experience, answerable to Chipman, who had none, and would have created an untenable situation in the Southern Party.[5]

Dr. Anderson's version of the meeting was quite different. Far from offering to resign, he told Stefansson either to take full control of, and responsibility for, the Southern Party or leave it alone. Furthermore, he (Dr. Anderson) had no authority to turn affairs over to Stefansson without the consent of his second-in-command, Chipman. And he knew Chipman would back him up, although he did not know that Chipman had already written in his diary " … if Anderson has any backbone he will either make Stefansson assume full control or absolutely none. No compromise can be satisfactory situated as we are."[6] Stefansson, however, refused to take full responsibility for the Southern Party. In Dr. Anderson's view Stefansson then had no right to interfere with his leadership and the operations of the Southern Party, as he was now proposing to do, having assured the scientists at both Victoria and Nome that once the Expedition left Nome the Southern Party would be solely under the direction of Dr. Anderson. He had been somewhat placated by Stefansson's verbal reassurances on those two previous occasions, but he was incensed now that Stefansson would go back on his word so quickly.[7]

From Stefansson's viewpoint, however, the apparent loss of the *Karluk* with its men, provisions and equipment had changed the fundamental conditions, and he was now simply exercising his authority as leader of the Expedition to obtain whatever he needed to carry out his exploration work. And, as he reminded Dr. Anderson, it was the exploration work that he, the Deputy Minister of the Naval Service and the Prime Minister of Canada regarded as the main purpose of the Expedition. Both Dr. Anderson and the Director of the Geological Survey of Canada thought differently at that time.

---

[5] Stefansson, 1921b, pp. 96–98. Dr. Anderson's copy of *The Friendly Arctic* carries the comment "purely imaginary" in his handwriting where Stefansson claims (p. 97, line 5) that he (Dr. Anderson) changed his mind. This book is housed in the CMN library.

[6] Chipman, 1916 (December 15, 1913).

[7] Anderson, 1921h (December 31, 1921).

Stefansson Creates his New Northern Party

Stefansson and Dr. Anderson parted without reaching any agreement. Dr. Anderson proceeded to Collinson Point, taking Captain Bernard with him, while Stefansson and Olsen continued on to the *Belvedere* where Stefansson spent a pleasant Christmas with the ship's Captain Steven F. Cottle and Mrs. Cottle. There he also arranged to have some of the Expedition's supplies freighted from the *Belvedere* to Martin Point for the use of his new Northern Party on its first ice trip. The sailor Olsen, instead of continuing on with Stefansson, returned to Collinson Point from the *Belvedere*.

Dr. Anderson justifiably objected to Stefansson's having many of the supplies from the *Belvedere* freighted to Martin Point for his new Northern Party because they had been purchased for the use of the Southern Party. In Esquimalt Dr. Anderson had tirelessly endeavoured to keep supplies destined for Stefansson's Northern Party separate from those intended for his Southern Party during the loading of the *Karluk*. At Nome some of the Southern Party's items intended for transfer from the *Karluk* to the *Alaska* and the *Mary Sachs* were instead retained by Stefansson on the *Karluk*, including most of the pemmican, with Stefansson arguing that the Southern Party could "live off the land." As a result, Dr. Anderson had to obtain additional supplies for his party at Nome, most of which ended up being shipped to Herschel Island on the *Belvedere*. That ship, however, like the *Karluk*, had failed to reach Herschel Island, and now Stefansson was helping himself to some of the Southern Party's supplies on it to restock his new Northern Party.

Before leaving Ottawa, Stefansson had been supplied by the government with a number of blank cheque-books for occasional emergency use. Having lost the service of the equipment and men intended for his Northern Party, he now started to issue cheque after cheque to purchase the supplies he needed and the services of whatever men and Iñupiat women seamstresses he required to get his newly planned Northern Party up and running. He had lost the *Karluk* but he was still determined to carry out the geographic plans for which it had been purchased.

Towards this purpose he left the *Belvedere* after Christmas and headed east for Herschel Island to obtain men, dogs, supplies and equipment. En route he stopped briefly at the camp of the trader "Duffy" O'Connor near the Alaska/Canada International Boundary. O'Connor had injured one of his legs while chopping wood a few weeks earlier and wanted to leave the Arctic, so Stefansson arranged to buy his entire stock of supplies, issuing him a cheque for the full price asked ($6,000).

Stefansson then proceeded some ten miles farther east to the trading camp of Matt Andreasen and purchased all of his supplies as well as his 50-foot schooner, *North Star*, for $13,000. Stefansson considered this well-designed little vessel ideal for his exploration work, because the shape of its bow permitted it to ride up on the ice, then break it by its sheer weight. None of the other three ships under his command was able to perform in that manner, nor were they sufficiently powered or strengthened in their prows to batter their way through the ice, which was the alternate method of dealing with thin ice. (Ice-breakers were unknown then in the North American Arctic, although Russia already had a few in operation.)

Thus, in amazingly quick order while travelling east from the icebound *Belvedere* to the Royal North-West Mounted Police (RNWMP) station at Herschel Island, Stefansson managed to obtain most of the supplies he needed for his new Northern Party. What he had purchased was, in fact, more than adequate for his anticipated needs. With the *Mary Sachs*, the *North Star* and supplies now at his disposal, he turned his attention to obtaining men and some special equipment.

He arrived at the barracks of the RNWMP at Herschel Island shortly before New Year's Day. Upon learning that a police patrol was leaving within a few days for Fort McPherson, he arranged to accompany it, for at Fort McPherson he could both dispatch his own mail south and collect incoming mail destined for members of the Expedition. He also intended to obtain two power boats and operators for the work he proposed being done by Chipman, J.R. Cox and J.J. O'Neill on the Mackenzie Delta early in the summer.

During his trip up the Mackenzie River with the police patrol, he chanced to meet Storker Storkerson, a Norwegian sailor whom he had known on the Mikkelsen-Leffingwell Arctic Expedition in 1906–1908. Storkerson was proceeding upriver to the place where his wife and young daughter were camped, bringing a load of supplies he had obtained from Matt Andreasen (before Stefansson bought Andreasen's entire outfit). Stefansson quickly persuaded Storkerson to join his new exploration party rather than continue trying to subsist by his trapping, in which activity Storkerson had had little success.

Living not far from Storkerson were Peder Pedersen and Willoughby Mason, with whom Stefansson then spent several days. Mason had a gasoline launch, the *Edna*, which Stefansson purchased for the geographer Chipman to use for mapping the Mackenzie Delta early in the summer. Stefansson also hired Pedersen, who claimed he was an engineer, to get the launch into shape and later to work with Chipman once the Mackenzie River Delta was free of ice. Stefansson had purchased another small launch at Christmas-time from Captain Cottle of the *Belvedere* for the use of the other geographer, Cox, on the Mackenzie River Delta.

There were too many channels in the delta to expect the two geographers to map all of them in the time available, but local people could provide useful advice on which channels were safer and deeper than others, and could direct them to the deeper channels for their mapping work. Only small motorized schooners and smaller craft including whale boats operated then in the delta during the brief navigation period each summer. This made it necessary to trans-ship at Fort McPherson supplies and other goods that the larger river-boats brought down the Mackenzie River destined for Herschel Island and a few other coastal localities. What Stefansson hoped to accomplish was the discovery and mapping of deeper channels in the delta that might be useable by the larger vessels coming down the river. Or alternatively, channels deep enough to accommodate ocean-going ships, permitting them to navigate for some miles up-river. Stefansson likely saw this mapping project as both a worthwhile way to keep several members of the Southern Party busy while they waited for the coastal navigation season to begin, and a means of improving his own image in Ottawa following the disastrous loss of his ship and men.

Stefansson's Irritating Letter to Dr. Anderson

From a temporary location miles up the Mackenzie River, Stefansson now wrote about his new plans to Dr. Anderson, giving his letter to Storkerson to deliver in person. In that letter he mentioned the survey work he wanted Chipman and Cox to do in the delta when the river became ice-free, and the arrangements he had made for motorboats and provisions for their use.

The geologist O'Neill was to accompany Chipman, with the local man, Peder Pedersen, whom Stefansson had hired to assist them, and map the geology of as much of the delta as he could observe. These were not assignments the officials in Ottawa had instructed the two geographers to undertake, but they were practical projects that might have proven highly beneficial to northwestern Canada's main transportation system and the people in that Western Arctic region.

Since the Southern Party was, through circumstances beyond its control, spending its first year in Alaska rather than in its assigned field area, one can readily understand why Stefansson assumed that the two geographers and geologist had no definite plans for spring field work. Thus his ideas to have Chipman, Cox and O'Neill investigate the various channels of the Mackenzie River Delta, his purchase of two power boats for that purpose, and his hiring of a boat engineer were all innovative and commendable actions on his part. However, he had carried them all out without any consultation with Dr. Anderson, and they were understand-ably regarded by Dr. Anderson as direct interference with the operation of his Southern Party.

Instead of proceeding directly to Collinson Point with Stefansson's letter for Dr. Anderson, Storkerson turned off and headed for Martin Point to take charge of the ice-trip activities, giving the letter to the sailor Olsen to deliver. Olsen delivered the letter to Dr. Anderson on February 10 when he reached Collinson Point.

Along with his "instructions" about the geographers' summer work, Stefansson included "directions" for Dr. Anderson to establish a base at Martin Point, just east of Barter Island and about forty miles east of Collinson Point. From that base Stefansson intended to commence his exploration trip north over the Beaufort Sea in search of undiscovered land. Such a trip could be regarded as either commendable or sheer madness since Stefansson had neither ship nor scientists, but he was determined to commence the Arctic exploration that he had come north to undertake. As preparatory activities, the "directions" in his letter required several of Dr. Anderson's men to make tents, overhaul sounding equipment, rate chronometer watches that Stefansson needed for his Expedition, move equipment Stefansson required by dog team to Martin Point and other such tasks. Stefansson wanted everything ready at the Martin Point camp by March 1 for him to start his first ice trip. It was evident from the tone of his letter that he expected Dr. Anderson and his men to respond promptly to his "directions," and to give Storkerson all necessary assistance and cooperation in establishing the base camp at Martin Point. It was not to be.

In his field notes the day after receiving Stefansson's letter, Dr. Anderson wrote:

> Made inquiries of members of the party in regard to volunteers for Mr. Stefansson's proposed ice trip. All members of the scientific [staff] are unanimously against having anything to do with it, first, because it is outside of the province of the plans provided for members of the Southern party, and would seriously interfere with work which the scientific men have all planned to do this spring … Secondly, they cannot see the advantage to be derived from going out on the ice in this region, near where the same thing has been tried by Mikkelsen and Leffingwell and ice conditions are well known …

> Other members of the crews do not consider that they have shipped for any such project. The Geological Survey men consider furthermore, that using the stores of the Southern party, which are on the "Belvedere," and upon which the Southern party is supposed to

subsist in part for the next year, in carrying out new plans of Mr. Stefansson's which are in contradiction to the avowed agreements made with the authorities at Ottawa and the individual members of the party when they joined the expedition, is in effect a misuse of Government property and stores ...

There is much dissatisfaction here at being asked to contribute a large amount of extra provisions and supplies of various kinds for the extra employees of the ice camp, because as we are fixed, on account of the *Mary Sachs* coming in with practically nothing in the way of provisions, we shall be running short of many things before the end of this year, to say nothing of next year, and we are not very certain of having them replaced next year.

The Polar Bear certainly has no supplies to sell and the Belvedere can have very little. In fact, Captain Cottle was obliged to use up some of our provisions which he was carrying as freight (viz. salt beef), on account of having the ship-wrecked crew of the Elvira thrown into his care ...[8]

From the above, it is apparent that, at least initially, few if any of the men at Collinson Point, scientists and crew alike, were willing to assist Stefansson to get his ice trip under way. Even stronger opposition was voiced over the use of supplies on the *Belvedere* to provision Stefansson's ice trip and to feed the host of Iñupiat he subsequently hired to work at Martin Point preparing his group for their trip. These provisions were intended for the use of the Southern Party. As events unfolded, after Stefansson reached Collinson Point early in March and presented his case, several of the scientists and crew did change their attitude and render assistance, even to being members of support parties. In his published account in *The Friendly Arctic*, however, Stefansson presented the picture that Dr. Anderson was almost alone in opposing his wishes, and described the entire incident as the "threatened mutiny."[9] His words and insinuations stirred up a mountain of resentment in Dr. Anderson and the scientists who were with him on the Expedition at that time.

Stefansson Returns to Collinson Point

After completing his reports to the government in Ottawa and other duties, Stefansson left Fort McPherson and started back to Collinson Point. About twelve miles west of the police barracks at Herschel Island, he met O'Neill, who was accompanied by the sailor Olsen and Captain Peter Bernard with two sleds heavily loaded with Expedition supplies. The three men were heading for Herschel Island. O'Neill, with the assistance of Olsen, was intending to map the geology of the Firth River, which emptied into the ocean between Herschel Island and the International Boundary, but first he planned to store his load of supplies at Herschel Island, discuss his trip up the Firth River with the police, and then proceed with his study of the geology of that river. Bernard, on the instructions of Dr. Anderson, was ferrying Expedition supplies from the *Belvedere* to Herschel Island, where he would store them for later use.

---

[8] Anderson, 1916a (February 11, 1914).
[9] Stefansson, 1921b, p. 123.

O'Neill handed Stefansson a typewritten letter from Dr. Anderson, which was the latter's response to the letter of instructions he had received from Stefansson. According to Stefansson, it stated that Dr. Anderson and his men refused to assist Stefansson or supply him with the resources and equipment he had requested. Dr. Anderson was said to have stated that his Southern Party needed these items for its own work, and that he regarded Stefansson's planned ice trip as primarily a publicity stunt of little scientific value, which would seriously risk the lives of both Stefansson and the men who would accompany him.[10]

Dr. Anderson recorded in his field notebook that he replied to Stefansson's letter on February 15, but for some unknown reason, unlike on other occasions, he did not copy his letter into his notebook.[11] His typewritten letter subsequently disappeared under strange and mysterious circumstances. In his book, *The Friendly Arctic,* Stefansson later claimed that Dr. Anderson had removed it from a padlocked box Stefansson left with the RNWMP at Herschel Island before embarking upon his ice trip,[12] but Dr. Anderson stoutly denied having done so.

Some months after his meeting with Stefansson, O'Neill wrote about the incident to a fellow geologist at the Geological Survey in Ottawa:

> We have been quite unanimous in backing Anderson's stand that the Southern Party has definite instructions to complete a certain work, and cannot agree to any proposition that would interfere with its chances of carrying out those instructions ... I told Dr. Anderson how I stood on the subject and also gave Mr. S[tefansson] a thorough insight into my views, when I met him at [*sic,* near] Herschel Island in February, together with some other things I thought he ought to know ...
>
> He was in a bad humor over letters I had brought him and made several rash statements, laying the law down in great style. I let him spiel away and kept my opinions until they were called for; then I wasn'[t] backward about telling him.[13]

After reading Dr. Anderson's letter Stefansson turned back angrily and accompanied O'Neill and his two companions to their destination, the RNWMP barracks at Herschel Island. While they walked together, Stefansson discussed Dr. Anderson's letter and his own trip plans with O'Neill, ultimately persuading O'Neill to hand over his chronometer, one of the few in the possession of the Southern Party. This accurately calibrated watch was absolutely essential to Stefansson, for without precise knowledge of the time of day, he could not determine his locations while on his ice trip and without that information, any bottom soundings he took or geographic positions he determined would have been almost meaningless.

At Herschel Island, having overridden Dr. Anderson's instructions to O'Neill regarding the retention of his chronometer and thereby depriving the Southern Party of its use for the rest of the Expedition, Stefansson now countermanded Dr. Anderson's instructions to Captain

---

[10] Stefansson, 1921b, pp. 112–114.

[11] Anderson, 1916a (Vol. II, February 15, 1914).

[12] *Op. cit.*, pp. 113–114, footnote.

[13] O'Neill, 1914 (July 4, 1914).

Bernard to freight supplies from the *Belvedere* to Herschel Island. Instead, after Bernard had stored his load of provisions, Stefansson ordered him to return with Stefansson to Collinson Point, and the two men set off at once. En route, Stefansson stopped briefly to discuss his ice-trip plans with Matt Andreasen, Duffy O'Connor, Captain Cottle and Hulin S. Mott (temporarily captain of the *Polar Bear* after Captain Louis Lane returned to the United States for the winter), when he reached each of their winter camps. All of them, he wrote later, independently offered him support in preparing for his trip. He met Storkerson at the trappers' log cabin occupied by Crawford, discussed arrangements and then continued on to Collinson Point, which he reached mid-afternoon on March 8. There, to his considerable dismay, he found that virtually none of his instructions to Dr. Anderson for the preparations for his ice trip had been carried out.

Greatly disturbed by what he regarded as outright rebellion, Stefansson distributed the mail he had brought from Fort McPherson,[14] and then called a meeting of the scientists and two captains that evening after they had finished their supper.

The Collinson Point Controversy

When all were gathered in the stuffy house, Stefansson talked of a wide range of topics for more than an hour. First he spoke of the aims of the Expedition, how most of the plans and arrangements for it had been his ideas and through his efforts, and his role in drawing up the instructions given each scientist at the start of the Expedition. He went on to claim that he had personally approved of all but one of the scientists (the exception being Wilkins), told of his successes in obtaining supplies and personnel for his new Northern Party, and then mentioned his disappointment in some of Dr. Anderson's actions and decisions on their previous expedition and how little biological work he had accomplished. He was particularly critical now of the Dr.'s opposition to his ice-trip plans—"raked me over the coals fore and aft before all hands, for not falling in immediately with all of his plans" was how Dr. Anderson described the verbal attack.[15] Stefansson also reminded the men that his [Stefansson's] exploration work was the prime purpose of the Expedition. He complained of the lack of cooperation of both Dr. Anderson and his men in helping him commence his ice journey, and Dr. Anderson's strong objections to Stefansson's requests for depth-measuring equipment, a chronometer, a sextant and other pieces of specialized equipment, which Dr. Anderson claimed were needed by the Southern Party to carry out the work assigned its members.

He told the men of his purchases of the supplies and equipment of both Duffy O'Connor and Matt Andreasen, including the latter's schooner, *North Star*. He had paid O'Connor $6,000 for his outfit and Andreasen $13,000 for his.[16] Objections arose from several members of the

---

[14] For some unexplained reason Stefansson withheld from Dr. Anderson for three days a letter from Mrs. Anderson, one Stefansson knew to be of extreme importance because he had opened and read it before reaching Collinson Point. In this letter Mrs. Anderson tells her husband of the birth and subsequent death a few days later of their first child, a boy, and begs him to return home to be with her (Mrs. Anderson to Dr. Anderson, November 7, 1913). There is no record of Stefansson having expressed his sympathy or condolences to Dr. Anderson over his loss.

[15] Anderson, R.M., 1916a (Vol. II, March 8, 1914).

[16] Chipman, 1916 (March 8, 1914).

Southern Party when he mentioned that he intended to use the *Mary Sachs* as well as the *North Star* for his new exploration plans. The Southern Party was now faced with carrying out three years' work in two years, they argued, and could make much better use of a second vessel than Stefansson, who would be the only scientist on his new Northern Party. A few among them indicated their resentment to him for interfering with their work, for the vagueness of most of his plans and for his demands for instruments they felt they could not spare. Chipman and Cox made it clear that if Stefansson took some of the equipment they needed for their mapping work they would leave the Expedition at Herschel Island and return to Ottawa, effectively scuttling the Southern Party. The Danish naturalist Johansen bluntly told Stefansson that he (Stefansson) had authority only over the *Karluk* and its personnel, and as it was now lost, he (Stefansson) might as well go home! Dr. Anderson insisted that Stefansson put in writing his plans for himself and for the Southern Party, and what he required to carry out his new activities. He also asked Stefansson for lists of the items Stefansson had purchased, as he (Dr. Anderson) would now have to account to Ottawa for them. Stefansson's acquisition of the O'Connor and Andreasen outfits and new personnel for both Northern and Southern Parties added many accounting and administrative tasks to the ones Dr. Anderson already had.

Dr. Anderson had remained quiet while Stefansson talked of what he proposed doing and what he had accomplished. Finally, after Stefansson had directed several pointed criticisms at him, he no longer could remain silent, and in a slow and deliberate voice informed Stefansson and the other men that he fully intended to follow the instructions he had been given by the government, and in keeping with those instructions was not prepared to offer any equipment or personnel to Stefansson that would severely handicap the operations of his Southern Party.

Wilkins was witness to the fray that evening, but remained a quiet observer, taking the position that as an employee of the Gaumont Company in London he was for all intents and purposes a guest with the Expedition. He did, however, mention in his diary that Stefansson had read them a letter from Rev. C.E. Whittaker, a Church of England missionary formerly in the Mackenzie River district, to a Native on Herschel Island, which warned him that most of the Expedition members were "bad men."[17] Chipman's diary added that Stefansson had shown him the copy of Rev. Whittaker's letter. It was apparently written in some manner the Native could understand, rather than in English, for Chipman wrote "The translation is that 'V.S. and some of those with him willfully and maliciously refuse to believe in the word of God; they are bad after your women; look out for them; and do not have anything to do with them.' " Chipman, a strictly raised Presbyterian in the late Victorian age, then added, "I am sure mother would be glad to know it."[18]

Cox kept his thoughts to himself, in keeping with his character. A few others—O'Neill, Jenness and McConnell—were absent. O'Neill had gone east to map the rocks along the Firth River and McConnell had gone west to Cape Smyth (Barrow) to collect Jenness. At some point during the evening Dr. Anderson voiced his willingness to step down as leader of the

---

[17] Wilkins, 1916 (March 8, 1914).
[18] Chipman, 1916 (March 12, 1914).

Southern Party and let Stefansson take full charge, as he seemed so intent on doing, but Stefansson ignored the offer.

Stefansson later described the meeting as "a rather tense two hours," with general agreement reached before eleven o'clock, but the diaries (or field notes) of Chipman, Dr. Anderson and Johansen indicate that it continued until about three in the morning. It is a wonder the men were not all asphyxiated in that small cabin with the men smoking their pipes and cigarettes, oil lamps giving off pungent fumes, and the wood stove burning during the seven hours the meeting lasted.

This meeting subsequently received considerable notoriety in 1922 after Stefansson described it in his popular book *The Friendly Arctic* as a "threatened mutiny" by the scientists at Collinson Point.[19] (See Chapter 23 for further details.)

Figure 17 Stefansson reading at Collinson Point camp, northern Alaska, March 13, 1914.

Photo: © Canadian Museum of Civilization, photo by George Hubert Wilkins, 50756

Calm Following the Storm

Looking back many years later, however, one has to admit that Stefansson pulled off an incredible feat when he revised his original plans after the disappearance of the *Karluk*, trekked hundreds of miles along the north coast of Alaska from Barrow to the Mackenzie River, and emerged with a schooner, new supplies and all the manpower he needed to carry out his exploration program—all within a few short months. His biggest obstacles had been the outright objections and attitudes of Dr. Anderson and his scientific colleagues at Collinson Point, but Stefansson had managed to persuade most of them to be at least minimally cooperative. In doing so, however, he had further alienated the friendship and aggravated the hostility of his former colleague, Dr. Anderson, and spent many thousands of dollars of the Canadian government's money.

Following the heated discussion when he first returned, Stefansson seemingly chose to largely ignore the hostility of most of the scientists at Collinson Point, or else was merely content with what he had accomplished. He had "borrowed" O'Neill's chronometer, had persuaded Chipman to lend him a sextant, and had obtained offers of assistance from the naturalist Johansen, the photographer Wilkins, Captain Bernard and several other

---

[19] Stefansson, 1921b, p. 123.

crew members in making preparations for his ice trip, all in spite of Dr. Anderson's opposition. Stefansson's new Northern Party at that point differed greatly from his original one, lacking the scientific talent and personnel, as well as the working space of the *Karluk*, but he was determined to succeed in his original plans to explore for new land in the Beaufort Sea. With any sort of luck, he might even still beat MacMillan to "Crocker Land," if it actually existed.

A forced calm enveloped the Collinson Point camp for the next several days. Stefansson typed lists,[20] as requested, and read while others undertook tasks preparing for their part in the pending ice trip. All was finally in readiness on March 16 for the start from Collinson Point, and after Wilkins had completed the ceremonial filming of Stefansson's departure in the lead of four teams of dogs and their heavily loaded sleds, the teams headed for Martin Point.

Figure 18 Departure of Stefansson's first ice party from Collinson Point, northern Alaska. Jennie Thomsen and daughter on left, Stefansson (in white parka with horizontal stripe), Cox (behind sled and wheel) and Chipman (to right of sled). Wilkins' movie camera ready to film the departure, March 16, 1914.
Photo © Canadian Museum of Civilization, photo by George Hubert Wilkins, 50762

---

[20] The Southern Party had two kinds of typewriter, a Hammond #12 and a Bijou portable (Anderson to Chief Accountant, 1914c). Stefansson later used a portable typewriter at Cape Kellett on Banks Island, so it appears that Wilkins may have taken the Bijou portable machine there in the summer.

THE SEARCH FOR "CROCKER LAND," 1914

10

Stefansson's new exploration plans called for him to proceed north, roughly along the meridian 143° W longitude to 76° N latitude, taking depth soundings regularly as he progressed. If the sea-ice over which he was travelling was drifting west, as he anticipated, he intended to return south to the Alaskan coast within two months, landing somewhere east of Point Barrow. From there he would head immediately east to the *Alaska* and *Mary Sachs* in Camden Bay. If the current did not carry him west and if he found no land as he progressed northward, he would turn east and land on either Prince Patrick Island or Banks Island, where he expected to meet the *Mary Sachs* or the *North Star* during the summer, in accordance with instructions he would soon send to Dr. Anderson and the captains of those two ships. And finally, if perchance he discovered new land, he would map it. If it was small in area he would return to Alaska before the summer, but if it was large, he would stay there and explore it for a year and travel to Banks Island or Alaska the following spring.

Following several days of preparation, he finally left Collinson Point for Martin Point on March 16, accompanied by eight men and four sleds. Travelling slowly because of their heavy loads, they covered the twenty-four miles by 8 p.m., arriving at the small cabin that the engineer Crawford occupied when he tended his fox-trapping lines. It was foggy the next morning as they set off again, but an Iñupiat couple who knew the way led them to their intended destination. After lunch, Chipman, Cox and Pedersen veered off to the southeast with their dogs and sled, and headed for the International Boundary post that marked the border between Alaska and Canada. Chipman intended to start his geographic mapping of Canada's Arctic Coast from that carefully surveyed site east to the Mackenzie River Delta.

Martin Point was a prominent sandspit jutting into the Beaufort Sea a few miles east of Barter Island. That island was so-named because it was the meeting place in earlier days for Alaskan Iñupiat from farther west and Inuvialuit from around the delta of the Mackenzie River to the east, when they wanted to trade with each other. When Stefansson and his men reached Martin Point they found a busy community of five tents, a dog barn, and assorted racks and caches, which had been recently erected by Storkerson and some men he had hired from the *Belvedere* and the *Polar Bear*. Those two ships were trapped in the ice farther east along the coast. Several Iñupiat women sat about the camp, hired by Storkerson temporarily as seamstresses. Among these was the robust, thirtyish widow named Pannigabluk, who had travelled extensively with Stefansson during his 1908–1912 expedition in the Arctic, and whom he had recently re-hired for his new Northern Party. With her was her four-year old son, Alex, widely rumoured at the time to have been fathered by Stefansson. Wilkins, for example, had heard such rumours and, after meeting Pannigabluk and Alex in January 1914, commented, "When looking down on the child from the side, there is a striking resemblance to V.S. to be

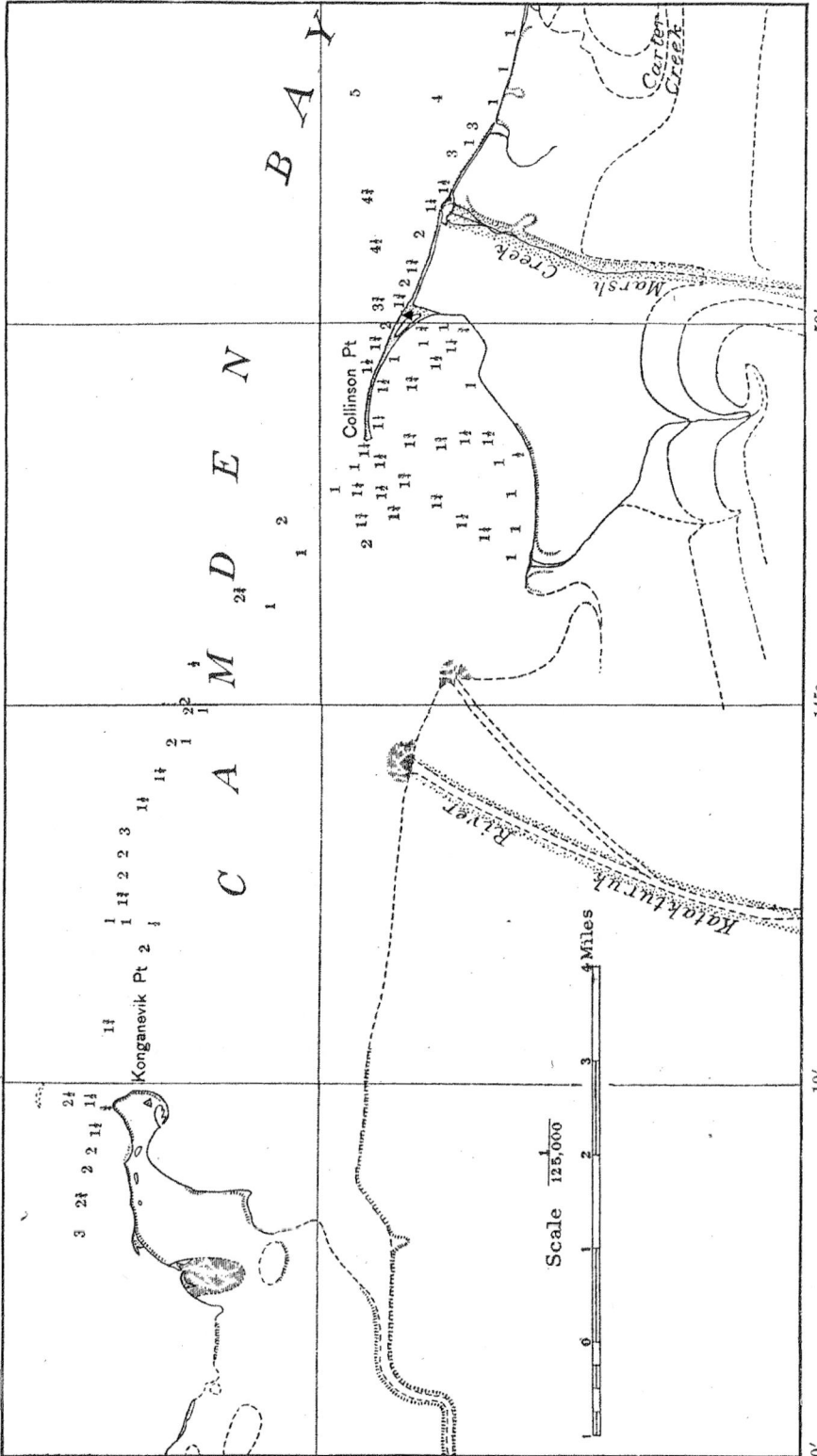

Map 6 Camden Bay and Collinson Point, northern Alaska, with a lagoon sheltered by a long sandspit. The Southern Party's base camp, a driftwood cabin (indicated by the symbol ▲), was located at the base of the spit. The *Mary Sachs* was pulled up on the shore near the cabin in 1913–1914. The *Alaska* was anchored in the deeper water of the lagoon, about a mile to the west. Water depths are in fathoms. (From Leffingwell, 1919.)

seen, and I believe that there is some truth in the common report along the coast that it is his child."[1]

Stefansson denied the rumours or remained silent on the matter, of course, but several decades later, a few years after his death, one of his long-time friends, Richard S. Finnie, reported having seen a birth registration for Alex in the record book of the Anglican Church at Aklavik, with Pannigabluk and Stefansson listed as the boy's parents. Unhappily, both the record book and the church were later destroyed by fire.[2]

The First Ice Trip

Stefansson intended to take provisions on his ice trip for only six to seven weeks, for he expected that he and his two companions would survive the rest of their trip on seal meat they obtained by hunting on the ice, and caribou meat after they reached land. Obviously, there was a considerable risk in such an undertaking—from ice breakup, injury, sickness or starvation. Most of the

Figure 19 Pannigabluk and her son, Alex Stefansson, at Martin Point, northern Alaska, March 18, 1914.
Photo: © Canadian Museum of Civilization, photo by George Hubert Wilkins, 50770

sailors and Iñupiat along the Alaskan coast who knew of his ice trip believed it was suicidal. Even Dr. Anderson, who was more familiar than any of the others with Stefansson's remarkable ability to survive in the Arctic, had little faith that the men and dogs could "live off the land" (or in this case, "live off the ice"), as Stefansson intended to do, and felt the risks were too high for the limited potential scientific findings of their trip. Those were two of Dr. Anderson's reasons for refusing to help Stefansson. After losing the use of the *Karluk*, however, Stefansson felt that it was essential that he succeed with this ice trip in order to retain the support and respect of both the Canadian government and the public. Additionally, he was determined to prove to the world that white men could "live off the land" in the Arctic just as the Inuit did. The latter, however, would not normally venture more than a few miles from land.

The two men willing to accompany Stefansson on his ice journey were Storker Storkerson and Ole Andreasen, both Norwegian sailors who had come to the Western Arctic many years before on whaling ships. When the Arctic whaling industry collapsed early in the 1900s, they had remained in the North as trappers and traders. The money Stefansson was willing to pay them was sufficient enticement for them to agree to accompany him, although within weeks they demanded a substantial increase.

---

[1] Wilkins 1916 (January 22, 1914).
[2] Finnie, 1978, p. 3.

Bad weather delayed the departure of the ice party from Martin Point for a few days. When it was finally ready to leave on the afternoon of March 22, Wilkins filmed Stefansson leading his men proudly seaward, followed by four teams of eager dogs pulling their heavily loaded sleds. Stefansson intended to use only one team of dogs and one sled for most of his journey, but for the first few days he was going to be accompanied by Wilkins, Johansen, Bernard, Aarnout Castel (a sailor he had hired from the *Belvedere*), Daniel Sweeney (also from the *Belvedere*), two Iñupiat men, and three additional heavily laden sleds. All told, the four sleds started out with 3,400 pounds of supplies and equipment, and Stefansson's party needed all the muscle power available to assist them over or through the hummocky ice ridges they soon encountered. After they had progressed a few miles beyond the near-shore ice ridges, Sweeney and two Iñupiat men (Ikey Bolt and Taliak) transferred their load to the other sleds and returned with their dogs and sled to shore, as Stefansson had planned.

Soon after Sweeney's departure, Stefansson's party came to the boundary between the land-fast ice and the moving pack ice. Here they camped, and Johansen took a depth sounding to the sea floor, obtaining a reading of eighty-four feet. Stefansson's secretary McConnell then arrived unexpectedly, travelling alone and on foot, bringing mail from Barrow for Stefansson and for some of the other men. He had reached Camden Bay on March 20, after collecting Jenness from Barrow, and upon learning that Stefansson's ice party had gone to Martin Point, expressed a desire to join it. Dr. Anderson had given him permission to do so, and he had managed to catch up to the ice party after several days of solo travel.

Problems

The next day Captain Bernard had an unexpected accident while crossing some hummocky ice, sustaining a serious cut to his forehead. McConnell stitched it up. Stefansson and three others then took Bernard back to Martin Point where someone could look after him while he recovered. They returned the next morning, bringing Crawford to replace Bernard. Later that day as the ice party slowly sledded seaward, it came upon open water in which seals were plentiful. Stefansson had Wilkins take several photographs of him seal hunting, one of which[3] attained considerable publicity over the next decade or two.

On March 27, five days north of Martin Point, Stefansson sent Wilkins and Castel back to shore with some equipment he considered no longer necessary, one sled and instructions to bring back several items needed by the ice party. Shortly before reaching the Martin Point camp, however, Wilkins and Castel ran into a raging snowstorm, which greatly hindered their arrival and forced them to remain at the camp for a day until the storm blew out. Later, when they tried to return to Stefansson's temporary camp, they encountered a wide lead of open water blocking their way and realized that Stefansson's party was somewhere to the north, adrift on a large ice mass at the whims of the westerly winds and the ocean currents, being carried many miles to the east.

The same raging snowstorm struck Stefansson's party on the ice. When the storm finally abated and the weather cleared, Stefansson could see mountain peaks well to the south in the

---

[3] Photograph 50776, CMC; also PA 214013, LAC. This photograph was used as the picture on the dust jacket of Stefansson's *The Friendly Arctic* (1921).

Endicott Mountains, which he knew from his previous years in the region were about forty miles east of Martin Point. He realized then that the winds had blown their ice mass well to the east, making it impossible for Wilkins and Castel to find them. The loss to his ice party of the assistance of Wilkins and Castel, whom he knew to be two good men, and of the several good dogs that pulled their sled was bad enough, but the sled Wilkins had taken to shore was the one that had attached to it a bag of spare ammunition, tools and special equipment required for repairing the sleds whenever they were damaged. It was a cruel and unexpected twist of fate, but Stefansson decided to continue northward notwithstanding.

Locating the Continental Slope

On April 4, Stefansson and Storkerson shot several seals in an open stretch of water and recovered six, a great boon to their food supply. Soon afterwards a polar bear made its appearance near their camp, attracted by the scent of the killed seals, and was quickly shot to death. The men cached the excess seal meat and bear meat for the support party to collect when they were returning to the shore. This seemed like a good idea at the time, but the support party, which consisted of all of the men except Stefansson, Storkerson and Andreasen, was later unable to find the cache and it was wasted.

As the ice party proceeded northward, the naturalist, Johansen, took depth soundings at frequent intervals, determining that the depth was increasing at the rate of just over six feet per mile. On April 5, however, his readings suddenly increased within a horizontal distance of two miles to nine hundred feet. This indicated quite vividly that they had reached the outer edge of the Continental Shelf, where the Earth's surface plunges to the oceanic depths.

Figure 20. Johansen taking the temperature of the sea water north of Martin Point, northern Alaska, March 25, 1914.
Photo: © Canadian Museum of Civilization, photo by George Hubert Wilkins, 50778

Stefansson was delighted to discover an abundance of seals in the open leads in the ice over the deeper water, for it was generally thought that they lived only in the shallower coastal waters. The likelihood of success of his trip over the ice was considerably increased by this discovery, for it was absolutely essential that he and his men obtain seals for food once their provisions were exhausted.

Departure of the Support Party

On April 7, Stefansson's party reached 70° 13' N latitude, 140° 30' W longitude, about fifty miles north of the Alaskan coast, and he decided to send his support party back to shore. Before they left, however, he took the time to write detailed instructions for Dr. Anderson, stating that in the likelihood that he and his two companions did not return to the Alaskan coast that spring, Dr. Anderson was to assume they had proceeded to the southwest corner of Prince Patrick Island or the northwest corner of Banks Island. Dr. Anderson was therefore instructed to send Captain Bernard and the *Mary Sachs* with a load of supplies from Herschel Island to Cape Kellett on the southwest coast of Banks Island, to establish a base depot there, or if ocean conditions permitted, farther north near Norway Island on the west coast of Banks Island. More important, Dr. Anderson was to send Wilkins with the smaller *North Star* to the west coast of Banks Island near Norway Island to meet Stefansson and his two companions, for this was the ship Stefansson intended to use for his exploration of the Beaufort Sea. In the event that he found neither men nor notes, Wilkins was to continue north to Prince Patrick Island to meet them. Stefansson also wrote similar instructions for Wilkins, and different ones for Jenness and Captain Bernard.

Stefansson was well aware that Wilkins had been loaned to the Expedition by the Gaumont Company for only one year and was expected to return to England that summer. By instructing him to bring the *North Star* to Norway Island or even farther north to Prince Patrick Island to pick up Stefansson's three-man exploration party, Stefansson was ordering Wilkins, for the benefit of the Expedition, deliberately to break his contract with the Gaumont Company.

When he completed his various letters, Stefansson gave them to McConnell to deliver when he reached shore. McConnell, Johansen and Crawford then started their fifty-mile journey to shore, taking with them the two weak sleds, and a thirty-day supply of food. They encountered numerous difficulties en route with open leads and high pressure ridges, but landed safely about thirty miles west of Herschel Island on April 16, from where they returned along the coast to Collinson Point.

Three Men on an Ice Raft

At 70° 13' N, Stefansson, Storkerson and Andreasen were standing on the ice about as far north as ships on the open sea had ever reached in the Western Arctic, with only their wits and guns to ensure their survival until they could expect to meet the *North Star* or the *Mary Sachs* some four months hence. As Stefansson later described it, "We were three men alone on the edge of the unknown."[4]

---

[4] Stefansson, 1921b, p. 162.

His party's situation had been duplicated once before, when two Norwegians, Fridtjof Nansen and Hjalmar Johansen, left Nansen's icebound ship *Fram* at 84° N latitude in 1895 and headed north on foot hoping to be the first men to reach the North Pole. They managed to get two degrees closer to the Pole (86° 1' N) before having to turn back and land on Franz Joseph Land. Their journey differed, however, as Stefansson noted, because "… they had food to carry them nearly or quite to land …," a tried method, and did not have to hunt until they were on land. "But we were facing the unknown part of the Arctic sea with a method not only untried, but disbelieved in by all but ourselves."[5]

Furthermore, it might be added, Stefansson's party was heading north not intent upon reaching the North Pole, but in search of land, new information and to test Stefansson's hypothesis that the seas could provide the food they needed while they were travelling on the ice. They had six strong dogs, a good sled, 1,236 pounds of supplies and equipment, and two rifles, each with about 160 rounds of ammunition. Stefansson estimated they had food for about thirty days. They were short of fuel (kerosene), however, because half of their supply had been on the sled Wilkins took to shore on March 27.

Were they brave or foolhardy to undertake such an adventure? Sailors from the two icebound ships *Polar Bear* and *Belvedere* thought them completely mad, and gave them little or no chance of survival. Indeed, they soon spoke of the two men with Stefansson as already lost. Foolhardy might be the best word for Stefansson, who seemed invincible and determined to become famous by his wits and his discoveries. The word greedy would apply better to Storkerson and Andreasen, to whom money was the main motivation. They had initially insisted upon being paid $5 per day to accompany Stefansson over the ice, which was more than double the local wages on shore. Then several weeks later, when they were far out over the ice, they had demanded that Stefansson increase their daily pay to $25 per day until they reached the ships, under threat of refusing to assist him. Stefansson was forced to agree.

Immediately after the departure of the support party, open leads, pressure ridges, gale winds and snow plagued the three men, allowing them to make very little progress at first. Then the winds abated and the temperature plunged to well below 0° Fahrenheit, and the men made better progress. With the colder weather, the snow became more suitable for the building of snowhouse shelters, and on April 15 Stefansson built his first one. Prior to that date they had used tents. They now discovered that they were progressing a little *east* of north, an indication that the ice was moving eastward, so changed their travel direction to a little west of north to try to remain along the 143rd longitude line.

By April 16 they had reached 71° N, and five days later, 72° N. They now saw few seals and the supply of food on their sled was diminishing rapidly. Because of breakage, they had only about 4,500 of their original 10,000 feet of sounding wire left. On April 22 the sounding wire failed to reach bottom, indicating for the first time that they were above water deeper than 4,500 feet and getting farther from land

---

[5] *Ibid.*

Drifting West

On April 25 the wind shifted and came from the east, which drove westward the ice upon which they were travelling, opposite to the direction in which they intended to travel. This meant, Stefansson reasoned, that they would encounter widening "leads of open water running north and south" parallel to the western shores of Prince Patrick and Banks Islands, which could prevent them from getting ashore.[6]

During the warmer daytime hours there was now more surface water on the ice, so Stefansson changed their schedule to travelling at night, although, of course, there was light from the sun for almost the full twenty-four hours each day. He also decided to head for Cape Prince Alfred on the northwest corner of Banks Island, as they had not progressed far enough north to aim for Prince Patrick Island.

The depths they were obtaining from their soundings, absence of sea birds and other factors clearly indicated that there was no major continent or even a small body of land nearby. It was obvious, therefore, that neither "Crocker Land" nor "Harris Land" lay in their immediate path, as had been thought by some Arctic "experts." It would have been very exciting to have found such land, but to find that none existed was still a definite contribution to the scientific knowledge of that part of the world, hence not entirely disappointing to Stefansson.

At this point Storkerson and Andreasen urged Stefansson to return to Alaska rather than head east to Banks Island, reasoning that they knew there were seals to the south and also their two ships to take them to Banks Island during the summer. They argued further that the Alaskan coast, while somewhat inhospitable, was familiar, and their survival would be more probable if they proceeded in that direction. Banks Island, on the other hand, was apparently uninhabited, almost unknown and might lack animal life upon which to exist. Stefansson, however, argued that in May, which would commence in just a few days, easterly winds would drive the ice off the Alaskan coast to the west, with large areas of open water between the onshore ice and the moving sea ice, making it impossible to reach shore without a boat. Furthermore, the likelihood was great of their being carried well west of Point Barrow.

Heading for Banks Island

And so they headed for Banks Island, but rather than travelling straight east, they followed a somewhat curved course, moving more northeasterly. By May 2 the sun remained above the horizon at midnight. On May 5 Stefansson's party used their lasts drops of kerosene. Their supply of food was also disturbingly low at that point, and they no longer saw seals in the leads they skirted. As a consequence, they went on half rations and hurried onwards, always travelling "at night" when it was cooler. On May 7 Stefansson sighted a seal in a lead, but it was much too far away to shoot. Still, it was a hopeful sign. On May 13 they shot two seals, but these promptly sank and could not be retrieved, leaving them more discouraged than before. Stefansson took the opportunity to explain to his two companions that seals will generally float if shot in salty water, but will sink if the surface water is relatively fresh from the melting of sea-ice, because the fresh water is less dense than the salty water. The dogs by this time were

---

[6] *Op. cit.*, p. 187.

being fed strips of bear skin, and the three men were on half rations, using for fuel pieces of fur cut from the two grizzly bear skins they had brought with them for bedding. Spirits were sagging, and even Stefansson began to question his belief about always finding food on the Arctic ice.

On May 15 they reached 74° N latitude, two degrees short of their intended northward target, and about three hundred miles north of the mainland. There were now many leads, which gave them more hope of finding much-needed seals, but also increased their problems of navigating landward. They desperately needed a westerly wind to drive the ice to the east and allow them to continue shoreward.

Stefansson now called a halt, and the three men waited by a lead for the much-hoped-for seals to appear. They did not have long to wait. Two seals poked their heads above the surface within two hundred yards of Stefansson and he shot them both. To his dismay both seals promptly sank. An hour or two later another one surfaced, and Stefansson carefully and deliberately shot it through the head. This seal floated, and soon both men and dogs enjoyed their first full meal in more than a week. After feasting they rested in their single Burberry tent, which led to their next mishap. Unexpectedly all three men were struck with painful attacks of snow-blindness, which cost them valuable travel time by having to remain in camp for the next two days.

Using a Sledboat

On May 20 they came upon a lead at right angles to their path, which was a quarter of a mile wide at its narrowest point. Suspecting that much time would be lost seeking a narrower opening to left or right along the lead, Stefansson decided to use their sledboat to cross to the opposite side. Unloading their sled, they spread their canvas cover on the icy surface. The cover measured eighteen feet by ten feet and was waterproofed with a thick coating of lard. Then they placed the sled on it, lashed two six-foot sticks (carried with them for this purpose) crossways to the sled, one at the front, the other at the back, and then a ski between the ends of both sticks on each side, making a frame. They then lashed the canvas cover to the frame. With this strangely constructed "boat," which they assembled within two hours, they were able to navigate carefully across the quarter-mile lead in two trips, taking three dogs and some supplies on each trip.

Drifting West Again

For the next several days they were stranded on a mass of floating ice that was several miles square, unable to cross to any adjoining ones because of choppy water stirred by a brisk wind. They used the time to shoot and recover several seals. The odour of the dead seals in turn attracted polar bears, five of which came to a sudden end after deigning to approach their camp.

Sextant readings now indicated the ice block on which they were camped was drifting westward, away from Banks Island, which was confirmed by depth soundings. These had registered 2,400 feet on May 20, and increased over the next few days until on May 29 they were again above water too deep for their sounding wire to measure.

Seizing an opportunity on June 5 to get across to the next ice mass, they moved their camp and some of their accumulated supply of meat in four sledboat trips. An increase of wind widened the lead while they moved across it, preventing them from making additional trips and forcing them to leave behind more than a ton of bear meat and seal meat. Once safely landed on the ice mass on the east side of the lead they loaded their sled and continued their journey ten miles northeastward, hoping for a west wind to reverse the ice drift and bring them nearer to their destination of Banks Island.

Land in Sight

For the next two weeks they trekked slowly northeastward, but the ice was moving southward, causing their actual direction to be southeastward. They took soundings more frequently now, for they encountered many leads. On June 11 they got a reading of 2,171 feet, on June 15 one of 1,137 feet. Soon they began to see a few birds—snow buntings, old squaw ducks and jaegers—which they knew indicated that land was not too much farther to the east. Greatly cheered by this observation but hesitant to be too optimistic, they pressed on, crossing from each ice mass to an adjoining one whenever possible. On June 22 they obtained a depth reading of only 162 feet, and the next morning they could see three low hills about twenty miles to the east. These later proved to be on Norway Island, by coincidence their intended meeting site with Wilkins. To reach it proved highly challenging, however, because of intervening pressure ridges and the amount of water lying upon the ice. They were ultimately forced to wade through some cold melt-water pools nearly up to their waists, with their dogs swimming and the sled floating behind.

Norway Island

At last, on the evening of June 25, ninety-six days after leaving Martin Point, they wearily climbed ashore on Norway Island. Stefansson estimated that they had travelled about seven hundred miles in the three months they had been on the ice. After weeks of little other than seal meat or bear meat to eat, his two companions welcomed the numerous signs they soon saw of caribou and Arctic hares, not to mention foxes, wolves, ptarmigan and other birds. True to his customary habit, Stefansson went off to explore the island, leaving the task of setting up camp and the collection of driftwood for their campfire to his two assistants.

Stefansson understandably felt very pleased with himself at this time for having arrived at Norway Island, more or less on schedule, exactly where he had instructed Wilkins to look for him. And contrary to the opinions of sailors and Iñupiat along the Alaskan coast, he and his two companions had survived in spite of overwhelming odds. Indeed, he frequently thereafter spoke of the beliefs, so far as he knew,[7] of Expedition members, police, traders and Iñupiat that the three of them had long since perished. After all, he had taken food and fuel to last only about forty days and had been unheard from for three months. He even assumed the Canadian government had given him up for dead.[8] Unfortunately, he had no way to notify anyone that he and his two companions had landed safely on Norway Island as planned.

---

[7] "So far as he knew" was a qualifying phrase Stefansson was adept at using.
[8] Stefansson, 1921a, p. 293.

BANKS ISLAND, 1914–1915

11

Leaving Storkerson and Andreasen to make camp, Stefansson climbed to the top of Norway Island to view the surrounding terrain with his field glasses and to obtain his first glimpse of Banks Island farther east. Three miles to the east lay frozen sea, indicating that the island he had landed on was only half the size it was shown to be on his old British Admiralty Chart. Beyond that, perhaps another three miles across the ice, lay what he now assumed to be Banks Island. Through his field glasses he perceived six white specks on a hillside there—caribou. He immediately started for the caribou rather than turning back to inform Andreasen and Storkerson of his plans, for he was well aware that the need to replenish the food supply for both his men and the dogs was urgent. By deliberately walking along the top of the hill he hoped they would see him against the skyline and understand where he was going.

Bernard Island and Bernard River

The distance to the caribou he had seen—seven miles he wrote later—proved farther than he had estimated. Several hours of brisk walking finally brought him within half a mile of them, as they grazed quietly on a broad grassy slope. After a brief period of careful stalking, he shot and killed the entire herd. He then spent the rest of the night skinning and dismembering the six carcasses, grateful for the existence of twenty-four hours of daylight at that time of the year. Once he had completed these tasks, he cached the skins and meat and then hiked eight miles back to his camp with the six caribou tongues, arriving on time for his two companions and himself to enjoy their first hearty breakfast in several weeks.

Once they had completed eating, the three men proceeded to the top of the highest hill on the island, where they erected a conspicuous beacon in the hopes of attracting the attention of Wilkins, whom Stefansson expected to arrive soon from Herschel Island with the *North Star*. Inside the beacon, Stefansson placed a record of their ice journey from Martin Point and also the information that he intended to spend the summer hunting on the mainland to the east in order to accumulate caribou skins and meat for the winter.

Stefansson then had Storkerson and Andreasen move their camp east to the locality where he had cached the six dead caribou. Once there and fully settled, Stefansson and Storkerson cooked and feasted upon boiled caribou heads. Such a delicacy did not happen to appeal to Andreasen, however, who fried and dined upon caribou steaks for which he lacked both salt and onions to improve the flavour. At some point while they were eating, Stefansson suggested to the disbelieving Andreasen that he ought to try eating the meat on the caribou's head, for it was certainly the tastiest part of the animal. Stefansson later reported that within a few days Andreasen joined the other two in eating the boiled heads, to which they added boiled briskets and ribs.

Stefansson soon determined that they were now upon a second island, larger and more fertile than the one on which the three of them had landed. It was, in fact, about eight miles long, with hills rising to a level of three or four hundred feet. Furthermore, it lay but a mile or so from what he now decided was the mainland of Banks Island. This island was not shown on his old Admiralty Chart. He had made his first discovery of new land, but curiously, he showed no special enthusiasm over the fact, even though finding new land was one of the major goals of his Expedition. Instead he merely observed, "The day after moving to the deer-kill we discovered we were on an island about eight miles in its longest diameter and three or four hundred feet high …"[1] Later he named it Bernard Island after Captain Peter Bernard of the *Mary Sachs*.

After killing yet another caribou and adding its meat to that already in their possession, Stefansson had his men move the camp a mile across the intervening ice to a sandspit on Banks Island close to the mouth of a wide river. He later named this river after Captain Bernard. Over the next few days Storkerson mapped the island they had discovered, managing in addition to shoot a bearded seal (*ugruk*) and several small ordinary seals, which he added to the supply of food and skins.

Safe on Banks Island at last, with plenty of food, water and driftwood, and no longer fearing that the ice beneath their tent might break up while they slept, Stefansson had time to reflect upon their success to date and to plan for the days ahead. He had good reason to feel satisfied. By his depth soundings he had determined the position of the Continental Shelf at two locations, one north of Alaska and the other west of Banks Island. He had learned something of the ocean currents in the Beaufort Sea. And he had proved the presence of seals in leads over much of the deep-water parts he had travelled, although they were admittedly more plentiful in the near-shore regions. He had sent instructions to Dr. Anderson in April to send two ships north that summer (the *Mary Sachs* to Cape Halkett on the southwest corner of Banks Island, and the *North Star* to Norway Island or near to it), so he could now anticipate the arrival of the *North Star* within days. He also envisioned using these ships to explore the Beaufort Sea that summer during the short navigation season that would soon be upon them, one ship northwest of Banks Island (from Cape Prince Alfred), the other northwest from Prince Patrick Island, then east to Isachsen Land and back to Prince Patrick Land.

Hunting Caribou Inland

Stefansson now resolved to accumulate a large store of caribou skins and meat for his new Northern (Exploration) Party, in contemplation of his fall and winter activities after Wilkins arrived with the *North Star*. The skins were for the sewing of new clothes when the ships arrived with several Iñupiat seamstresses. And should the ships fail to arrive, Storkerson and Andreasen were capable sewers and could keep Stefansson and themselves adequately dressed.

For the first half of July they hunted near their camp on the coast of Banks Island opposite Bernard Island, because Stefansson wanted to be sure that Wilkins found them. Later in the

---

[1] Stefansson, 1921b, p. 235. The Expedition's members frequently used the word "deer" for caribou.

month he and Storkerson, the more experienced of his two companions, trekked inland, primarily to explore but also to hunt, leaving Andreasen to guard their supplies at the coast. The two hunters set up a temporary hunting camp on a high hill about fifteen miles inland from the coast, from where Stefansson could see through his field glasses the condition of the Beaufort Sea around Bernard and Norway islands. From that inland camp they hunted for the next three weeks, and within a short time had killed forty-two bull caribou and dried more than half a ton of back fat, as well as hams, shoulders and other fleshy parts, and a good quantity of skins. During that time they lived on the caribou heads and back bones, and fed the internal organs to their dogs. There were plenty of heather-like plants with which to make fires for cooking, and Stefansson found the grassy, flower-strewn interior of the island with its sparkling icy streams and few mosquitoes almost idyllic. Most of the streams could be forded if one went well inland from their outlets, so they were not a major impediment for travel.

Stefansson considered caribou fat especially good to eat, so he intended to shoot bulls, whose back fat was well developed at that time of year. The fat was needed for both food and winter candlelight, should Wilkins and the *North Star* fail to appear.

The Muskoxen Problem

A matter that soon puzzled Stefansson was the number of bleached muskoxen bones and partial skeletons he observed during his wanderings in the countryside, for not a live muskox was to be seen. He ultimately concluded that they had been plentiful at some time, but had been killed off by Copper Inuit from Victoria Island in the latter half of the 1800s after they discovered the abandoned British naval ship HMS *Investigator* in Mercy Bay, on the north coast of Banks Island. That ship had taken shelter there in 1851, while searching for Sir John Franklin's Expedition. After being prevented by ice from leaving the bay for the next two years, its crew abandoned the ship and crossed M'Clure Strait to another British ship wintering on the south coast of Melville Island, and reached England in 1855. Inuit from Victoria Island crossing to Banks Island in search of game later discovered the abandoned ship in Mercy Bay, and for many years thereafter made regular pilgrimages to it to obtain iron, wood and other salvageable items. Stefansson concluded that these Inuit had killed all the muskoxen over a period of years, which seemed a reasonable explanation for why no live muskoxen were seen on Banks Island by Stefansson or by any other member of the Canadian Arctic Expedition between 1914 and 1917.

During the nearly one hundred years since Stefansson was last on Banks Island (in 1917) muskoxen evidently crossed the ice to that island from Victoria Island and thrived, for today they number in the thousands.[2] Stefansson disliked the word "musk-ox," claiming that the animal did not have a musk odour. He preferred instead to call these survivors of the Ice Age by their Latin name, *ovibos*, or alternatively, "polar oxen."[3] His arguments notwithstanding, the term "muskoxen" remains in current usage for these shaggy northern dwellers

---

[2] David R. Gray, 1987, p. 23, and personal communiqué, 2007.
[3] Stefansson, 1921b, pp. 238–241.

Towards the end of July, Stefansson decided to make a quick trip from his temporary hunting camp to the coast in order to see what the coastal navigation conditions were like. Before departing he instructed Storkerson to cover the drying meat during times of rain or fog, to spread it out to dry upon the sun's return, and to protect it from gulls, ravens, foxes and wolves. During his trek to the coast he came upon several bull caribou, but being alone, on foot and lacking a means of transporting a lot of skins and meat, he deliberately refrained from shooting any of them until he was close to the coast. At that point he spotted a particularly fine specimen among a small herd and quickly felled it, retrieving about forty pounds of back fat, which he then carried on his back for the rest of the way to the coastal camp.

At the camp he found Andreasen well and content, but full of stories about wolves attempting to steal the meat that was stored alongside and his having to use a number of cartridges from their limited supply of ammunition to scare off the wolves. In reality, it later surfaced, Andreasen had wasted most of the ammunition shooting at geese to satisfy his craving for food other than their normal stock of seal and caribou meat.

Where was the *North Star*?

Stefansson examined the ice conditions to seaward as soon as he reached the coast, hoping to see the *North Star* approaching with Wilkins and its supplies and personnel of sailors, seamstresses and dogs. He thought that perhaps he might also see the *Mary Sachs*, if there was sufficient open water for it to manoeuvre safely past Cape Kellett. As commander of the Expedition, he assumed that his instructions to Wilkins and Dr. Anderson about the two ships would be carried out to the letter. What he saw now as he looked seaward was a wide band of ice along the shore (shore-fast ice), grounded in the shallow waters that occurred along much of the western side of Banks Island. To seaward the sea-ice was breaking up, and he could see widening leads of open water. He hoped that the shore-fast ice, which was noticeably melting and rotting away, might soon be dislodged by the right combination of high tides and easterly winds to permit a vessel to navigate close to shore. The *North Star* should certainly be working its way north any time now along one of the leads. Time and again he looked south in hopes of seeing the sails of that ship, only to be disappointed.

Hunting Farther Inland

Stefansson then decided that it would be relatively safe to leave the supplies of skins and meat at the shore camp, and to take Andreasen and the dogs inland to assist with the hunting. Before doing so, however, he and Andreasen constructed a rack of driftwood on which to place the skins and drying meat to keep them out of reach of the ever-hungry wolves or foxes. He knew that polar bears could easily get at the meat, but he had seen little evidence of them in the area, so felt the caribou and seal meat and skins would be safe if left unattended. When all of the meat and skins were safely piled on the rack, Stefansson and Andreasen set off on the fifteen-mile trek inland to rejoin Storkerson.

Stefansson learned to his dismay, upon arriving at the hunting camp, that during his absence wolves had succeeded in consuming much of the meat he and Storkerson were drying. They would now have to obtain a lot more. Taking Andreasen with him this time, Stefansson hunted farther inland for several days with considerable success, constantly making note of the terrain while he was looking for caribou. Storkerson was left to guard the hunting camp. During this period, Stefansson frequently thought about Wilkins and the *North Star*, and by the middle of August, with still no sign of either it or the *Mary Sachs*, was growing uneasy. Why had no ship come into view, for offshore conditions appeared to be excellent? He eyed the coast daily thereafter with his field glasses.

South to Cape Kellett

Finally, on August 27, with no ships in sight, Stefansson decided to lead his men south to Cape Kellett. Before doing so, however, he had them dig a deep pit at the hunting camp, line it with stones, set their dried meat in it, and then cover it with stones. They certainly could not transport all that food on their backs, and cached underground in this manner it should be quite safe from foxes, wolves and polar bears. It seemed sensible to try to preserve the meat rather than wastefully abandoning it. The cache would be a valuable supply to draw upon, Stefansson reasoned, when he returned north later to commence his next exploration over the Beaufort Sea.

Accordingly, the three men and their dogs started south on foot for Cape Kellett on September 1, following along the coast, hoping to spot one or other of the two ships at anchor in one of the many bays along their route. Stefansson went ahead by three or four miles, travelling on the hilltops, occasionally leaving little temporary monuments along the way, in some of which he placed messages for his two companions. Storkerson and Andreasen carried the heavy camp equipment on their backs, while the dogs carried three or four days' supply of food, all of which slowed their progress.

Now and then, Stefansson descended to the shore and stuck pieces of driftwood on end to mark the shoreline as an aid if they travelled this coast again after the snows came. His frequent examinations of his old British Admiralty Chart to determine his location revealed that the chart contained many inaccuracies. This was not really surprising, however, for the coastline had been drawn originally from the deck of the British naval ship *Investigator*, when M'Clure had sailed along that coast in 1851. To the extent possible, Stefansson marked corrections on the chart, which included several deep embayments and a number of islands.

After they had progressed southward for a few days, all three men were forced to travel well inland from the coast to avoid encountering the many bays that cut deeply into the island's west coast. This deprived them of driftwood for fire-making, but they found ample amounts of dried willow and heather, which served almost as well. What proved more serious, however, was the absence of game, and their supply of dried meat quickly approached exhaustion. Fortunately, just when their situation looked serious, Stefansson sighted a small herd of caribou and shot four. This provided them with more than enough meat for several more days. To avoid unnecessary wastage, they again dug a hole, lined it with stones and placed the well-chilled leftover meat in the hole as an additional cache for later use, marking the locality with a monument.

Just north of Terror Island, which Stefansson found to be correctly positioned on the Admiralty Chart, they came upon a large bay and valley that was not shown on their map, even though the bay continued inland for many miles. The bay forced them to veer quite a distance inland in order to find a place where they could cross it. Stefansson later named this deep coastal indentation Storkerson Bay after his capable travelling companion.

On September 10 Stefansson reached the top of a hill from which he could examine the prominent, hook-shaped sandspit known as Cape Kellett, and some of the shore east of it. Now at last he would find out if the ships had reached the cape. Carefully he scanned the distant horizon through his field glasses, and his spirits sank. There was absolutely no sign of human activity. His instructions in April had called for a base camp to be established at Cape Kellett during the summer, but there was no camp and no ship. Discouraged by his discovery, he erected a prominent beacon and left a note within it instructing his companions to make camp on the nearby beach. Without a ship, he and his two companions would have to contemplate either wintering on Banks Island or heading for Victoria Island by dog-sled once the intervening ocean had frozen over, thence across to the mainland and on to the Mackenzie River—a frightfully long journey.

That evening the trio was pretty despondent, for they had set their hopes on finding both ship and camp at Cape Kellett. Unknown to them, however, two men were at that precise time just a few miles inland searching for the three of them. Their paths would soon cross.

Sighting of a Ship

The next morning, September 11, Stefansson set out for Cape Kellett on the chance he might find a beacon or message there. En route he came upon a small creek where he spied the fresh imprint of a man's boot in the adjoining mud. With renewed spirits, he signalled to his companions behind him and forged ahead. Half a mile farther, he came upon another fresh footprint, the pattern of which he recognized as being that of the footwear worn by members of the Canadian Arctic Expedition. He searched the elongate, curved cape again with his field glasses. Once again he was disappointed, for he could see no beacon on the sandspit, nor for a visible distance east of it. Undaunted, he continued onward. His diary records what happened next: "I got to the top of a hill from which I saw the tips of two masts. I could hardly believe my eyes—somehow it seemed unnatural to find a ship in Banks Island where it ought to be."[4]

Hastening shoreward in fear that the vessel was moving away, he reached a point where he could see the entire ship and stopped short. It was not sailing off as he had feared. Instead, it was sail-less, immobile and pulled up on the beach. And it was not the *North Star* that he had expected to see but the *Mary Sachs*! Her cargo was scattered on the beach nearby and several men were busily constructing a house.[5]

Stefansson quickly approached the men. From time to time one of them glanced his way but then went on with his work. The engineer James Crawford was the first to recognize him, but not until Stefansson was close enough to speak. For a moment Crawford stared at him in

---

[4] Stefansson 1918 (September 11, 1914).

[5] The house was located on the shore about eight miles east of Cape Kellett. Little evidence of it remains today (D. R. Gray, 2010, personal communication).

disbelief, then called out to Captain Peter Bernard that Stefansson had reappeared. The men hastily gathered around Stefansson, all questioning him at the same time. After a few moments, the captain insisted Stefansson have something to eat, and Crawford and Karl Thomsen went off to find Storkerson and Andreasen and to help them with their dogs and packs.

Over coffee, bread and butter, Stefansson told his story, then listened to Captain Bernard's account of bringing the *Mary Sachs* there from Herschel Island. Presently "Levi" Baur appeared, whom Stefansson had first met in 1906. A sailor on the *Belvedere* when Stefansson had started on his ice trip in March, Baur was now cook for the crew of the *Mary Sachs*. Crawford and Captain Bernard had been startled by the unexpected appearance of Stefansson, but Baur was wide-eyed in disbelief, for he had been certain that Stefansson could not possibly have reached Banks Island, nor survive if by some chance he had managed to reach it.

Catching up on the News

Stefansson then inquired about who had come on the *Mary Sachs* and was given all the names. On asking why Baur was the ship's cook, he learned from Captain Bernard

Figure 21. Stefansson at Cape Kellett, Banks Island, six months after heading north from Martin Point, Alaska, over the Beaufort Sea, September 13, 1914.

Photo: © Canadian Museum of Civilization, photo by George Hubert Wilkins, 50869

of the tragic suicide at Collinson Point in April of Andre Norem, the *Mary Sach's* former cook. Norem's death was just the first of several deaths of Expedition members Stefansson was soon to hear about.

Next he heard about the *Karluk's* fate. Captain Bernard told him of the sinking of that ship and the safe arrival of all of the men on the northeast coast of Wrangel Island (the captain had his facts wrong about this). Then he mentioned Captain Bartlett's lengthy journey to seek a rescue expedition, trekking with an Eskimo companion from Wrangel Island to the Siberian coast and then hundreds of miles along that coast to East Cape, and then travelling by ship to Nome, Alaska. The most recent news was that two U.S. revenue cutters and a Russian ice-breaker had been dispatched to rescue the people on Wrangel Island. It was not until the next summer that Stefansson learned the full story of the Wrangel Island rescue and about the eleven deaths among the *Karluk's* original twenty-five survivors.

Stefansson then mentioned that he had issued instructions in the spring for the *North Star* to be sent to Banks Island and asked why this had not been done. The captain told him that he did not know the reasons for the change in plans, but the *North Star* had gone east from Herschel Island with Dr. Anderson's party. Had the captain a report or letter of explanation from Dr. Anderson, Stefansson then asked? He had neither a report nor a letter, was Captain Bernard's prompt response. Stefansson was infuriated by Dr. Anderson's apparently deliberate defiance of his written "instructions" to send Wilkins and the *North Star* to Norway Island, because without it, and with the *Mary Sachs* beached, he could not carry out the exploration plans he had recently formulated.

Dr. Anderson's Quandary

Unbeknownst at that time to either Stefansson or Captain Bernard, Dr. Anderson had received official orders at Herschel Island in August 1914 from his employer, the Director of the Geological Survey of Canada—a major branch within the Department of Mines—not to risk sending any of his Southern Party men—*and especially not Chipman*, who was second-in-command of that party—with a ship to Banks Island to search for Stefansson. At the same time, he had received a second letter, this one from the Deputy Minister of the Department of the Naval Service, which included instructions to send an Expedition *under the command of Chipman* in search of the missing Stefansson party. Obviously the two government departments had not consulted each other before sending those letters.

The conflicting orders placed Dr. Anderson in an exceedingly difficult position, for a number of reasons: (1) As an employee of the Department of Mines, he was responsible to that department and not to the Department of the Naval Service. Stefansson, on the other hand, dealt with the Department of the Naval Service, the senior of the two government departments, which had made him leader of the Expedition, but he was not one of that department's paid employees (at his own insistence), which puts in question to what extent he was responsible to it. (2) Stefansson's written instructions to Dr. Anderson in the spring had called for the dispatch of both the *Mary Sachs* and the *North Star* to Banks Island for the use of Stefansson's new Northern Party, but no one at that stage had any idea where Stefansson and his two companions were, or even if they were still alive. (3) Stefansson's instructions had stipulated that Wilkins was to bring the *North Star* with supplies, crew and several Natives to Banks Island, but Wilkins was an employee of the British Gaumont Company and was only seconded to the Expedition for one year. Any extension of that time without the company's permission would be a breach of his contract. (4) Dr. Anderson had concerns over the reliability of Captain Peter Bernard to carry out Stefansson's instructions to establish a base camp with two years' supplies at Cape Kellett, especially if Stefansson failed to appear. (5) Dr. Anderson was under clear and definite instructions from the Director of the Geological Survey not to disrupt the intended program of his Southern Party to meet requests for the Northern Party. He also knew that if any one of the scientists under his direction took command of the *Mary Sachs* to enable Stefansson to undertake new exploration work, his Southern Party would not be able to complete the field work it was expected to carry out. There was an added factor, in that regard: Chipman was the only scientist who might have been able to take charge of such a relief operation, but he was not prepared to do so, because it was not part of the field assignment he had agreed to come north to carry out. And then, (6) the *North Star* was not large enough to carry all the personnel, equipment, dogs and two years' supply of provisions required by Stefansson's new party.

Stefansson, of course, had no way of knowing what was going on with the Southern Party nor of the new instructions from Ottawa. And Dr. Anderson had been too occupied with attending to preparations for getting his own Southern Party east to Coronation Gulf as quickly as possible that he had no time to write a report or letter to Stefansson. The conflicting instructions from Ottawa, poor communications between Northern and Southern Parties, and confusion at the small settlement of Herschel Island at this time all contributed to a significant increase in the misunderstandings and ill feelings between Stefansson and Dr. Anderson.

News about the *Karluk*

The *Herman* arrived at Herschel Island on August 9 with news of the *Karluk* and its survivors. Dr. Anderson learned from its Captain C.T. Pedersen (the same man who had negotiated the purchase of the *Karluk* for the Expedition a year earlier) that ships had been dispatched to Wrangel Island to rescue the *Karluk* survivors. As a consequence, Dr. Anderson concluded there was no longer any need for him to send a second ship to Banks Island to bring back the *Karluk* survivors, as per Stefansson's instructions. An extra vessel could be of considerable assistance to the Southern Party as it struggled to complete its intended activities in the Coronation Gulf region in two thirds of the originally allotted time. Naturally, Dr. Anderson had no knowledge that Stefansson was alive and safe, nor of the new exploration plans he had formulated over the summer. Nor was the information that Captain Pedersen imparted about the *Karluk* and its survivors available to either Stefansson or the government officials in Ottawa when they wrote their respective orders to Dr. Anderson in the spring of 1914.

It is difficult to comprehend in today's world of almost instant communications just how infrequent and uncertain the communications were between the government officials in Ottawa, the members of the Canadian Arctic Expedition at Herschel Island and with Stefansson himself in the summer of 1914. The instructions Dr. Anderson received from Ottawa early in August had been written early in the spring based upon news and information sent by Stefansson early in January, a seven- to eight-month period during which much had happened in the North. Dr. Anderson was forced to make several major decisions hastily at Herschel Island early in August 1914 concerning ships and men, based upon the information he possessed at the time. One of these decisions, which resulted in considerable criticism later from Stefansson, was to send only the *Mary Sachs* to Banks Island. It was reached, following discussions between Dr. Anderson and his scientific companions, based upon the facts that the *North Star* was too small to carry enough provisions and people to meet Stefansson's needs for two years, and no government man was available to take charge of the *Mary Sachs* to be sure it went to Banks Island.

At that point, Dr. Anderson asked Wilkins to assume charge of the *Mary Sachs* instead of the *North Star* and to take it to Banks Island, and also to take on the responsibility of finding the missing Stefansson men or what happened to them. By agreeing to undertake those tasks, Wilkins effectively brought about his immediate loss of employment with the Gaumont Company.

Thus freed of its need to proceed to Banks Island, the *North Star* became available for use by the members of the Southern Party, which cheered them immeasurably. As events unfolded, however (see Chapter 12), Stefansson prevented them from getting much use of it by sending Wilkins to the Southern Party's headquarters in the spring of 1915 with orders to bring the schooner to Banks Island.

Stefansson, meanwhile, was safe at the new camp near Cape Kellett on the southwest coast of Banks Island, critical of the changes made in the instructions he had written in April and making no allowance for other events necessitating those changes. Some months after learning that his instructions to Dr. Anderson had not been carried out, he initiated a series of complaints about the latter's actions in letters to the Deputy Minister of the Naval Service. Needless to say, while he was complaining over the apparent disobedience to his instructions by Dr. Anderson, Chipman, and one or two of the other scientists on the Southern Party, Stefansson seemed totally unconcerned that some of his own actions reflected his own disdain for governmental instructions, principles and procedures. He had little tolerance for any of his men who gave independent instructions that conflicted with his own. His expectations that the scientists on the Southern Party should respond dutifully and without discussion to his "orders," in spite of their commitments to other activities (and allegiance to a different government department), reflected an attitude on his part that created major hostilities between him and several of the scientists both while they were in the North and later.

When Stefansson's party reached Cape Kellett, Wilkins and Billy Natkusiak were inland looking for both caribou and signs of Stefansson. Less than a day after Stefansson left his last note for Storkerson and Andreasen by a beacon, Wilkins discovered the beacon and recognized Stefansson's handwriting on the note. He also suspected that it had been written very recently, so hastened back to his hunting camp to round up Natkusiak and return to the ship. Together they arrived early in the morning of September 13, to find Stefansson and his two companions sound asleep. After the three weary explorers awakened, Wilkins noticed immediately that none of them looked emaciated from their five-month ordeal, as he had imagined they would be. Indeed, they looked healthier and fatter than he remembered them!

Discussions between Stefansson and Wilkins

Stefansson and Wilkins conversed for much of that first day, the former asking about all that had happened during his absence. He especially wanted to know why his plans for the *North Star* had been changed, why that vessel went east with the Southern Party and why it took with it certain equipment he had wanted, especially a heavy sled and the Southern Party's depth-sounding wire. Wilkins answered all of the questions to the best of his knowledge.

According to Wilkins, Dr. Anderson had dispatched a well stocked ship to search for Stefansson, as instructed by the government officials in Ottawa. Wilkins had led the search party because he was the only scientist with knowledge of marine engines.[6] He had been

---

[6] Wilkins had studied both mechanical and electrical engineering in Australia in his teens, and in the spring of 1914 had completely overhauled the engine of the *North Star.*

ready to sail for Banks Island with the *North Star* early in August as Stefansson had instructed, but Dr. Anderson did not feel confident that Captain Bernard would follow Stefansson's four-months-old instructions, take the *Mary Sachs* to Banks Island and establish a base camp at Cape Kellett without someone like Chipman or Wilkins in command. As the *North Star* could not carry sufficient provisions and people for a two-year search, Dr. Anderson had asked Wilkins to take command of the *Mary Sachs*.

Wilkins would much rather have returned to London as his company had instructed him to do, but out of duty and obligation, had agreed to lead the search for Stefansson on the chance that the three men were still alive. On August 6 he wrote in his diary:

> I can but hope for little reputation in the photographic world by continuing with the expedition, but there are the three men on the ice who if not actually depending on some boat reaching Banks Island for their lives, would suffer great hardships and accomplish very little towards the success of the expedition after this summer; and it was agreed by each of the members of the scientific staff that I was the only logical member of the expedition to command the relief ship … It means breaking my contract with the Gaumont Co[mpany], and the only compensation I hope for is to be able to bring help to the missing ice party.[7]

Stefansson soon compensated him by writing to Deputy Minister Desbarats, asking him to put Wilkins on the payroll of the Department of the Naval Service at the salary of $1,500 per year, a level equal to that of Chipman (and three times that of Jenness, McKinlay or Beuchat). He also made Wilkins second-in-command of his Northern Party.

After Stefansson heard about all the changes that had been made to his spring plans at Herschel Island, he discussed his new plans. Wilkins had some doubts about them. "V.S. discussed several plans for the coming season, some of them I think too extensive to be carried out by the men we have and the resources available. However, we hope to accomplish a great deal and I trust we will not be disappointed."[8] Stefansson was more optimistic now and confident that valuable exploratory work could still be done. "Now, with most of our best men and resources gone, it had become a matter of individual prowess," he wrote. "We had to show that by adapting ourselves unaided to local conditions a few could do the work of many."[9]

Actually, Stefansson's activities thereafter focused almost entirely on exploration because, having lost his scientists and their equipment, he could no longer oversee studies in oceanography, marine biology, geology and geography. In addition, operating thereafter in unpopulated regions, he was forced to abandon most of his own ethnographic studies, especially those of the "Blond Eskimos." Indeed, it is fair to say that the loss of the *Karluk* helped bring about changes at this time in the careers of both Stefansson and Wilkins.

---

[7] Wilkins, 1916 (August 6, 1914).
[8] *Op. cit.* (September 13, 1914).
[9] Stefansson, 1921b, p. 278.

Plans for a New Base Camp

Stefansson's latest plans called for the establishment of a new base camp near Norway Island, from which exploration farther north could be undertaken. The *North Star* with its shallow draft could have moved their supplies from the base camp near Cape Kellett with little effort, for its prow had been shaped and constructed to ride up on thin ice and break it with its weight. The larger *Mary Sachs*, alas, was not designed to do that, and its two 30-HP gasoline engines lacked the strength to break through any but the thinnest ice. Additionally, one of its propellers had been damaged by ice en route to Banks Island.

With the *North Star* somewhere in the Coronation Gulf region, Wilkins suggested to Stefansson that they try to launch the *Mary Sachs* right away so that they could move the camp north from the Cape Kellett area. Stefansson dismissed this suggestion for some unexplained reason, and instead instructed Crawford to get the launch *Edna* ready to ferry the supplies north. Cape Kellett was ice-free at the moment, and from Cape Kellett north to Norway Island was also ice-free near shore, as Stefansson had observed on his way south. He was confident, therefore, that the launch *Edna* could serve their needs adequately by making several trips, although it was not suitable for use in ice. He was not aware, however, of the accident that had befallen the launch en route to Cape Kellett.

The *Alaska*'s engineer, Daniel Blue, had put the launch *Edna* into good condition at Herschel Island early in August after Chipman had found it mechanically unreliable when he tried to use it for his survey of the delta of the Mackenzie River. However, during Wilkins' trip to Cape Kellett with the *Mary Sachs*, the *Edna* partly capsized and filled with water near Baillie Islands while it was being towed behind the *Mary Sachs*. Somehow Wilkins' men managed to empty the water from it and hoist it on board the *Mary Sachs*, but its engine had been soaked with salt water and was in all probability damaged.

Crawford soon found that the engine of the *Edna* needed a considerable overhaul, and by the time he got it operating safely some ten days later, the opportunity to move the camp had passed. The weather had cooled and new ice was forming on the sea.

The new ice notwithstanding, Stefansson asked Wilkins to take Storkerson and his family with supplies and equipment up the coast a dozen or so miles. There they could establish a small winter camp, hunt caribou and trap foxes. Storkerson, happily re-united with his family at Cape Kellett, hoped to accumulate a good supply of fox furs before the winter, which he could sell as compensation for his loss of wages while the *Mary Sachs* was inactive.[10] Stefansson also thought Wilkins could move a load of supplies to Norway Island on the same trip. Meanwhile, he intended to finish writing about his ice trip and accompany Wilkins, but in typical fashion, changed his mind at the last minute and went hunting instead.

---

[10] Most of the ship's crew were laid off once the navigation season ended and given permission by Stefansson to hunt foxes and other animals for their furs, for which they hoped to receive reasonable compensation. Deputy Minister Desbarats of the Department of the Naval Service in Ottawa thought differently when he learned of Stefansson's arrangement, and notified him by letter in the spring of 1914 that all skins obtained by members of the Expedition belonged to the government. Wilkins carried that letter to Cape Kellett on the *Mary Sachs* that August, where Stefansson received it.

Wilkins, Crawford and the Storkerson family left Cape Kellett on September 21, but encountered repeated problems with the *Edna's* engine, and also with slush ice and a light headwind after they rounded Cape Kellett. A band of shore ice forced them to tie up to an ice floe that evening. Overnight the fresh sea-ice thickened, stopping them from any further progress. Wilkins, Storkerson and Crawford had to unload the launch, cache the surplus freight that had been destined for Norway Island, pull the launch onto the shore and walk back to Cape Kellett. They left the Storkerson family where they landed, about twenty-five miles north of the main camp.

A Successful Hunting Trip

During Wilkins' absence, Stefansson had taken stock of the state of the main camp's provisions and decided they needed to lay in a supply of fresh meat for the winter. He knew from previous experiences that caribou were at their fattest and tastiest in the early autumn. So when Wilkins and Crawford returned, Stefansson, Wilkins and Natkusiak went inland to hunt. Natkusiak had just returned from several days of unsuccessful hunting, but Stefansson was confident that they would find caribou.

Almost exactly a year had passed since Stefansson and Wilkins had left the *Karluk* to hunt caribou. On that occasion Wilkins was not given any opportunity to hunt with Stefansson. He was again being offered the opportunity to go hunting with Stefansson and looked forward to learning this time from the latter's experience. However, his hopes were short-lived, for once they had camped, Stefansson asked Wilkins to look after the camp and dogs while he and Natkusiak hunted.

Fog and blowing snow interfered with their hunting for several days, but finally Natkusiak killed fifteen caribou. They promptly moved their camp closer to the scene of the killing, and all three men tackled the task of gutting, skinning and cutting the meat of the dead caribou. When that was done, Stefansson went after other caribou, telling his companions to move the meat and skins from any caribou he killed to their camp. Wilkins had yet to hunt with Stefansson, but he did manage to shoot two foxes for a collection of fox skins that he was starting for the National Museum in Ottawa. Stefansson soon returned with the news that he had killed another twenty-three caribou, so the process of gutting, skinning, cutting up the meat, and moving skins and meat to their hunting camp took place all over again. It was a constant race to protect the dead animals from the ravages of the hungry wolves and foxes that were soon attracted by the scent of the kill, and to gut and skin the carcasses before the meat spoiled. Spoiled meat was unfit for human consumption, although it could sometimes be fed to the dogs.

By October 11, Stefansson and Natkusiak had killed forty-one caribou while Wilkins had not even had the opportunity to hunt his first one. Finally, his wish was fulfilled two days later, when he downed one of the caribou in a group from which Natkusiak killed an additional seven.

By mid-October the hours of daylight were noticeably diminished. Stefansson suddenly decided to return to the base camp near Cape Kellett and took Natkusiak and the dogs with him, leaving Wilkins alone to guard the supply of meat from nearly sixty caribou that they had killed. Natkusiak returned three days later with a note from Stefansson asking Wilkins to continue hunting with Natkusiak until November 5. Wilkins in turn asked Natkusiak to build a snowhouse so that they could be more comfortable than in their tent while they waited for Stefansson's return. Thereafter, they saw few caribou, but shot or trapped several foxes to add to Wilkins' collection.

Stefansson returned to the hunting camp early in November, bringing Thomsen and two sleds and dogs. Then began the task of moving the accumulated supply of caribou meat. Some of it was to go to Storkerson's camp on the coast for their use, the rest to the base camp. They undertook the move in stages over several days. Stefansson, Wilkins and Natkusiak finally reached the base camp in time for a special dinner in celebration of the American Thanksgiving. Their eight-week hunting trip had netted them enough meat for the men and dogs for many weeks.

A Handy Whale Carcass

By mid-November the sun had disappeared below the horizon for several weeks. Nevertheless, Stefansson decided to go east to De Salis Bay, where he expected to encounter a number of "Blond Eskimos" from Victoria Island. He had met them in Prince Albert Sound in 1911, at which time he understood them to say that they spent part of the winter at De Salis Bay. Accordingly, he and Natkusiak headed east, but Stefansson returned unexpectedly a few hours later with the news that a dead whale lay partly buried in sand and snow about ten miles down the coast. This could supply food for their dogs for the entire winter. Polar bears, wolves and foxes were already feasting on it, Stefansson added, so that Wilkins might get some good photographs, in spite of the lack of sunshine. Thomsen promptly went off to the whale carcass with a number of traps, and he and Natkusiak spent the night catching and skinning foxes. Stefansson then returned to the whale carcass, and for several days he and Thomsen ferried whale meat to the base camp to ensure that there was enough dog food for the winter.

Stefansson and Natkusiak finally set out again for De Salis Bay on December 21, immediately after eating a Christmas dinner held four days early to accommodate their travel plans. It was a very low-key event. They expected to meet the Victoria Island Inuit at De Salis Bay, and Stefansson hoped he could induce them to return with him to the base camp so that Wilkins could photograph them. Wilkins was left in charge of the base camp. Every day he spent many hours carefully skinning and preserving the fox skins he had accumulated, a good number of which had been collected for him by Captain Bernard and Thomsen. These two men were on wages during the winter, so were not permitted to keep the foxes they trapped or shot. Andreasen, Crawford and Storkerson also trapped foxes and accumulated their skins, but by arrangement with Stefansson, they expected to sell their furs in lieu of receiving wages during the winter season. By the time Wilkins had finished his messy, bloody and smelly skin-cleaning task, he had seventy-one fox skins and seven caribou skins, all carefully labelled as to age, sex, size, and date and place of killing, a valuable original collection from Banks Island. These specimens now form part of the collections of the Canadian Museum of Nature in Aylmer (Gatineau), Quebec.

Wilkins also undertook the first systematic tidal studies on Banks Island, taking readings near Cape Kellett every four hours for a week right after Christmas. He took these readings in the shelter of a small snowhouse observatory he had constructed over a hole in the ice one hundred feet off shore.

Stefansson and Natkusiak returned towards the end of January without seeing any Inuit, although they had gone around the southeast coast of Banks Island beyond De Salis Bay and even crossed over Prince of Wales Strait to Victoria Island in search of them. His efforts to bring the "Blond" Inuit to their Cape Kellett camp for study and filming had failed.

Plans for Second Ice Trip

Stefansson now talked with Wilkins of his next plan—a second ice trip over the Beaufort Sea. It called for his men to ferry large amounts of supplies from the camp near Cape Kellett 150 miles north along the west coast of Banks Island, then across M'Clure Strait to the southwest corner of Prince Patrick Island, where they would establish a new base camp. From there, Stefansson and a small party of men would head northwest over the Beaufort Sea in search of land, the location of the Continental Shelf and any other oceanographic information they could obtain.

Stefansson asked Wilkins to lead the advance party of men and supplies he wanted moved up the coast to the new base camp early in February. Eager to play a useful role in the exploration work, Wilkins agreed to undertake the tedious and challenging task of sledding the supplies north.

FIRST DISCOVERY OF LAND, 1915

12

Preparations for freighting supplies to the northwest corner of Banks Island and possibly farther on to Prince Patrick Island occupied the men at Cape Kellett for more than a week. Several of the dogs had recently succumbed to a strange sickness that sometimes proved fatal to dogs in the North, which resulted in there being only two good dog teams and one poor team available for the freighting. Stefansson wanted Wilkins to set off immediately, saying that he intended to follow within a few days. Wilkins had some misgivings, however, having already experienced Stefansson's tendency to change his plans, or to be considerably delayed. Nevertheless, Wilkins had finished preserving all the skins in his collection for the National Museum in Ottawa, and was ready and willing to head north with some supplies. This time he fully expected to have an active part in Stefansson's exploration party.

Freighting Supplies to Northwestern Banks Island

Wilkins, Thomsen and Natkusiak got underway on February 9, with two heavily loaded sleds and Stefansson's revised instructions to establish the new base camp near Cape Prince Alfred. They made slow progress owing to the heaviness of their loads, thick fog and temperatures well below 0° F, but succeeded in travelling the fifteen miles to Andreasen's trapping camp the first day and ten more miles to Storkerson's camp the following day. By February 16 they were temporarily camped between Bernard Island and the coast of Banks Island. There they took the time to pick up some of the caribou meat Stefansson had cached fifteen miles inland during the previous summer. Fog and cold continued to impede their progress, but on February 21 they reached their intended destination at Cape Prince Alfred, having travelled more than 175 miles in the twelve days since they left Cape Kellett.

Open water lay just west of the cape, preventing the start of any ice trip from that location. Wilkins therefore decided that he and his men would continue northeast along the coast until they reached the closest part of Banks Island to Prince Patrick Island. Pressure ridges close to shore forced them to keep on the seaward side of the ridges. On February 26 Wilkins' party reached Cape Wrottesley, and there he decided they would build a large snowhouse and leave some of their supplies, for that seemed to be the best locality from which to cross the eighty-two-mile-wide M'Clure Strait after Stefansson arrived. After completing these two tasks, they turned back to meet Stefansson. He had not reached Cape Prince Alfred when they arrived, so they continued on south.

On March 3, after feeding their last rations to their dogs, they headed for a snowhouse they had built on February 16 on the north bank of the Bernard River in the lee of Bernard Island. There they found Stefansson and Storkerson, newly arrived from Cape Kellett. All of Stefansson's dogs suffered from cuts inflicted by sharp, thin slivers of ice, and needed time to recuperate.

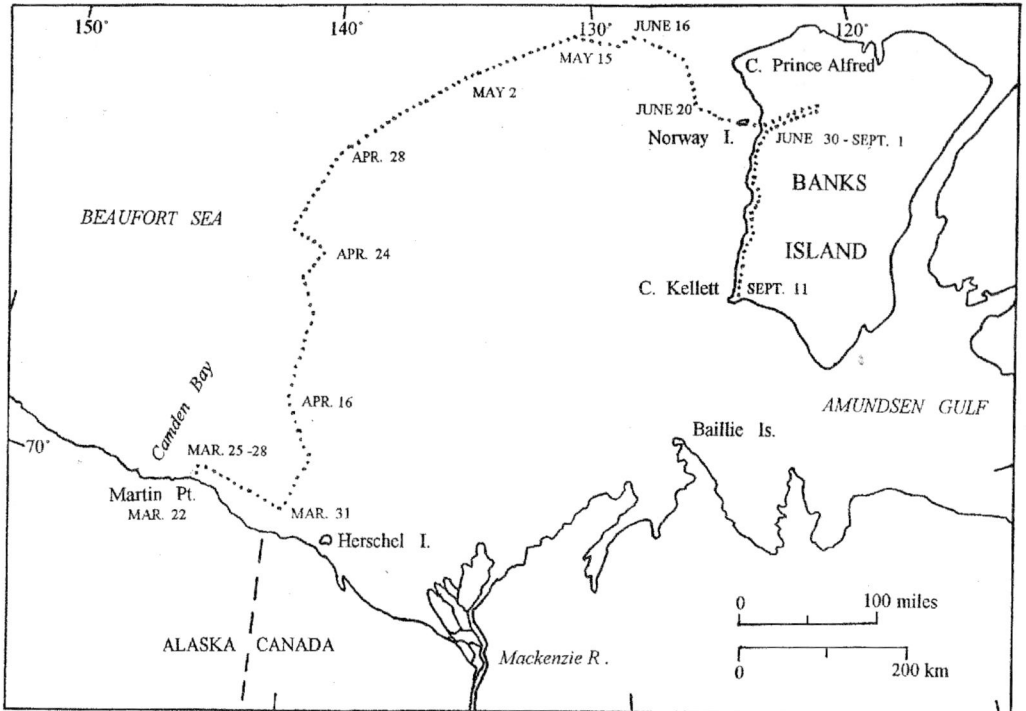

Map 7 Route of Stefansson's 1914 ice trip on the Beaufort Sea. (After a map in Stefansson's *The Friendly Arctic*, 1921, facing p. 140.)

Andreasen and Crawford arrived four days later from Cape Kellett with the third and final sledload. A quick inventory revealed that most of their coal–oil supply had leaked during the freighting, so Stefansson sent Thomsen and Storkerson back to Cape Kellett for another twenty gallons, which he deemed sufficient for his ice trip. Stefansson, Andreasen, Natkusiak and Crawford then headed north on March 11, ferrying two loads of supplies to their camp at Cape Prince Alfred.

Once again Wilkins was left behind in charge of both the camp and several of the dogs whose feet were still in poor condition. Storkerson and Thomsen returned from the base camp near Cape Kellett on March 25, reporting to Wilkins that they had been deathly ill for a few days en route after eating polar-bear liver. Three more dogs had died of the strange canine sickness at the Cape Kellett base camp, they reported, leaving Captain Bernard with only three working dogs. Additionally, the cook Levi Baur had been making "hooch" (an alcoholic liquid) from prune juice, which did not bode well. Storkerson and Thomsen rested for a day, then continued on to Cape Prince Alfred with the coal oil and other supplies. Wilkins, alone and without books to help pass the time, found the days tedious, the nights long. He hunted caribou on two days and succeeded in shooting two. These he had Andreasen and Crawford bring to his camp after they returned from Cape Prince Alfred. Then on April 1 they closed the snowhouse and spent two days moving the last of the supplies north.

Problems with the Men

Stefansson drew Wilkins into a serious discussion about employee problems soon after the latter arrived at Cape Prince Alfred. Crawford had refused to go on the ice trip if he was required to live on only seal meat or caribou meat, and also refused to accompany Stefansson if he decided instead to go to Coronation Gulf to get the *North Star*. Additionally, Thomsen and Storkerson refused to go on the ice trip with Stefansson if Crawford was sent back to Cape Kellett, for they were concerned about the safety of their wives and children at the cape if Crawford got together with Baur, because of the alcoholic prune juice that Baur made and Crawford's inclination to drink heavily. Wilkins knew of the drinking problem with Crawford, for the latter had been frightfully drunk on the *Mary Sachs* on the voyage from Herschel Island to Banks Island.[1] And then there was Natkusiak, who refused to go far out on the sea-ice. Stefansson was understanding about the latter, aware that with their spiritual beliefs no Inuit were willing to venture far out on the sea-ice.

Stefansson was therefore in a quandary. He wanted to lead the ice trip to be sure he was witness to any exciting discoveries, but he also wanted to go to Coronation Gulf to learn of the location and progress of the Southern Party, to take reports and mail there so that they could be sent south during the summer navigation season and to bring back the schooner *North Star*.

He finally decided that his best solution was to have Wilkins return to Cape Kellett with Crawford and Natkusiak and to continue from there to the Southern Party's base camp somewhere around Coronation Gulf. Wilkins would be able to photograph Copper Inuit at the Southern Party's main camp, would act with Stefansson's authority to claim the *North Star* and would then proceed to Herschel Island. There he would terminate Crawford's employment and leave him, then pick up supplies and mail for the Northern Party, and sail the *North Star* up the west coast of Banks Island and establish a base on Prince Patrick Island, assuming, of course, that the route north was not blocked with ice.

For the immediate present, however, Stefansson needed the assistance of all seven men to start his new ice trip. In the five weeks since Wilkins' initial arrival at Cape Prince Alfred, the open water then visible off shore had frozen solid, and ice travel was now possible.

Start of the Second Ice Trip

On April 5, the men and dogs started over the ice of M'Clure Strait northwest from Cape Prince Alfred, stopping to camp after progressing eight miles. At that point Stefansson had Wilkins photograph their snowhouse and sleds on the ice. Wilkins also took a series of pictures to show two men stretching a well-greased tarpaulin over an empty sled to make a "sledboat," and then pictures of them using this unusual device to transport themselves and several dogs across an open-water lead.[2]

Two days later Wilkins said farewell to the exploration party (Stefansson, Storkerson, Thomsen and Andreasen) and, with Crawford and Natkusiak, started for Cape Kellett. They had a difficult journey south, almost running out of food, but finally reached the base camp near

---

[1] Jenness, 2004, pp. 142–150.
[2] Several of the resulting photographs appear in Stefansson's *The Friendly Arctic* (1921b).

Figure 22 Storkerson and Natkusiak preparing a sledboat in order to cross an open lead northwest of Banks Island, M'Clure Strait, April 7, 1915.

Photo: © Canadian Museum of Civilization, photo by George Hubert Wilkins, 50885

Cape Kellett ten days later. Following a week's respite they commenced their five-hundred-mile journey to the Coronation Gulf region in search of the Southern Party and the *North Star*. Before telling of Wilkins' assortment of experiences on that journey (see Chapter 20), however, it seems more fitting to return to Stefansson's important exploration activities.

Over the Ice

After the departure of Wilkins and his two companions on March 7, Stefansson mustered his three Norwegian companions—Thomsen, Storkerson and Andreasen, one more man than he had on his ice trip the previous year—and continued his journey seaward with the dogs and two heavily laden sleds. Because of the lateness of their start, they encountered a great deal of open water, necessitating delays to assemble and disassemble their two sledboats. Periodically they took depth soundings at a lead, initially obtaining readings of 600 to 1,200 feet. After a time, however, their 4,500-foot sounding wire no longer reached the bottom, and they knew that they had passed the Continental Slope and were over deeper ocean.

The four men deliberately travelled northwesterly, but the drift of the ice was to the southwest, partly counteracting their efforts and seriously retarding their progress. Thus, by the middle of May they had only reached 76° 40' N latitude, which was the same latitude as the southern part of Prince Patrick Island, although they were about one hundred miles west of it. At that point Stefansson decided to return to shore to complete the mapping of the west coast of that island.

In the two months they had just travelled over the sea-ice, they had accomplished several things: they had obtained an assortment of useful depth soundings, found no land and ascertained a strong southwesterly drift of the sea-ice off the western end of M'Clure Strait. This drift was caused by a current that continued south along the west coast of Banks Island, then turned west along the north coast of Alaska, flowing between two and fifteen miles per hour.

Their journey back to land brought them to the southwest corner of Prince Patrick Island near Land's End, which they reached on June 4. By then they had used up almost all of their fuel and dog food, and were faced with two choices: to continue exploring, hoping to find these necessary commodities, or to give up and head south for Cape Kellett. Stefansson had no intention of giving up, for he knew with reasonable certainty—as the British explorer Lieutenant G.F. Mecham had not known sixty-five years earlier when he originally explored Prince Patrick Island—that there were likely to be seals in leads off shore even if he failed to find caribou and fuel on the island. The seal meat would nourish both men and dogs, and the seal oil would provide them with cooking fuel. Parts of Prince Patrick Island still remained unmapped since Mecham's time, and Stefansson was determined to complete that mapping on this ice trip. To his great relief, his three companions agreed to go on.

Mapping Prince Patrick Island

Stefansson's examination of the west coast of Prince Patrick Island now revealed numerous errors in the outline of the shoreline on his old British Admiralty Chart, as well as the existence of several small islands, which had not been mapped in 1853 by either Mecham or Captain Francis Leopold M'Clintock. However, his efforts to correct and complete the mapping of the island's west coast proved highly difficult because of two unexpected factors: snow and ice covered the gentle seaward slope of the land, obscuring the actual shoreline, and foggy weather obscured everything else.

Stefansson also spent several days searching the western part of Prince Patrick Island without seeing a single caribou or any other mammals, for their supply of food was becoming alarmingly low. At the end of that time, the need for dog food had become desperate, so he led his three companions several miles off shore to the edge of the land-fast ice. There, to their great joy, they found numerous seals swimming and splashing in the open water. They needed several seals a week to feed their dogs and themselves. Stefansson had learned from experience that if he waited for the seals to poke their heads above water within a range of fifty feet, then shot them in the head, the seals would usually float. To retrieve the dead seal they would then cast an Inuit-type *manak*—a series of hooks on a baseball-sized piece of wood, all attached to a long line—beyond the dead seal and pull it towards them, hooking the seal and enabling them to pull it to the water's edge where they could retrieve it. In this manner they shot and retrieved several seals in short order, thereby obtaining sufficient fuel and food for several days.

Working slowly northward along the west coast of Prince Patrick Island, Stefansson and his companions reached Cape Leopold M'Clintock at the island's northernmost part on June 15. The Admiralty Chart showed two tiny islands (islets) immediately northwest of Cape Leopold M'Clintock. He and his men proceeded to one of them to make camp temporarily,

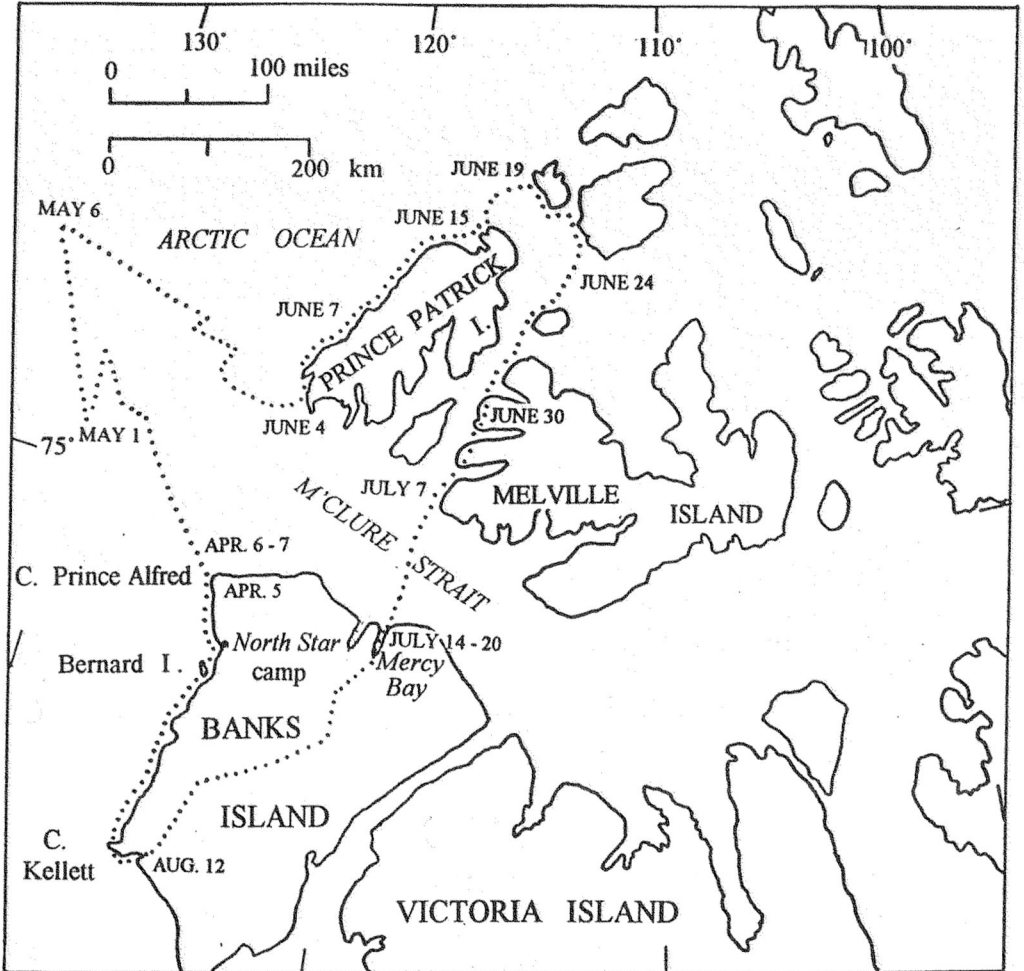

Map 8 Route of Stefansson's 1915 ice trip on the Beaufort Sea. (After a map in Stefansson's *The Friendly Arctic,* facing p. 292.)

expecting to find seals in the area. In this they were not disappointed and soon had enough seal meat for several meals. Stefansson then noticed a small monument-like object on a point on Prince Patrick Island and headed back to examine it. Within the piled-up stones he discovered a small papier-maché cylinder, roughly the size of a shotgun cartridge, which he brought back to his camp. Within the cylinder lay a piece of paper left by Captain M'Clintock, by coincidence on exactly the same day of the year, June 15, but sixty-two years earlier, in 1853. M'Clintock had written on the paper that he had examined about 150 miles of shore to the southeast and was about to sled back to the south side of Melville Island where his ship, the HMS *Intrepid,* lay. His subsequent 1,400-mile sled journey without dogs (the British Navy considered man-power superior to dog-power for pulling sleds)[3] from Melville Island to Prince

---

[3] This belief may be why the two Scotsmen, Dr. Mackay and Murray, refused to take any dogs when they left the *Karluk,* heading for Siberia (Bartlett and Hale, 1916, p. 123; Niven, 2000, p. 147).

Patrick Island and back to his ship was certainly one of the greatest of all Arctic journeys. In discovering M'Clintock's note, Stefansson was satisfied that he had completed the mapping of the northwestern part of Prince Patrick Island left unfinished by Mecham and had connected with the farthest north reached by M'Clintock along the island's eastern coast. It was an important locality to obtain latitude and longitude readings.

Two days later Stefansson instructed his men to head straight north into a region in which his Admiralty Chart showed no land. Meanwhile, he returned to Cape Leopold M'Clintock, rebuilt the cairn and placed an account of his own journey in it for someone else to discover. He then followed after the other three, striking north. Before catching up to them, he checked some features M'Clintock had marked on his map as "Polynia Islands" and decided that they had simply been shore leads. Three reefs marked on his map proved to be reefs, all right, but a feature M'Clintock had marked as "Ireland's Eye" was nowhere to be seen.

Discovery of New Land

Stefansson travelled about fifteen miles, slowly following his three companions northward over the ice, when he suddenly saw them stop and set up their tent. For a few minutes he watched them through his field glasses. During that period, Storkerson walked a little ahead of the other two, climbed a hummock and examined the surroundings with his field glasses, then waved to his two companions. It could mean only one thing, thought Stefansson. Storkerson had found land. Storkerson had indeed found land, although their chart showed none north of Prince Patrick Island. Stefansson hurried towards their tent. A short time later they all stood on the hummock and looked at the land that Storkerson had seen. Satisfied that it really was there, they retired to their tent and celebrated their discovery with a "feast" of biscuit crumbs mixed with malted milk. It was not what one could call exciting fare, but it was different from seal meat.[4] Then, because it was already early morning, they sought to get some sleep before exploring the new land.

That afternoon (June 18, 1915), they headed east-northeast for the land they had seen. After travelling about a mile and a half they spotted a seal hole and took a depth reading—414 feet. Stefansson recorded that there was a strong current at that locality, running a little west of north. Seven and one-half miles from Prince Patrick Island they came to a small low gravel island, one of several they noticed between the temporary campsite they had left and the new land. Another five miles brought them to the shore of the new land. The two sleds landed east of where Stefansson stepped ashore, for he had noticed a prominent hill and was heading for it in order to get a good view over the surrounding countryside. By the time he reached its summit, however, he was partly enveloped in fog and could make out very little.

Visibility the next day proved no better, which prevented Stefansson from obtaining good solar readings for latitude and longitude. He had, however, been successful in getting

---

Stefansson, however, stated in an appendix in *The Friendly Arctic* entitled "The story of the *Karluk*," which he described as a condensation of a lengthy document prepared by John Hadley, that Bartlett flatly refused to provide Dr. Mackay's party of men with any dogs, telling him at the time, "Not one dog; if you go off and leave us you play dog yourself" (Stefansson, 1921b, p. 710).

[4] Stefansson, 1964, p. 192.

good latitude and longitude readings at Cape Leopold M'Clintock, so felt that his compass bearings would allow him to plot the location of the land reasonably accurately upon a map.[5]

While on the new land Stefansson took note of the nature of its terrain, plant life, and its animal and bird life. Snow still lay on the coastal beaches, but grasses, mosses and lichens were plentiful elsewhere. Caribou tracks were plentiful, too, and he saw an ivory gull attempting to catch a lemming before his men scared off the gull. Snow buntings and larkspurs were also present.

Stefansson now became concerned over the lateness of the season and his need to reach Banks Island in time to make contact with one or more captains of whaling ships operating along that island's southern coast. Having discovered new land, he was determined to get the news to the outside world as quickly as possible, which he could only do by giving letters to a whaling-ship captain to mail. He therefore decided to turn eastward on the following day rather than to continue mapping northward along the western coast of the land they had just discovered.

Before doing so, however, he had his men heap some gravel into a three-foot mound and then placed a note in it stating that he had "hoisted the flag of the Empire" and had "taken possession of the land in the name of His Majesty King George V on behalf of the Dominion of Canada."[6] He also stated that Storkerson had sighted the land early in the morning of June 18 from a position fourteen miles due west, and that Ole Andreasen had been the first one to walk on it early on June 19. His note ended with his comment that all were well and that from this locality he and his companions would proceed east along the coast for some distance, depending upon the extent of the land, then head south to Melville Island and the Expedition's headquarters near Cape Kellett. They would continue to travel during the night because the temperature was cooler than during the daytime and the travel conditions were accordingly better.

On June 20 Stefansson and his three companions started southeastward along the coast of the new land. Stefansson followed the shore to record its shape, while his men and the two sleds maintained a straight course from point to point. About noon the men reached what Stefansson thought was a deep and wide bay, for they could not see its northern end. Here they camped to permit Stefansson to obtain solar readings for latitude and longitude. He then walked inland and climbed to a level of about 175 feet according to an aneroid barometer he carried, from where he could see Prince Patrick Island to the southwest and two smaller islands east of it.

The next day the three assistants crossed the sixteen-mile-wide "bay" with the two sleds and camped, while Stefansson proceeded deeper into the "bay" heading towards a prominent, snow-capped hill or mountain well in the distance on the "bay's" east side. As the weather was at last perfectly clear and because it happened to be the longest day of the year, he resolved to climb the mountain in order to obtain a superior view of the surroundings. Reaching it turned out to be a difficult twenty-mile walk, for the warm sun had softened the snow,

---

[5] The reliability of some of Stefansson's data was later questioned in an exchange in the *Geographical Journal* between geographers James White (1924) and F.A. McDiarmid (1923).

[6] Stefansson, 1921b, p. 330.

rendering the turf wet and muddy. Nevertheless, he finally reached its summit, about 1,500 feet above sea level. It proved to be the most conspicuous landmark he found on any of the new lands he discovered during his several years in the Arctic. He named it "Leffingwell Crags" in honour of the American geologist Ernest de Koven Leffingwell, the co-leader of the first Arctic expedition with which he had been involved.

From this mountain-top Stefansson gazed in all directions. Since he had come from the west, he wasted little time examining the land in that direction. Fog obscured his view to the north and northwest. To the northeast and east, however, lay high hills and low mountains for as far as he could see. To the south he could barely make out the camp where his companions now slept. He took a compass bearing on the camp to guide his direction there, then shortly before noon started for the coast.

Hours later when his men awoke and found Stefansson still absent—he had by then been separated from them for twenty-four hours—they became alarmed and decided to hitch up a dog-team to go in search of him. Before they had set forth, however, they were much relieved to see him approaching the camp. His journey from the mountain back to the camp took him seven-and-a-half hours.

Stefansson had now been without sleep for more than twenty-four hours, so needed to rest. Before he lay down, however, he instructed Storkerson and Thomsen to take an unloaded sled and map the coast to the southeast as far as they could go and return, leaving themselves time to have a short sleep before they all started south for Melville Island that same evening. Storkerson found that the land continued to the southeast for more than thirty miles, which indicated to Stefansson that they had discovered a sizeable body of land.

After awakening, Stefansson prepared a list of reasons justifying his decision to discontinue so hastily his exploration of this new land and return to Cape Kellett. Evidently he anticipated being criticized for his decision and wished to be well prepared with answers. One of his reasons was his desire to have time to spend a week or two excavating an old Aboriginal village, evidence of which he had noticed at Cape Kellett. This, however, he failed to find the time to do.

British-born zoologist-explorer Tom Manning subsequently examined the site briefly while conducting geographical, biological and oceanographic studies around Banks Island for the Defence Research Board in Ottawa between 1951 and 1953. Archaeologist Robert McGhee later led a small party from the National Museum of Canada to the site in the 1960s. From his excavations McGhee determined that the early dwellers of the site had been whale-hunting people, now referred to as the Thule people. These people had spread across the Arctic from Alaska to Greenland between 1000 and 1500 AD.[7]

Return to Banks Island

On the evening of June 22, Stefansson and his men started their long journey southward. Their exploration on this second ice trip had taken them almost to 78° N latitude. From the new land they headed south for Eight Bears Island, each of them in turn leading the way ahead of the dogs. And each man, when it was his turn to lead, found it helpful to wear the one pair

---

[7] McGhee, 1978, pp. 83–85.

of skis they had brought with them, for their snowshoes sank into the soft wet snow in the warmer weather, impeding their progress. They were also grateful for their specially made sleds with their "toboggan bottoms," which prevented the runners from sinking more deeply into the snow. Stefansson later commented, "I have no doubt that but for the toboggan bottoms our progress would have been even slower than McClintock's."[8] The latter had passed along the same stretch of sea-ice in 1853 while searching for the lost Sir John Franklin Expedition. M'Clintock and his men were on foot, however, and sank deeply into the soft snow with every step, whereas Stefansson and his men were on snowshoes or the one pair of skis.

Periodically Stefansson's men found a seal's breathing hole and stopped to use it to ascertain the depth of the ocean beneath them. They also occasionally shot a seal for food and fuel. Seals were in sufficient numbers that they could count upon killing one every few days, which allowed them to travel without burdening their sleds with unneeded seal meat. A wet snowstorm greeted them when they reached Eight Bears Island, thoroughly soaking their fur garments and footwear, and emphasizing the urgency of their reaching Banks Island quickly before it became impossible to cross the ice on M'Clure Strait.

The following day they headed for Emerald Island after Stefansson had obtained its direction visually from a high point on Eight Bears Island. The men drove their sleds past the island while Stefansson walked over it. The small amount of grass and moss on the island did not seem, in Stefansson's view, to justify Captain M'Clintock's name for it. They then proceeded southwards along the west coast of Melville Island, which they found to be marked by deep gorges, sheer cliffs and bold headlands, vastly different terrain from the rolling prairie-like grasslands of much of the west coast of Banks Island. Snow still covered the rocky terrain along the west coast of Melville Island, with bare patches only here and there.

On June 30, on the southwest side of Melville Island, Stefansson encountered two bull muskoxen, the first of these shaggy creatures he had seen. He shot them both, for he had reached an area where seals were difficult to obtain and was in need of replenishing his food supply.

With the arrival of July, Stefansson grew increasingly worried about reaching Cape Kellett in time to contact any of the whaling ships operating in that region. He reckoned that he needed to arrive there no later than August 10, and commenced writing reports for the government in the back of his diary and letters telling of his activities, which he hoped to send out with the ships.

Expecting his party to cover the ninety miles across M'Clure Strait to Banks Island in four days, he and his men started across the strait on July 8 with only that number of days' supply of food. The ice they encountered was extremely cut up, however, with much surface water, forcing the dogs to swim in places and the men to wade. On July 13, they ate their last food—a piece of sealskin with some attached blubber—when about seven miles from Banks Island. Eight hours later they reached a place within six hundred feet of Banks Island, but were stopped by a lead of similar width and forced to assemble their sledboat. After crossing the lead by this means, they landed on the east side of Mercy Bay, and Stefansson promptly decided to replenish their food stock. As a result, he and Thomsen went off in search of caribou, leaving

---

[8] Stefansson, 1921b, p. 341.

the other two to prepare the camp. Six hours later, they returned with two caribou and a hare, only to learn that in their absence Storkerson had killed a seal. They then feasted for the first time in some while.

Stefansson later observed in *The Friendly Arctic*, "Mercy Bay is one of the historic places of the North."[9] Robert M'Clure and his crew from the HMS *Investigator* had been icebound there from 1851 to 1853 during their search for Sir John Franklin's Expedition. They had then been rescued by a British Navy search party in the spring of 1853 and led across the ice in Viscount Melville Sound to Dealy Island, where Captain Kellett's ship HMS *Intrepid* had wintered, and from there they ultimately reached England safely. Of the abandoned *Investigator* Stefansson found no sign, apart from an ice anchor and a grappling hook, both well rusted. Likewise there was little remaining of the monument erected by M'Clure's men, or the food depot he had left on the west side of the bay. There was, however, a small pile of soft coal, which was largely overgrown with grass and many pieces of wood, chiefly barrel staves, the oak wood of which had proven too hard for the Copper Inuit years earlier to carve and use.[10]

Stefansson and his men remained at Mercy Bay for five days, cutting up their worn and waterproofed sled tarp in order to make back-packs for their dogs to carry supplies overland to Cape Kellett. They also cached their two sleds and some of their equipment to reduce their own carrying loads to manageable levels.

Hiking across Banks Island

On July 20 the four men and their thirteen dogs started their more than two-hundred-mile trek across Banks Island. Within eight miles they came upon a large, previously unknown, north-flowing river, which Stefansson later named the Thomsen River after one of his companions. Descending from the upland into the valley, they followed the river's winding course upstream seeking a shallow place where it could be forded. This they did not find, so they returned to the higher ground. Stefansson then went on ahead of the others, serving as an advance guide, examining the countryside and hunting as he hiked along. On the eighth day inland, he came to a place in the river where it turned sharply to the east and there, to his great relief, they were all able to cross to its western side.

Travelling during the cooler night-time hours, they slowly progressed across the island, covering about ten miles each day. Along the way, Stefansson occasionally noticed rings of stones that indicated the former locations of Native skin tents, the stones having held down the bases of the tents. When the Natives took their tents elsewhere, the rings of stone—tent rings—remained. Bones of muskoxen and geese lay scattered about the tent rings, indicating the kind of food the Native people had eaten. Here and there he also came upon large piles

---

[9] *Op. cit.*, p. 359.

[10] The remains of the HMS *Investigator* were located in July 2010, in twelve metres of frigid Arctic water approximately 150 metres off the west shore of Mercy Bay. Scientists from Parks Canada, using side-scanning sonar from an inflatable Zodiac boat, made the discovery shortly after the ice moved out of the bay. The discovery was promptly reported on the Internet as well as by the newspapers.

of bleached muskoxen bones, like the piles he had seen near the island's west coast the previous summer, indicating that a mass killing had occurred there sometime in the past.

Stefansson had fully intended to use his sextant regularly to establish the latitudes and longitudes of his locations as he crossed the island, which would provide him with the data he needed later to plot the route he had followed. In order to do this, however, he had to take his latitude readings exactly at noon, a time of day when he was likely to be asleep after walking all night. Unfortunately, he had inadvertently left his alarm clock with the sleds at Mercy Bay, which meant that he either had to remain awake until noon, or hope to awaken in time to take the midday reading on the sun. He chose to depend upon awakening in time, for he was very much in need of sleep each morning when they made camp. He took time sights for longitude around seven in the morning before going to sleep and the midday reading for latitude whenever he managed to waken on time.

On August 7, when still thirty to forty miles northeast of the base camp near Cape Kellett, Stefansson's party came upon a family of four Copper Inuit from Minto Inlet, Victoria Island. They consisted of the father, Kullak,[11] his wife, Neriyok, their daughter, Titalik, age about ten and facially tattooed, and their son, Herona, age about six. All were camping in a skin tent and catching molting geese for food. They told Stefansson that they had met Wilkins, Crawford and Natkusiak on the ice in Prince of Wales Strait in the spring (when those three men were heading southeast in search of the headquarters of the Southern Party), and from them had learned about the camp near Cape Kellett. Kullak unexpectedly presented Stefansson with a pair of white sealskin slippers, asking in return that Stefansson make sure that his wife, who was pregnant, would produce a son and both would be healthy. Stefansson accepted the gift reluctantly, for his doing so presented him with quite a predicament. If the child was a girl, both parents would be angry with him; if the child or mother or both died, he would be blamed for murdering them, and Kullak might seek to kill him in retaliation. But if he refused the gift, Kullak would take offence and blame him for whatever went wrong.[12]

From Kullak, Stefansson now learned that the people at the base camp were short of fresh meat and that Kullak had herded a flock of molting geese to the camp to help them. Hearing of the plight of the men in the base camp, Stefansson went after some caribou and killed four, instructing his men to skin them, then have the dogs carry the fresh meat to the base camp near Cape Kellett, leaving their equipment behind for later retrieval. He then set off alone for the base camp, arriving the next day, August 9.

---

[11] Kullak is Stefansson's spelling, which I have used because of his great knowledge of the Native language and sounds. Wilkins spelled it Kudlak (Wilkins, 1916 [May 9, 1916]; Jenness, 2004, p. 392, note 7).

[12] A few months later Kullak's wife died shortly after giving birth to a daughter. The daughter survived and was adopted by another Inuit woman. The following spring Wilkins visited a snowhouse village at the mouth of Minto Inlet to photograph the Copper Inuit in their winter habitat, and came upon Kullak among the villagers. Kullak was by then a very angry man, eager to get even with Stefansson for "murdering" his wife. Kullak might have taken his revenge on Wilkins instead, if the latter had not acted on the advice of some other members of the village and departed before matters reached a crisis (Jenness, 2004, pp. 322–323).

Stefansson had now safely completed his second ice trip, had discovered new land, the extent of which he could not yet say with certainty, had taken numerous soundings, had made geographic corrections to the coastlines shown on the British Admiralty Chart and could happily report no illness, accident, or loss of men or dogs. He had, however, left his two sleds and assorted equipment at Mercy Bay, so that he and his three companions could walk across Banks Island from north to south in time to reach Cape Kellett before August 10. They had only twenty days to make the trip and make it they did, on August 9—the first white men to do so. They had been absent for about 160 days, and had covered a total distance during that time of about two thousand miles.[13] Once back at the base camp, Stefansson still had a little time to write his reports and letters before any whaling ships chanced to come into view and spot them.

Gale winds blew the following two days. During the afternoon of the second day, August 11, a schooner unexpectedly came in sight from the southeast and anchored several miles west of the camp in the shelter of the sandspit at Cape Kellett. Stefansson hastened along the shore to meet it. Although he was disappointed that it was not the *North Star*, he readily recognized the ship as Captain Louis Lane's *Polar Bear*. Captain Lane had been cruising for whales and the gale winds had driven him to the cape in search of shelter. As Stefansson later reported in *The Friendly Arctic*, several men, including Captain Lane, soon got into a whale-boat and headed for the shore to meet him. Suddenly someone in the whale-boat recognized Stefansson, which resulted in cries of amazement that he was alive and well. As soon as the boat touched the beach Captain Lane jumped out and raced up to shake Stefansson's hand. Others quickly followed, giving Stefansson the most enthusiastic welcome of his life. Captain Lane explained that Stefansson was supposed to be dead. Then they invited him back to their ship for a hearty meal.[14]

---

[13] Stefansson, 1921a, p. 298.

[14] *Ibid.* This may be one of Stefansson's fantasized interpretations, so criticized by Dr. Anderson and McKinlay when he expressed them in *The Friendly Arctic* (1921b).

ENTER THE *POLAR BEAR*, 1915 13

At Cape Kellett, Stefansson needed a ship to undertake the exploration he still hoped to accomplish, which is why he had sent Wilkins to Coronation Gulf in April to bring back the *North Star*. However, it was already early August and there was no sign of either Wilkins or the *North Star*. Then the schooner *Polar Bear* unexpectedly sailed into view at Cape Kellett, offering Stefansson new opportunities and new ideas. He would discuss those ideas with the ship's captain, his long-time friend Louis Lane, after going on board, but not before he had a chance to catch up on the latest news of the *Karluk* and other matters.

Captain Lane first told Stefansson of the rescue of the twelve survivors of the *Karluk* from Wrangel Island the previous September. Then he mentioned that there was no trace of the eight men who had disappeared attempting to reach Wrangel Island after the *Karluk* sank. More startling to Stefansson, however, was the news that Europe had been at war for a year, a world-wide war involving the British and French Empires against the German-Austro-Hungarian Empires. Stefansson was greatly relieved to hear that the United States was not officially involved in the war, probably because that meant his activities in the Arctic should still attract interest in the newspapers in that country. Personal publicity was vitally important to him, as he confessed some years later in *The Friendly Arctic*, where he mentioned with evident satisfaction how stories (even if untrue) expressing his dramatic reaction to the news of the war were published in every American newspaper and also in many European papers. Before heading for the Arctic, he had employed a newspaper-clipping service to supply him with information of that kind.[1] It charged him five cents for each newspaper clipping that mentioned his name.

Captain Lane then told Stefansson that he had heard nothing of the *North Star*, but there had been an unusually early spring along the Arctic coast, with open water around Baillie Islands for the past month, so he thought that the *North Star* should have put in an appearance. As it happened, Captain Lane was familiar with ice conditions in Amundsen Gulf because he had hunted whales there, but was not familiar with the conditions farther east in Dolphin and Union Strait. He probably did not know, therefore, that the eastward-drifting ice from Amundsen Gulf could delay ship navigation at Bernard Harbour for weeks after navigation was possible around Baillie Islands.

Captain Lane then mentioned that the Hudson's Bay supply vessel, *Ruby*, was bringing in supplies for both Northern and Southern parties, and that the Southern Party's *Alaska* lacked an engineer, because its engineer, Daniel Blue, had died of scurvy and pneumonia during the winter.

---

[1] Stefansson, 1921b, p. 377.

An Expensive Arrangement

The news of the *Alaska's* trouble promptly raised the question in Stefansson's mind of how supplies would reach both the Northern and Southern Parties from Herschel Island. On an impulse he asked Captain Lane if he could charter the *Polar Bear* to go to Herschel Island and attend to his urgent supply problems there.

In the past Captain Lane had made his living by hunting and killing bowhead whales. These whales had yielded large amounts of whale oil, which commanded a good price in the United States, where it was used for lighting, soaps, engine lubricants and other purposes. In addition, baleen (also known as whalebone) from the whale's head was used for umbrellas, brushes, the stays in women's corsets and a variety of other commercial objects. Cheap petroleum-based products replaced the need for whale oil after 1906, and whaling ships pretty well vanished thereafter from the Beaufort Sea and Amundsen Gulf. To remain in business, Captain Lane had been forced to become a trader. In some years, however, after completing his trading activities at Herschel Island, he would spend a few days hunting bowhead whales in Amundsen Gulf between Herschel Island, Baillie Islands and Banks Island. That was what he had been doing in 1915, which is how he happened to appear at Cape Kellett on the afternoon of August 11, shortly after the overland arrival there of Stefansson and his three companions.

Captain Lane knew that Stefansson possessed a thick government cheque-book, so when Stefansson asked to charter his schooner, he responded by saying it was available for $1,000 per day. Making money in that manner was certainly easier and less messy than killing bowhead whales. His price was a truly outrageous amount in 1915, but Stefansson nevertheless agreed to pay it, although he refrained from mentioning the amount later when he wrote about his encounter with Captain Lane in *The Friendly Arctic*.

With Stefansson and his three companions (Storkerson, Andreasen and Thomsen) safely on board, the *Polar Bear* crossed Amundsen Gulf to the tiny settlement of Baillie Islands, where Stefansson learned that the *Alaska* had already gone west to Herschel Island. There was no news of Wilkins or Dr. Anderson, so he left a note for Wilkins to proceed directly to Cape Kellett with the *North Star*, await his arrival there for a week, then head north with supplies for a new base camp. (Wilkins and the *North Star* reached Baillie Islands two days after Stefansson left that note.) The *Polar Bear* then sailed west to Herschel Island, where it anchored near the *Alaska*, the Hudson's Bay Company supply ship *Ruby* and four smaller schooners.

At Herschel Island Stefansson reported that he was welcomed as a man returned from the dead, for no news of him had reached that settlement since he commenced his ice trip north of Martin Point months earlier. A short time later he learned that he would be unable to obtain the supplies he sought from the *Ruby* for several days. Each day he retained the use of the *Polar Bear* increased his charter bill by $1,000, so he arranged to purchase the *Polar Bear* from Captain Lane for $20,000, plus an additional $10,000 for its cargo of trade goods. The transaction also required that Stefansson make a ship available for Captain Lane, so that the Captain, the many furs he had accumulated through trading during the summer and some of his crew could return south. Stefansson therefore purchased the small schooner, *Gladiator*, from its owner, Fritz Wolki, for an additional $6,000.

With the use of the *Polar Bear*, Stefansson envisioned that he could now expand his dreams for exploration by establishing a base camp on Melville Island, from where he could more easily reach the still unexplored Far North. Then he envisioned an even more startling scheme. Once his northern explorations were completed, he would sail the *Polar Bear* east to the Atlantic Ocean, and thus be able to claim that he was the first person to sail through the Northwest Passage from west to east. His craving for public acclaim seemed insatiable.

While at Herschel Island, Stefansson kept exceedingly busy. He oversaw the loading of his Northern Party supplies onto the *Polar Bear*, purchased dogs for his Northern Party, wrote reports for the Canadian government, hired engineer J.E. Hoff from the *Ruby* for the *Alaska*, and arranged with Captain Alexander Allan, the owner of the newly arrived schooner, *El Sueno*, to take the Southern Party supplies from the *Ruby* to that Party's base camp, thereby sparing the *Polar Bear* from that task, for the *Alaska* had already gone east to Bernard Harbour. He also sought and hired four Iñupiat men as hunters and packers (Alingnak, Emiu, Illun and Pikalu) and eight Iñupiat women as seamstresses (Guninana, Ikiuna, Ikugana, Kutok, Violet Mamayauk, Pukalook, Pusimmik and Uttaktuak) to sew the clothes for his men at Cape Kellett and on the *Polar Bear*. "Seamstresses such as these we need so badly," Stefansson commented, "that we are willing to engage along with them comparatively useless husbands and families of several children."[2] Additionally he hired as first mate on the *Polar Bear* his old friend John Hadley, who had survived the sinking of the *Karluk* and had just reached Herschel Island as the ice pilot on the Hudson's Bay Company schooner, *Fort Macpherson*. This gave Stefansson access to Hadley's story of what happened to the *Karluk* and its personnel, which he encouraged Hadley to write down.[3]

Once all his business was completed at Herschel Island, Stefansson hastened back to Baillie Islands on the *Polar Bear*, followed by the *Gladiator*, hoping to find Wilkins there with the *North Star*. Instead, he found a note, dated August 16, in which Wilkins reported that he was proceeding to Cape Kellett as instructed.

Stefansson was then unexpectedly delayed at Baillie Islands for a few days while he awaited the arrival of the small auxiliary schooner *Atkoon*. At Herschel Island he had encountered a young Church of England missionary, Rev. Herbert Girling, who, with two assistants from southern Canada (G. Eldon Merritt and William B.E. Hoare), was intending to take the overloaded *Atkoon* east to establish a Church of England mission among the Copper Inuit. Knowing something of the navigation difficulties the missionaries would face without an

---

[2] Stefansson, 1921b, pp. 390–391.

[3] *Op. cit.*, pp. 704–730. McKinlay informed me in 1980 that Hadley kept a diary only after the *Karluk* survivors reached Wrangel Island. Years earlier McKinlay had written Dr. R.M. Anderson "… I know that Hadley kept no diary on board the ship and did not commence keeping one until he reached Wrangell [*sic*] Island. That diary he handed to me …" (McKinlay, 1922b). With Hadley's permission McKinlay took that diary to Scotland in 1914, and ultimately deposited it in the National Library of Scotland. It was roughly written and its contents frequently differ from the account in Stefansson's book. "It is merely a brief summary of events on Wrangell [*sic*] Island, and contains no comment of any kind on either Captain Bartlett or Stefansson" (McKinlay, 1922b). Stefansson stated (1921b, p. 704) that the published account in his book is his condensation of Hadley's lengthy version, written by Hadley in 1917–1918. Later McKinlay suggested "I am of the opinion that the author was not Hadley but Stefansson" (McKinlay, 1925). That seems probable in view of the literary quality of the Hadley account in *The Friendly Arctic*.

experienced Arctic sailor on board, Stefansson had lent Rev. Girling two seasoned men—Hadley and a knowledgeable local Inuk named Illun—to navigate the *Atkoon* safely past the hazardous mouth of the delta of the Mackenzie River as far as Baillie Islands. As that vessel took longer than expected to reach Baillie Islands, the delay prevented Stefansson from getting to Cape Kellett before Wilkins and the *North Star* started north. With the *North Star* not available to turn over to Captain Lane as he had planned, Stefansson agreed to Captain Lane taking the *Gladiator*. Some of the crew members from the *Polar Bear*—those who were not willing to become part of Stefansson's Northern Party—then transferred to the *Gladiator*. Then, Ole Andreasen suddenly left the *Polar Bear* and climbed on board the *Gladiator*, having decided to leave Stefansson's Expedition at this time to start his own trading business with the money Stefansson had paid him.

Captain Lane then departed with the *Gladiator* on its long journey out of the Arctic, well satisfied with his financial arrangements with Stefansson. Ten years later, he told Mrs. R.M. Anderson, "Sure I am a friend of his. Why shouldn't I be? I got $52,000 out of him in 1915 up at Herschel Island, and so I like him $52,000 worth, but he is a liar just the same ..."[4] The Captain evidently prided himself on being "crooked" and regretted that he had not asked more for his ship.[5]

Northern Expansion

The size of Stefansson's Northern Party was now increased by ten of the *Polar Bear's* crew, including four young Americans—Herman and Martin Kilian, E. Lorne Knight and Harold Noice. Stefansson appointed the Englishman John Hadley as its first mate; the Australian, William Seymour, also a seasoned sailor, as second mate; and Seymour's Iñupiaq wife, Ikugana (Anna), was one of the women in Stefansson's seamstress group. The ship's captain, Henry Gonzales, was originally from Portugal. Other crewmen are listed in Appendix 1.

At Cape Kellett an unexpected change in the direction of the wind to the northwest drove the pack ice shoreward along the west coast of Banks Island and prevented Stefansson from following Wilkins north with the *Polar Bear*. He therefore decided to try to reach Melville Island by way of the east side of Banks Island and Prince of Wales Strait. With the added power of the *Polar Bear's* engine, he hoped to get north of Prince of Wales Strait into Viscount Melville Sound, which both Captains Robert M'Clure and Richard Collinson had failed to accomplish in 1850 and 1851 respectively with their sailing ships. However, heading east from Cape Kellett on September 3, the *Polar Bear* encountered heavy ice as soon as it entered Prince of Wales Strait. Proceeding cautiously, it worked its way slowly northward in some open water along the Victoria Island coast to a location just north of Deans Dundas Bay, near Armstrong Point. There it became enclosed in the ice.

Armstrong Point

Further progress by ship now became out of the question, so Stefansson had Hadley, a competent carpenter, oversee the construction of a house on the shore, using lumber brought on

---

[4] Mrs. R.M. Anderson, 1925 (February 4, 1925).
[5] Diubaldo, 1978, p. 119.

the nearby icebound *Polar Bear*. By September 21, the house was completed and everyone moved in. As there was no turf available nearby for insulation, the men packed snow against the walls of the house for insulation, but it proved quite unsatisfactory, and the house at Armstrong Point was very cold that winter.

Most of Stefansson's new employees devoted the next few weeks to getting settled, killing seals for skins and food, and collecting driftwood, which was scarce. They shot a few caribou during the initial days, but thereafter saw very few, for the season was too late for the caribou to be in that part of Victoria Island.

Late in September, Stefansson led a party of six Iñupiat south along the coast and established a temporary sealing camp just north of Deans Dundas Bay. Two weeks later he sent Storkerson and three of the ship's crew north along the ice in Prince of Wales Strait to take tidal observations and to establish a small food depot at Peel Point for later use. When those tasks were completed, Storkerson's support team of Harold Noice and Karsten ("Charlie") Andersen returned from Peel Point to the *Polar Bear* camp. Meanwhile, Storkerson and Herman Kilian headed east to commence the mapping of the north coast of Victoria Island, in rapidly diminishing daylight, from the easternmost location mapped by Lieutenant Robert Wynniatt, one of Captain M'Clure's officers from the *Investigator*, in the early 1850s.

During the latter half of October Stefansson visited a snowhouse village of Copper Inuit at the mouth of Minto Inlet, where he traded for ethnographic material and wrote down several stories and beliefs he heard. He was greatly disappointed to discover that Pammiungittok, a very old man in this village, whom Stefansson had met in Prince Albert Sound in the summer of 1911, could no longer recall details of his visit as a very young boy in 1852 to Captain Richard Collinson's ship, the *Enterprise,* in Walker Bay (Minto Inlet). During their previous meeting, the old Inuk had told Stefansson about the discovery and plundering by his people of Captain Robert M'Clure's ship *Investigator* in Mercy Bay, Banks Island. On the present visit, however, Stefansson obtained no new information from him. Some of the villagers recalled visiting Captain Charles Klengenberg's ship when he wintered near Bell Island on the southwest coast of Victoria Island in 1905–1906, while most of them remembered Captain Billy Mogg's ship in Walker Bay in the winter of 1907–1908, but these events were of less interest to Stefansson.

Before leaving the village, Stefansson persuaded two young Copper Inuit men, Nutaittok and Taptuna, to return with him to his camp so that he could observe their reactions to the strange things they would see and hear—the ship, house, windows, stoves, cooking and eating utensils, foods, and music from a gramophone. The two men duly examined the new objects with much curiosity, finding special interest in items whose usefulness they were able to comprehend. Their chief interests, however, centred on knives, pots and the window glass. The gramophone held their interest only briefly, little of the white man's food appealed to them apart from weak tea, and they found the house cold and damp, quite unsuitable for fur clothing.

Early in November Stefansson prepared to travel from the *Polar Bear* camp to the base camp near Cape Kellett, a journey of nearly three hundred miles, to obtain the heavy sleds that, in August, he had asked Captain Bernard to make, as well as to obtain news of Wilkins and the *North Star*. Before leaving, however, he wrote a letter for Storkerson, instructing him,

Figure 23 The *Polar Bear* camp south of Armstrong Point, Prince of Wales Strait, Victoria Island, June 1, 1916.
Photo: © Canadian Museum of Civilization, photo by George Hubert Wilkins, 51181

after he returned to the *Polar Bear* camp from the north coast of Victoria Island, to journey to Mercy Bay on Banks Island and bring back the two good sleds Stefansson and Storkerson had left there during the summer. After that had been accomplished, Storkerson was to prepare for the next ice trip, which would be in the spring to survey the land they had discovered north of Prince Patrick Island. Stefansson also hoped to establish a new base camp well north of the *Polar Bear* camp to permit exploration even farther north in 1916.

In preparation for his journey to Cape Kellett, Stefansson demonstrated how to construct snowhouses (called *iglu* or *igloo* in Inuktitut, the language of the Inuit across North America) for the benefit of both the new Iñupiat members of the Expedition, and the white sailors he had hired at Herschel Island. Snowhouse building was an art practised only by the Inuit in the Central and Eastern Arctic at the time, where driftwood and whale bones (the customary building materials of the Western Arctic) were scarce or non-existent.

On November 16 Stefansson left the *Polar Bear* camp for Cape Kellett, accompanied by Noice, Martin Kilian, the Alaskan Iñupiaq Emiu, two sleds and nine dogs. They spent a week or more at Ramsay Island waiting for the approach of the full moon to illuminate their way, because the sun no longer rose above the horizon at that latitude so late in the year. Then on December 1 they continued on to Cape Kellett, which they reached towards the middle of December. Once there, Stefansson learned to his satisfaction that Wilkins and the *North Star* were settled in a small bay about twenty miles north of Norway Island, that is, about 140 miles to the north of Cape Kellett. There Wilkins and his men had built a comfortable camp alongside the beached schooner, using oil drums, canvas, and the spars and beams from the ship. Some while later, after he had spent a few days in Wilkins' camp, Stefansson commented, "Wilkins had here the most comfortable and the most sensibly arranged of our three winter houses."[6] It was also located just where Stefansson had wanted it to be a year earlier in 1914, when he completed his first ice trip. Now, while he was at Cape Kellett, he wanted to discuss his future plans with Wilkins, so he sent Thomsen and Emiu north to bring Wilkins back to the base camp near Cape Kellett. They returned with both Wilkins and Castel on January 1, 1916.

Over the course of the next few days, Stefansson and Wilkins discussed an assortment of activities for the Northern Party, and Stefansson developed several new and distinct exploration plans for the spring. These included two ice trips, one continuing the exploration of the Beaufort Sea northwest from Cape Prince Alfred, the other returning north of Prince Patrick Island by way of Melville Island to complete the mapping of the land discovered the previous spring and to explore farther afield. Stefansson would keep his base camp near Cape Kellett, with his reliable friend Captain Peter Bernard remaining in charge, and have a second base where the *Polar Bear* was icebound on Victoria Island.

Figure 24 The Storkerson family at the *Polar Bear* camp south of Armstrong Point, Prince of Wales Strait, Victoria Island, June 1, 1916.

Photo: © Canadian Museum of Civilization, photo by George Hubert Wilkins, 51180

---

[6] Stefansson, 1921b, p. 457.

MORE DISCOVERIES OF LAND, 1916

14

Harold Noice, Lorne Knight and Charlie Thomsen were the first to initiate Stefansson's plans for his third ice trip. They left the Cape Kellett camp on January 6, 1916 for the *Polar Bear* camp. Darkness, blizzards, dog fights and other incidents marred their journey, proving as Noice later wrote, Stefansson's saying that "an adventure is a sign of incompetence."[1] Arriving at the *Polar Bear* camp on February 1, they learned that the handsome Captain Henry Gonzales and the bright-eyed young Inuk, Violet Mamayauk, from Baillie Islands had been married a few days earlier. The Captain exercised his rights as ex-officio magistrate and clergyman to perform the ceremony himself.[2] Additionally, Storkerson had been to Mercy Bay with Captain Gonzales and brought back to the ship the two sleds left there the previous summer, the *Polar Bear's* second engineer, J.J. Jones, had died of heart failure, and Hadley had received a nasty wound in his arm as a result of an unpleasant encounter with a polar bear. Thomsen then handed Stefansson's instructions to Storkerson, which told him to transport a load of supplies from that camp around the north coast of Banks Island and meet Stefansson somewhere near Cape Prince Alfred. Storkerson's supplies with those being transported north from Cape Kellett by Wilkins and Stefansson would ensure that Stefansson was well prepared to spend several months that spring investigating the land he had discovered the year before and continue his search for new land towards the end of January.

It was already February 1, however, when Storkerson was handed Stefansson's instructions, too late to take supplies to Stefansson on the northwestern coast of Banks Island. Instead, Storkerson reasoned, he would head north to Melville Island with the supplies and intercept Stefansson en route. As a safety measure before starting north, however, he sent Herman Kilian and Palaiyak to Mercy Bay with a note for Stefansson telling of his change of plans.[3]

Freighting Supplies North

Two days after the departure of Noice, Knight and Thomsen for the *Polar Bear* camp, on the coldest day of the winter, Wilkins left Cape Kellett with Castel and Martin Kilian for the *North Star* camp, bearing instructions from Stefansson to continue up the west coast of Banks Island to Cape Prince Alfred. Wilkins had left Natkusiak alone at Cape Prince Alfred before Christmas with instructions to catch seals for dog food for the ice trip, and for the three parties—those of Wilkins and Stefansson from Cape Kellett and Storkerson from the *Polar Bear*—who were

---

[1] Noice, 1924, p. 92.
[2] *Op. cit.*, p. 96.
[3] Ashlee, 2008, p. 133.

all expected to converge on Natkusiak's camp at Cape Prince Alfred. As usual, Stefansson was delayed and did not head north from Cape Kellett until January 23.

After eleven disagreeably cold and windy days with little daylight, Wilkins and his two companions reached the *North Star* camp on January 19. Wilkins was pleased to find that Peter Lopez and his native wife, Uttaktuak, whom he had left in charge of that camp, were both well. After two days of rest, the men loaded two sleds and then Castel, Martin Kilian and Lopez headed north for Natkusiak's camp. Wilkins remained at the *North Star*, expecting Stefansson to arrive within a few days. While he waited, he wrote in his diary "I can't imagine what prevents V.S. from coming, but as I have never known him to do what he proposed to do except once (and that was by accident), I will not expect him until he comes …"[4] Kilian and Lopez returned to the *North Star* camp on January 26 as per Wilkins' instructions, but both men were badly frostbitten. Castel had remained behind at Cape Prince Alfred with Natkusiak.

Stefansson finally reached the *North Star* camp on February 5, accompanied by several Iñupiat, which made it necessary to build an additional snowhouse to accommodate everyone. Much to Wilkins' annoyance, Stefansson now informed him that he intended to abandon the *North Star*, on the excuse that it was a year late reaching the locality where he had wanted it. Instead he would freight all of the supplies to Melville Island where he would settle Lopez

Figure 25 Men of the Northern Party at Cape Kellett, Banks Island, taken by flashlight. Back row (l. to r.): Knight, Thomsen, Castel, Noice; front row, M. Kilian and Captain P. Bernard, January 1, 1916.
Photo. © Canadian Museum of Civilization, photo by George Hubert Wilkins, 51098

---

[4] Wilkins, 1916 (January 26, 1916).

and Alingnak and their wives (Uttaktuak and Guninana) in a new base camp from which Stefansson could explore farther north. After the weeks of effort Wilkins had devoted to rendering the *North Star* seaworthy in the spring of 1914 and then sledding five hundred miles to Bernard Harbour to retrieve it in the summer of 1915, he was understandably upset to hear that Stefansson had no further use for the schooner. Nor did he like Stefansson's new plans to sled all the supplies to Melville Island.

By an unexpected twist of fate, Stefansson did not have the mail Wilkins had anxiously waited for all summer, having kept it on the *Polar Bear* when he reached Cape Kellett from Herschel Island, and it was now at the *Polar Bear* camp on Victoria Island. Wilkins was concerned about the condition of his elderly father, who had been in poor health when last Wilkins had received any mail. He was also worried about his mother. He had had no mail from Australia in well over a year, and now wanted to get to the *Polar Bear* camp to read it. He also wanted to return to England to see if he still had a job with the Gaumont Company. He therefore informed Stefansson he would need a team of dogs to return to the base camp of the Southern Party at Bernard Harbour so that he could leave the Arctic with the scientists there in the coming summer. (The Canadian government had sent instructions to both Dr. Anderson and Stefansson to return from the North in the summer of 1916, and the Southern Party intended to follow those instructions.) Having discovered new land, however, Stefansson now intended to remain longer than the planned three years in the North, and sought in vain to persuade Wilkins to help him carry out his new plans.

Where was Storkerson?

By early February there were sufficient supplies at the *North Star* camp on the west coast of Banks Island for Stefansson's third ice trip over the Beaufort Sea. Storkerson was to be in charge of the trip and head northwest from the Gore Islands near Cape Prince Alfred in February, but there was still no sign of him. Stefansson's instructions to Storkerson in January were for him to reach Cape Prince Alfred by February 10 after picking up the two sleds that had been left at Mercy Bay. Stefansson had no way of knowing, however, that his instructions did not reach Storkerson until February 1, by which time it was too late to get to Cape Prince Alfred with a load of supplies from the *Polar Bear* camp.

When Storkerson had not appeared by February 20, Stefansson became uneasy. To start an ice trip northwest from the Gore Islands at this late date would be not only dangerous but also a duplication of Stefansson's 1915 trip. Stefansson therefore abandoned his plans for further exploration of the Beaufort Sea, deciding instead to pursue his northern explorations by way of Melville Island. Towards that purpose he asked Wilkins and Castel to start freighting supplies from the *North Star* along the north coast of Banks Island to a site from which it would be suitable to cross M'Clure Strait to Melville Island. He was determined to establish a base on that island whether or not the *North Star* or the *Polar Bear* could reach it.

Wilkins agreed to take charge of the freighting activities and left the *North Star* camp with Castel on February 21 for Natkusiak's camp at Cape Prince Alfred. While his men were busy ferrying supplies, Stefansson obtained Iñupiat folklore and other ethnographical information from Alingnak's wife, Guninana, whom Stefansson found exceptionally well versed in such matters. Finally, on March 2 he, too, left the *North Star* camp, followed soon afterwards by Lopez, who closed the *North Star* camp and joined the others farther north.

At Natkusiak's camp Wilkins and Castel joined Martin Kilian and the Iñupiaq Emiu (who had the nickname "Split-the-Wind" because he enjoyed driving fast dog teams). Leaving Castel to take care of the camp, the other four headed slowly northeastward along the coast. Wilkins hunted a short distance inland, as Stefansson customarily did, while the other three men drove the two sleds along the coast. Over the next few days he killed several caribou, which provided ample fresh meat for both men and dogs. Castel arrived on March 4 with new instructions from Stefansson, who by then had reached Natkusiak's camp at Cape Prince Alfred.

Stefansson caught up to Wilkins on March 12, by which time Wilkins had sent Castel and Kilian ahead to look for signs of Storkerson, now a month overdue. Castel and Kilian returned after finding no sign of Storkerson. They had, however, discovered a wide and very long bay flanked by steep cliffs, which Stefansson subsequently named Castel Bay.

Much disturbed by the continuing lack of news of Storkerson, Stefansson decided to go ahead of Wilkins to Mercy Bay with Natkusiak and Emiu, looking for signs of Storkerson. On March 31 Natkusiak and Emiu arrived back at Wilkins' camp with a note from Stefansson, which stated that he had found a letter from Storkerson at Mercy Bay. The letter revealed that Storkerson had retrieved the two sleds in the fall and then decided to forego Stefansson's ice trip because of circumstances at the *Polar Bear* camp. Instead he intended to prepare for the trip from that camp north to investigate the "New Land." Storkerson's letter also mentioned that he had been unsuccessful in completing the mapping of the unfinished section of the north coast of Victoria Island before Christmas, but had discovered a major mountain range extending inland from that island's north coast. Stefansson later named this mountain range the Shaler Mountains after Nathaniel Shaler, his geology professor at Harvard University a decade earlier.

After receiving this news about Storkerson, Wilkins, Natkusiak and Emiu moved to Mercy Bay, where they waited with Stefansson for the slower-moving sled parties still en route freighting supplies east from Cape Prince Alfred. While they were thus delayed, Stefansson contemplated trying to reach the *Polar Bear* camp before Storkerson set off for the "New Land" and Wilkins thought about his long-awaited mail that was at the *Polar Bear* camp.

A Change of Plans

Suddenly Stefansson announced that he wanted his party to move directly across the strait to Liddon Gulf on Melville Island. He suspected that Storkerson was already on Melville Island or beyond, and wanted to overtake him. This was Stefansson's exploration trip, and he did not want Storkerson to receive the acclaim that he sought for himself for the discovery of new land. Wilkins had other plans. He wanted to head for the *Polar Bear* camp in order to retrieve his previous summer's mail, then proceed to a Copper Inuit camp near the mouth of Minto Inlet and photograph their daily activities. Stefansson's sudden change of plans extinguished those of Wilkins, who complained in his diary that he had spent nine months hunting and moving freight for the exploration party, but "when it comes to the time of the year when I can do some work in my own department I am begrudged a few days."[5] He consoled himself by thinking that he could obtain some good photographs of muskoxen, which Stefansson had told him were plentiful on Melville Island. That idea also would soon be dampened by Stefansson's unpredictable changes in plans.

The distance across M'Clure Strait to Liddon Gulf is more than eighty-five miles when crossed diagonally. Stefansson and his companions left Mercy Bay on April 5 and made the crossing in eight days. When half-way across the strait, Stefansson called a halt and had Wilkins photograph him and several others building a snowhouse, photographs he later made good use of in some of his many publications. On the eighth day near Cape Ross they came upon a group of three roofless snowhouses and a note left by Storkerson. Cape Ross is a steep elongated cliff of stratified rocks, which mark the entrance to Liddon Gulf. The note stated that he had gone farther north on April 4. A polar bear approached the camp just as Stefansson's party arrived, evidently thinking the intruders might make a fine meal or two, but was sighted and shot by Natkusiak before it caused any problem.

Figure 26 Minnie Guninana, wife of Alingnak, and one of Stefansson's best informants on folklore and language, and Uttaktuak, wife of Peter Lopez, at *North Star* camp on the northwest coast of Banks Island, February 21, 1916. Photo: © Canadian Museum of Civilization, photo by George Hubert Wilkins, 51100

The Polar-Bear Liver Test

The next evening Stefansson persuaded his men to join him in an experiment to test an old Inuit taboo—eating polar-bear liver. He suspected that the taboo against eating it was simply one of the many Native superstitions he had learned about. His men divided the liver into five roughly equal portions and cooked each portion either well done or slightly underdone according to their individual tastes. All was well for about six hours after they finished eating. Then Emiu became violently sick to his stomach, with a raging headache and slight fever, which lasted several hours. Martin Kilian and Natkusiak became similarly ill a while later. Next to react was Wilkins, who developed a slight headache, but attributed this at the time to

---

[5] Wilkins, 1916 (April 4, 1916).

snow-blindness. Castel also developed only a slight headache, while Stefansson claimed no ill effects whatsoever. In each case the symptoms lasted for several hours. Storkerson and Thomsen had had a similar experience after consuming polar-bear liver a year earlier, and had been ill for several days.

In Stefansson's view, the test gave definite credence to the Native taboo by demonstrating that the eating of polar-bear liver could result in painful after-effects. Years later, laboratory studies in the United States provided an explanation: the liver of the polar bear is unusually high in vitamin A, which the bear acquires from the seals it eats. Human ingestion of too much of this vitamin can create scurvy-like symptoms that could prove fatal.[6]

More Changes of Plans

Stefansson had failed in his initial efforts to overtake Storkerson. He now hastily modified his plans, announcing that he and three men would re-cross M'Clure Strait to Russell Point on Banks Island, where there was a cache of supplies. Two of the men, Castel and Emiu, would bring back a load from that cache to the abandoned snowhouse camp in Liddon Gulf where they were currently located and wait there, while Stefansson and Wilkins would continue on south to the *Polar Bear* camp near Armstrong Point. There Stefansson could discuss plans with Captain Gonzales to bring the *Polar Bear* to Melville Island during the summer and establish

Figure 27 Snowhouses built by Storkerson's men at the base of cliffs in the east side of Liddon Gulf, Melville Island, April 13, 1916. This is the earliest known photograph of Melville Island.
Photo: © Canadian Museum of Civilization, photo by George Hubert Wilkins, 51146

---

[6] Rodahl, K., and Moore, T., 1943; Rodahl, K., 1949.

a camp in Liddon Gulf while Wilkins photographed the Minto Inlet Inuit, who would be camped near the ship. Once again Stefansson's change of plans irritated Wilkins, because this newest plan deprived him of the opportunity of photographing muskoxen.

Leaving Martin Kilian, who still suffered from the after-effects of the bear liver, to guard the snowhouse camp in Liddon Gulf, Stefansson, Wilkins, Castel and Emiu started south on the afternoon of April 17, retracing their trail across M'Clure Strait. That evening, as they were getting ready to camp in one of their former snowhouses fifteen miles off shore, Martin Kilian's brother Herman drove up with a dog team and two Iñupiat men, heading for the *Polar Bear* camp. Kilian had news that Storkerson was freighting supplies to a base in Hecla and Griper Bay on the northwest side of Melville Island, in preparation for exploring and mapping the new land discovered the previous year.

This news disturbed Stefansson, and once again he changed his plans. Instead of proceeding to the *Polar Bear* to discuss the coming summer's plans with Captain Gonzales, he decided he would now turn back immediately and try to overtake Storkerson. He briefed Wilkins on his latest change of plans and said he would draw up his summer plans for Wilkins to take to Captain Gonzales. By now thoroughly frustrated with the frequency in which Stefansson changed his plans, Wilkins replied that he wanted to pursue his own photographic interests and then leave the Arctic in the summer to return to England. He felt that Stefansson had enough men to carry out his further exploration work, which was mainly sledding and freighting, and he, Wilkins, was no longer needed for that purpose. He would transfer his things to Herman Kilian's sled and go south with him to the *Polar Bear* camp, photograph the Inuit encamped around the ship and also those at the mouth of Minto Inlet, and then head for Bernard Harbour and the headquarters of the Southern Party. The main problems with this plan were, however, firstly, that he would have no interpreter while he was among the Inuit—he had counted on Stefansson filling that role—and secondly, he would lack a sled and dogs to get to Bernard Harbour.

Stefansson may have tried to persuade Wilkins to remain with the Northern Party, for he later wrote in *The Friendly Arctic*: "We discussed the possible necessity for his taking command of the *Bear* to bring her to Melville Island the summer of 1916, but agreed that her present crew were quite capable of doing that and would probably use their best endeavors in that direction."[7] Stefansson may have suspected that Captain Gonzales would not carry out his written instructions and hoped that his offer to Wilkins would persuade him to stay in the Arctic. However, it is also conceivable that Stefansson merely invented that argument later for inclusion in his book, because Wilkins' diary makes no mention of any such discussion between them. In reality, at the time, the two men simply parted with a cordial handshake, not to meet again until sometime after World War I, when they renewed their friendship.

Wilkins Starts South

After loading his possessions on Herman Kilian's sled, Wilkins accompanied him to the *Polar Bear* camp, which they reached on April 26, and where he finally received his long overdue mail. His mail included a letter containing the sad news that his father had passed away many months earlier. Up until that point Wilkins had been wavering between returning to England

---

[7] Stefansson, 1921b, p. 487.

or staying with the Northern Party. That news finally convinced him to leave the Arctic, for he felt it more important than ever that he see his elderly mother again, now that she was alone.

At the *Polar Bear* camp, Storkerson had done well in overseeing the preparations for the spring exploration work before heading north. The women had been making seal-skin boots, deer-skin socks, mittens, *attigis* (parkas), sleeping bags and drill snow-shirts, while the men had been repairing dog harnesses, weighing rations of dog food and man food, which they sewed into canvas sacks, and repairing the sleds. Several of the men had taken provisions and made a cache at Russell Point on the northeast coast of Banks Island, about sixty miles north of the *Polar Bear*. Several of the dogs had died from dog-sickness, while others had been badly injured or died from a horrendous fight that ensued after Captain Gonzales had tied twelve of them to a sled in tandem. That fight happened while he was on one of the three journeys Storkerson had attempted to make to Mercy Bay to get Stefansson's two sleds.[8]

On May 1 Wilkins undertook a sled trip of several days' duration, going south from the *Polar Bear* camp to Minto Inlet to photograph and trade with some of Stefansson's so-called "Blond Eskimos" in their snowhouse village. This was, after all, one of the main reasons his company had sent him on the Expedition. On the ensuing days he obtained still shots of the Inuit and their snowhouses as well as some moving-picture film showing how they caught seals through holes in the ice. He also had an opportunity to examine the facial and other features of many of the villagers and saw no evidence to justify their being called "Blond Eskimos." His observations provided additional evidence for his colleague, the anthropologist Jenness, who a few years later refuted Stefansson's idea that there was a link between these Copper Inuit and the long vanished, early Viking settlers in western Greenland.[9] Conflicting scientific evidence notwithstanding, Stefansson's idea had wide and lingering appeal, especially in the United States, for one still sees the term "Blond Eskimos" in print occasionally today.

On his trip to Minto Inlet, Wilkins was accompanied by Silas Palaiyak, a young Mackenzie Delta Inuvialuk (as an Inuk from that locality is called),[10] and Anna (Ikugana) Seymour, the Iñupiaq wife of Bill Seymour, the *Polar Bear's* first mate. During their visit to the village the three of them had a tense encounter with Kullak, the Inuk Stefansson had met the previous summer on Banks Island, whose wife had been pregnant then but had since died following childbirth. It so happened that Kullak blamed Stefansson for her death and sought revenge for that tragic event on him or, as Palaiyak and Anna suspected, on Wilkins as a fellow evil white man. On the urgent advice of his two frightened companions, Wilkins hurriedly vacated the village and returned with them to the *Polar Bear* camp.

Wilkins then spent several days hunting caribou with three companions on Banks Island. They had considerable success and returned to the *Polar Bear* camp with a useful supply of caribou skins and meat. At the camp, they found Storkerson, Martin Kilian and Illun newly arrived from "First Land" (or "New Land," which Storkerson had discovered the year before,

---

[8] Noice, 1924, pp. 98–99.

[9] Jenness, 1923, p. B46.

[10] Palaiyak had been with Stefansson during his 1908–1912 expedition. Stefansson had found him at Herschel Island in the summer of 1915 and hired him for the Northern Party.

and which was later named Brock Island after Dr. Reginald Brock, the Director of the Geological Survey of Canada from 1908 to 1914). Stefansson had sent them back south with instructions for Storkerson to bring his wife and family, Pannigabluk and her son, Alex, and Pikalu and his wife north to Melville Island to hunt and sew during the summer and lay in food supplies for the following winter. Later Storkerson was to return south to complete the mapping of the northern coast of Victoria Island.

On June 2, Wilkins and Palaiyak started south from the *Polar Bear* camp for the Southern Party's base camp at Bernard Harbour, which they reached without major incident two weeks later. While at Bernard Harbour, Wilkins took many photographs of Copper Inuit as he waited for the scientists to complete their investigations.[11]

Catching up to Storkerson

Meanwhile, back in April 1916, after Wilkins said goodbye to Stefansson and started south from M'Clure Strait to the *Polar Bear*, Stefansson and his party headed in the opposite direction. They quickly reached Liddon Gulf, from where they followed Storkerson's sled trail northwards across the isthmus that separated Liddon Gulf from Hecla and Griper Bay. On the north side of the isthmus Stefansson discovered a note from Storkerson that stated he had passed there about a week before. After following Storkerson's trail along the west side of the bay to Cape Grassy, Stefansson found that the trail headed seaward, so he did likewise. A short while later Stefansson's foot broke through some thin ice and he injured an ankle. He continued walking on it until it became so painful that he was forced to ride on his sled. Near Eight Bears Island his party encountered Thomsen and Illun travelling south with a light sled to meet them. Thomsen handed Stefansson a message saying Storkerson was at a certain unnamed point on the "New Land." Stefansson promptly sent Emiu with his empty sled and fast dogs north with instructions for Storkerson to wait for him where he was.

Stefansson later gave the name "Cape Murray" to their meeting place on "New Land," after the oceanographer James Murray, who had lost his life after surviving the sinking of the *Karluk*. Storkerson had determined the latitude and longitude for the cape before Stefansson's arrival. Stefansson reached the cape on May 3 and stayed with Storkerson's party for three days to rest his dogs and his ankle, and also to formulate his next plans. With Wilkins gone, Stefansson appointed Storkerson his right-hand man, instructing him to return south to Melville Island, make contact with the hunters Alingnak and Lopez and their wives, and establish a winter base at a suitable location on the east side of Liddon Gulf. The Inuit were to remain at this base and kill muskoxen and caribou during the summer and to dry their meat when the animals were in their best condition.

Once the winter base was established in Liddon Gulf, Storkerson was to continue southward with Martin Kilian and Illun to the *Polar Bear* camp, taking Stefansson's instructions to Captain Gonzales to bring the ship to Liddon Gulf during the coming summer (1916) if navigational conditions made it possible. If the captain was unable to move the *Polar Bear* to Liddon Gulf, he was to remain at his present camp near Armstrong Point and send word of his actions to

---

[11] Readers interested in reading a fuller account of Wilkins' Arctic activities are referred to Jenness (2004). For a well-written biography of Wilkins' fascinating life, see Nasht (2005).

the winter camp at Liddon Gulf early in 1917. After providing the captain with his instructions, Storkerson was to bring his own wife, Uiniq Elvina, and their two young daughters, Martina and Aida, north from the *Polar Bear* to the winter camp and oversee the meat-drying operations and accumulation of seal oil from seal fat, which would be needed for cooking fuel. He was also to search for layers of lignite or other coal-like substances like the ones they had observed on the north side of Banks Island, to ensure they had an adequate supply of heating fuel during the following winter. Finally, if time permitted, he was to finish his mapping of the north coast of Victoria Island. In this manner, Stefansson tactfully removed Storkerson from any further exploration in that northernmost region, and the resulting acclaim that would go with it.

Stefansson spent the next few days writing out instructions for his men. On May 5 he had Castel, Noice and Karsten Andersen start north from Cape Murray to explore the west coast of the "New Land." As his ankle was still not well enough to do much walking, he sent Natkusiak inland to hunt and report on the nature of the land. For perhaps the first ten miles Natkusiak found the interior generally level, with mud, sand or gravel where it was not snow-covered. In this level region he saw no caribou. Beyond it the land rose to several hundred feet, and in that region Natkusiak saw evidence of the presence of numerous caribou, but no living animals.

Two days later (May 7) Storkerson started south to Lidden Gulf as Stefansson had instructed him to do, accompanied by Thomsen, Martin Kilian and Illun. Thomsen had received Stefansson's permission to continue south to Cape Kellett to bring his family north to the winter camp Storkerson would oversee in Liddon Gulf.

The Tragedy of Thomsen and Captain Peter Bernard

Thomsen had said he was willing to remain with the Expedition provided that he could go to Cape Kellett and bring back his wife, Jennie, and their two children (their daughter, Annie, and their one-year old son, whose name seems to have gone unreported). Stefansson agreed to this plan provided that Thomsen did not linger at Cape Kellett but returned north immediately, and brought back a new sled made by Captain Bernard. On his way south, additionally, Thomsen was to pick up bird and mammal specimens that Wilkins had collected on Banks Island and left at the *North Star* camp, and transport them to Cape Kellett for Captain Bernard to box for shipment south to the National Museum of Canada in Ottawa. (Sadly, Thomsen failed to collect these specimens, and they were never recovered.) On his return trip to Melville Island, he was to travel by way of the *Polar Bear*. If he was too late in the spring to cross M'Clure Strait, Thomsen was to summer at Mercy Bay. There he could lay in a supply of seal oil and dried caribou meat, using fuel from a "coal mine" they had seen there recently, and cross the strait to Liddon Gulf when it was safe to do so, soon after the New Year. Failure to follow these instructions of Stefansson may have cost Thomsen his life several months later.

Thomsen left Cape Murray to go south to Cape Kellett on May 7, 1916. Months later Captain Gonzales brought the news north to Stefansson[12] that Thomsen had spent the summer

---

[12]   Captain Gonzales sledded north to see Stefansson on Melville Island in February 1917, bringing news he had obtained from Jim Crawford, the former engineer of the *Mary Sachs*. Crawford had come to Minto Inlet with the schooner *Challenge* in the summer of 1916 to trade with the Inuit.

with his family at Cape Kellett instead of promptly returning north, as Stefansson had stipulated. Also that Thomsen and Captain Bernard had started north for Melville Island by way of the west coast of Banks Island late in the fall of 1916, again defying Stefansson's instruction to return north by way of the east coast and the *Polar Bear.* They had with them nearly five hundred pounds of mail and two new sleds for Stefansson. There had been no further news of them.

The following spring, Stefansson sent Castel and Karsten Andersen south to Cape Kellett from Melville Island to get the *Mary Sachs* ready to leave the Arctic. They were to cross to Banks Island from Liddon Gulf and follow the west coast of Banks Island, searching en route for signs of Thomsen and Captain Bernard.

Castel and Karsten Andersen left Storkerson's party at Melville Island for Cape Kellett early in May 1917. At Mercy Bay they found two sleds, one being the one Thomsen had taken from Melville Island a year before, the other being a new one made by Captain Bernard. A note signed by the two missing men and dated December 22, 1916, was attached to one of the sleds. It stated they had run out of food and left a cache twenty miles NNE on the ice in M'Clure Strait, and with one sled, the large bag of mail and half their original dogs—the rest having died—they were starting back south. They discovered Thomsen's body in the snow two-days' journey west of Mercy Bay, but never found the body of Captain Bernard. They found no evidence of his reaching the *North Star* camp on his southward journey, nor did they locate the bag of mail. In a touching epithet, Stefansson later wrote: "He [Captain Bernard] and Thomsen were lost in a brave attempt to do what they thought was best and most con-ducive to the success of the expedition."[13]

Castel and Andersen continued on to Cape Kellett, arriving there in June to discover two strangers, Otto Binder and August Masik, who had been left by Jim Crawford and the *Challenge* near the cape in the summer of 1916 to trap foxes, following which Crawford proceeded with his schooner to Minto Inlet to trap and trade with the Inuit. They explained that the Kilian brothers had arrived in November with instructions for Captain Bernard, waited several months for his return, and then led the few remaining Expedition members, all of them Inuit, back to the *Polar Bear* camp. Castel and Masik, with the help of Andersen and Binder, then repaired the *Mary Sachs* and launched it early in August, ready to sail west when Stefansson arrived with his party from the north.

Continuing the Exploration Northward

On May 8, 1916, Stefansson, Natkusiak and Emiu started north from Cape Murray on Brock Island with two sleds, on the lighter one of which Stefansson rode as a passenger because of

---

He had encountered Captain Pedersen on the *Herman,* who had just been to Cape Kellett and mentioned that Thomsen was intending to start north for Melville Island late in the fall of 1916.

[13] Stefansson, 1921b, p. 654. Curiously, although speaking well of Bernard here and on earlier occasions, as mentioned in several places in *The Friendly Arctic,* Stefansson later stipulated that Bernard was not to receive any proceeds from the sale of the Northern Party's collection of fox furs, many of which he had caught, because of unsatisfactory service. By then, of course, Bernard was dead, so that his widow received none of the benefits due her (Anderson, 1923f.)

his ankle. After travelling for a little more than thirty miles, they overtook Castel's party, which had followed the coast of the "New Land" north from the cape (Cape Murray), but then had discovered that the coast turned east.

The next day they all followed the coast to the east and discovered after twenty miles that it turned south. From this location they could see the mountain Stefansson had called "Leffingwell Crags" the previous year, rising about five miles beyond what appeared to be an

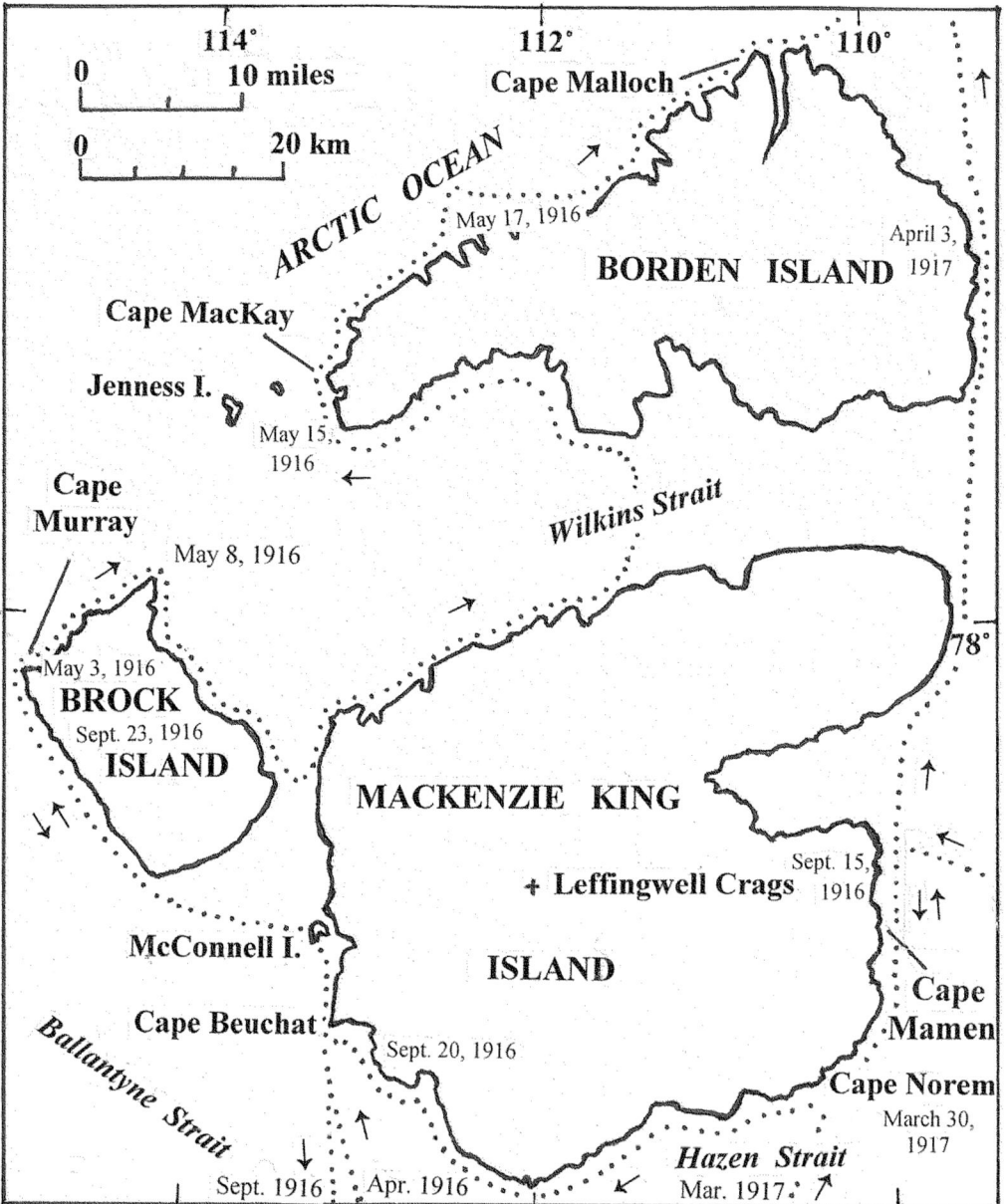

Map 9 Brock, Mackenzie King, and Borden Islands, which were mapped in 1916 and 1917 by V. Stefansson and assistants of his Northern Party. He named several geographic features on these islands after members of the Canadian Arctic Expedition.

ice-covered channel. Stefansson realized at this time that the channel divided his "New Land" in two, and that "Leffingwell Crags" lay on a separate island. To this second island he later gave the name Borden Island after Sir Robert Borden, the Canadian prime minister who had authorized the financing of his Arctic Expedition.

Stefansson and his men then sledded across the channel, which a year earlier he had thought to be a long bay, and continued northeastward along the coast of the adjoining island (Borden Island). Fog soon obscured their view much of the time, rendering their progress both difficult and uncomfortable. Continuing along the low-lying, treeless coast—the men with the two heavy sleds travelling from point to distant point while Stefansson with his lighter sled investigated the indentations between them—they soon realized that they were going in a westerly direction, then a while later, southwesterly. By May 15, Stefansson realized that they had just travelled around the southern end of a peninsula and were again heading northerly along the west coast of the new land. Fog greatly obstructed their view of the terrain, but it appeared to contain many low rolling hills.

Snow-blindness now added to their discomfort, at times forcing them to halt for two or three days to rest their eyes. Their amber-coloured sunglasses had been left on the *Karluk* and they had but one good pair of goggles with them. This generally was assigned to the person selecting the trail.

In 1947, as a result of the first aerial photographs made of the Arctic Islands region, Stefansson's "Borden Island" was in turn discovered to be two large islands. Stefansson's name, Borden Island, was retained for the northern island of the two. The southern island was then given the name Mackenzie King Island after Canada's prime minister at that time (in 1947). Thus, what Stefansson had originally called "New Land," proved to be three large islands.

Stefansson Changes his Plans Again and Again

Although Stefansson had rested his ankle for three weeks, it was still not fit for walking on, so he was again forced to change his plans. Turning over most of the provisions to Castel and Noice, which Stefansson estimated would last the two men and their nine dogs for thirty days, he instructed them to map the coast northeast and east until it turned south, then strike northeast for Cape Isachsen on the northwest side of Ellef Ringnes Island. Once there, they were to take location readings and leave a record of their progress and observations, then head north over land-fast ice in search of land, possibly "Crocker Land." Stefansson had planned to do this exploration himself, but feared his injured ankle might fail him. Noice recorded in his diary on May 20:

> If we find land we are to explore that side which faces the open water (the west shore). We are to return to Melville Island by the 10th of July—if circumstances permit. If possible we are also to determine whether the New Land (Borden) is connected with Findlay Island; and also to ascertain if Findlay Island, Paterson Island, and King Christian Land are in reality one island.[14]

---

[14] Noice, 1924, pp. 116–117.

If instead of land-fast ice north of Cape Isachsen they encountered east-moving floe ice, they were to follow it towards Axel Heiberg Island for as far as it appeared safe to do so, and then head back south to Melville Island where they could easily obtain muskoxen to replace their depleted food supplies.

Castel and Noice left with their assignment on May 21. For the next few days Stefansson, Natkusiak, Andersen and Emiu hunted seals to feed themselves and their dogs. Then, with Stefansson's ankle duly rested, they followed after Castel and Noice, taking depth soundings periodically as they crossed from the "New Land" to Cape Isachsen, at the northern end of Ellef Ringnes Island. They reached that cape on May 31, a few miles ahead of Castel and Noice, whose heavy load prevented them from making rapid progress in the soft snow.

Castel and Emiu greatly taxed Stefansson's patience one day by discussing their delight at eating canned sardines, which were Emiu's favourite food, and boiled potatoes, which were Castel's preference.[15] Fearing that their comments would negatively influence the positive attitudes of Noice and Andersen towards their meat diet and disrupt the morale of his men, Stefansson decided to send Castel, Natkusiak and Emiu back to Melville Island. For their route he instructed them to proceed from Cape Isachsen southward to the northwest corner of King Christian Island, a body of land shown on their old British Admiralty Chart to lie south of Ellef Ringnes Island. Stefansson wanted to know if it was a separate island or the eastern part of his newly named Borden Island. If it proved to be a separate island, they would have established that a sea passage existed between Findlay Island and Borden Island. In that case, Castel and his companions were to follow down the west coast of Findlay Island and then strike west for Cape George Richards on the northeastern tip of Melville Island, from where they would proceed to Hecla Bay and Liddon Gulf and rejoin Storkerson's party, which was then busy drying meat at a base camp established there.[16]

Castel's party departed on June 4. Stefansson, Noice and Andersen then continued the summer's exploration to the northeast of Cape Isachsen along the edge of the land floe, where Stefansson expected the seals to be plentiful. His ankle was still very tender, but he determinedly donned snowshoes each day and walked as much as he could.

Discovery of "Second Land"

Fog and giant blocks of crushed ice, forming ridges tens of feet high, greatly hampered their daily progress over the ice. On June 13 Andersen suddenly sighted land, much to the joy of

---

[15] Stefansson, 1921b, p. 505. It is interesting how people's observations can differ so greatly on the same topic! Noice (1924, p. 123) recorded that peas were Emiu's favourite dish while baked potatoes were Castel's. Although Stefansson was one of the most patient men Noice had ever known, "He believed that no one on an exploring expedition should ever express a longing for any food which he was accustomed to eat when in civilization and which he was then without; furthermore, he thought that everyone should agree with him that a diet of meat alone was as agreeable as any other" (*Op. cit.,* pp. 122–123).

[16] Noice (1924, p. 124) stated that Castel was to head due south from [Cape] Isachsen, map the east coast of Borden Island and then continue on to Melville Island. Castel, he commented, was the only man among them who was capable of mapping that as-yet unmapped part of the island.

all three men. They reached it the following day, and at a little after midnight of June 15 Stefansson let Noice take the first step ashore. Unable to undertake a lengthy walk inland because his ankle still pained him, Stefansson sent Noice to explore as far inland as he could manage. Andersen wandered along the beach and near-shore, seeing what he could find. Snow covered some of the terrain; the rest was gravel with little vegetation and no visible bedrock. Low rolling hills occurred near the coast, a gentle dome lay farther inland. Andersen discovered small pieces of petrified wood on hilltops two hundred feet above sea level.

It is somewhat ironic that Stefansson allowed one or other of his assistants to take the first steps on each new island discovered. He was thus neither the first to see them nor the first to step on them, yet he was the person later credited with their discovery.

To mark their discovery of this previously unknown island on which they were camped, slightly north of the 80th latitude, they built a three-and-one-half-foot-tall stone cairn on the northern tip of the island, a mile east of their camp, and deposited within it a can containing a written record of that event.[17] Stefansson stated on this record his intention to map its eastern coast, then determine the extent of Findlay Island and spend the summer on "First Land" (Borden Island).

Stefansson gave this unknown island the temporary name, "Second Land, because it was the second island of considerable size to be discovered by us."[18] Later he gave it the formal name, Meighen Island, in honour of Arthur Meighen, the man who was Minister of the Interior for Sir Robert Borden from 1917 to 1920, then succeeded him briefly in 1920–1921 as Canada's ninth prime minister, just about the time Stefansson was completing his book, *The Friendly Arctic*. The island was roughly three-cornered, about forty miles in diameter, and "the most nearly barren land

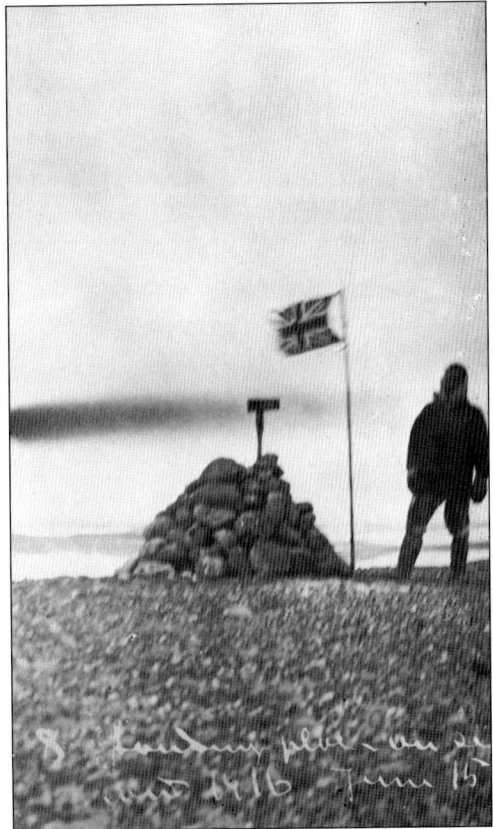

Figure 28. Stone cairn erected by Stefansson's men at landing site, British flag to claim land for Canada, and Harold Noice on Meighen Island (which Stefansson initially called "Second Land"), June 15, 1916.

Photo: © Canadian Museum of Civilization, photo by Vilhjalmur Stefansson, 50688

---

17 The message was recovered in 1955, when the Geological Survey of Canada undertook a major reconnaissance-geology mapping project—known as "Operation Franklin"—which examined most of the Arctic islands with the aid of helicopters, fixed-wing aircraft with balloon tires and ships.

18 Stefansson, 1921a, p. 299.

I have seen in the Arctic,"[19] with only a small amount of grass and some lichens and mosses. However, he claimed to have seen three types of gulls, one or two types of sandpipers, and many Hutchin's geese. In fact he considered the island to be rather a paradise for the Hutchin's goose.[20] There were no mammals, however, forcing Stefansson's party to live on the seal meat already on their sled, and several goose eggs they collected.

On June 17 Stefansson sent Noice and Andersen to sled around the coast to the north while he hobbled across the island, a distance of ten miles. He later remarked on the scattering of sea shells all across the island and the presence of beach gravels, which he correctly interpreted to be the result of land rising in recent times, evidence similar to what he had noted elsewhere in the High Arctic region. Stefansson's interpretation was an early recognition of what later became recognized as large-scale land rising relative to sea level (post-glacial uplift) following the gradual disappearance of thick ice masses covering much of Canada some ten thousand to twelve thousand years ago.

The next day, as he stood on the north coast of the island, he noted the existence of a smaller, low-lying island a short distance to the north, but did not bother to name it. He also observed several ivory gulls nesting on a small reef that lay between the two islands, and on impulse walked to the reef and collected a nest with two eggs for museum biological specimens.

Standing at that far northerly latitude, about 1,200 miles from their base camp near Cape Kellett, or about 950 miles from the *Polar Bear* camp in Prince of Wales Strait, the three men must have felt an intense sense of quiet and isolation, yet deep satisfaction and pride at what they had managed to discover and accomplish.

Heading South

"By June 22nd the sun had gone as far north as it intended, and so had we," commented Stefansson.[21] It was time to return to Melville Island. Before heading south, however, he and his two companions took turns examining the horizon through field glasses for signs of the towering mountains that Peary had believed he had seen in 1906, the land to which he had given the name "Crocker Land." They saw nothing. When Stefansson was finally satisfied that such a land simply did not exist, an interpretation that has since proven correct, they started south.

By June 28 they reached the southernmost tip of "Second Land" (Meighen Island) and headed due south across the frozen sea for Amund Ringnes Island, which they sighted two days later. Melting temperatures were producing much wet snow, slush and standing water, making it very difficult for the dogs to haul the sled and forcing the men to wade much of the time. They landed on the west coast of Amund Ringnes Island on July 2 and called a halt there because they were all suffering from snow-blindness. Andersen especially was in great pain.

Three days later they resumed their journey southward along the west side of Amund Ringnes Island. As they progressed, Stefansson made significant corrections to the shape of

---

[19] Stefansson, 1921b, p. 522.

[20] A very small, short-billed subspecies of the Canada goose (Taverner, 1940, pp. 80–81; Godfrey, 1966, p. 49). The name is no longer used.

[21] Stefansson, 1921b, p. 523.

the coastline as it was depicted on the old British Admiralty Chart. He also hunted some distance inland, for they were in need of food. His men, meanwhile, followed along the coast with the sled and dogs, keeping a straight line from one point to the next. On July 12 Stefansson shot three caribou, which alleviated their food shortage briefly. Two days later they crossed Hassel Sound to Ellef Ringnes Island, periodically taking depth soundings as they moved across the fifteen-mile-wide channel. Then they continued southward for another fourteen or fifteen miles.

MacMillan Got There First

Near the south end of Ellef Ringnes Island, Stefansson's attention was drawn to a mound that included several pieces of board. The mound was on the top of a nearby point. Drawing closer he recognized the familiar red tins of the Underwood brand of man pemmican, which was produced by the American manufacturer that had supplied his own Expedition. The boards carried lettering that indicated they had come from boxes of condensed milk. There were also empty rifle cartridges of American manufacture lying around. Stefansson realized immediately that his rival, Donald MacMillan, had been here recently, for being a disciple of Peary, MacMillan considered condensed milk an essential food item along with pemmican, hard bread and tea. At that location he was exploring as far from his base camp in Greenland as Stefansson was from his on Banks Island.[22]

> I knew that Donald MacMillan's Crocker Land Expedition had its base at Etah in north Greenland, but I hardly expected them to be working down in this vicinity. Yet the distance from his base at Etah to the west coast of Prince Patrick Island by way of the sled route is no greater than the distance from our base at Cape Kellett to the north end of Meighen Island, so that if both of us went equally far from home our fields of work would overlap by two hundred miles.[23]

A few moments' search led to the discovery, in a used can Stefansson's men extracted from mud, of a neatly written note on letter paper bearing the heading "Crocker Land Expedition." It revealed that MacMillan had paused briefly at this site on April 23, 1916, while returning to Etah from Findlay Island. With him were three Inuit, who were doing his hunting, and he reported the killing of thirteen bears, thirteen seals, sixteen hare, two ptarmigan and thirty muskoxen. He also had lost eight dogs out of forty-seven (three from illness, three from exhaustion on the trail and two killed by bears). As for MacMillan's beacon, having been made chiefly of chunks of hard frozen earth on top of a box, it had softened during the summer and collapsed to only a fraction of its probable original size. Stefansson removed MacMillan's record from the tin, replacing it with a copy that he made, together with a record of his own, and with his men rebuilt MacMillan's cairn with mud and empty cans, taking the boards for fuel.

Realizing the significance of the occasion, he later commented "It was East meeting West for the second time in arctic exploration, the other case being that of McClure [sic] and Kellett

---

[22] For more on MacMillan's expedition, see MacMillan (1918).
[23] *Op. cit.*, p. 528.

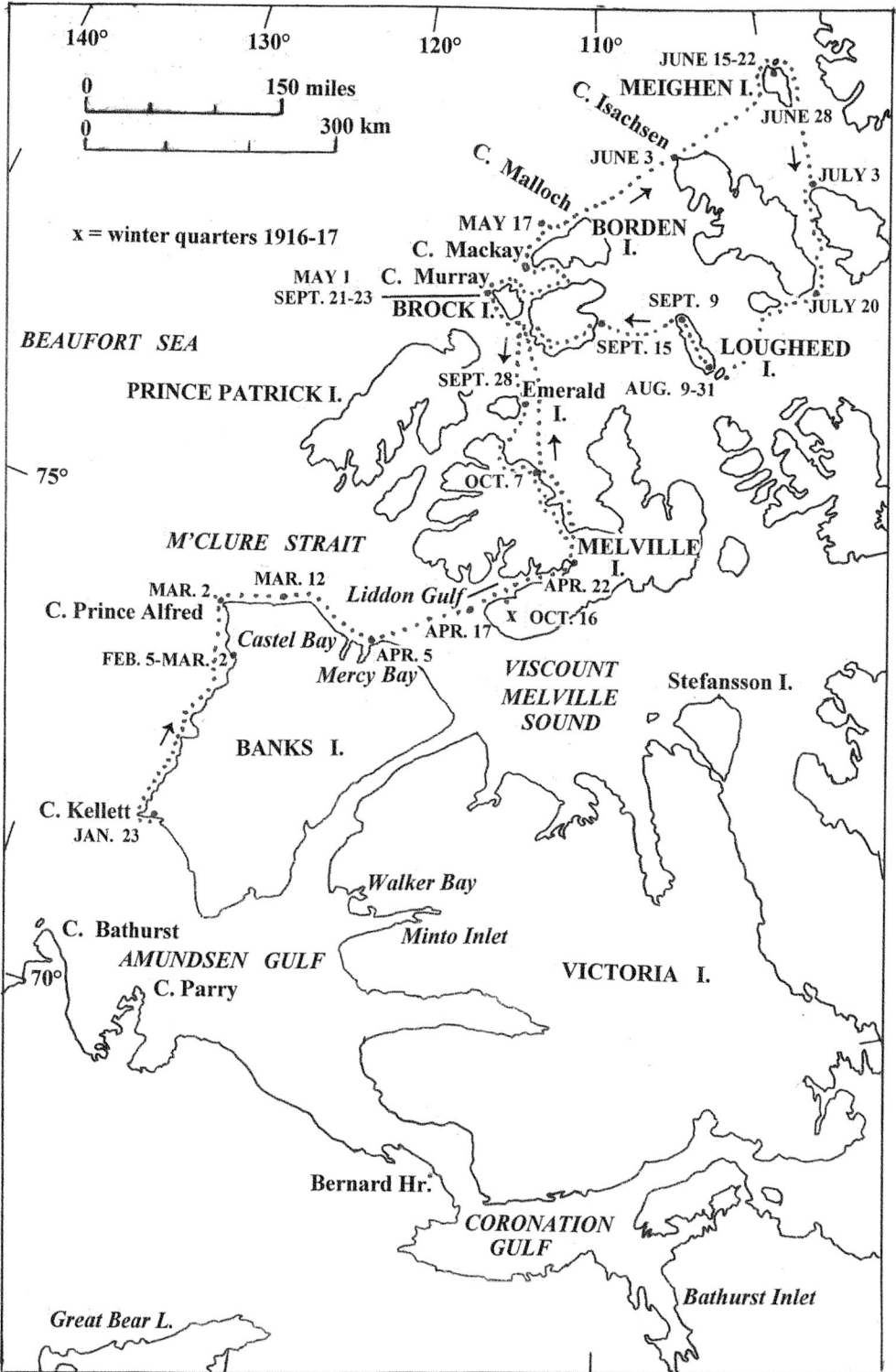

Map 10   Route of Stefansson's 1916 ice trip. (After a map in Stefansson's *The Friendly Arctic*, 1921, facing p. 450.)

at Melville and Banks Islands in 1853."[24] Stefansson further remarked that this occasion was also a meeting of two different methods of exploration.[25] MacMillan was the sole white man on his exploration party, with three Inuit hunters, a great number of dogs pulling several sleds well laden with provisions, specifically pemmican, condensed milk, hard bread and tea. Stefansson, in contrast, was travelling at this time with two white men, no Inuit, one sled with seven dogs (an eighth dog had died) and was obtaining food from the sea or land (seals, caribou and birds) en route. Of course one of Stefansson's personal objectives was to prove to one and all that a man could "live off the land" in the Arctic. It took Noice, however, to state clearly one of the fundamentals for living in that manner: "One of the first rules to be observed in living off the country, especially if one is entering unknown territory, is to take the fewest possible number of men and dogs."[26]

Noice undoubtedly learned that valuable lesson from Stefansson, yet the latter chose to give the general impression to the public that any number of people could live in the manner he advocated. Stefansson also avoided admitting that the limited animal population could not sustain a large number of predatory humans for very long.

Stefansson's party left Ellef Ringnes Island and headed a little south of west across Danish Sound to determine whether the land identified on his British Admiralty Chart as "Findlay Island" was the same as King Christian Island on Otto Sverdrup's more recent map. Findlay Island had been discovered by Lieutenant Sherard Osborn in 1853, while searching for the Sir John Franklin Expedition, whereas King Christian Land was seen by the topographer Captain Gunnar Isachsen from Ellef Ringnes Island in 1901 while he was a member of the exploration party led by Norway's Otto Sverdrup. Stefansson soon discovered that the large island Sverdrup had mapped as King Christian Island was at best a much smaller island.

Travel conditions worsened appreciably as Stefansson and his two companions headed over the sea–ice towards Findlay Island. The men were again frequently forced to wade, and the dogs had to swim while towing the partly floating sled across narrow channels between one ice island and the next. On August 1 they were subjected to drizzling rain. The next day they sighted land to the southwest and headed for it. What they had seen initially proved on closer viewing to be several islands. They camped on the ice, and then Stefansson followed a shore lead for half a mile and landed alone on one of the islands, where he soon succeeded in shooting seven caribou bulls. His companions then brought their sled to where the caribou lay. In due course they loaded several hundred pounds of boneless caribou meat on it and headed north along the ice off-shore for a larger island. Before they reached it, however, a hickory fender at the front of their sled broke after an especially hard shock, when the front end of the heavily laden sled dove from one ice island into the next. Forced to cross water in order to land, and lacking a well greased tarpaulin with which to make a sledboat, they ingeniously inflated four sealskins into air bladders to serve as floats, which took them several hours. In this way they managed to reach shore safely.

---

[24] *Op. cit.*, p. 529.

[25] *Ibid.*

[26] Noice, 1924, p. 123.

Summer on "Third Land," 1916

Once ashore, on an island of well-vegetated, rolling hills, Stefansson decided to spend the balance of the summer of 1916 there. His sled was damaged, and further travelling on the ice was too risky for several more weeks. Leaving Noice to guard the camp, he and Andersen explored the island, finding it to be roughly forty-five miles long by twelve miles wide. As it was not shown on his old Admiralty Chart, he tentatively assigned the name "Third Land" to it. Later he named it Lougheed Island after Sir James Alexander Lougheed, leader of the Conservative Party in the Senate, and Minister of the Interior and Minister of Mines in Arthur Meighen's cabinet in 1920–1921.

From his examination of Lougheed Island, Stefansson concluded that it harboured about three hundred caribou, but no wolves or mosquitoes. There was plenty of grass, fine for the caribou, which Stefansson described as "the fattest for the season that I have seen anywhere,"[27] but there appeared to be no combustible vegetation, which was a big problem for his party. (The following year Stefansson came upon a sizeable "coal mine" when he returned to the same island.[28]) Fortunately, they had brought numerous boards with them from the MacMillan cairn on Ellef Ringnes Island. These, duly splintered and smeared with caribou fat, served their fuel needs fairly well during their four-week stay on the island.

From the tops of hills on various parts of the island, Stefansson was able to see and record the compass bearings on points on Bathurst Island, King Christian Island and Borden Island, which would serve as helpful guides to map-plotting later on. However, he was unable to see Melville Island. While they were camped on Lougheed Island, he and his companions took tide observations every ten minutes for thirty hours, which added useful data to the observations he or others had made in scattered localities around the Western Arctic.

Disappointment at "Borden Island"

Snow fell late in August, so on September 3 Stefansson, Noice and Andersen headed north along the west coast of Lougheed Island. They reached the northwest corner of that island on September 8 and the next day started across the sea-ice towards "Borden Island." Their ultimate destination was Cape Murray on Brock Island, where Stefansson expected to find Natkusiak encamped in accordance with instructions given him earlier. Stefansson now planned to use Cape Murray as a base from which to commence his fourth ice trip in the spring of 1917. He led the way while they crossed the ice, constantly probing the snow and ice with his staff to ensure their safe passage. Noice described the conditions they encountered in this fashion: "It was hard work crossing the strait. There was so much soft snow on the ice that the sledge at times sank down to the toboggan bottom and the dogs got stuck. Their feet found no hold on the glass-like ice under them."[29]

---

[27] Stefansson, 1921b, p. 542.

[28] Stefansson, 1921a, pp. 301–302. The material Stefansson called "coal" varied from tree trunks to deposits of lignite. They had seen deposits on Amund Ringnes Island, Lougheed Island, Melville Island and Banks Island.

[29] Noice, 1924, p. 169.

The distance across the ice to "Borden Island" was about fifty miles, which they covered in seven days, reaching that island's east coast on September 15.[30] Instructing his two companions to take the sled southward along the coast, taking note of its shape while they progressed, Stefansson employed his customary procedure of travelling parallel to the coast but some distance inland, which permitted him to obtain a geographic impression of the land, to hunt caribou and still return to the coast towards evening to rejoin the others. On September 20, however, he changed his routine and followed the coast, coming upon a very large log after travelling about three miles. In the treeless Arctic region the discovery of such an object is highly unusual. Andersen and Noice measured it as seventy-five feet long, with a circumference of twelve feet.[31] At the time, however, their only interest was evidently in its potential as firewood, a little of which they obtained by breaking off the tips of some roots. Today, I suspect, persons would wonder what kind of tree it had been and its possible source and age. It had certainly not come from anywhere on the Canadian Arctic islands, unless perhaps it had come from one or other of the fossil forests now known on Axel Heiberg Island or Ellesmere Island. If that were the case, however, it would be readily recognized by the kind of tree and its condition. If it had drifted to its present location from the North American mainland, it would have had to have come from hundreds of miles inland from the coast and then hundreds of miles north, a seemingly impossible happening. No trees of such dimensions are found today anywhere near the North American Arctic coast. The probable source of this unusually large log would therefore appear to have been Siberia.

Stefansson had expected to find a cache of skin boots, ammunition and other equipment somewhere on the southwest coast of what he at the time thought to be "Borden Island," left there by Castel as per his instructions. Instead, he found nothing, not even a note of explanation. Likewise, no one was at Cape Murray nor was there a message there when they arrived, the revelation of which upset Stefansson so much that he made no entries in his diary for almost a week. Faced with no cache from which to replenish his supplies, increasing darkness and concern over what might have happened to Castel and Natkusiak, Stefansson concluded that he had no option but to abandon his planned spring ice trip northwest from Cape Murray and head south to Melville Island.

South to Melville Island

The autumn equinox was upon them, with its fog and dropping temperatures, as well as unsuitable snow conditions for making snowhouses. Consequently, Stefansson and his companions were forced to use their double tent for shelter. With inadequate fuel to heat it, they were terribly cold and damp, which added to their general discomfort and misery. On September 23, they headed for Melville Island, passing Emerald Island in a snowstorm. Their progress was also unduly slow because of deep snow, and the 130-mile trek took them a week. The dogs had to be fed worn-out boots for lack of other nourishment. They finally reached

---

[30] In reality, they reached the east coast of the island later named Mackenzie King Island near its southeast point.

[31] Noice, 1924, p. 172. The tree-trunk had a diameter of about four feet. This had obviously once been part of a very large tree.

the east side of the northwest peninsula of Melville Island on October 2 and camped, completely out of food. The next day "Dame Fortune" smiled upon them, for Stefansson sighted and killed two muskoxen, which provided them with a good supply of fresh meat for both men and dogs. This was most timely, for they were almost immediately engulfed in a snowstorm of several days' duration.

Natkusiak's Camp near Cape Grassy

When the snowstorm finally passed, they continued southeast for several days along the coast of northwest Melville Island. Near Cape Grassy Stefansson discovered Natkusiak in a comfortable camp, with a good supply of lignite coal nearby for fuel. At the camp were Alingnak, his wife Guninana, and their foster daughter, Topsy Ikiuna, whom Natkusiak had married ("Native-style") during the past year. Any ill feelings Stefansson might have held because Natkusiak and his people were located here rather than at Cape Murray, where he had instructed them to establish their camp, were quickly quenched by the warmth of the fire blazing in the open fireplace in their muskox-skin camp and the realization that a valuable source of fuel (in the form of lignite) occurred nearby, so far north. Natkusiak's reason for failing to camp at Cape Murray was that his heavily loaded sled had been damaged beyond repair on the rough ice nearby, forcing him to spend the summer at Grassy Cape.

Stefansson and his two travel-weary companions stayed briefly at Natkusiak's camp to rest and to have their footwear repaired by the two women. During that time Stefansson concluded that this locality was actually a far better outfitting base for a spring ice trip than Cape Murray.

Castel's Report

A report from Castel awaited Stefansson at Natkusiak's camp. When he, Emiu and Natkusiak left Stefansson, Noice and Andersen at Cape Isachsen on June 4 they had headed for the northwest corner of King Christian Island, but found no land where it was shown on their map. In foggy weather, they turned west and came upon land where none occurred on their map. Thinking it was King Christian Island but misplaced on the map, Castel left a cairn with a message. Stefansson later credited Castel with landing on the north end of Lougheed Island, making Castel and not Stefansson the discoverer of that island.[32] Castel encountered Vasey Hamilton Island but not Markham Island while heading for Cape George Richards on the northeast corner of Melville Island.

Castel then decided it was too late to head for "Borden Island" and leave supplies there, as Stefansson had instructed him to do, so instead proceeded to the base camp in Liddon Gulf, where he found Storkerson. Storkerson had then outfitted Natkusiak, who with the Alingnak family had started for Cape Murray. En route their overloaded sled had been badly damaged and they had therefore ended up at Cape Grassy.

The Base Camp in Liddon Gulf

A week later, after proceeding southward from Natkusiak's camp and crossing to Liddon Gulf, Stefansson chanced upon Storkerson and several of his men, and joined their base camp near

---

[32] Stefansson, 1921b, p. 563.

Cape Ross. Storkerson's people had killed ninety muskoxen, twenty-seven seals and two or three polar bears during the summer, and dried a very large quantity of meat from them, exactly as Stefansson had wanted. Lacking anything with which to keep the meat off the ground for drying, all suitable meat had been laboriously removed from the skeleton, sliced thin and spread out on stones to dry.[33] The Iñupiat women in his camp had been sewing clothing and footwear as well, which also pleased Stefansson greatly.

Storkerson had established the Melville Island base camp on the east side of Liddon Gulf near Peddie Point, where bituminous shale capable of burning lay exposed. Here was constructed a house of muskox hides "… with a main floor space of twelve by twenty-eight feet and an additional sleeping alcove about eight feet by eight."[34] The men then created a stove and stove pipes from empty cans, which proved adequate for the camp's cooking needs.

There was no sign in Liddon Gulf of Captain Gonzales and the *Polar Bear*, however, although the captain had stated in a letter to Stefansson his intention to bring the ship north during the summer of 1916. Storkerson suggested that Gonzales had probably crossed to Winter Harbour, on the south coast of Melville Island many miles east of Liddon Gulf, so Stefansson sent him there, together with Castel and Emiu, to find out. They returned two weeks later with the report that they saw no sign of the *Polar Bear*, but at Winter Harbour they had found a frame house containing a note as well as a considerable amount of supplies left by Captain Joseph Bernier in 1910. The note, signed by Captain Bernier on behalf of the Canadian government, stated, "This cache is intended only for shipwrecked crews in case of bare necessity."[35] In spite of that stipulation, Storkerson had brought some of the supplies back to Stefansson's Liddon Gulf base camp.

The arrival of items such as flour, powdered milk, dried vegetables, butter, jam, coffee, hard bread, potatoes, tobaccos (both smoking and chewing) and baking powder created unexpected problems at the Liddon Gulf camp. There had been few complaints about their food, Stefansson commented, when only meat and broth were available, but soon after the arrival of the supplies from Winter Harbour, they had "… all the troubles of a boarding house."[36] More cooking time was required, which consumed more fuel than usual and made the house uncomfortably hot. The Native women complained over having to rise earlier to prepare breakfast, some of the men ate more butter than others considered reasonable, others grumbled that coffee should be served every meal and some individuals wanted more of the depot supplies while others urged they be carefully conserved. Stefansson listed still other complaints, all arising from the arrival of the food supplies from Captain Bernier's cache, and constituting behavioural traits unwelcome in a camp where the people were isolated together for a lengthy period of time.

Storkerson reported the presence of some hardwood and iron at Winter Harbour, which could prove useful for repairing the sleds, and also brought back some kerosene and a lantern and spare globes, which would be of considerable use to Stefansson's travel plans.

---

[33] *Op. cit.,* 1921b, p. 567.
[34] *Op. cit.,* 1921b, p. 568.
[35] *Op. cit.,* 1921b, p. 577.
[36] *Op. cit.,* 1921b, p. 578.

At the end of October, Storkerson, Castel, Noice and Emiu left the hidden Gulf base camp and headed for Cape Grassy with two heavily laden sleds of dried meat and seal oil, starting the transfer of provisions to a more northerly base in preparation for Stefansson's intended explorations the next spring. En route they intended to pick up some kerosene and food supplies that they had left near the foot of Liddon Gulf a few days earlier while on their way back from Winter Harbour. At the same time, Andersen and Lopez were delegated to bring to the Liddon Gulf camp the meat of thirty-eight muskoxen that some of the Inuit hunters had killed recently. Meanwhile, Stefansson remained in the Liddon Gulf base camp near Cape Ross, busily writing reports of his recently completed exploration and jotting down ethnographical information he obtained from Guninana (Alingnak's wife). He also planned the activities for his Northern Party for the spring and summer of 1917.[37]

Seventeen members of Stefansson's Northern Party and about fifty dogs spent the winter of 1916–1917 on Melville Island.[38] "We lived in houses of musk-ox skins and burned coal in sheet-iron stoves which we made of tin cans that we happened to have with us. For light we used musk-ox tallow. We were able to clothe ourselves suitably with caribou skins that had been secured during the summer, and we spent the winter in comfort and the best of health."[39]

---

[37] While at Herschel Island in August 1915, Stefansson received clear instructions from G.J. Desbarats, Deputy Minister of the Naval Service, to terminate his Expedition in the summer of 1916. He chose to ignore those instructions, however, reasoning that the government would surely approve his continuation beyond that date after it learned of his discovery of new land in 1915. He then avoided getting any mail in 1916 by staying on Melville Island for the summer. A large bag of mail intended for him and members of his Northern Party reached Banks Island on the *Herman* in the summer of 1916, but was lost when Captain Peter Bernard perished along with Thomsen trying to take it by sled to Stefansson on Melville Island. Stefansson's decisions and actions at that time appreciably increased the cost of his Expedition and greatly embarrassed senior government officials and one or more cabinet ministers. Several years after his return south Dr. Anderson commented in some of his correspondence with American friends and acquaintances that Stefansson deliberately remained out of touch in the Far North until after the war ended in order to receive the acclaim and publicity he sought (see, for example: (1) Anderson (1921c); (2) Anderson (1921e); and (3) Anderson (1922d).

[38] Stefansson, 1921b, p. 581.

[39] Stefansson, 1921a, p. 302.

STEFANSSON'S FINAL EXPLORATIONS, 1917

15

Stefansson intended to complete his exploration in the High Arctic during the spring and summer months of 1917. Accordingly, at the end of February that year he instructed several teams of men to move dogsleds loaded with supplies one hundred miles from his base camp on the south coast of Melville Island near the mouth of Liddon Gulf to Cape Grassy on the north coast. Then just as he and Storkerson were about to head for Cape Grassy, Captain Gonzales arrived by dogsled from the *Polar Bear*. Stefansson immediately asked him why he had not brought his ship to Liddon Gulf the previous summer (1916) as instructed. The captain's response was that Prince of Wales Strait had remained blocked with ice throughout the summer, preventing his ship from travelling north.[1] The *Polar Bear* and its crew had moved since then from their camp near Armstrong Point to a site farther south in Walker Bay on the south side of Minto Inlet, where fresh water, caribou and Arctic char were readily available. He also brought news, albeit many months old, that the Southern Party had left the Arctic in 1916, and that Thomsen was intending to return to Liddon Gulf from Cape Kellett by way of the west coast of Banks Island in November with two new sleds and the mail received during the summer of 1916.[2]

Gonzales had obtained this latter news from Crawford, the former engineer on the *Mary Sachs,* whom Wilkins had left at Baillie Islands in August 1915. Crawford had returned from Nome to Victoria Island in the summer of 1916 with the schooner *Challenge*, which he and his new partner, Leo Wittenberg of Nome, had purchased from Captain Lane. They were wintering in Minto Inlet about half a day's dogsled journey from the *Polar Bear*, where

---

[1] Naming Lorne Knight as his source of information, Harold Noice later claimed that Captain Gonzales could have reached Melville Island with the *Polar Bear* in the summer of 1916 because the Prince of Wales Strait had been free of ice. Why he did not do so, in defiance of Stefansson's orders, was apparently because of the following drama, one that Noice described as The Ladies' Mutiny. All of the Inuit women at the *Polar Bear* camp wanted to return to their homes. When the wives of Gonzales and Seymour learned that they would be sent back to Herschel Island from Victoria Island with the families of Illun and Pikalu while their husbands would accompany Stefansson on the *Polar Bear* from Melville Island east to the Atlantic Ocean and south to Montréal, they set up such a scene of weeping and wailing, in which they were quickly joined by other Inuit wives, that Gonzales had relented and wintered instead at Walker Bay, where fresh water and salmon were available (Noice, 1924, pp. 242–243; see also Ashlee, 2008, pp. 145–146). Later Captain Gonzales destroyed the *Mary Sachs* to ensure that he escaped from the Arctic before Stefansson reported his disobedient and destructive actions.

[2] Stefansson, 1921b, p. 599.

they were trapping foxes and trading with the Inuit with the intent of making a fortune.[3] They were not connected in any way with the Canadian Arctic Expedition at this time, but would soon be briefly but actively involved with Stefansson's Northern Party. Gonzales was unaware when he presented his news to Stefansson that Thomsen had failed to reach Stefansson's camp.

Preparing for the 1917 Ice Trip

Storkerson and four men left Liddon Gulf with four loaded sleds for Cape Grassy on March 3. Stefansson and Emiu followed two days later, soon catching up to the Storkerson party. The temperature was between -50° and -60° F, and the travelling was extremely uncomfortable, especially when blizzards enveloped them. Stefansson considered it his most uncomfortable winter trip.[4]

At Cape Grassy he found the meat supply low, so stopped to hunt more muskoxen, but sent Castel, Noice, Andersen, Knight and Pikalu ahead to "Borden Island" with three sledloads. He then followed, after shooting a muskox, reaching Cape Mackay on the west side of "Borden Island" on March 30. Several days later he sent back three Inuit men—Illun, Pikalu and Ulipsinna, all originally from the *Polar Bear*—with two sleds and twenty-nine dogs, and instructions to take Alingnak and the Iñupiat women at Natkusiak's Grassy Point camp back to the main camp in Liddon Gulf.

On April 7 Stefansson joined forces with Castel's advance party at the edge of the ice off the northwest coast of "Borden Island" and headed seaward. Periodic soundings thereafter as they progressed farther from land consistently indicated depths of about 1,460 feet.

Sending the Support Party South

On April 16 Stefansson sent Castel and Storkerson south with several men and various instructions. They were to travel together to Liddon Gulf, then Castel and Karsten Andersen would cross M'Clure Strait to Mercy Bay and follow the west coast of Banks Island south to Cape Kellett, looking for signs of Thomsen. After reaching Cape Kellett, Castel and Captain Pete Bernard were to fix up the *Mary Sachs* and in due course launch it. Late in August they were to load everyone at Cape Kellett on board and proceed to Herschel Island and then to Nome, where they were to notify the Canadian government of their actions.

Natkusiak and his young wife, Topsy Ikiuna, together with her adoptive parents, Alingnak and his wife, Guninana, were to accompany Castel and Andersen as far as the schooner *North Star* on the west coast of Banks Island. Natkusiak was keen on obtaining the schooner and Stefansson was prepared to sell it to him.[5] Natkusiak and his in-laws then spent the next four years trapping Arctic foxes on Banks Island, the first two of which were around the *North*

---

[3] A Seattle newspaper in the fall of 1919 reported that Crawford returned south with a thousand fox skins, which he hoped to sell for $50,000. (Anderson, 1920a (January 7, 1919 [*sic* 1920a])).

[4] Stefansson, 1921b, p. 602.

[5] The *North Star* was turned over to Natkusiak in place of $2,000 of wages. Some supplies were included (Natkusiask file, RG42, Vol. 471, File 82-2-5, LAC).

140°   130°   120°   • APR. 24   100°

↑ ↓

0   150 miles

0   300 km

ELLEF RINGNES I.

APR. 17 •↘

..... Stefansson's route 1917
— — — Storkerson's route 1917

BORDEN I.   MAY 13-27
APR. 3

BROCK I.

PRINCE PATRICK I.   MAR. 30
JUNE 13

—75°   MAR. 22-25

M'CLURE STRAIT   **MELVILLE I.**

BEAUFORT SEA   MAR. 5   Dealy I.   JUNE 28-JULY 4
x   Winter Hr.
JULY 6

↙ **VISCOUNT MELVILLE SOUND**

BANKS I.   JULY 27   Stefansson I.

JULY 14

C. Kellett   JUNE 6
AUG. 23

JULY 31
x = winter quarters of
Polar Bear 1916-17

C. Bathurst   Walker Bay
Minto Inlet

—70°   **AMUNDSEN GULF**   **VICTORIA I.**

Bernard Hr.   **CORONATION GULF**

Great Bear L.   *Bathurst Inlet*

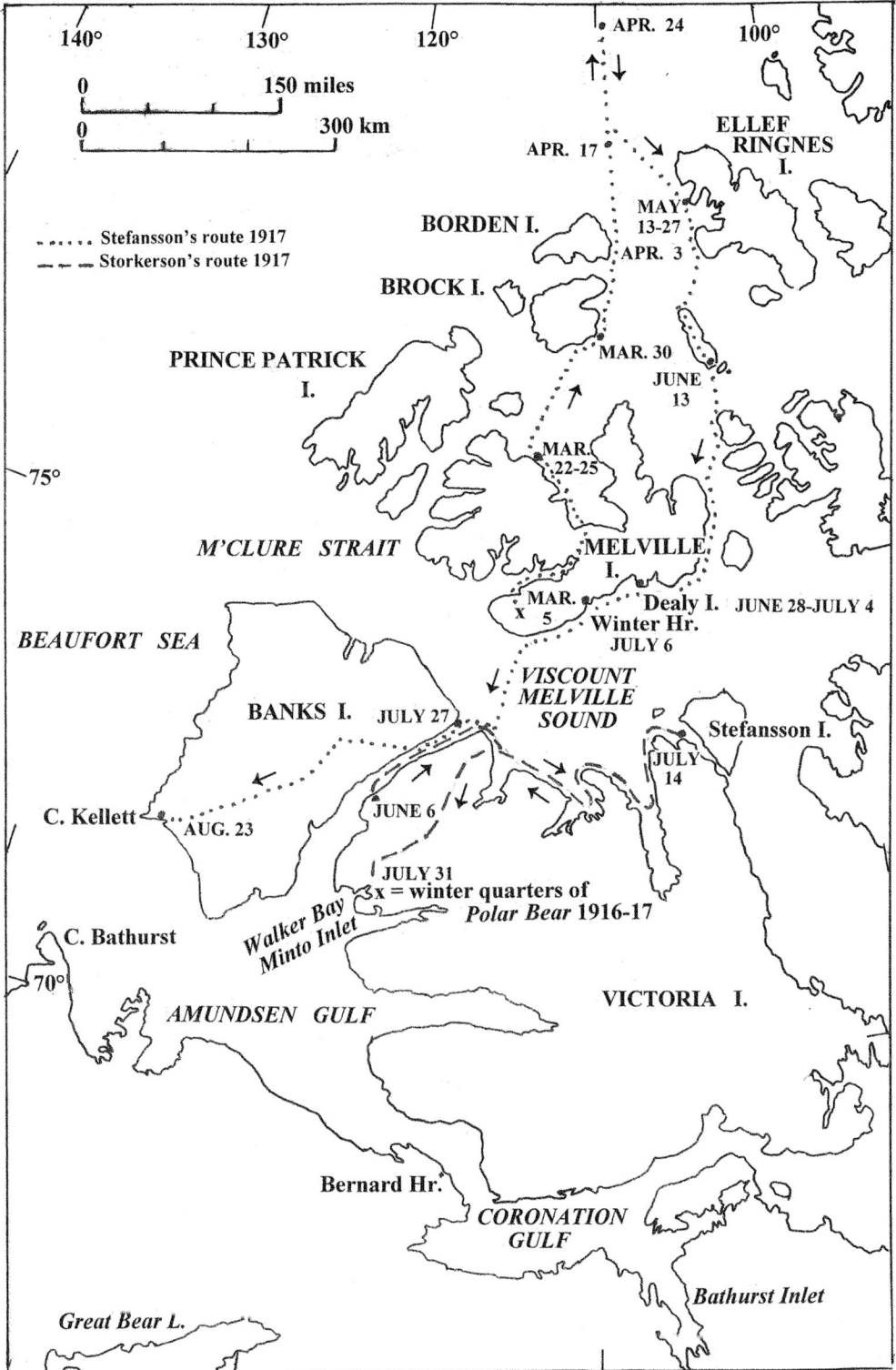

Map 11 Route of Stefansson's 1917 ice trip. (After a map in Stefansson's *The Friendly Arctic*, facing p. 450.)

Star site followed by two years near Cape Kellett. In 1921 they moved on the *North Star* to Baillie Islands and settled there for more than a decade. A storm destroyed their schooner there in 1928.[6]

Meanwhile, Storkerson was to continue with Captain Gonzales and all the Inuit from Liddon Gulf camp to the *Polar Bear* camp, where the captain was to supply him with a party of men and equipment to complete the mapping of the northeast coast of Victoria Island. When that task was done, Storkerson was to return to the *Polar Bear,* and it was then to proceed to Cape Kellett and assist if necessary in launching the *Mary Sachs.* Meanwhile, Stefansson and his three companions would try to reach Cape Kellett on foot before August 20, 1917, that is, before the two ships left Cape Kellett for Herschel Island and Nome.

These instructions were intended to wind up the explorations of Stefansson's Northern Party. All seemed straightforward except Stefansson's plans for himself and his three remaining men (Knight, Noice and Emiu). In *The Friendly Arctic,*[7] he suggested four fanciful options. If he found new land he and his three companions might spend a year on it, or a year on the ice, or might go east by way of Ellesmere Island to Greenland and come out on a trading vessel, or might cross Victoria Island to Bernard Harbour and go from there south by way of Great Bear Lake. All reflected his own interests, of course, and his absolute belief that he and his three companions could survive on the meat of seals, polar bears, caribou or muskoxen wherever they went, that they would have enough ammunition and would suffer no equipment damage, illness or injury. He was determined to show one and all that white men were capable of "living off the country" in the North.

The Third Ice Trip

On April 16, after the support party had departed, Stefansson, Knight, Noice and Emiu, with two teams of dogs and sleds, continued their north-northwesterly exploration from "First Land" (Borden Island). The depth-sounding readings they took as they progressed seaward deepened to 1,997 feet about one hundred miles from land, and then shallowed slowly to 1,612 feet at 140 miles.

Stefansson had just started to wonder if the soundings indicated land in the direction they were travelling when Noice and Knight complained of feeling ill. From their responses to his questions, Stefansson quickly realized they showed symptoms of scurvy, symptoms similar to those shown by Andersen two weeks earlier. Both Andersen and Noice, Stefansson now learned, had refrained from eating fresh meat during the summer, in defiance of Stefansson's instructions. Instead they had eaten dried meat and groceries they had obtained from Captain Bernier's depot when they were repairing sleds at Winter Harbour. Knight had fallen in with

---

[6]   Condon, 1996, pp. 100–102. Billy Natkusiak moved his family from Baillie Islands back to Banks Island in 1933, then the following year to the west coast of Victoria Island, remaining chiefly in the Walker Bay area before moving to Holman (now known as Ulukhaktok) in 1938 or 1939. Because of his early years on Banks Island (originally called Banks Land in 1820 by Lieutenant W.E. Parry of the Royal Navy), Natkusiak was often called Billy Banksland.

[7]   Stefansson, 1921b, p. 612.

their eating habits after he joined them in Liddon Gulf. Stefansson knew that the vitamin C of fresh seal meat could relieve their ailments, but the ice conditions all around their present location—pressure ridges and old ice—were unlikely to attract seals.

The serious condition of his two men and the absence of seals forced Stefansson to turn back on April 26 from about 80° 30' N latitude, 113° W longitude, and head for the nearest land, which was Cape Isachsen, about 125 miles away. They reached that destination on May 11, after chopping laboriously with their pick-axes for several hours to get through a new pressure ridge a short distance off shore. With their supply of food down to three days and no likelihood of seeing any seals, it was imperative to find caribou, for Noice was barely able to walk and Knight was almost as badly crippled, although both men tried their best to be useful. The next day Stefansson and Emiu shot and killed an entire herd of twenty-three caribou. For the next two weeks, they remained at an inland camp, which they called "Camp Hospital," during which time the two patients made a noticeable recovery, initially in spirit, then more slowly physically.

By May 27 Noice and Knight were well enough to move on, so they all started south for Lougheed Island. They reached that destination four days later, after being delayed both by thick fog and attacks of snow-blindness. As in the previous year Stefansson encountered many caribou, but unlike the previous year, there was now an equal number of wolves on the island, which made the caribou wary and hence more difficult to hunt. Nevertheless, Stefansson managed to kill several caribou, and with the addition of the fresh meat the men continued south to the site on the island where they had camped the previous year. En route they passed a large deposit of lignite coal and paused to load as much of it as they dared onto their two sleds.

They remained at this camp long enough to obtain solar observations on their location, which would permit a check on their readings of the previous year. Then on June 13 they continued south, this time heading across Byam Martin Channel towards the east side of Melville Island, taking depth soundings along their line of progress between Lougheed Island and Melville Island. Stefansson intended to reach Melville Island just west of Bradford Point. As they neared that point Stefansson was pleased to notice that both seals and the tracks of polar bears were unusually plentiful. They still had a good supply of caribou meat, however, so continued on their way.

After passing Bradford Point they rounded the southeast coast of Melville Island and on June 25 came upon a rock cairn, about five feet high, built by members of Captain Bernier's crew from the CGS *Arctic* in 1909, shortly after his visit to Winter Harbour, according to a note they found in a glass fruit-jar.[8]

Dealy Island and Winter Harbour

Three days later Stefansson and his companions reached the depot Captain Kellett had left at Dealy Island in the spring of 1853, during his search for the Sir John Franklin Expedition. This depot was marked by a large pyramidal monument of black-painted boulders on top of the highest hill on the island. A tall pole rose like a mast from the middle of the boulders. Nearby at the base of a steep cliff was a large stone house, still largely buried by a huge snowbank. The

---

[8] Noice, 1924, p. 238.

house's original wooden roof had caved in under the weight of drifted snow some years after its construction, and had been replaced by Captain Bernier's men in 1909, but this roof, too, had collapsed under the weight of the snow. Great oaken barrels, stacked in three tiers, lay within the house, their contents identified by large printed letters. According to their labels, some contained flour, others sugar, peas, molasses and other food items, much of it unfit for consumption. Nevertheless, Emiu managed to salvage some flour from one barrel and sugar from another, which he considered edible, although the flour was lumpy and smelling a bit sour, and prepared a meal of palatable pancakes. Knight and Noice then mixed sugar with unsweetened chocolate they found in another barrel and made fudge. Curiosity resulted in the men opening still other barrels. "Some contained heavy wool sweaters; others fine brass-buttoned, scarlet-coloured and satin-lined broadcloth pea-jackets; still others had brightly coloured, fancifully designed mittens. There were also barrels of long leather sea-boots, felt shoes, knitted underwear."[9] The men each took a pair of mittens, a sweater and a pea jacket.[10]

Emiu was unexpectedly ill the next day, which Stefansson attributed to overeating rather than food poisoning, but it forced them to remain at Dealy Island until July 4. Then they continued to the west, stopping for two days at Captain Bernier's house near Winter Harbour to enable Stefansson to determine their exact location by means of good sets of solar observations.

Across Viscount Melville Sound to Banks Island

From Winter Harbour they started across Viscount Melville Sound, expecting to strike Banks Island near John Russell Point. Conditions were unpleasant during the crossing, however, with deep-water channels locally forcing the dogs to swim, the sleds to float, and the men to wade in icy water, waist deep. To add to their discomfort, it rained heavily on July 21. Sharp ice needles sliced the men's footwear and lacerated the feet of the dogs. Somewhere en route, Stefansson's normally accurate sense of direction went astray, for instead of arriving on Banks Island they ended up east of Peel Point on the north coast of Victoria Island, far to the east of their intended destination. Turning west they continued on to Peel Point and crossed Prince of Wales Strait to Banks Island. A shore lead about two hundred yards wide halted them as they approached that island. Their sled canvas was full of holes and hence of no use to make a sledboat, and they had been forced to feed their seal pokes to their dogs weeks earlier near Cape Isachsen, so they faced a new challenge to reach the shore. As Noice described their dilemma,

> … we had therefore no means of crossing this lead except by using a cake of ice as a ferry. Fortunately there was a loose cake handy which, although not very large, yet was large enough for carrying about a third of our load. The Commander [Stefansson] and I ferried the first load across together, using shovels for paddles, and when we got close to the beach poled our ice-raft along with our tent-poles. Then, while I returned for a second load, he went to look for a good camp-site. Knight and Split [-the Wind or Emiu] made the next trip, and then Split came back with the raft and we moved the remainder ashore.[11]

---

[9] *Op. cit.*, p. 239.
[10] Stefansson, 1921b, p. 631.
[11] Noice, 1924, p. 247.

Knight showed Stefansson a natural harbour close to where they landed, a harbour originally reported by Storkerson. Stefansson later named it "Knight Harbour."[12] Nearby Stefansson found a brass cylinder in which was a folded document signed by M'Clure in April 1851, before he turned his ship back south and sailed around the west coast of Banks Island to finally winter in Mercy Bay. Stefansson built a cairn at the site, removed M'Clure's original document from the brass cylinder, and left in it a copy together with a report of his own work. M'Clure's note stated that he had seen the waters of Viscount Melville Sound northeast from that locality in 1850, which observation later entitled him to the rich British Admiralty reward for being the first to discover the Northwest Passage.[13] On his own report Stefansson stated that he and his men intended to trek overland to Cape Kellett.

Last Trek across Banks Island

Stefansson now left his two well-used sleds and other items on the northeast coast of Banks Island for they were too heavy to transport overland.[14] On July 28, he and his three companions donned their back packs, tied smaller back packs on their dogs, and started south along the shore of Prince of Wales Strait, with Noice making notes and sketching the shape of the coastline. This activity slowed their progress too much, however, so that after five days Stefansson led them inland. Time was fast running out for them to get to Cape Kellett so that they could depart on the *Mary Sachs* or the *Polar Bear*.

By August 13, Stefansson reckoned they were still about eighty miles northeast of the base camp near Cape Kellett. Impatient to reach the camp, he walked on ahead of the others, soon becoming enveloped in a thick fog. After more than twenty-four hours of walking, by his own account, he reached the coast several miles east of the camp. As he examined it through his field glasses, he recognized the shape of a ship, but it had only one mast so it did not appear to be either the *Polar Bear* or the *Mary Sachs*.

Treachery at the Cape Kellett Base Camp

Stefansson approached more closely and suddenly realized that the ship was indeed the *Mary Sachs*, but her foremast was gone, as were her sails, and her wheelhouse was resting about three hundred feet along the shore. There were two strangers nearby, whom Stefansson soon learned were Otto Binder, an American, and August Masik, a Russian. Both were hunters from Nome,

---

12 Stefansson, 1921b, p. 399; Jenness, 1997b, p. 39.

13 Stefansson, 1921a, p. 302.

14 Thirty-seven years later, in 1954, a helicopter airlifted the better of Stefansson's two sleds to the Canadian ice-breaker, HMCS *Labrador*, and in due course the sled reached Ottawa. Years later, the sled—plus other items left by Stefansson—was displayed at the National Archives at Ottawa and then at the Prince of Wales Museum in Yellowknife. It was ultimately returned to Ottawa in 2007 and resides in the Canadian Museum of Civilization (CMC) in Gatineau, Quebec. In 2008 a replica of the original sled was put on display in the main foyer of that Museum. Stefansson's second sled was retrieved from Banks Island by an oil company in 1971, and it too resides now in the CMC, but is in need of restoration (David R. Gray, personal communication, 2008).

who had come east from Alaska on the schooner *Challenge* with Jim Crawford and set up a hunting camp at De Salis Bay, but shortly afterwards moved to Cape Kellett. From them Stefansson now also learned much troubling news.

Firstly, there was tragic news about the deaths of his two stalwart men, Captain Peter Bernard and sailor Charles Thomsen. Stefansson had reluctantly sent Thomsen back alone from Melville Island to Cape Kellett in the spring of 1916 to collect his wife and two children, and bring them north without delay to the base camp in Liddon Gulf. Thomsen carried with him instructions from Stefansson for Captain Bernard to build more sleds and to remain at Cape Kellett. Thomsen had arrived safely at Cape Kellett early in the summer, but decided to remain there with his wife and children instead of returning immediately as Stefansson had instructed him to do.

The trading ship *Herman* reached the cape sometime in August 1916 and supplied Captain Bernard with the material he needed to build two sleds, nearly five hundred pounds of mail, and sundry other materials. Together Captain Bernard and Thomsen decided several weeks later that the mail and the two newly completed sleds should be taken to Stefansson. Ignoring the latter's instructions that Captain Bernard should remain at Cape Kellett, they started north in October 1916 for Liddon Gulf, with their dogs pulling the two loaded sleds and towing a third one. They travelled by way of the west coast of Banks Island. Bernard was an experienced Alaskan traveller, but apparently grossly underestimated the difficulties they would encounter on such a journey.

The following May, Castel and "Charlie" Andersen found two sleds just west of Mercy Bay while they were heading south for Cape Kellett. One of these they recognized as the one Thomsen took south from Melville Island the previous year, the other was a new one of the type built by Captain Bernard. A note dated December 22, 1916 was attached to one of the sleds and indicated that the two missing men had left a cache on the ice of M'Clure Strait, twenty miles north-northeast of Mercy Bay, had the mail with them, but were out of food, and half of their dogs had died. Sadly, they were less than one hundred miles from the camp on Liddon Bay, which was their destination, but for some reason were trying to return the much greater distance to Cape Kellett. Some miles west of Mercy Bay, Castel found part of a new sled with axe marks indicating it had been chopped to reduce its length and weight, and seven miles west of Cape M'Clure he discovered the snow-covered, frozen remains of Thomsen. After ascertaining that Thomsen had frozen to death Castel and Andersen replaced the body on the land, and then covered it to protect it from scavenging animals. Castel then ordered Natkusiak and Alingnak to bury the body in the spring when the ground had thawed. Captain Bernard evidently continued westward alone, leaving a trail here and there for another sixty to seventy miles, then there was nothing. No trace of his body or of the mail he was so faithfully taking with him was ever found. "Thus died two of the expedition's best men," wrote Stefansson a few years later.[15]

Secondly, there was the destruction of the *Mary Sachs*. As Stefansson later pieced the details together, when Castel and Andersen reached Cape Kellett in June 1917, the only persons at the camp were Binder and Masik. Mrs. Thomsen and her two daughters and some other Inuit

---

[15] Stefansson, 1921b, p. 653.

had left the camp near Cape Kellett for the *Polar Bear* camp with the Kilian brothers a month earlier, after Captain Bernard failed to return from taking his sleds to Melville Island.

From June to August, the long-time sailors, Castel and Masik, helped by Andersen, spent many hours restoring the seaworthiness of the *Mary Sachs*, which included caulking its seams, scraping and painting the ship, and overhauling its rigging. Binder, a mechanic, overhauled the two engines and repaired the propeller shafts. They had then managed to partly launch the vessel on August 6, in preparation for moving it three miles east along the shore to the shallow lagoon they called "Baur Harbour" (after W.J. "Levi" Baur, the cook of the *Mary Sachs*), to ensure its safety from stormy weather while they awaited Stefansson's arrival. Once the *Mary Sachs* with its six-and-a-half-foot draft was safely anchored in the lagoon, Castel intended to take depth soundings of the bay partly enclosed by the curved sandspit at Cape Kellett, in accordance with his instructions from Stefansson. That project was to ascertain if the cape sheltered a harbour that was deep enough for ships larger than the *Mary Sachs*. Unfortunately, Castel was unable to carry out this work.

On August 7, Captain Gonzales arrived with the *Polar Bear*. Castel informed him that he, Masik, Binder and Andersen had been getting the *Mary Sachs* ready to take Stefansson and his men back to Alaska, and were intending to wait for them until August 25, in accordance with Stefansson's instructions. Castel then asked Captain Gonzales to use the *Polar Bear* to pull the *Mary Sachs* off the beach. This Captain Gonzales proceeded to do. The work on restoring the *Mary Sachs* continued over the next few days. On August 11, the restoration was finally completed, so Captain Gonzales, as the senior officer present, took it for a trial trip and then examined it inside and out. Later he told Storkerson that he found it in good condition, "… the strongest built vessel he had seen of her size …"[16]

Details of what happened next remain somewhat vague, but as Noice and Ashlee recounted them,[17] Captain Gonzales, exercising his senior officer status, suddenly told Castel and Andersen to bring their possessions on board the *Polar Bear,* and then he and some of his men ran the *Mary Sachs* broadside on the beach and rendered her unserviceable. This they did by cutting down her foremast and sawing it into firewood for the *Polar Bear*, removing the sails, and then taking the tools from the engine-room. Captain Gonzales also told Binder and Masik they could break up the ship and use its timber to build themselves a house. He then left a supply of groceries, ammunition and traps, some of which were for Stefansson's small party of men if they appeared, but he left no sleds, primus stoves, nor writing materials and few books, taking all such items away from the camp. Storkerson, Castel and Hadley hastily wrote letters for Stefansson, which they left with Binder, in which they stated that they had protested strongly to Captain Gonzales over the wanton destruction of the *Mary Sachs*, but to no avail.[18] Then the *Polar Bear* departed. Captain Gonzales' actions certainly appeared like a deliberate act of sabotage in an effort to hinder and delay Stefansson's departure from Banks Island.

Thirdly, there was news of Storkerson's mapping activities on Victoria Island commencing early in June 1917. Together with two members of the *Polar Bear* crew (Martin Kilian and

---

[16] Storkerson, 1917.

[17] Noice, 1924, p. 260; Ashlee, 2008, pp. 156–157.

[18] Stefansson, 1921b, pp. 658–660.

Adelbert Gumaer), he had undertaken to complete the mapping of Victoria Island's north coast. Captain Gonzales had told him that he intended to sail from Walker Bay on August 1, and threatened to leave Storkerson and his men behind if they were not back to the ship by then. As a result, Storkerson succeeded in mapping only part of the remaining unmapped coastline before having to hurry back. He finally reached Walker Bay on August 3 fearing the worst, but to his great relief found the *Polar Bear* still there.

After hearing all this troubling news from Binder and Masik, Stefansson pondered his next move. It appeared he would have to make the long trek back to the northeast coast of Banks Island with Emiu, Knight and Noice, retrieve his two sleds, return along the east side of Banks Island and cross Amundsen Gulf (when it became sufficiently frozen over) to Baillie Islands, and proceed from there west to Fort McPherson and over the mountains to Dawson. From Dawson they could continue by water south to Vancouver.

Rescue

On August 26, however, while still at the base camp contemplating what to do, Stefansson suddenly noticed a schooner approaching from the southeast. Rescue seemed at hand. Ironically, it was the *Challenge*,[19] with Jim Crawford, the engineer on the *Mary Sachs* from 1913 to 1915 whom Stefansson had dismissed in the summer of 1915. With Crawford was his current trading partner, Captain Leo Wittenberg. Having spent the winter of 1916–1917 near the *Polar Bear* camp in Walker Bay, they knew of Captain Gonzales' plans to leave the Arctic, and were intent upon scavenging his abandoned Expedition supplies. They had headed first for the cache that the *Polar Bear* had left at Armstrong Point in Prince of Wales Strait, but were prevented from reaching it because of ice conditions in the strait. Then they had come to Cape Kellett, hoping to salvage what was left there. They anchored at Cape Kellett, and Crawford and one of his sailors walked back to the campsite, intending to assess what items they could take. Great was their surprise to discover Stefansson and some of his men, all in need of transportation, for they had expected to find the camp abandoned.

Stefansson soon arranged to purchase the *Challenge* from Crawford for $6,000 plus what was left of the stripped *Mary Sachs* and all the supplies at Cape Kellett. Crawford and some of his crew then moved into the Expedition's camp with the intent of trapping foxes for the winter,[20] and Binder, Masik, the Stefansson party and Wittenberg sailed for Herschel Island. Masik became the *Challenge's* new captain, being the most competent sailor among them, and Binder became its engineer. The sun shone brilliantly on their first day, the Beaufort Sea was calm, and the *Challenge* soon left Cape Kellett far behind.

---

[19] The *Challenge* was a gasoline auxiliary schooner of about 60 tons, formerly a Japanese sealer, which had been confiscated by the U.S. government for poaching and then sold. Leo Wittenberg of Nome purchased it, stocked it with "a gang of seven beach combers and a few natives" and Crawford as second-in-command, and headed for Minto Inlet to establish a camp there to trade with the Copper Inuit. They stopped briefly at Herschel Island en route to Victoria Island (La Nauze, 1917).

[20] ???.

Confrontation with Captain Gonzales

Towards evening the *Challenge* encountered thick ice, the western edge of which ran northwest and southeast. With the ice came fog, and Stefansson had the ship turn southeast and cruise slowly along the margins of both the ice and the fog. The following morning the fog suddenly lifted, and the crew was startled to see the *Polar Bear* less than two miles away.

Stefansson's account in *The Friendly Arctic* of the events is markedly tempered,[21] so most of the following details are from Noice's version of what happened.[22] According to Noice, Wittenberg ran up a distress signal on the *Challenge* to lure the *Polar Bear* closer. Captain Gonzales saw the distress signal, recognized the *Challenge* and promptly headed towards it. Stefansson then went below deck, and Noice, Knight and Emiu concealed themselves so that they would not be recognized and forewarn the men on the approaching ship. When the *Polar Bear* drew close, one of the *Challenge's* crew hailed Captain Gonzales, instructing him to tie up to the ice, and then added "Stefansson is aboard and wants to speak to you." Captain Gonzales and Seymour promptly conversed on the *Polar Bear*, unsmiling. Then the captain ordered his helmsman to steer slowly towards the ice. As the bow of the *Polar Bear* touched the ice, Storkerson leaped over its rail onto the ice and tied that ship to an ice-boulder. The crew of the *Challenge* followed suit with their ship.

Captain Gonzales then came on board the *Challenge* and was ushered down to the cabin, where Stefansson was seated at the chart-table. The two men conversed behind a closed door for the next fifteen minutes. Then the captain returned to his ship, displaying somewhat less than his customary swagger.

After Captain Gonzales had departed, Stefansson told his curious crew the various reasons the captain had given him for destroying the *Mary Sachs*. It was unseaworthy; he needed the firewood from the foremast for his galley stove; and he had taken all of the sleds from the camp in case the *Polar Bear* was crushed by the ice trying to cross to the mainland. Furthermore, he did not have time to tow the *Mary Sachs* to the nearby sheltered harbour, and the ship would constantly leak and the two men living there (Masik and Binder) would not have the time to keep pumping out the water. Captain Gonzales then added that four of his officers had agreed that the ship was not seaworthy. Stefansson was little impressed by the man's arguments and stated that he was going to ask Storkerson, Hadley and Castel about what had happened before deciding what action he would take.

Stefansson then went over to the *Polar Bear* and took charge, while Castel came from that ship to take charge of the *Challenge*. Both ships then got under way for Baillie Islands. Deliberately employing all of the canvas he felt safe to hoist, Castel soon had the *Challenge* literally and deliberately sailing circles around the slower *Polar Bear*; "... nothing makes a captain madder than that," commented Noice.[23] Of course, the *Challenge* happened to be one of the swiftest ships in the Arctic. Each time it swept past the stern of the *Polar Bear*, Castel taunted Captain Gonzales by holding out a rope in mock offer of a tow. Fog finally enveloped

---

[21] Stefansson, 1921b, pp. 664–665.

[22] Noice, 1924, pp. 263–266.

[23] *Op. cit.*, p. 265.

the ships that evening, and they lost contact with each other. The *Challenge* reached Baillie Islands three days later, whereas another three days passed before the *Polar Bear* finally made its appearance. It had stopped for a while at Booth Island en route.[24]

Captain Gonzales' real reason for ordering the destruction of the *Mary Sachs* remains unknown. Stefansson subsequently suggested that the captain feared his pay would be withheld for his failure to take the *Polar Bear* to Melville Island during the summer of 1916. As a consequence, he may have thought that by preventing Stefansson from reaching a telegraph office for many months, Gonzales could reach Victoria, explain his actions plausibly to the Canadian government and collect his wages before Stefansson had any opportunity to prevent it.[25] In any case, when the *Polar Bear* reached Baillie Islands Stefansson discharged the captain and put him ashore, along with his wife, Mamayauk, and several Inuit. The only punishment he meted out to the captain at the time for his malicious actions was to cancel his pay "… from the time of his refusal to bring the ship north to Melville Island …" in the summer of 1916.[26] Stefansson later urged G.J. Desbarats, the Deputy Minister of the Naval Service in Ottawa, to charge Captain Gonzales with deliberately destroying a government vessel and related charges, but nothing came of the matter.[27] Quite likely the government officials preferred to avoid the embarrassment that would have resulted from the public revelation of the gross expenditures incurred by Stefansson and other controversial Expedition matters during the war years.

Noice's Later Exploration and Anthropological Activities

Meanwhile, the young sailor Noice, though not yet twenty-one, had decided he wanted to remain in the Arctic and follow in Stefansson's footsteps. An opportunity arose while the *Polar Bear* and the *Challenge* were anchored at Baillie Islands, so he asked to be discharged and paid for his service with the Expedition. Stefansson obliged him, and soon afterwards Noice teamed up with Otto Binder and A.A. Carroll, a Texan known to Stefansson who happened to be at Baillie Islands representing a fur-trading company. Together the three men bought the *Challenge* from Stefansson, intending to head east into Coronation Gulf, where Binder and Carroll would obtain furs through trading with the Inuit. Noice, on the other hand, intended to continue Stefansson's Arctic work, first by completing the mapping of the northeastern part of Victoria Island that Storkerson had been unable to finish (Noice failed to undertake this project), and then by a study of the Inuit on the eastern end of Victoria Island, collecting furs, weapons

---

[24] Donohue, 1922.

[25] Stefansson, 1921b, pp. 659–660.

[26] Noice, 1924, p. 266.

[27] Stefansson stated in this letter that he had gone before a magistrate in Victoria, B.C., in November 1918 with the testimonies of various members of the crew of the ill-fated *Mary Sachs* concerning the vessel's destruction by the orders of Captain Gonzales. He was evidently informed that he needed Storkerson's sworn testimony, and asked if the statement he had obtained at Herschel Island, declared before Inspector Tupper of the RNWMP, would suffice. He was informed that it would. The Canadian Department of Justice later considered the charges, but by then Gonzales was sailing on the American ship *Herman* and outside of Canadian jurisdiction (Stefansson, 1919c [December 10, 1919]).

and other artifacts from them (which he succeeded in doing). Having only a high-school education, Noice was quite unqualified for such undertakings. But in his youthful naivety he was confident that he had learned sufficient ethnology from Stefansson in the two years he had been attached to his Northern Party to carry them out well.

Noice spent the winter of 1917–1918 at Pearce Point, where he discovered and excavated the ruins of prehistoric Eskimo dwellings of earth, stone and whalebone. Among these ruins he discovered fragments of pottery, which greatly interested archaeologists, who later traced the occurrence of similar pottery fragments in early ruins from Alaska eastward. In 1918, Noice sold his share of the *Challenge* when its crew refused to go east, and proceeded there on his own, spending the next two years with the Copper Inuit on southeast Victoria Island learning their language, and observing their dances and religious seances. In the summer of 1921 he reappeared in the western part of Coronation Gulf with many furs and Inuit artifacts, then proceeded west to Herschel Island with Pete Norberg on the *El Sueno*, and from there south to the United States. He made $4,000 for the furs he had acquired, then apparently sold his Inuit artifacts to a dealer (John G. Worth) who sold some of them to two museums, where they are now housed—the Heye Museum in New York City (now the National Museum of the American Indian, which is a part of the Smithsonian Institution), and the Field Museum in Chicago. He also sold at least 250 objects to the Royal Ontario Museum. He then aroused some publicity during the next two years by supporting Stefansson's contentions that there were Inuit in the central Arctic with blond characteristics,[28] and that man with a knowledge of hunting and travelling, and a supply of ammunition, "… can sustain himself indefinitely,"[29] which was the essence of Stefansson's "living off the country" claim.

In 1923 Noice took charge of the *Donaldson* at Nome, Alaska, on Stefansson's instructions, and sailed to Wrangel Island to leave supplies and several Native settlers as part of Stefansson's scheme to take possession of the island for Canada or Britain, and to bring back the five young people forming a party that Stefansson had sent to the island in 1921. Noice deposited the Native settlers as instructed (they were arrested and removed from the island in the summer of 1924 by some men from a Russian ice-breaker), and returned to Alaska with Ada Blackjack, the lone survivor of the 1921 party. The news of the loss of all four men from his 1921 expedition to Wrangel Island resulted in considerable publicity unfavourable to Stefansson, during the course of which he and Noice had a major falling-out.[30]

The schooner *Challenge* was wrecked on the mainland coast of Amundsen Gulf sometime shortly after Noice sold his share of it.[31]

The *Polar Bear* Fails to Leave the Arctic in 1917

After completing his activities at Baillie Islands, including the sale of the *Challenge* to Messrs. Binder, Carroll and Noice, Stefansson headed the *Polar Bear* west to Herschel Island, with

---

[28] Anonymous, 1923d (*Ottawa Citizen*, May 1, 1923). Noice had come to Ottawa to hear a lecture by Stefansson.

[29] Anonymous, 1922b (*Honolulu Star-Bulletin*, February 24, 1922).

[30] Noice, 1924 (December 24, 1924).

[31] Stefansson, 1921b, p. 669.

Figure 29 Inuit women and children at Walker Bay, Victoria Island; (l. to r.): Alex and Pannigabluk Stefansson; Annie, Jennie and little boy Thomsen; Martina, Aida and Elvina Storkerson, July 1916.

Photo: © Canadian Museum of Civilization, photo by John Hadley, 57024

William Seymour as its new master, arriving there September 7. While there, Stefansson wisely took additional supplies on board the *Polar Bear* in case it and its crew encountered adverse ice conditions and were forced to winter before reaching Barrow. In addition he paid off some of the *Polar Bear's* crew, including Seymour, whom he had promoted to first officer after the *Challenge* made contact with the *Polar Bear*. After Seymour and his wife left the ship, Stefansson made Hadley the new master of the *Polar Bear*. Castel and Masik then became first and second mates, respectively.

Stefansson kept the *Polar Bear* at Herschel Island for what seemed to its crew an unduly long and needless time, during which the weather was good. Finally, on September 12 he issued the orders for the *Polar Bear* to head west for Nome, lightly loaded and aided by easterly winds. It is highly probable that it could have left the Arctic that summer had Stefansson wanted it to do so, for "It was common knowledge in the north in 1917 that they had three weeks of good open water after they left Herschel Island and could have easily gone out …"[32] But Stefansson had heard from the RNWMP at Herschel Island that the

---

[32]   Anderson, 1921e (November 19, 1921).

Figure 30 *Polar Bear* tent camp on Barter Island, northern Alaska, with men pulling the schooner upright using a windlass and cable attached to one of the masts, July 4, 1918.

Photo: Reproduced with the permission of Natural Resources Canada 2011, courtesy of the Geological Survey of Canada, John Hadley, 63564

war in Europe was far from over, and knew that if he returned to the United States at that time it would arouse little interest. Besides, he still had one more ice trip he wanted to undertake. So now, despite receiving clear government instructions from Ottawa to leave the Arctic that summer, he deliberately ignored them[33] and delayed unduly his departure from Herschel Island.

The following evening the *Polar Bear* anchored in a harbour on the west side of Barter Island. Early the next morning, according to Stefansson's account,[34] the winds shifted to a southwest gale and within minutes drove the ship broadside onto a sandbank. The winds in turn caused the sea level near the shore to rise several feet, which allowed the waves to push the ship farther onto the sandbank. Then suddenly both the winds and the sea level dropped, leaving the schooner on her side in a mud bank. Lightening her load of cargo alone would not have been sufficient to refloat her. In addition to lightening her load, the men needed to keep her anchors to windward, and wait for another southwest gale, which might not happen

---

[33] It is noteworthy that Stefansson, in *The Friendly Arctic*, accused Dr. Anderson, Chipman and other members of the Southern Party with disobedience to *his* instructions at Collinson Point, even using the naval term "mutinous" to describe their behaviour, yet remained silent about having ignored the instructions he received from the Department of the Naval Service to leave the Arctic in 1916. Dr. Anderson and his men acted in accordance with the clear instructions from their government department in 1914, whereas Stefansson deliberately ignored his instructions in 1915 and again in 1917, a far more serious offence. He may have felt no guilt over this, however, for he was not being paid by the government for his services, at his own request.

[34] Stefansson, 1921b, pp. 672–673.

again for weeks. Before that could happen, however, freeze-up occurred, and the *Polar Bear* was caught on the sandbank for the winter.

Early in 1918, Dr. Anderson received a letter from the *Polar Bear's* first mate, Bill Seymour, at Herschel Island reporting that Stefansson had discharged him and withheld a year's pay from him "for some reason whitch He wont explane. He is going to leave it to the Dep Minister [Desbarats] to Decide ..." Seymour asked Dr. Anderson to write on his behalf to the Deputy Minister [of the Naval Service] to help him get the money he was owed.[35] The letter then stated that the *Polar Bear* was wintering between Martin Point and the Southern Party's winter quarters at Collinson Point so that Stefansson could remain another winter in the Arctic.

> We had a verry late season Heare the Polar Bare left Herschell Island on Her way out about the 12th of Sep Just about the time she Left the Easterly weather set in and blew strong East gales for three weeaks which I ame shure cleared the road for Him but we all new that He did not want to get out in fact He told me so at Bailey Island that the Publick would not take aney Notice of Him or the Expedition while the war is going on.

Dr. Anderson evidently believed Seymour, mentioning a short time later to his friend J. Eugene Law at the University of California, Berkeley: "Stefansson stayed up there until 1918, and according to testimony of his own ship's officer [Seymour] and Officers of the R.N.W.M.P. up there, said he would not go out while the war was on, because people would not take notice of his expedition."[36] Dr. Anderson also wrote to General A.W. Greely that the *Polar Bear* could easily have gone out in the summer of 1917, but instead had "... managed to get stuck at Barter Island in a favourable place to make another 'great ice trip.' "[37]

Seymour had several other startling comments. "I herd that He was going to bring a charge of Mutiny against the whole bunch of your outfit and you the ringleader ..." This warning was later borne out in 1921 by the charges made in Stefansson's book *The Friendly Arctic*, which aroused widespread newspaper publicity. Clearly, Stefansson had been thinking about making these charges for many months. Some months after his book appeared, however, he admitted to the Deputy Minister of Mines, Dr. C. Camsell, that perhaps he had been too strong with the words he had used in his book and blamed the usage on his publisher.[38]

Seymour also remarked: "He got rid of the Capt at Bailey Island and me at Herschell so He could put lots of blame on us and take some off Him Self ..." Stefansson withheld one year's pay from each man for not waiting with the *Polar Bear* for Stefansson and his party at Banks Island and for destroying the *Mary Sachs*.[39]

---

[35] Seymour, 1917 (December 24, 1917).

[36] Anderson, 1921c (March 26, 1921).

[37] Anderson, 1921e (November 19, 1921).

[38] Anderson, 1923d (May 11, 1923).

[39] Dr. Anderson forwarded a copy of Seymour's December 24, 1917 letter on March 22, 1918 to G.J. Desbarats, Deputy Minister of the Naval Service, Ottawa, for his information and to make a case for the payment to Seymour of the money owed him. Desbarats' reply noted Seymour's "...

And lastly, Seymour also informed Dr. Anderson that Stefansson was "… going to make another Ice trip take his Departure from the Midway Islands[40] about the first of feb. and wind up on the Coast of Sibiria if He Gets Thire."

This was quite possibly the first indication Dr. Anderson had received that Stefansson was contemplating initiating a radical ice journey early in 1918, using sleds and dog teams and large drifting floes of ice. Rather than using a drifting ship for base and supplies, as Fridtjof Nansen had done twenty years earlier, he would use a sled party on drifting ice and "live off the country" (or more exactly live on food from the sea) after his initial supplies became exhausted. In September 1893, the Norwegian explorer Nansen had deliberately allowed his well-strengthened ship, *Fram*, to become ice-bound near the New Siberian Islands north of eastern Siberia (and well west of Wrangel Island), following which it drifted westward to emerge north of Spitzbergen. "This was a plan the carrying out of which I had had in mind for several years," wrote Stefansson later, "indeed, ever since we had found on the Martin Point trip the abundance of animal life in the sea."[41] By the kind of journey Stefansson now envisioned, he would be able to prove both the westward drift of the polar ice, which the voyages of the *Jeannette* and the *Fram* seemed to indicate, and the existence of plenty of life in the ocean well beyond the coastal regions.

Seymour's comments suggest that Stefansson deliberately remained in the Arctic in 1917 to serve his own vanity and pocketbook, knowing that he would not get the publicity he sought back in the United States and Canada while the war was still raging in Europe. Instead he simply dreamed up his new project on the Beaufort Sea to justify his actions, and disregarded the orders from the Canadian government to return that summer.

Another crew member from the *Polar Bear* subsequently provided a more detailed account of the *Polar Bear's* 1917 activities, which confirmed Seymour's claims. This was Peter Donohue, who wrote the account for Captain Joseph Bernard in November 1922, who in turn forwarded a copy to Dr. Anderson. It reveals clearly that Stefansson had no intention of leaving the Arctic that summer:

> At Herschel Island we stopped more than a week. None of us seemed to know what we were delayed for. We were all anxious to get out so as to go around Point Barrow before it froze up … The weather was good, season mild, and no ice in sight, but still no

---

complaint regarding part of his pay which had been held back by Mr. Stefansson …" but responded that "… on the representations which I have received I am inclined to think that the Captain [Gonzales] and the Mate [Seymour] are liable to arrest for the destruction of a government ship and for abandoning their Chief in the Arctic Regions." He had therefore telegraphed Stefansson to "… take the necessary legal action against this man (i.e. arrest and charge Seymour) if the Mounted Police considered the evidence would warrant such action" (Desbarats, 1918 (March 27, 1918). The police evidently did not interpret the incidents as the Deputy Minister did, and no action was taken against Seymour.

40 The Midway Islands referred to here are two off-shore-bar islands about ten miles north of Prudhoe Bay, northern Alaska, and five miles west of Cross Island, from which Storkerson's 1918 ice trip actually commenced.

41 Stefansson, 1921b, pp. 674–675.

order was given to sail. Some suggestions were made to Stefansson about the delay, but he said there was plenty of time, and they were not ready. Apparently the only reason for causing the delay was to load on some provisions and supplies, of which they took enough for a year.

To us it seems very strange to take on board such an amount of supplies as we should have reached Nome in 8 or 10 days had we sailed on. It became very evident that it was the intention of the leaders to be frozen in somewhere between Herschel Island and Point Barrow. At last we sailed out. Things went well until we reached Barter Island. Here a few pieces of ice lay grounded on the beach and sand bar. It was just a little before dark when we got abreast of the island, and to our surprise we saw the vessel being steered at the western end of the island. We were sure then convinced that it was their intention to winter there. We were sure that the ice was no excuse for us to go there. We anchored close between the sandspit and the island.

There was plenty of water. During the evening the wind shifted and the vessel swung on her anchors, stern grounded on the sandspit. It was not grounded very hard. Would have been very little trouble to get her off, but nothing was done. She lay quiet and was not hurting herself. The next morning the wind was blowing from off the Island.

We ran an anchor, set the foresail and in less than an hour we had the vessel afloat. It was a great surprise to me when this fall I read in "The friendly Arctic" by Stefansson, that the cause of the last year's stay at Barter Island had been by the mischief of the *Polar Bear* having ground on the sandspit and the next morning the tide had receded and left her three feet short of her floating line and that caused them so much delay that they were obliged to winter there.

It surprised me for it is the truth that it never took us over an hour to get her afloat and after we were afloat we would have been all ready to sail as everything was in shape. We lay here for a whole week without seeing any ice except for a few grounded cakes that lay along the coast. Finally after a week there came a light frost and a little snow. It seems it was all that the commander was waiting for, for right away an order was given to build a house there and prepare for winter. And then for more than a week after we had started building the house there was no ice in sight at sea, although the ground was slightly frozen up. For two or three weeks after we had come to Baillie [*sic*, Barter] Island I am convinced that there was no ice and the vessel would have had no trouble to get by even on the sail, without power …[42]

Shortly thereafter, however, Stefansson became desperately ill.

---

[42] Donohue, 1922.

STEFANSSON'S SKIRMISH WITH DEATH, 1917–1918

16

Stefansson's New Ice Trip

Soon after the *Polar Bear* was driven aground at Barter Island, Stefansson produced what appeared to be a new exploration plan, although it may have been one he had contemplated for several years. He had reflected frequently in the past upon the routes across the Arctic drifted by Lieutenant De Long's icebound *Jeannette* in 1879–1881 and Nansen's icebound *Fram* in 1893–1896. Now he could add the route of his own icebound *Karluk* in 1913–1914, as he understood it to have been from Hadley's account. Combining these, he envisioned an almost continuous trans-Arctic, east-to-west ice drift from Barter Island to Spitzbergen (a Norwegian archipelago 360 miles north of Norway).

During the past three winters he himself had ventured onto many drifting masses of ice, from which he had been able to shoot seals and polar bears for food and fuel, and had obtained geographic locations, ocean depths, ocean-current directions and other useful oceanographic information. Why not, he reasoned, establish himself on a large Beaufort Sea ice mass, live on seal meat and polar-bear meat, and see if the ice will drift west across the Arctic from North America past Siberia to Spitzbergen? Now that would make newspaper headlines!

He reasoned that a scientific party with dogs and sleds that was stationed on drifting Arctic ice would have the ability to move its location from time to time, should such moves prove advantageous or necessary, in sharp contrast to the limited mobility of the scientific personnel on an icebound ship. With the *Polar Bear* assured of remaining aground for the next eight or nine months at Barter Island, he had been presented with a splendid opportunity to carry out such a plan. Any additional food supplies he would need were probably available at Herschel Island, but he would need new sleds and dogs, because he had left his sleds on Banks Island and had gotten rid of his dogs on his way west.

He was confident he could obtain dogs from persons living on the Mackenzie Delta, and from Ole Andreasen, who was now running a trading post at Shingle Point, on the mainland between Herschel Island and the Mackenzie River Delta. As for sleds, he knew that among the men on the *Polar Bear* who had joined the crew at Herschel Island were Anthony Shannon, who was a competent iron worker, and Peter Donohue, an excellent carpenter. With suitable material, these two men could build the sleds he needed for his ice-drift project.

Accordingly, a few weeks later when sled travel again became possible, Stefansson and Storkerson proceeded east to Herschel Island and the Mackenzie Delta in search of dogs, material to build sleds, and other supplies. In this mission they proved surprisingly successful. Storkerson then spent the next several weeks freighting their newly acquired supplies from Herschel Island and two or three small trading posts west to the *Polar Bear*. Soon the men at the ship were busily constructing new sleds. Stefansson, meanwhile, was able to buy good dogs in the delta and upriver at Fort McPherson, engage several Inuit for the early part of the journey, and persuade his former ice-trip assistant, Ole Andreasen, to join the planned ice journey. The ice-trip preparations were progressing beautifully when adversity struck.

Serious Illness

After two years of isolated survival among the Arctic islands, far from civilization's multitude of germs, Stefansson had initially caught a bad cold in August at Baillie Islands. No sooner had he recovered than he caught a second one and then a third one. He had largely ignored these inconveniences at the time, but when one day in January 1918 he suddenly felt weak as he reached Ole Andreasen's trading post at Shingle Point, he was forced to stay and rest for three days. He then continued westward towards Herschel Island, but grew increasingly weaker, and was finally forced to ride on his sled. After he spent an uncomfortable night in an Inuit house en route, his host thought it imperative to transport him by sled to Herschel Island, where he might obtain medical attention. Upon his arrival at the RNWMP barracks on Herschel Island, the police immediately sent for Mr. W. Henry Fry, the Church of England missionary, who was the most medically knowledgeable person on the island. Mr. Fry also happened to have a thermometer. Stefansson's temperature was found to be above 104° F, and so he was promptly put to bed, where he spent the next several weeks, cared for by the police with the assistance of Mr. Fry, and with food frequently sent over by Mrs. Fry.[1]

During some of the time he was ill, a young Loucheaux First Nations boy attended to his needs. Stefansson had hired the boy weeks before when he was seeking to buy dogs in the delta region, but did not give his name. The boy assisted him faithfully during his subsequent illness and accompanied him on at least part of his journey to the hospital at Fort Yukon.[2]

Meanwhile Storkerson was at the *Polar Bear* camp on Barter Island, miles to the west, in charge of preparations for Stefansson's lengthy ice drift. After his convalescence had extended from days to weeks, Stefansson sent a message to Storkerson, asking him to come to see him at Herschel Island. When Storkerson finally arrived, Stefansson discussed with him the necessity of postponing their ice-journey plans. Two weeks later, however, believing he was almost restored to good health, he sent word to the *Polar Bear* to have everything ready for the ice journey. Then he had a relapse.

One of the policemen stationed at Herschel Island, Constable Alexander Lamont, had about that same time come down with what was diagnosed as typhoid fever and was seriously ill in a room across the hall from Stefansson. Tragically his condition worsened, and in due course he died. The news of his death did not exactly elevate Stefansson's spirits. It was then realized that Stefansson's initial illness had likewise been typhoid fever. Now his medical symptoms indicated he had developed pneumonia. He was transferred at this point from the

---

[1] In a strange twist of fate, the very Church of England missionaries whose social and spiritual teachings to the Aboriginal population Stefansson had been questioning for some years were the persons who nursed him back to health in the winter of 1917–1918 at Herschel Island and at Fort Yukon.

[2] This boy was Abraham Stewart.

police residence to the house of Leo Wittenberg, where he had frequent visits from Mr. Fry and Mr. Harding, the man in charge of the local Hudson's Bay Company post.

Having pneumonia meant that his recovery would take weeks, and it was now February. Stefansson therefore wrote again to Storkerson at Barter Island, instructing him to take command of the spring exploration operations.

A few weeks later three First Nations men from the vicinity of Rampart House arrived at Herschel Island to trade. Stefansson immediately saw an opportunity to be transported south four hundred miles to a hospital at Fort Yukon, the most northerly hospital on the North American continent at the time. The police felt the journey too great a risk for Stefansson and told him they would not permit his making it. Though much disappointed, Stefansson continued to insist that arrangements be made for taking him to Fort Yukon. Subsequently his condition worsened as he developed pleurisy, and Inspector J.W. Phillips, the police officer in charge, finally agreed to let him go. It was perhaps a case of letting the doomed man have his last wish, if not a secret preference to have him die away from the police post.

Stefansson left Herschel Island for Fort Yukon in the first week of April on a sled specially prepared with bed springs for his comfort. Constable J. Brockie accompanied him, along with Mr. Fry, two Aboriginal men (Naipaktuna and Saryoak), and Abraham Stewart. Constable Brockie was instructed to accompany Stefansson wherever he was needed, but Mr. Fry turned back on the second day, and the three other men were paid off at Old Crow, Yukon on April 15.[3] Along the way Stefansson reportedly enjoyed the kind of food he had been accustomed to for the past several years—meat and fish, instead of a liquid diet. Sometimes it was boiled, the rest of the time it was raw and frozen.

At the mouth of the Crow River Stefansson stayed briefly with a trader, whose wife was a trained nurse. This permitted Constable Brockie and his assistants to return to Herschel Island. Stefansson then hired local men to transport him the rest of the way to Fort Yukon. His party reached Archdeacon Stuck's St. Stephen's Episcopal Hospital at Fort Yukon on April 27, by which time he had gained thirty pounds and, remarkably, felt well enough to walk into the building on his own strength! There, in addition to the typhoid, pneumonia, pleurisy and alimentary trouble he had been fighting since January, Stefansson was diagnosed with neurasthenia.[4]

After a three-month convalescence at the hospital, Stefansson proceeded up the Yukon River by river-boat to Dawson and Whitehorse, then by rail south to Skagway, and by ship to Vancouver and Victoria. In Victoria he by sheer chance met the crew of the newly arrived *Polar Bear* and paid them their final wages. While there he also went before a magistrate seeking to bring charges against several members of the crew of the *Mary Sachs* and others, including

---

[3] Stefansson, 1918 (April 15, 1918).

[4] Anderson, 1921h (December 31, 1921). Dr. Anderson commented dryly, in this letter to the newspaper man J.J. Underwood, that Stefansson seemed to have recognized the "dynamite" in admitting he had neurasthenia (a type of neurosis, usually the result of emotional conflicts), hence refrained from mentioning that ailment in *The Friendly Arctic*. This letter also mentioned that during his illness, Stefansson begged canned milk from the Frys, which Mrs. Fry had been saving for her two infants. Dr. Anderson had received this information from Mr. Fry.

Captain Gonzales, for their part in the deliberate destruction to that vessel. However, many of the persons involved were American citizens and no longer in Canada, so the charges seem to have been set aside.[5] From Victoria Stefansson went east by train to Ottawa to report to the Canadian government. A while later he proceeded south to New York, thus ending what turned out to be the last of his Arctic journeys.

---

[5] RG42, Vol. 473, File 84-2-13, LAC.

STORKERSON'S DRIFT ON THE ARCTIC ICE, 1918

17

Stefansson's serious illness forced him to turn over the command of his ice-drift project in February 1918 to Storkerson, who had been freighting supplies from Herschel Island to Barter Island since the previous November. Over the winter of 1917–1918 Stefansson's men at Barter Island had built sleds, packaged rations, made camping gear, tents, canvas sledboat covers and sledboat frames, overhauled and packed arms and ammunition, and completed numerous other tasks necessary to equip the men about to venture north over the Beaufort Sea. Iñupiat seamstresses, meanwhile, kept busy making clothing and footgear. For the important task of making clothing and footwear, Storkerson hired all of the Inuit women in the vicinity who were willing to sew. These included Dora Brower, who was a step-daughter of Charlie Brower of Barrow, Elvina Storkerson, Pannigabluk, Jennie Thomsen and Mamie Mamayauk, wife of Ilavinirk.[1]

Stefansson's departure from Herschel Island early in April for the hospital in Fort Yukon had removed him from the leadership of his intended ice trip. As a result, Storkerson was forced to shoulder the entire responsibility for both the ice trip and winding up the affairs of the Expedition, and the various men and women who had been employed at Barter Island. He did not see Stefansson again for a year and a half.

Stefansson's plans called for the ice-drift party to start from Cross Island, northwest of Barter Island, and proceed north along the 148° W longitude for about three hundred miles to 74° or 75° N latitude, where they were to camp. They would thus follow approximately the route taken by Ejnar Mikkelsen and Ernest de Koven Leffingwell in 1907, when Storkerson was one of the members of their exploration party. But Stefansson wanted Storkerson to attain a position about one degree of latitude farther north of that reached by Mikkelsen and Leffingwell. They were then to drift with the ice, take soundings, determine ocean currents, take solar observations for location and collect meteorological, zoological and oceanographic information. This was a pretty tall order for men with little scientific training. The ice-trip members commenced freighting their supplies to Cross Island about February 20. By March 11 all was in readiness for them to start their adventure.

Stefansson's original plan, had he led the ice party, was to spend a year adrift on the ice, but none of Storkerson's men was willing to join him for that length of time on the ice. Storkerson was thus forced to change Stefansson's plans without having the benefit of first discussing the matter with him. As he revised them, Storkerson's plans called for the ice party to travel north to about 77° N latitude, then east to Prince Patrick Island and from there south to Banks Island and Cape Kellett, reaching there in time to be picked up by a whaling ship the following summer.

---

[1] Ashlee, 2008, pp. 166–167.

Useful information would undoubtedly still be obtained, but the change in plans defeated Stefansson's original purpose of obtaining information on the annual drift of the polar ice and testing the idea of a trans-polar westward movement of the Arctic ice from the Arctic Archipelago.

Storkerson's Ice Trip Commences

Led by Storkerson, the ice journey commenced from Cross Island on March 15, 1918, with twelve men, fifty-six dogs, eight sleds, and about eight thousand pounds of provisions and equipment.[2] Stefansson could hardly have claimed that his men would be "living off the country" with that amount of provisions and equipment. Their camp that evening, some thirteen miles northwards, was on westward-drifting sea ice. Slowly they worked their way through the dangerous rough ice that lay between the shore-fast ice and the main pack ice. By March 25 they were about sixty miles north of the Alaskan mainland. On April 4, Storkerson sent three men from his support party, led by Herman Kilian, with two sleds and nineteen dogs, back to the *Polar Bear* camp at Barter Island, because their assistance was no longer required. They also took with them Storkerson's initial reports of progress for Stefansson. The ice party's location at that date was about 105 miles north of Cross Island at 72° N latitude and 147° W longitude.

On April 8 a very wide lead of open water halted the ice party, a lead much too wide to risk crossing with sledboats. The men set up camp nearby. That night the easterly winds, which had blown since their departure from Cross Island, increased to a gale, accompanied by much snow. Storkerson soon realized that the old floe on which they were camped was drifting rapidly northwestward. This strong drift reminded him of the westerly drift of the ice he had encountered in 1907 while with the Mikkelsen-Leffingwell expedition, and also of the westerly drift of the ice in April 1914 when he was with Stefansson north of Martin Point.

At this point Storkerson explained to his men that this strong westerly drift made it unlikely that they could now reach Prince Patrick Island and then proceed south to Cape Kellett as they had originally intended. He then called for volunteers to drift with him on the ice for a year and carry out scientific studies. In essence he was assuming the role of Stefansson and again asking his men to carry out Stefansson's original ice-drifting project. Four men and a boy volunteered: Masik, Gumaer, Knight, Martin Kilian and Fred Wolki, the last just a boy. The other men flatly refused, so Storkerson decided to send them back to shore, along with the young Wolki, as soon as the strong southeasterly winds diminished.

On April 14, the weather improved sufficiently to permit Storkerson to obtain solar readings, which indicated their position as 73° 3' N latitude, 148° 32' W longitude. This placed them about 190 miles north of the Colville River delta. The second support party, led by Castel and consisting of Andersen, Emiu, one other man, and Wolki, now started south for the coast with twenty dogs and one sled. They took with them Storkerson's latest report for Stefansson and personal letters. Storkerson and four men (Gumaer, M. Kilian, Knight and Masik), sixteen

---

[2] Storkerson's exploration party in March 1918 included the following men: Andersen, Castel, Gumaer, H. Kilian, M. Kilian, Knight, Masik, Fred Wolki, Emiu and three other unidentified men.

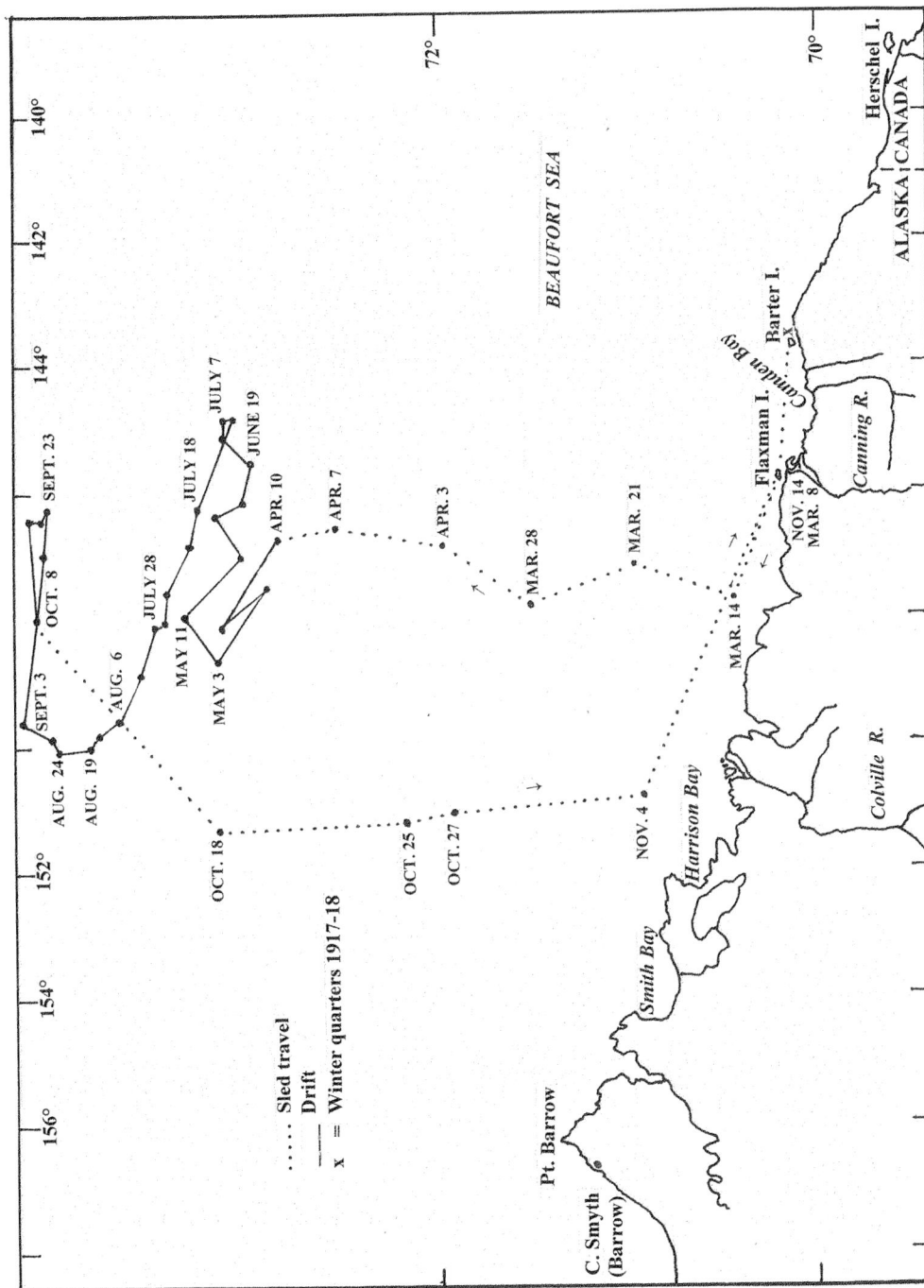

Map 12 Route of Storkerson's 1918 ice drift north of Alaska. (After a map in Stefansson's *The Friendly Arctic*, 1921, in pocket, back cover.)

dogs, one thousand rounds of ammunition, and 101 days of rations for both men and dogs remained at the ice-floe camp to drift at the whims of Mother Nature for the next year.

Stockpiling Meat

Storkerson's first priority was the accumulation of a stockpile of seal meat, fat and polar-bear meat, which necessitated moving their camp some three miles east to a better hunting site. He then assigned the task of hunting to his men, while he carried out the scientific activities and kept a record of the results. By the middle of June the men had accumulated about three tons of deboned meat from forty-two seals and four polar bears, which Storkerson calculated would be sufficient to last them until the middle of the summer. He knew that seals killed in the leads in mid-summer generally sank in the fresh water layer that overlay the salt water, and were therefore far more difficult to retrieve than during other seasons. To preserve their ammunition, therefore, the men refrained from hunting seals between mid-June and mid-August. Thereafter the increased fat on the seals, plus changes in seawater conditions resulting from the winds and from reduced melting, meant that the seals again tended to float after they were shot and could generally be recovered. During their eight months on the sea-ice (April 8 to October 9, 1918), Storkerson and his men shot a total of ninety-six seals and six polar bears.

Storkerson estimated that the ice floe on which they drifted that summer was more than fifteen miles long and eleven miles wide. As the smaller floes drifted faster than the large one they were on, they nearly always had open leads to east or to west, depending upon which way the winds blew. These open leads were their potential seal-hunting sites.

To Storkerson's surprise, the ice floe on which he and his men had camped did not maintain the westward drift that he had anticipated. Soon after the departure of Castel with the second support party on April 14, their ice floe commenced drifting steadily east! On April 14 they had been located at 148.5° W longitude; on July 14, they were located at 144.5° W longitude. Then the winds shifted and they drifted northwest for the next six weeks to 151° W longitude. A further change of the winds then caused them to drift northeasterly, to the most northerly position they attained, at 74° N latitude, on September 3. Thereafter until they started for shore on October 9 their floe drifted back and forth in a zigzag fashion. Storkerson estimated that they had drifted somewhat more than two miles per day during the April-to-October period, for a total of about 440 miles in all. Depth soundings taken over the summer supplied much new data. Their deepest reading was just over fifteen thousand feet at a position about ninety miles north of Alaska.

Storkerson's Unexpected Illness

Late in August Storkerson had an unexpected attack of asthma. It was an ailment from which he had not previously suffered, so he had no experience in how to deal with it. When the affliction worsened late in September, he decided that it was best to head for shore, for he was growing increasingly concerned that he might become unable to lead his inexperienced men through the winter. He was also having problems maintaining discipline and congeniality among his men, who had grown restless from homesickness. His loss of sleep because of his constant wheezing did not improve matters either.

After 184 days of drifting, Storkerson and his party packed up and started south on October 9 from 73.9° N latitude. On November 5 they reached land-fast ice just north of Harrison Bay (by coincidence, not far from the position of the *Karluk* when Stefansson left it to go hunting in September 1913), and turned east along the coast. At Flaxman Island, Storkerson obtained a confirming solar reading, for the location there had been well established previously by the U.S. geologist, Ernest de Koven Leffingwell, between 1906 and 1914. In the intervening years Leffingwell had returned to the United States, hence was not at his camp at the time of their visit. Storkerson's party then continued east to Herschel Island, which they reached late in November. They had been on the Beaufort Sea ice for 238 days. They did not stop at Barter Island on their way east, of course, because the *Polar Bear* had sailed south during the summer, and the residual Inuit assistants had scattered.

Storkerson remained at Herschel Island until the spring of 1919, at which time he accompanied Inspector Phillips of the RNWMP to Fort McPherson. From there he proceeded south in the summer by river steamer to Fort McMurray, and continued onward from there to Banff, where by pre-arrangement he met Stefansson, and together they finalized his observations.

Was Storkerson's sled journey north of Alaska between March 17 and November 5, 1918 worth the time, effort and cost? And what useful information resulted from it?

From Storkerson's depth-sounding data, Stefansson now knew that deep ocean lay north of the Continental Slope off Alaska for at least two hundred miles. Consequently, a body of land known as "Keenan Land," which had been shown to lie north of Alaska on a National Geographic Society map in 1912, simply did not exist. From Storkerson's data on ice movement, Stefansson had proof that a permanent current previously thought to exist in the Beaufort Sea between 72.5° N and 74° N latitudes did not exist. Storkerson had proved that a party of men could live for many months on a large detached ice mass in the Arctic Ocean and obtain valuable scientific information. And Storkerson had found that seals and polar bears lived in the Beaufort Sea hundreds of miles north of the coast, confirming Stefansson's claim that the Arctic Sea was "not as inhospitable as people think." The geographical value of those four findings alone made the ice trip an outstanding success.

It is perhaps worth adding as well that the drifting of Storkerson's ice party on the Arctic Ocean for months in 1918 was a forerunner to later more professional scientific studies in polar oceanography. Some sixty years later, for example, Russian scientists established a floating scientific laboratory on an Arctic ice island and drifted for a year. Ice-drift research in the Arctic continues to this day.

Disposal of the *Polar Bear*

Late in June 1918, with Storkerson and his men somewhere on the Beaufort Sea, the crew of the *Polar Bear*, under the command of the carpenter Captain John Hadley, got to work salvaging their schooner, which was lying on her side and still frozen in the ice and mud. First, the crew set off carefully planted explosive charges to loosen the ice that imprisoned her. Then the men pumped the water out of her, for she was submerged almost to the level of her main deck. When that was accomplished, they used a wooden windlass to pull the schooner upright. Their next task was to refloat her. This required waiting for a higher water level, which in turn depended upon a wind change. The wind ultimately changed direction, the water level rose, and the men succeeded in refloating the ship. They then cleaned up its interior and exterior, reconditioned the engine and readied her to sail west to Nome. Tragically, on July 22, 1918, Pipsuk, an Iñupiaq from Nome who had been obtained from the schooner *Challenge* in August 1917, drowned after his kayak capsized while he was tending the Expedition's fish nets.[3]

The *Polar Bear's* subsequent voyage from Barter Island to Nome early in August proved uneventful. Instead of Nome, however, she was taken to Saint Michael, some miles farther south, where she was hauled onto the shipway, braced and repaired in preparation for her sale.[4] Jafet Lindeberg of Nome bought her the next year for $5,000.[5]

The completion of Stefansson's Arctic Expedition in 1918 brought a close to his days as an Arctic explorer, but not to his interest in the polar regions. During the rest of his life, he actively wrote about the Arctic, accumulated a very large library of Arctic literature, oversaw the compilation and publication of a valuable Arctic encyclopedia, lectured and consulted on the Arctic, and stimulated many young persons to follow similar interests. His reputation as a person who roamed almost at will anywhere in the Arctic, month after month without communication of any sort with the outside world, remains unmatched in today's technologically linked world.

---

[3] Pipsuk's body was recovered and he was buried near the east end of Barter Island on July 23, 1918, with an appropriate headstone (Baur, 1918). The incident was mentioned only in the diaries of Hadley, Castel and Baur (each of which is in Library and Archives Canada), and evidently escaped the notice of Dr. Anderson some years later when he was preparing the list of men whose lives had been lost on the Expedition for the memorial plaque.

[4] Hadley, 1918 (August 1918).

[5] Stefansson had paid $20,000 for the schooner in 1915. Dr. Anderson suspected that Stefansson had deliberately reported the *Polar Bear* as unseaworthy to enable his friend Lindeberg to purchase it from the Canadian government at a greatly reduced price (Anderson, 1921h [Dec. 31, 1921]).

PART 4
DR. ANDERSON'S
SOUTHERN PARTY

NORTHERN ALASKA, 1913 18

The Canadian Arctic Expedition was publicized during its organizational phase as having the largest number of scientists (twelve including Stefansson) ever taken on a Polar expedition. Three of these scientists (assistant topographer Bjarne Mamen, geologist George Malloch and oceanographer James Murray), all members of Stefansson's Northern Party, perished tragically within a year of going north. Two others, members of Dr. Anderson's Southern Party, the French ethnologist Henri Beuchat, and the Scottish meteorologist and magnetician W.L. McKinlay, were sent to Herschel Island on the *Karluk* because of the limited accommodations of the *Alaska* and *Mary Sachs,* and thus were carried off with it when it was trapped by the ice. Beuchat perished later with the oceanographer Murray; McKinlay survived, but had no opportunity to carry out any of his intended work with the Southern Party. Whatever scientific information these five might have accumulated during the first few months was lost with the sinking of the *Karluk* in January 1914. Neither McKinlay nor Stefansson prepared reports for the government's official printed record, the former because he had no data deserving a report, the latter because he chose not to write one.

The six remaining scientists with the Southern Party were an international group, consisting of an American, a Dane, an Englishman, a New Zealander and two Canadians. Their principal task was to map the geology and geography of a hundred-mile-wide band of coast along the central Canadian Arctic mainland between Cape Parry (on the west) and Bathurst Inlet (on the east), with special attention being given to the copper-bearing rocks known to exist in the region. These activities were to be accompanied by ethnographical and archaeological investigations of the Copper Inuit on the mainland and on Victoria Island, as well as biological studies of marine life, mammals, birds, plants and insects. These men completed the field tasks they had been assigned, returned south in 1916, and wrote and published many of the government reports that today comprise the official printed record of the Canadian Arctic Expedition 1913–1918.

The Scientists of the Southern Party

The six scientists who carried out the studies for the Southern Party were: K.G. Chipman (Canada) and J.R. Cox (England and Canada), both geographers; J.J. O'Neill (Canada), geologist; Diamond Jenness (New Zealand), ethnologist; Fritz Johansen (Denmark), botanist and marine biologist; and Dr. R.M. Anderson (United States), zoologist and executive head of the Southern Party. The ethnologist Henri Beuchat (France) and magnetician/meteorologist William McKinlay (Scotland) by misfortune were on the *Karluk,* as was the Australian photographer and engineer, G. H. Wilkins.

Stefansson had expected McKinlay and Wilkins to somehow join his Northern Party after a year with the Southern Party. However, they both quite sensibly informed Stefansson in Nome in July 1913 that they would make no effort to join his Northern Party unless they had specific information as to where it was located, and a ship was available to transport them there. It is not known what Stefansson's answer was on that occasion, but at that stage of the operations, Stefansson had no idea where his Northern Party would be in a year's time, nor was he prepared to send a ship to fetch the two men. He had clearly stated his views on that subject in a letter to Dr. Brock, the Director of the Geological Survey of Canada, several months before. The scientists, he wrote, would be taken to the base camp, but from there they were on their own.

"From that base there will be no question of being taken to other places … every man is expected to get to wherever he wants to be within 400 miles of a base. We will provide Eskimo companions but the white men will have to walk … and to depend upon themselves."[1]

Among this group of scientists lay seeds of disparate allegiance. Chipman, Cox and O'Neill were employed by the Geological Survey of Canada within the Department of Mines. Dr. Anderson was employed by the National Museum of Canada, which was then under the administration of the Geological Survey of Canada, and Jenness was on a three-year contract with the National Museum. Johansen was employed by the Department of the Naval Service. Beuchat, McKinlay and Wilkins also answered to the Department of the Naval Service. Fortunately, little conflict arose because of this, but Johansen and Wilkins were instigators of tense moments on a few occasions in the second and third years.

Failure to Reach their Destination

As mentioned in Chapter 1, the two small schooners, *Alaska* and *Mary Sachs*, after leaving Nome, Alaska, in July 1913, were forced by unusually severe ice conditions to seek shelter at Collinson Point in Camden Bay later in the summer, three hundred miles east of Barrow, northern Alaska. This was the first summer since 1888, in fact, that not a single vessel from the Pacific Coast reached the small Arctic settlement at Herschel Island, the Expedition's initial destination. As members of a Canadian government expedition, the men on the *Alaska* and the *Mary Sachs* had no business spending time on the northern coast of Alaska, but navigation conditions in the summer of 1913 left them no option.

Nine vessels were trapped in the ice along the Western Arctic coast in 1913, six of these along the Alaskan north shore. None left the Arctic that year. The *Karluk* was one of these, as we have seen, having been caught in the ice north of Cross Island. The steam-whaler *Belvedere* and four smaller gasoline schooners, *Mary Sachs, Alaska, Polar Bear* and the *Elvira*, were icebound along the ninety-mile stretch of coast between Collinson Point and the Alaska-Canada boundary. The *Elvira* sank shortly after getting caught in the ice, and its crew were taken on board the *Belvedere* for the winter. Two additional schooners were trapped in ice about ten miles east of the Alaska-Canada boundary: the *Anna Olga* and the *North Star*, and an additional gasoline schooner, the *Teddy Bear*, wintered east of the Mackenzie River Delta. Of these, all but the *Karluk, Mary Sachs, North Star, Alaska* and the *Elvira* left the Arctic the following summer (1914).

---

[1] Stefansson, 1913b (March 1, 1913).

Collinson Point

When it became apparent that his two schooners could not reach Herschel Island, Dr. Anderson and his Southern Party sought shelter in Camden Bay, more specifically in a lagoon within that bay that was protected by the elongated sandspit known as Collinson Point. At the neck of the sandspit was a vacated log house measuring about twenty feet by thirty feet, which would serve nicely as their base camp. It had been well built of driftwood in 1911 by an American trader named Duffy O'Connor, who just a few weeks before Dr. Anderson's arrival had conveniently abandoned his house and moved east to Demarcation Point. Expedition members soon pulled the *Mary Sachs* partly on shore near the vacant house, and anchored the larger *Alaska* about half a mile away in the lagoon. From materials on the ships the men soon rigged up bunks and shelves in the house, then used driftwood to build a barn to shelter the sled dogs.

The scientists spent most of the fall of 1913 learning to adapt to Arctic life—the bitterly cold temperatures, the few hours of daylight, the frequent blizzards, the need to collect driftwood regularly for fuel, the necessity of wearing fur clothing and fur footwear for warmth, learning how to travel with dogsleds and taking scientific observations at low temperatures.

They constructed one snowhouse in which they operated a tide-measuring machine and then a second one as a specially designed observatory from which they could take solar and stellar readings. These snowhouses provided protection from the winds, but in the Arctic cold, a man's breath can quickly create a film of frost on the lenses and metal parts of their surveying instruments. Some of the scientists hunted in the mountains south of their camp in search of mountain sheep and caribou, but had little success. Chipman and Cox prepared a detailed

Figure 31 The CAE house at Collinson Point, northern Alaska, occupied by the Southern Party during the first winter in the Arctic, February 1, 1914.

Photo: © Canadian Museum of Civilization, photo by Rudolph Martin Anderson, 38703 topographic map of the lagoon at Collinson Point and the adjoining ten square miles of tundra. Their sounding of the lagoon indicated a depth of only seven feet at its entrance, thereby limiting the size of vessels it could shelter. Parts of the lagoon were deeper, and the *Alaska* chose one of the deeper spots to drop anchor and winter.

Stefansson's Unexpected Arrival

Stefansson arrived unexpectedly at Collinson Point from Barrow on December 14, bearing news of his separation from the *Karluk* since late in September. Travelling with him were his male secretary McConnell, the photographer Wilkins and Ikey (Angutisiak) Bolt, an Iñupiaq from Point Hope, Alaska, all of whom appeared a day later than Stefansson. The latter had intentionally left them at a cabin twelve miles west of the Southern Party's winter quarters in order to have the opportunity to present his version of the separation from the *Karluk* without any disagreement from either Wilkins or McConnell (as both men commented in their diaries at the time). When Stefansson first entered the house, some of the scientists were lying in their bunks reading, writing or playing games of chess, cribbage or checkers. He promptly scolded them severely, telling them that they ought to be out working or exercising in order to be in suitable condition for more rigorous work when the sun returned. Of course, little scientific work of the kind to be done by the Expedition scientists could be done without daylight, but Stefansson felt they should be actively undertaking camp chores, such as gathering and chopping wood, fetching water, repairing equipment, attending to the dogs, checking the condition of the schooners and related activities. His unexpected appearance and sharp words had the immediate effect, understandably, of lowering his popularity with the men at the camp even more than it had been after the confrontational meetings at Victoria and Nome. Luckily Dr. Anderson and a few of his men were away hunting in the mountains at the time, so missed the unpleasantness of the occasion.

Stefansson remained at Collinson Point for only a few days, just long enough to give his dogs a rest and for him to assess the local situation. During that time he explained to the scientists still at the camp that Jenness was spending the winter with an Iñupiat family near the mouth of the Colville River and that he himself was having to make new plans, for the men who comprised his Northern Party were on the *Karluk*, adrift within a large mass of Arctic ice, and hence were no longer available to him. On December 17, he headed east to the *Belvedere*, taking with him Louis Olsen (a sailor from the *Alaska*), saying he intended to start organizing a new Northern Party and would be back soon. He subsequently spent Christmas at the *Belvedere*, sent Olsen back to Collinson Point, then went on to Herschel Island, not returning until early in March.

Christmas 1913

A few days before Christmas Day, their first in the Arctic, some of the men sought ways to make that day more festive. Wilkins rooted through the warehouse cache and emerged with a motion-picture projector and some films with which he planned a showing after the Christmas dinner. Chipman and Johansen created a Christmas tree using a small shrub, which they decorated with tinfoil from cigarette packages. McConnell went west to Flaxman Island to bring the American geologist, Ernest de Koven Leffingwell, to the camp for the holidays.

Leffingwell brought with him a bottle of whisky, which proved to be the only alcoholic beverage in the camp. Others wrapped personal items as gifts for one or other of their colleagues, and the cook Charles Brooks planned a remarkably lavish Christmas meal.

The festivities were unexpectedly marred, however, when the assistant cook, Andre Norem, failed to return on Christmas Day after a visit to the trapping cabin of the *Mary Sachs'* engineer, Jim Crawford, twenty-four miles to the east. Dr. Anderson then dispatched a sled party to a cabin ten miles to the east of their base camp, which the *Alaska's* Captain Otto Nahmens had been using as a haven for his trapping activities, because Crawford said that on the previous day he had left Norem at Captain Nahmens' cabin. The sled party returned without finding any sign of the missing man. A search party then set out immediately, but darkness soon forced it to return. All knew that Norem had been suffering from bouts of despondency of late, so his disappearance increased their concerns.

In spite of worries over Norem, the men enjoyed a Christmas dinner fit for royalty. Their menu consisted of soup, stuffed roast duck with gravy, cranberry sauce, mashed potatoes, string beans, Christmas pudding, mince pie, dates, chocolates, fruit and coffee. Several Alaskan Iñupiat from nearby trapping camps also shared in the festive meal. It took three sittings at the table to feed all of the people present, but the cook Brooks somehow managed to satisfy everyone. After the meal most of the men smoked cigarettes or cigars, which certainly must have fouled the air in their crowded house. Afterwards the men passed around candies and exchanged assorted gifts, many of which had been supplied the previous June by several ladies' organizations in Victoria, British Columbia. Wilkins then showed moving pictures, the subject matter of which was never disclosed. It is probably safe to state that that was the first film showing anywhere on the northern Alaska coast. His projector was operated manually and illuminated by an acetylene lamp.

The next morning almost all of the men at the base camp set out to search the surroundings for Norem. A few hours later Cox and Wilkins found him wandering in a disoriented manner about two miles west of the camp. Although he had been exposed to the elements without food or fluids for two days and nights, in temperatures well below freezing, somehow he had survived with little evident physical harm.

The days passed quietly between Christmas and New Year's Day. Some of the men made brief forays to gather driftwood logs from the shore for use in building a tunnel under the drifted snow from the house to the warehouse and the dog shed. Others went off to collect loads of whale meat for the dogs from a whale carcass beached near Flaxman Island some thirty miles to the west. On New Year's Eve the men toasted the success of the Expedition with a hot toddy of lime juice, followed by a snack of biscuits and cheese, then cigars. Johansen, whom the others sometimes called "Prof" because of his frequent inclination to lecture them, then entertained them with an assortment of his stories, of which he was said to have many.

Several of the scientists undoubtedly reflected on the months they were wasting as they waited on foreign soil to reach their assigned field areas, with little prospect of accomplishing much the following year except to get established in the area where they were to work.

NORTHERN ALASKA TO BERNARD HARBOUR, 1914 19

The cold and darkness of January forced the men to remain close to their base at Collinson Point until the sun finally started to appear above the horizon in the latter half of January. By February plans were well advanced for starting their spring field work east from the Alaska-Canada boundary. It was awkward being located more than one hundred miles west of that boundary, for it required them to make long sled journeys before they could even commence their intended work. Their plans called for the reconnaissance mapping of the Firth River from its mouth upriver to the International Boundary line, both geographically and geologically, and then the geological reconnaissance of Herschel Island, and geographical mapping of the coast from the International Boundary east for about one hundred and fifty miles to the Mackenzie River Delta. Accurate geographical locations would be carried eastward from Demarcation Point, where Canadian surveyors just two years previously had established and marked the boundary between the two countries at the coast by means of an international monument and benchmark. The men would use their dogs and sleds for transportation until they completed their work to the Mackenzie Delta. Stefansson hoped to obtain mechanized water transport for the use of the two geographers and the geologist once navigation was possible. With such equipment, the scientists were to examine as much of the delta as time would permit, and meet the *Alaska* and *Mary Sachs* at Herschel Island after these ships were free to start their journey east to the Coronation Gulf region. It was hoped that the delta work would reveal one or more channels sufficiently deep to accommodate coastal vessels that wished to proceed upriver, or alternatively would permit river steamboats to proceed downriver to the coast, thereby avoiding the necessity of transferring the cargo that was intended for coastal settlements to smaller vessels at Red River or Fort McPherson.

O'Neill's Mapping of the Firth River[1]

The first to leave Collinson Point was the geologist J.J. O'Neill. On February 18, he started east with Captain Peter Bernard as his dog-driver, the sailor Louis Olsen as assistant, and two heavily laden sleds. Dr. Anderson had asked Bernard and Olsen to pick up certain Southern Party supplies from the *Belvedere* and freight them to the Mackenzie River Delta for use during the early summer's work there.[2] O'Neill intended to ascend the Firth River (then locally known as the Herschel Island River), mapping the geology along its course to the International

---

[1] O'Neill's activities are interpreted from his field notes in Library and Archives Canada, Ottawa, and from his published account (O'Neill, 1924, pp.10A–15A).

[2] There were seventy-five tons of CAE supplies on board the *Belvedere* (Stefansson to Desbarats, 1913g [October 18, 1913]). These were the supplies bought by Dr. Anderson in Nome for the

Boundary in order to obtain a cross-section of the geology from the coast inland to the Endicott Mountains.[3] When that project was completed he would examine the geology of Herschel Island and then go east and link up with Chipman and Cox for the mapping of the Mackenzie River Delta. He was also to be at Herschel Island early in August to connect with the *Alaska* and the *Mary Sachs* for their journey east to the Coronation Gulf region.

En route to Herschel Island the O'Neill party stopped briefly at Martin Point, then at the camp of the *Polar Bear's* crew, at the *Belvedere* (where they were joined by Storker Storkerson and Bill Seymour and their dogsleds), at Edward (Duffy) O'Connor's trading post near Demarcation Point, and finally at Martin Andreasen's camp.[4] On February 28, O'Neill's party met Stefansson and Peder Pedersen about fifteen miles west of Herschel Island. Stefansson was on his way back to Collinson Point after being upriver at Fort McPherson. O'Neill handed Stefansson a letter from Dr. Anderson responding to the instructions Stefansson had earlier sent him (Dr. Anderson) from Fort McPherson. Following a brief conversation about the situation at Collinson Point, Stefansson decided to return to Herschel Island with O'Neill for further discussions (as was mentioned in Chapter 9), so sent Pedersen ahead to Matt Andreasen's camp, where he would join him later. Stefansson had further discussions with O'Neill at Herschel Island and then left for Collinson Point on March 2, taking Bernard with him.

O'Neill had developed an unpleasant blister on one foot and injured a knee en route to Herschel Island, so welcomed a brief respite there, where he spent some of his time reading and playing games of cribbage with RNWMP Inspector Phillips.[5] He did, however, study the geology of the island briefly early in March, ascertaining that it was formed of deeply dissected marine clay containing the shells of Pleistocene fossils. Cliffs of silty mud, forty to fifty feet high, occur around the coast of the island. Two sandspits nearly connect the island with the mainland, while a third one, at the island's southeastern corner, is curved and makes a sheltered haven, known as Pauline Cove, deep enough for schooners, whaleboats and other small vessels.

O'Neill left Herschel Island on March 12, accompanied by Olsen and a man named "Siberia" Mike, who was originally from East Cape,[6] and had been hired by Stefansson to assist O'Neill. With Mike were his wife, Sis, and their young son, Georgie Mike. Together they headed for the Firth River, but scarcely had they progressed a few miles up that river when they were beset by gale-force winds and blowing snow, which lasted for three days. Upon awakening on the morning of March 17, O'Neill could not light his candle in their tent. Alarmed by this, he sought to exit the tent and discovered that the tent was completely covered with snow, which was smothering those who were inside it. It had been a narrow escape. O'Neill and Olsen then dug their way out of the tent as quickly as they could, using a frying pan as a shovel. Their

---

use of the Southern Party after Stefansson informed him he intended to retain most of the supplies on the *Karluk* for the use of his Northern Party.

[3] The name Endicott Mountains was a general one used in the early 1900s for all of the mountains oriented east-west along the northern Alaskan coast.

[4] Martin Andreasen was more commonly known as Matt.

[5] O'Neill, 1916 (March 4 and 5, 1914). His field notes indicate that he played a good many games of cribbage over the course of the summer.

[6] East Cape, now known as Mys Dezhneva or Dezhneva Cape, is the most easterly point in Asia.

shovel was somewhere in the drifted snow where Mike's wife had left it the night before, and she could not recall where that was. Next they dug out their sled and dogs, and in due course moved their supplies from the deep snow to the windswept frozen river. That accomplished, they all returned to Herschel Island, leaving their nearly buried tent behind.

When they returned to their river camp on March 19, having obtained another shovel and a tent from the police, they were without Mike's wife and baby son, both of whom remained behind at Herschel Island with bad colds. O'Neill instructed Olsen and Mike to set up their new tent and then to retrieve the old tent, which was covered with nine feet of snow. While the two men were thus occupied, O'Neill began his examination of the rocks that were exposed here and there along the sides of the river valley. All were bedded sedimentary rocks, such as sandstones, shales and limestones. He made notes on the rocks he examined, their direction (strike) and inclination (dip), both of which he determined using a special geological compass. He also noted the location and nature of contacts between dissimilar rocks, and the presence or absence of fossils. He continued his geological studies for a week, travelling farther and farther each day from his camp with the use of the sled and dogs for transportation. On March 25, he reached a site fifty miles upriver, where a surveyed line marking the International Boundary between Canada and Alaska crossed the river. This was as far as he

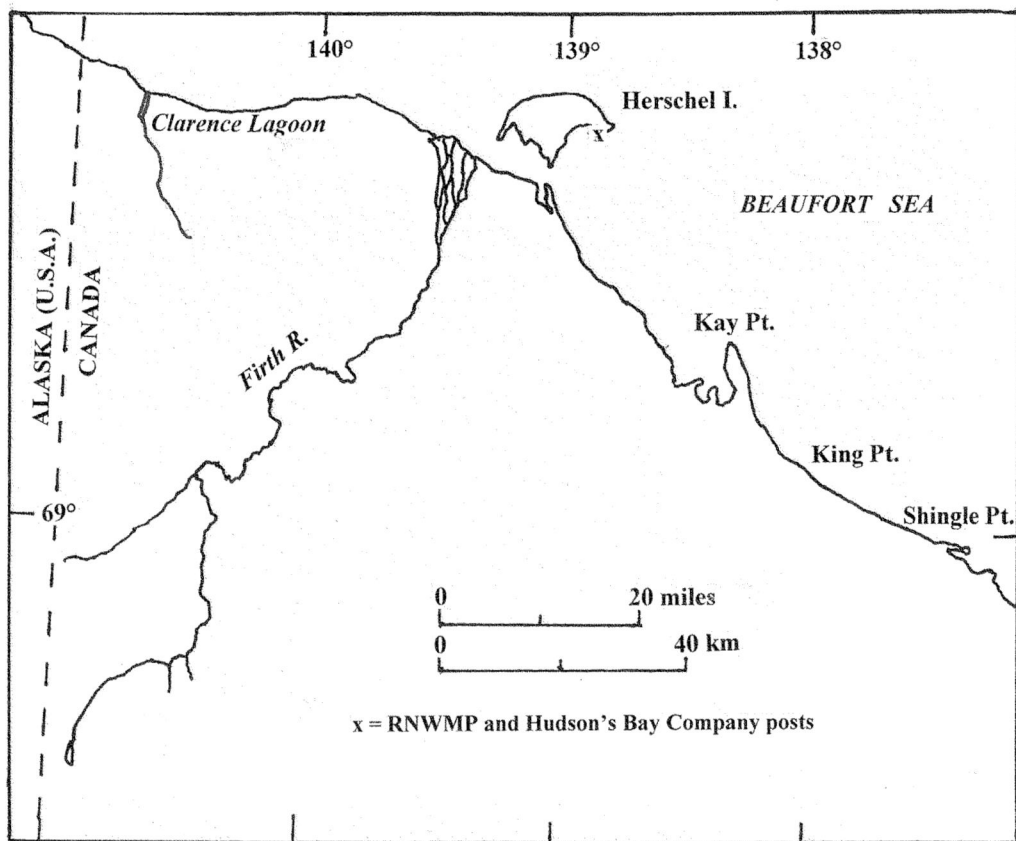

Map 13 Herschel Island and nearby Firth River, which Cox and O'Neill explored and mapped in 1914.

was expected to map, so he turned back. For the next few days he examined closely a few other rocks exposed along the river, searching with limited success for fossils with which he hoped to determine the age of the enclosing rocks. The marine fossils he found later proved to be Middle Jurassic in age, hence were about 150 million years old, deposited during the extended era when dinosaurs roamed the Earth.[7]

Once his Firth River project was completed, O'Neill and his two assistants packed their camp gear and returned to Herschel Island, reaching the police barracks on March 31. There they encountered Chipman, Cox and Peder Pedersen, who had just arrived from Collinson Point with the intention of mapping the Mackenzie River Delta as soon as it was free of ice.

Cox's Mapping of the Arctic Coast[8]

The geographers[9] Chipman and Cox had left Collinson Point on March 16, following the coast east. When they reached the boundary monument at Demarcation Point (an unmanned Canada-United States border point), they stopped to take a series of solar readings to rate their chronometers at this carefully surveyed site on the 141st W meridian. Cox stopped again at this boundary two years later when the Southern Party was leaving the Arctic so that he could get a new series of time sights at the same location to check with their previous readings. All of their subsequent location readings to the east would be related to the benchmark monument that was erected here in 1912. Their instructions from Dr. Anderson were then to map as much of the Mackenzie River Delta as time permitted before the end of July, at which time they were to return to Herschel Island to rejoin the *Alaska* and *Mary Sachs* and head for Coronation Gulf.

Commencing at the International Boundary, Cox mapped the configurations of the coast east to the mouth of the Firth River, where Chipman waited for him. For his mapping Cox used the age-old surveying method of pace-and-compass traversing to determine the directions and distances between prominent coastal features, then carefully sketched the shape of the coastline between the two features. From the mouth of that river the entire mapping party crossed the ice to Herschel Island on March 31, which they made their base until they could work on the Mackenzie Delta.

O'Neill joined them a few days later. Cox and O'Neill then conferred at length about what O'Neill had just seen along the Firth River. Two days later, accompanied by a local Inuk named Dick, with the latter's dogs and sled and four hundred pounds of Cox's supplies and equipment, Cox headed back to the Firth River. Over the next two weeks he traversed, sketched and made notes on about seventy-five miles of the river. Much of the river course was cut through canyons that were about one hundred and fifty feet wide, with walls one hundred to two hundred feet high. About thirty-one miles above its mouth the river flowed through a narrow winding pass in the British Mountains, a site locally known as "The

---

[7] Fossil, rock and mineral specimens collected by O'Neill were stored by the Geological Survey of Canada for some years, but appear to have been discarded.

[8] Cox's activities are detailed from his field notes (in Library and Archives Canada, Ottawa) and his brief published account (Chipman and Cox, 1924, pp. 17B–18B).

[9] Their official designation on the Expedition was topographer.

Blowhole" because the winds funnelled through it with howling ferocity. The adjoining mountains rose to elevations of about eight thousand feet. Cox, Dick and the dogs had a great deal of difficulty sledding through that pass on April 8 because of the slipperiness of the wind-swept ice on the river and the force of the driving winds.

When about fifteen miles beyond the south end of the pass, they encountered running water covering the entire frozen river bed. Cox climbed a nearby hill to ascertain the extent of the problem and saw that the water flowed from a large lake some miles upriver and that the water covered the ice on the river channel for as far as he could see. It made further progress impossible for him and so he was unable to get to the International Boundary line. He therefore took a longitude reading where he was and headed back to the mouth of the river, reaching the police post at Herschel Island on April 14. There he paid Dick for his seventeen days of service, his dogs and his sled at $5 per day. Dick had been recommended to Cox by RNWMP Inspector J.W. Phillips because he was about the only Inuk on Herschel Island then willing to work on Sundays. Most of the other Inuit on the island had been persuaded by the local Church of England missionary, the Reverend C.E. Whittaker, that Sundays were strictly days of rest and prayer. This was the same Rev. Whittaker who had also warned the local Natives against having anything to do with members of the Expedition, for they were a godless bunch and inclined to abuse the Native women. Rev. Whittaker and Stefansson were, understandably, not on the best of terms.

Lacking a dog team, sled and assistant, Cox now had to wait at Herschel Island until Olsen finished helping Captain Bernard with the freighting of supplies from the *Belvedere* to Martin Point for Stefansson. Olsen finally became available at the end of April. Cox needed his assistance to complete his coastal survey from the Firth River one hundred miles east to Escape Reef, which was at the western mouth of the Mackenzie River Delta, five miles east of Shingle Point. They encountered no major difficulties in carrying out that activity, and on May 1 Olsen left Cox alone with his summer's supplies at Escape Reef and returned with their sled and dogs to Herschel Island. Cox cleared the snow out of a cabin recently vacated by trapper Samuel "Scotty" McIntyre and moved in. He then spent the month of May alone, repairing the hull and engine of a launch on the beach about a mile away, which Stefansson had purchased earlier in the spring for $1,000 from a prospector named Willoughby Mason.[10] The launch was for Cox's summer use on the delta.

During May, Cox was visited briefly by Dr. Anderson and two members of the Herschel Island detachment of the RNWMP. Then two Mackenzie River Delta Inuit families (Inuvialuit) set up camp nearby. One of these, Roxy Memoganna and his wife, Monica, with their four children, arrived at the end of the month. Roxy had been hired by Stefansson to help Cox survey the delta, and planned to leave his family at Escape Reef for the summer while he was with Cox. Meanwhile, Cox finished repairing the launch about the end of May and gave it a test run, which proved satisfactory. He was now ready to commence his survey of the west branch of the delta just as soon as the ice cleared out of it.

---

[10] Auditor General Report, 1920.

The Mackenzie Delta Activities of Chipman and O'Neill

Meanwhile, Chipman, O'Neill, Olsen and Pedersen left Herschel Island on April 5 with two heavily laden dogsleds, heading for Pedersen's cabin on the Mackenzie River Delta. Their route took them along the mainland coast and up the Moose River, following the regular dogsled route to Fort McPherson. The Moose River was the most westerly of the navigable channels on the delta. They were thus conveniently able to make use of a series of police

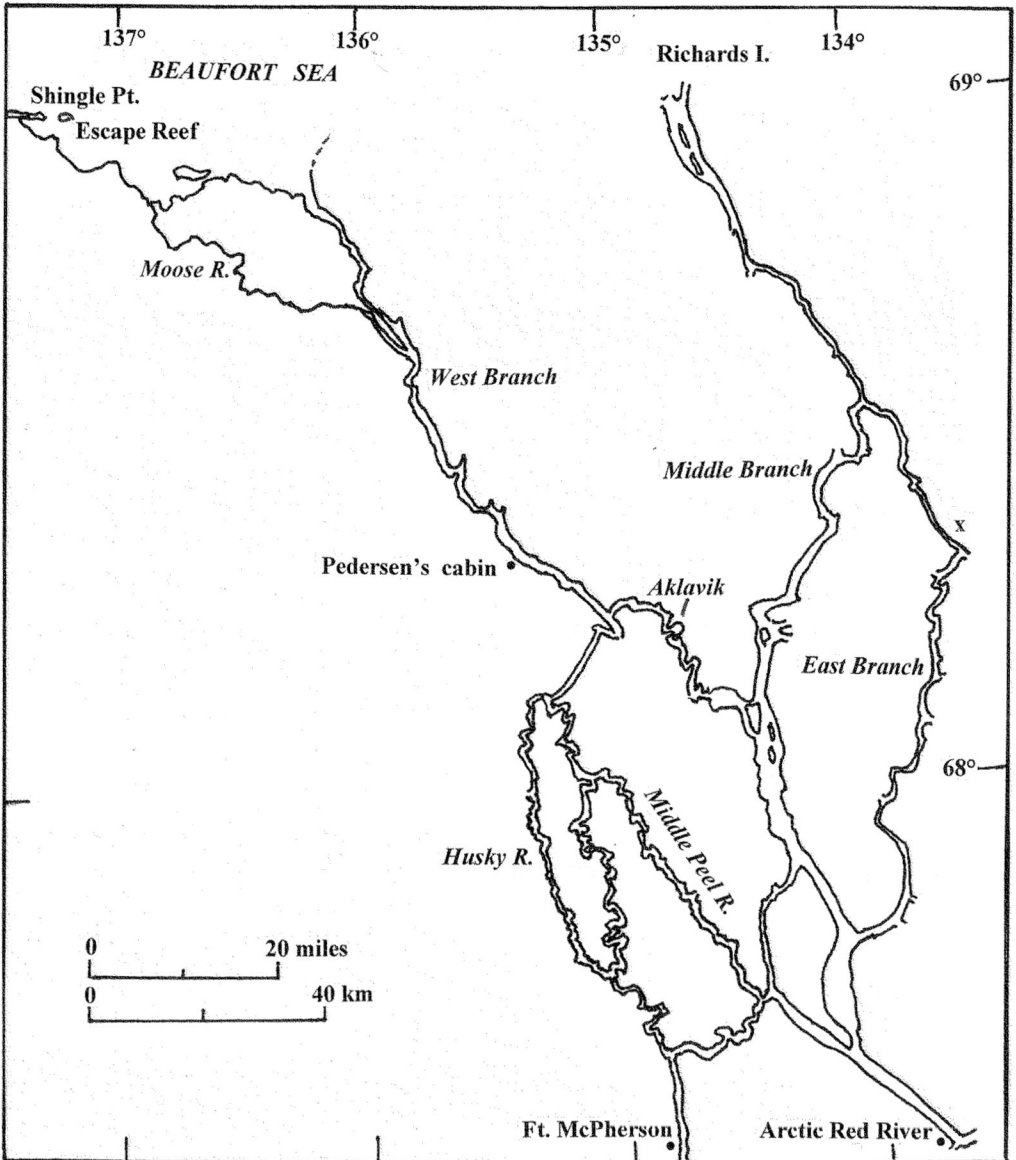

Map 14 Mackenzie River Delta, as mapped by Chipman, Cox and O'Neill in 1914. (Based on an insert map in Chipman and Cox, 1924.) The location of the town of Inuvik, which was founded many years later (1958), on the east branch of the Mackenzie River east of Aklavik, is indicated by an "x".

cabins located every twenty-five miles or so along the route. With Olsen and Pedersen driving the dogs, Chipman and O'Neill observed the terrain and hunted hares and ptarmigan on both sides of the river, obtaining food for the dogs. On April 12 they came upon the camp of Mrs. Elvina Storkerson, the eldest daughter of the Danish trader, Captain Charles Klengenberg, who was living there with her two young daughters while her husband was on the ice trip with Stefansson. Chipman left her the caribou skins and supplies that Stefansson had asked him to take to her. Mrs. Storkerson took advantage of their brief meeting to complain to Chipman that she and her two girls were always hungry, although Stefansson was paying a local Native $10 a month to look after her and her children.

Two days later (April 14) Chipman's party reached Pedersen's cabin, which was on the west branch of the Mackenzie River a dozen miles northwest of its junction with the Aklavik branch, and there they settled down to wait for the spring breakup of the river. This did not occur for six weeks. Nearby was Pedersen's power launch *Edna*, which Stefansson had purchased in January when he hired Pedersen to assist Chipman and O'Neill in their survey of the east branch of the delta. At the same time Stefansson had arranged for the use of Pedersen's dog team, sled and whaleboat. Stefansson had realized that if Chipman and Cox each had his own launch for transportation, they could work independently and hence accomplish twice as much as if they worked together. With any luck they would find one or more deep-water channels in the delta that would permit vessels larger than whaleboats to operate in the lower part of the river.

Commencing on April 20, Chipman, O'Neill and Pedersen trekked west of the river for a few days to the Black Mountain region to ascertain the geological sequence of the rocks on its slopes. They climbed to an elevation of fourteen hundred feet on Round Mountain, but found further progress unsafe underfoot at that time of year. O'Neill described the bedded rocks and collected a few fossils from them, for little was known of the geology of the region at the time. Meanwhile, Chipman viewed the delta, took several photographs, and then they returned to their base camp.

Mrs. Storkerson arrived on May 10 with her children, brought upriver by some Natives, who then continued up the river. Chipman helped her to get settled in a cabin near the Pedersen cabin and provided her with basic food supplies. Two weeks later she moved across the river to join some Natives there in hopes of finding more fish and game.

On May 28 Pedersen put his launch *Edna* in the water, but found that it leaked badly and needed considerable repairs. He and O'Neill managed to seal the hull, then O'Neill spent several days trying unsuccessfully to overhaul the launch's seemingly temperamental engine. Disgusted by their failure to get the *Edna* operating satisfactorily, Chipman and O'Neill finally decided to use Pedersen's whaleboat instead. On June 13, they sailed it down the middle branch, charting part of the river's course, and then followed the east branch of the delta northwards to the south end of Richard Island. Adverse winds prevented them from continuing to their intended destination, Kittigazuit, so they turned back and proceeded upriver to Arctic Red River to meet the steamer bringing in government supplies and mail to the Expedition. Sailing upriver against the current proved difficult at times, and it was frequently necessary for someone to walk along the shore pulling the whaleboat with a long rope. This activity together with Chipman's need to stop periodically to take compass directions, magnetic declinations, and solar readings for latitude and longitude delayed their progress. Swarms of mosquitoes also plagued their every move. Chipman and Pedersen finally left O'Neill at Arctic Red River to receive the supplies that they expected to arrive for the Expedition on the river steamer, and arrange for their transportation down the river to Herschel Island. They would, in the meantime, survey the Peel River to Fort McPherson. At Fort McPherson they were joined by Cox on July 3.

Cox's Mackenzie Delta Activities

Commencing on June 19 and ably assisted by Roxy, his Inuvialuk guide, Cox charted the west (or Aklavik) branch of the Mackenzie River from Escape Reef at the coast to the mouth of the Peel River. Fortunately his launch, which was jovially dubbed "Wild Bill," operated satisfactorily, enabling them to navigate the west branch of the river, take frequent depth measurements, and record a number of latitude and longitude readings by means of which Cox could later plot the course of the river's west branch on the map. After connecting with the mouth of the Peel River they proceeded to Fort McPherson to rejoin Chipman.

In this manner, within a period of less than four months, Chipman and Cox successfully completed the three major geographical surveying tasks Stefansson had set out for them the previous December. Firstly, Cox had mapped the course of the Firth River from the coast upriver for more than fifty miles almost to where it intersected the International Boundary line. He also mapped the coast from the International Boundary monument and benchmark west of Herschel Island east to Escape Reef at the mouth of the west branch of the Mackenzie River Delta.[11] Then, while Cox surveyed the west branch of the delta, Chipman surveyed the east branch and some connecting channels, permitting them to tie their surveys in with those done earlier by other surveyors working down the river.

The work by Chipman and Cox established that both east and west channels of the Mackenzie River Delta were sufficiently deep, although the river channels shifted from year to year, to accommodate ships drawing six to six-and-a-half feet of water, somewhat larger than those which had used the river previously, but were insufficiently deep for either the river steamers or large ocean-going vessels wanting to proceed upriver from the ocean. O'Neill, meanwhile, had carried out reconnaissance geology along the Canadian portion of the Firth River, on Herschel Island, and along the eastern branch of the Mackenzie River.

Jenness' Barter Island Ethnographic Activities

After two months with the Iñupiat families of Aksiatak and Aluk near the mouth of the Colville River, the ethnologist Jenness ran out of food late in January and was forced to sled to Barrow, taking the Hopson boy with him. During his two months with these two families he had diligently studied their language, activities, beliefs, games, folklore and physical characteristics, and had managed to acquire a working knowledge of their language. At Barrow for the next nine days he studied a collection of northern Alaskan Iñupiat artifacts that Stefansson had purchased earlier from Charles Brower for the National Museum of Canada,[12]

---

[11] No bedrock was exposed in the two hundred miles of coast.

[12] The Canadian Museum of Civilization possesses "extensive Stefansson/Barrow collections," which Dr. David Morrison catalogued in the 1990s (D. Morrison, personal communication, 2010).

and worked on the extensive notes he had made during the previous months. McConnell then arrived by sled from Collinson Point and took Jenness back there with him early in March. During April and May, Jenness consolidated his winter's notes, took brief charge of the Collinson Point camp and then sledded east to trader Duffy O'Connor's camp at Demarcation Point, where he made an inventory for Dr. Anderson of the goods Stefansson had acquired through purchase from O'Connor.

Late in May, in response to instructions sent him by Stefansson from somewhere on the Beaufort Sea late in April, Jenness moved to Barter Island and for the next two months carefully excavated three old Inuit ruins that marked the ancient trading sites of the Mackenzie and northern Alaskan Inuit. Despite near-freezing temperatures, he succeeded in gathering a large collection of artifacts for the National Museum of Canada. Helping him with his digging were two Inland Alaskan Iñupiat, Aiyakuk and his step-son Ipanna.

Fifty years later radiocarbon tests dated the artifacts he had collected at about 1500 AD. His Barter Island collection of artifacts was one of the first ones made systematically of the whale-hunting "Thule people," who spread east across the northern Arctic from Alaska to Greenland between 1000 and 1600 AD. The importance of his collection increased immeasurably in the mid-1950s after U.S. Army personnel obliterated his three archaeological sites on Barter Island during the hasty construction of a military station and airport.[13] His collection is thus not just the best one, but the only one from Barter Island.

Johansen's Spring Biological Activities

There was little outdoor work that the marine biologist and naturalist, Fritz Johansen, could do during the fall and winter months, but in the spring he made collections of plants, insects and fish in the vicinity of Collinson Point, and somehow managed to rear several species of local insects, enabling him to study their life histories. He also collected data on ocean temperatures and tide movements at both Collinson Point and Martin Point.

Late in March he accompanied Stefansson's first ice trip from Martin Point over the Beaufort Sea for two weeks as part of the support party, during which time he took frequent measurements of the ocean temperatures and ocean-floor depths. And then in spare moments between March and June of 1914 he wrote four articles in Old Danish (to continue the three he had written in 1913) as part of a series of twelve articles on the activities of the Expedition that he ultimately sent to the Danish newspaper, *Politiken*. Of course, these publications contravened the publication ban Stefansson had imposed on all of the scientists, but Johansen was a non-conformist and may have thought that the ban applied only to English publications.[14]

---

[13] This construction work was on one of more than sixty North American Distant Early Warning (DEW-line) stations that the American military constructed every fifty miles along the 69th latitude from Alaska across Canada to Greenland between 1954 and 1957 to guard the United States and Canada against possible long-range bomber attacks from the Soviet Union.

[14] Johansen may have arranged to have the funds paid by the Danish newspaper for his articles to go directly to his wife and small daughter in Denmark.

Dr. Anderson's Activities

Dr. Rudolph M. Anderson, the executive head of the Southern Party as well as its zoologist, had to spend a great deal of his time planning and coordinating the activities of the scientists and sailors in his party, their supplies and equipment, writing reports about the activities and accomplishments of his Southern Party for the government, and other related administrative duties. Although he had tried unsuccessfully to settle with Stefansson at a meeting in Victoria early in June the issue of their separate authorities for the Southern and Northern Parties, problems resulting from conflicting instructions (or "orders" as Stefansson called them) occurred all too frequently during the first winter in the Arctic. Stefansson, for example, had arranged to pay the captains and crews of the two schooners only for the navigation season rather than retaining them on salary year-round. These men had Stefansson's approval to hunt and trap foxes and other small mammals during the many winter months, with supplies provided by Dr. Anderson's party. Stefansson had also assured these men that they could sell any furs they accumulated by their trapping activities, although this arrangement was strictly contrary to instructions from the Department of the Naval Service in Ottawa. In fact, it led to considerable unrest, which Dr. Anderson was expected to deal with. Orders Stefansson later sent or gave to these men while they were under Dr. Anderson's charge represented further interference of what Dr. Anderson properly regarded as his own jurisdiction.

When camp conditions permitted him relief from his administrative tasks, Dr. Anderson made brief dogsled trips into the Alaskan mountains to the south, hunting mammals and birds to add to the collections of the National Museum of Canada. Over the course of the winter he obtained, cleaned and prepared specimens of fifty-two species of birds and thirteen species of mammals from northern Alaska, and in the spring gathered nests and eggs of many species of breeding birds. Although these were from northern Alaska and not Canada, the movements of birds and animals in those northern regions were not governed by political boundaries, and many if not most of the specimens collected were representative of Canadian species as well as American ones. Their collection also contributed to the understanding of the geographical ranges and seasonal appearances of these birds.

In May Dr. Anderson received instructions from Stefansson, from somewhere on the Beaufort Sea, to use the *Mary Sachs* to transport to the Coronation Gulf region any supplies for the Southern Party that could not be loaded on the *Alaska*, after which it was to bring additional supplies to Cape Kellett at the southwest corner of Banks Island and establish a base camp there for Stefansson's Northern Party. The *Mary Sachs* would thereafter serve Stefansson as a supply vessel, and if the men on the *Karluk* showed up on Banks Island it would be available for their scientific work.

Wilkins' Spring Activities

George Hubert Wilkins, loaned to the Canadian Arctic Expedition to serve as its official photographer for one year by the British Gaumont Company in London, had lost his company's expensive photographic equipment when the *Karluk* was carried off in the storm in September 1913. From the American geologist Leffingwell at Flaxman Island he obtained a Graphlex camera and film in November, by which time there was no sunlight to take outdoor pictures. In February he was able to buy a bulky, late-model Williamson 35-mm moving-picture camera with some equipment and 7,100 feet of film from John Clark, the photographer from the schooner *Elvira*, which had been crushed by the ice and sank east of Collinson Point. To operate this camera the photographer had to aim it and turn the handle simultaneously. After a few trial tests, Wilkins took his first moving pictures with it to show the departure of Stefansson's ice-trip party from Collinson Point on March 16, 1914. This was followed by brief segments of film of the ice-trip party leaving Martin Point, then of a camp on drifting ice, and finally several feet of Stefansson catching a seal. After becoming unexpectedly separated from Stefansson's ice-trip party, Wilkins sledded west to Barrow with Billy Natkusiak in the hope of filming the annual near-shore bowhead whale hunt, a local Iñupiat activity that was fast disappearing. He returned to Collinson Point without having seen a single whale.

In May 1914, Wilkins received instructions from Stefansson to proceed to Clarence Lagoon, a dozen miles east of Duffy O'Connor's trading camp, to ready the schooner *North Star* for service. Stefansson had purchased this trim little schooner from the trader Matt Andreasen in March, along with his supplies. Wilkins was to load it with supplies for two years, hire sailors and native seamstresses, and sail to Norway Island on the west coast of Banks Island, where Stefansson expected to meet him. Stefansson intended to use the *North Star* as an exploration ship thereafter, making use of its specially designed bow, which would ride up on the ice, permitting the weight of the schooner to break the over-ridden ice. Neither the *Alaska* nor the *Mary Sachs* was able to deal with the ice in that manner.

In June Wilkins journeyed to Clarence Lagoon, where he teamed up with sailor Aarnout Castel to renovate the engine and paint the *North Star*. Early in August, when near-shore navigation became possible, Wilkins moved the *North Star* to Herschel Island and prepared to proceed to Banks Island and commence his search for Stefansson and his two companions. With him were seamstresses Pannigabluk, Mrs. Thomsen and Mrs. Storkerson, with their respective small children, numerous dogs, and sailors Thomsen, Castel and Natkusiak. Pannigabluk subsequently decided to remain with her brothers in the delta region and left the ship with her son.

Spring Activities of Stefansson's Secretary, McConnell

In February, Dr. Anderson dispatched Stefansson's secretary, McConnell,[15] west with a sled and team of dogs to pick up Jenness at the Colville River and bring him east to the Collinson Point camp. His trip took McConnell to Barrow, where he found Jenness on February 19, and together they arrived at the base camp on March 20 after a rather trying journey east. McConnell promptly set off in pursuit of Stefansson's ice-trip party, taking the mail he had gathered at Barrow for him. He caught up with them a few miles off shore on March 23, just in time to provide medical assistance to Captain Peter Bernard who had suffered a bad head injury. On April 7 he left Stefansson's ice-trip party and returned to the coast and Collinson

---

[15] Burt McConnell was Stefansson's secretary and press-agent in 1913. In order to have him on the government's payroll, Stefansson had taken him north for one year under the title Assistant to the Meteorologist, although he never actually did any meteorological work. McKinlay was the Expedition's official meteorologist. Nevertheless, McConnell later claimed he was meteorologist of the Expedition (Anderson, 1922b [February 22, 1922]; see also McConnell [1915b, p. 672]).

Point. Dr. Anderson, faced with having to keep McConnell occupied, sent him east to Demarcation Point to make an inventory of the supplies at Duffy O'Connor's trading camp, which Stefansson had purchased earlier. He completed this task rather unsatisfactorily, then continued east ten miles to read and relax at the trading camp of Captain Matt Andreasen at Clarence Lagoon.

McConnell had been hired by Stefansson as his secretary on a one-year contract, and when he was no longer needed by Stefansson or by Dr. Anderson, he became a free agent. His contract expired on June 1, after which time he stayed with Captain Andreasen until August, occasionally hunting but mostly reading. Then he bought passage to Nome on the schooner, *Anna Olga,* which had wintered in the ice at the mouth of Clarence Lagoon. After reaching Nome he managed to participate in the rescue of the *Karluk* survivors from Wrangel Island, as was mentioned in Chapter 8.

Other Southern Party Activities, Spring and Summer 1914

Meanwhile, the despondency of the *Mary Sach's* cook, Norem, at Collinson Point worsened over the winter months, and on April 17 the troubled man shot himself in the head at the entrance to the Southern Party's house. Wilkins and Thomsen cleaned up the remains and laid out the body, adequately chilled for preservation, awaiting the return of Dr. Anderson, who was on a hunting trip up the Hulahula River. After a few days, however, the men at Collinson Point decided to hold a brief funeral service and then buried Norem in a simple wooden coffin near the camp, marking his grave with a small wooden cross.[16]

The sea-ice finally moved far enough off the coast to let the two ships, *Alaska* and *Mary Sachs*, leave Collinson Point on July 25, 1914, and edge their way east along shore towards Herschel Island. After minor delays en route caused by the ice, the *Alaska* reached Herschel Island on August 5, the *Mary Sachs* the next day. Wilkins had been waiting for them with the *North Star* at Herschel Island since July 24.

Changes of Plans at Herschel Island

Chipman, O'Neill and Pedersen arrived at Herschel Island in Pedersen's whaleboat from the Mackenzie River Delta about 1 a.m. on the morning of August 6, the first to reach the island from the east that year.[17] They brought with them the mail that had been sent down the Mackenzie River from Ottawa and elsewhere. Cox followed shortly afterwards in his launch, along with more than two dozen other small craft.

The newly arrived mail included two important letters for Stefansson from G.J. Desbarats, the Deputy Minister of the Naval Service, both dated April 30, 1914. Acting in accordance with an arrangement made previously between himself and Stefansson, Dr. Anderson read Stefansson's mail. In one letter Desbarats informed Stefansson that he was not to interfere with the program set out for the Southern Party by taking any of its men for his new Northern Party.[18] In his second letter, Desbarats stated that specimens collected by members of the

---

[16]   Wilkins negative 50807, April 19, 1914, LAC.
[17]   Chipman, 1916 (August 6, 1914).
[18]   Desbarats, 1914a (April 30, 1914).

Figure 32 Southern Party men enjoying Christmas dinner at Collinson Point, northern Alaska, December 25, 1913. (l. to r.): Captain P. Bernard, McConnell, Johansen, Chipman, Leffingwell, O'Neill, Dr. Anderson, Cox and C. Brooks.
Photo: © Canadian Museum of Civilization, photo by George Hubert Wilkins, 50701

Figure 33 The three CAE schooners at Herschel Island, Yukon Territory; (l. to r.): *North Star*, *Alaska* and *Mary Sachs* (behind *Alaska*), August 10, 1914.
Photo: © Canadian Museum of Civilization, photo by John Johnston O'Neill, 38668

Expedition were the property of the government. Skins of animals killed by members of the Expedition that were not required for clothing for the members or by the zoologist as scientific specimens should be held by the Expedition's storekeeper,[19] and their disposal would be decided upon by the Minister of the Naval Service. This latter ruling related directly to the skins collected through trapping by the sailors Storkerson, Crawford, Nahmens, Peter Bernard, Andreasen and Thomsen during that first winter. Stefansson later ignored these special instructions on the grounds that these men were not actually in the government's employ when they trapped the animals, thanks to Stefansson's arrangement to pay them only during the navigation season.[20]

There was also a letter dated May 7, 1914 for Stefansson from Dr. Brock at the Geological Survey which contained the following blunt comments:

> The disappearance of the KARLUK puts an end to the northern expedition, except what you may be able to accomplish yourself.
>
> I think it would be a fatal mistake to weaken or interfere with the programme of the Southern Party in planning your northern trip.
>
> While the northern work was the main object of the expedition, the Southern Party was made strong and assigned to do important work so that if anything happened to the northern plans, the work of the Southern Party would amply justify the expedition, and make it as a whole, a success. The disappearance of the KARLUK is an irreparable blow to the Northern plans. The thing to do now it seems to me is to make certain of the success of the Southern expedition, which has now become the paramount issue.
>
> Your own northern trip, while necessary and important, should not I think upset the Southern party's plans.[21]

In the mail received that August Dr. Anderson also received a letter from the Deputy Minister of the Naval Service. His letter instructed him to send a ship and search party, headed preferably by Chipman, to look for Stefansson and his men. However, along with that letter came one to Dr. Anderson from Dr. Brock, which stated that *under no circumstances* was any member of the Southern Party to be spared to look for Stefansson. These conflicting instructions placed Dr. Anderson in an awkward situation. Whatever his decision, he was not prepared

---

[19] Chipman had been designated as storekeeper for the Expedition. He was also given the task of collecting customs duties for the Canadian Customs Department from U.S. ships trading at Herschel Island or elsewhere along the Canadian Arctic coast.

[20] In 1917 Stefansson sold all the accumulated skins to the Hudson's Bay Company before starting for Nome, then submitted accounts to the Department of the Naval Service showing that he had spent some of the resultant money re-equipping his Northern Party for its final ice trip after the *Polar Bear* went aground at Barter Island. Documentation pertaining to this activity is located in RG42, Vol. 481, File 84-2-42, LAC.

[21] Brock, 1914.

Figure 34 Group photograph of some CAE personnel and others at Herschel Island, Yukon Territory. Identified behind and to the right of Jennie Thomsen and her daughter Annie (both in furs) are Cox (behind Jennie), Chipman, Dr. Anderson, Jenness (with beard but no hat) and O'Neill (second from right), August 7, 1914.
Photo: © Canadian Museum of Civilization, photo by George Hubert Wilkins, 51435

to release his most experienced man, Chipman, to take charge of the *Mary Sachs* for a year or two to search for Stefansson and his two companions on the chance that they were still alive after an absence of almost five months. Nor, as it happened, was Chipman willing to undertake that challenge. Yet Dr. Anderson could not simply abandon the three missing explorers to their fates, for he personally felt that it was probable they were still alive.

Wilkins was ready to sail from Herschel Island with the *North Star* to meet Stefansson, as Stefansson's April letter to him instructed. But the arrangements also called for the *Mary Sachs* to proceed to Cape Kellett after caching some supplies for the Southern Party and to establish a base there. On that part of the plan Dr. Anderson lacked confidence that Captain Bernard would actually carry out the plans without there being someone in charge on board who was responsible to the Canadian government. Recalling that Stefansson wanted the *Mary Sachs* at Banks Island, he decided to ask Wilkins to take command of that vessel instead of the *North Star*, as it could carry more supplies than the *North Star*, should it be necessary for Wilkins' search party to spend more than a year on Banks Island. This decision later angered Stefansson, who had developed finite exploration plans that called for the use of the *North Star*, plans that the *Mary Sachs* was not designed to undertake. And without the *North Star* Stefansson would have to cancel some of his new plans, about which, of course, Dr. Anderson had no knowledge.

Although Wilkins was still employed by the British Gaumont Company at the time and thus not officially a member of the Expedition, he had considerable knowledge of marine engines and had shown himself to be a reliable and responsible person. He was also a spare hand, for whom the government and Dr. Anderson had no specific planned activities. He was thus the obvious choice to take charge of the search ship. As Wilkins commented in his diary, "… it was agreed by each of the members of the scientific staff that I was the only logical member of the expedition to command the relief ship."[22]

In the same August mail, however, was a letter for Wilkins from his employer, the Gaumont Company, instructing him to return to England as soon as possible. He was reminded that the company had made an agreement with Stefansson to lend him to the Expedition for only one year. This letter forced Wilkins to make a difficult decision. Either he would obey the instructions from his employer and return to England or he would break his contract in order to search for the missing Stefansson ice party. His made his decision quickly. So great was his concern for the three missing men that he felt he must take charge of the search expedition and proceed immediately to Banks Island, even if it cost him his job with the Gaumont Company.

Once Wilkins, Captain Bernard and the crew of the *Mary Sachs* agreed to have Wilkins in charge of the ship, the Expedition members gathered to transfer the supplies, equipment and people from the *North Star* to the *Mary Sachs*. That left no one available to take the *North Star* to Banks Island, but Dr. Anderson saw no reason other than Stefansson's April instructions for the *North Star* to go there now. Furthermore, as no one could say for certain whether Stefansson was alive or not, the small schooner could be put to more useful purposes by the Southern Party. Dr. Anderson decided, therefore, with the agreement of his fellow scientists, to take the *North Star* to Coronation Gulf, but needed someone who could operate the schooner's engine. Cox, a mining engineer with considerable mechanical aptitude, was the most suitable man for that job, so Wilkins showed him how to run the *North Star's* engine. Having spent days overhauling and putting it into good running condition in June and July, Wilkins was well versed with the idiosyncracies of the *North Star's* engine.

News of the *Karluk*

Early the next morning, August 9, 1914, the SS *Herman* of San Francisco arrived fortuitously at Herschel Island under the command of Captain C.T. Pedersen. Then regarded as the most knowledgeable and capable of the ship captains in the Western Arctic, Pedersen was the man who had selected and purchased the *Karluk* in San Francisco for Stefansson the previous year and then delivered it to Esquimalt, British Columbia. He was also the man who, in May 1914, had taken Captain Bartlett from Emma Harbor to Saint Michael, with the news of the loss of the *Karluk*. From Captain Pedersen Dr. Anderson and the Expedition members now learned of the sinking of the *Karluk*, of the survivors reaching Wrangel Island, of Captain Bartlett's remarkable sled journey from Wrangel Island to Emma Harbor, his transport from there to Saint Michael south of Nome on the *Herman* with Captain Pedersen, and of the arrangements to rescue the survivors from Wrangel Island. However, Captain Pedersen did not know if the rescue ships had reached the survivors. It was not, in fact, until

---

[22] Wilkins, 1916 (August 6, 1914).

November of the following year, in 1915, that the Southern Party men would hear about the successful rescue of the survivors from Wrangel Island and about the deaths of eleven of the *Karluk* personnel.

Wilkins and the *Mary Sachs* Start the Search for Stefansson

By August 11 Expedition members had completed the transfer of goods, equipment and personnel to the *Mary Sachs* from the *North Star*, and it left Herschel Island on its rescue mission to Banks Island, cheered on by whistle blasts from the other vessels and the dipping of flags. On board were ten people: Wilkins, Captain Peter Bernard, engineer James Crawford, cook W.J. "Levi" Baur, Billy Natkusiak, sailor Charles Thomsen, Mrs. Jennie Thomsen and her young daughter, Annie, and Mrs. Elvina Storkerson and her young daughter, Martina. The schooner carried a supply of provisions, distillate, coal oil, dogs and sleds sufficient for two or more years. When the *Mary Sachs* reached Baillie Islands, Wilkins left a note for Dr. Anderson stating that it had broken one of its two propellers as a result of hitting some ice while en route there, and that there had been some dissent and drunkenness among his crew, but in spite of that he was heading for Banks Island on August 19. (Activities about the *Mary Sachs* and its passengers are in Chapters 11 to 17.[23])

Figure 35. Departure of the heavily loaded *Mary Sachs* from Herschel Island for Banks Island, to enable Wilkins to search for Stefansson and his two companions, Herschel Island, August 10, 1914.

Photo: © Canadian Museum of Civilization, photo by Kenneth Gordon Chipman, 43249

The *Alaska* Heads East

Dr. Anderson left Herschel Island with the *Alaska* on August 17, heading for a safe harbour he hoped to find somewhere on the mainland coast in the Coronation Gulf region in which he could establish his base camp, but with no specific destination in mind. On board were three other scientists (Chipman, O'Neill and Johansen) and the crew (Captain Daniel Sweeney, who was an

---

[23] For more details on the problems that beset the *Mary Sachs* en route to Banks Island, see Jenness (2004, pp. 145–153).

experienced sailor obtained from the *Belvedere*; cook James "Cockney" Sullivan, originally from London, England, whom Dr. Anderson had just obtained from the *Herman*; engineer Daniel Blue and assistant engineer Siberia Mike, recently a helper for O'Neill; and two native hunters, Ikey (Angutisiak) Bolt and Silas Palaiyak.) Sweeney had replaced Captain Otto Nahmens and Sullivan had replaced Charles Brooks after both Nahmens and Brooks voluntarily left the Expedition at Herschel Island. Palaiyak, a young Mackenzie River Delta Inuvialuk, had worked with both Stefansson and Dr. Anderson during their 1908–1912 expedition. Mike's wife, Sis, and their young son, Georgie Mike, were also on the *Alaska*.

The *Alaska* reached Baillie Islands on August 21, after minor delays caused by ice conditions near Kay Point. By chance, the trading schooner *Teddy Bear* was anchored there, the captain of which was Joseph (Joe) Bernard, a nephew of Captain Peter Bernard on the *Mary Sachs*. He had been trading with the Copper Inuit in Amundsen Gulf and Coronation Gulf for five years and was en route to Nome with a load of furs and Copper Inuit artifacts.[24] During their brief conversation, Captain Joe Bernard informed Dr. Anderson about ice and navigation conditions to the east, and sketched for him a map of the best harbour between Pearce Point and Coronation Gulf.[25] The captain's advice and map proved to be of considerable help to Dr. Anderson.

The *Alaska* left Baillie Islands on August 22 and anchored without incident two days later in the small sheltered harbour Captain Joe Bernard had recommended on the mainland coast northwest of Cape Krusenstern.

The *North Star* Heads East

The *North Star* left Herschel Island a day after the *Alaska*, that is, on August 18, having been delayed by the need to reorganize its cargo. On board were Cox, Jenness and Aarnout Castel, a Dutch sailor formerly on the *Belvedere*. They reached Baillie Islands a few hours after the *Alaska*, where Cox borrowed Palaiyak from the *Alaska* to help with the operation of the *North Star* for the rest of the voyage. The *North Star* then left Baillie Islands on August 22, several hours later than the *Alaska*, after waiting for Castel and Palaiyak to obtain a considerable quantity of meat and blubber for dog food from a dead whale beached near the tiny Inuit settlement.

The *North Star* proceeded cautiously across Franklin Bay to Cape Parry, thence across Darnley Bay to Cape Lyon. A little farther to the east the ship's crew found the camp they had been seeking and cached provisions and coal oil for the use later of members of the Southern Party. That camp belonged to Captain Christian Klengenberg[26] and was on a beach

---

[24]   He later sold his furs and artifacts to the University of Pennsylvania (MacBeth, 1923).

[25]   This sketch map is in CMNAC/1996-077, Series A—R.M. Anderson Collection, Box 67, Folder 25, Archives, CMN.

[26]   Danish-born Christian Klinkenberg Jorgensen, known to most of the Canadian Arctic Expedition members as Captain "Charlie" Klengenberg, became a rather notorious pioneer trapper and trader in the Amundsen Gulf and Coronation Gulf regions in the 1910s and 1920s. He ceased using his family name Jorgensen while in the Arctic. Stefansson and Dr. Anderson generally used the correct Danish spelling of his name (Klinkenberg), but Klengenberg himself, as well as his descendants

in a small cliff-lined bay at Pearce Point. It was occupied at the time by two of Klengenberg's children, teen-aged son Patsy and daughter Etna, who had fox traps set in the surrounding area. Klengenberg's oldest daughter Elvina (also known by her Iñupiaq name Uiniq or its derivative, "Weena," which was what her father called her[27]) was the wife of Stefansson's assistant, Storker Storkerson. Klengenberg was a Danish sailor, married to an Alaskan Iñupiaq woman, Gremnia Kemnik,[28] and was at the time trapping and hunting with the rest of his family near the bottom of Darnley Bay.

The *North Star* then motored slowly from Pearce Point eastward along the coast, its men ever alert for unexpected shoals. Fog enveloped it as it approached Stapylton Bay and from there on, greatly hindered its progress. On August 27, however, the ship's crew heard the barking of dogs and a gunshot in reply to their rifle shots and realized they were probably close to their destination. Dr. Anderson and Chipman soon emerged from the fog, paddling a canoe, and guided the *North Star* safely through a narrow passage between the mainland and a small gravel island into the inner harbour.[29]

Bernard Harbour

Once the *Alaska was* safely anchored in the inner harbour, Dr. Anderson and Chipman decided to establish their base camp there, as Captain Joseph Bernard had recommended. Next they chose the site on which to construct their base camp, a short distance in from a small sandy beach on the south side of the inner harbour. With those basic matters settled, all hands set about unloading the *Alaska*. It was not an ideal location, but the harbour offered safe shelter to the two schooners, and its location near Victoria Island and also near the centre of the mainland region that the scientists were to investigate seemed advantageous for their intended work. Furthermore, Dr. Anderson knew that Captain Bernard had spent one winter there safely.

One of Dr. Anderson's first activities while the unloading was under way was to ascertain the availability of adequate fresh water and fuel. A lake within a mile offered a good if inconvenient supply of fresh water. There was some driftwood along the coast nearby, but insufficient for their needs over two years. Realizing that the amount of coal they had brought with them

---

in Arctic Canada, ultimately adopted the Anglicized spelling "Klengenberg" (Ashlee, 2008 p. 47, footnote 27).

[27] Ashlee, 2008, p. 55, footnote 67.

[28] She was called Gremnia by Klengenberg's biographer, Tom McInnes (1932). Anthropologist and author Ashlee (2008, p. 56, footnote 71) learned that her Iñupiaq name was Kemnik from Klengenberg's second daughter, Etna, and the latter's husband, Ikey Bolt. A detailed Klengenberg genealogy compiled by a living descendant, Helen Klengenberg, which I found on the internet in 2009, gives Klengenberg's wife's name as Gremnia Qimnik.

[29] The gravel island is approximately one mile long, but less than twelve hundred feet wide and thirty feet above sea level at its highest elevation, and almost blocks the narrow entrance to the inner harbour. It was named Teddy Bear Island in 1960 by N.G. Gray, the Canadian government's Chief Hydrographer, after Captain Joe Bernard's schooner, which wintered there with its owner in 1912–1913.

would not last them the two years either, Dr. Anderson decided to return with the *Alaska* to Baillie Islands to retrieve the supply that had been left there for them. Before that could be undertaken, however, it was necessary to replace the propeller on the *Alaska*, which had been broken after the *Alaska* left Herschel Island. The replacement proved more difficult than expected, so that it was September 7 before Dr. Anderson and the *Alaska* headed west.

Dr. Anderson's decision to return to Herschel Island provided him with an opportunity to inform the government officials in Ottawa by mail of the location of his Southern Party's base camp for the next two years. However, he lacked a name for the location. The British Admiralty Chart he was supplied with showed names for several nearby coastal features, names such as Chantry Island, Stapylton Bay, Liston and Sutton Islands, Dolphin and Union Strait, all named by Sir John Richardson after his remarkable reconnaissance of the Arctic coast in 1848 from the mouth of the Mackenzie River to the Coppermine River in search of Sir John Franklin and his men. There were no official names in 1914, however, for the body of water behind Chantry Island or the several small bays and points along the nearby coast, including the sheltered harbour where the Southern Party was creating its base. After consultation with all of the scientific men in his party, Dr. Anderson proposed the name "Bernard Harbour" in honour of Captain Joseph F. Bernard, in a summary report for the government of his party's activities for 1913–1914.[30]

In 1920, Stefansson wrote Desbarats, the Deputy Minister of the Naval Service, urging the rejection of Dr. Anderson's name "Bernard Harbour," which Stefansson assumed had just been proposed. His reason seemed justified at the time, for he pointed out that the name "Bernard Harbor" (with the U.S. spelling for harbour) had previously been used by the American geologist Ernest de Koven Leffingwell for a bay on Barter Island, northern Alaska, where the same Captain Joe Bernard had spent the winter of 1909–1910.[31] Stefansson advocated instead that Dr. Anderson's name "Bernard Harbour" be changed to "Fort Bacon," a name used by the Hudson's Bay Company for the post it established near the Expedition base camp in 1916, a few weeks after the departure of the Southern Party. He also claimed that he personally saw the harbour in the spring of 1910 before Dr. Anderson saw it, when he passed it while sledding along Dolphin and Union Strait. "I was struck with the favorable appearance of the harbor," he wrote in his letter to Desbarats, "but had no means of ascertaining the depth of water."

Desbarats asked Dr. Anderson to comment on Stefansson's claim. Dr. Anderson in turn sought the comments of the Expedition's geographer Chipman, who replied that he had already protested to E. de K. Leffingwell over the latter's use of the name Bernard Harbor in his Alaskan report (Leffingwell, 1919) on the grounds of prior usage by Canada in the 1915 report. The U.S. Geographic Board subsequently withdrew its use of the name for the Barter

---

[30] Anderson, R.M. 1915. Dr. Anderson originally proposed the name "Bernard Harbour" for the small harbour sheltered by the island now identified as Teddy Bear Island, but it was changed a few years later to define the larger bay sheltered by Chantry Island (O'Neill, 1924, geological map). The change was undoubtedly made with the approval of both Chipman and Dr. Anderson.

[31] Stefansson, 1920 (September 7, 1920). Leffingwell used the name "Bernard Harbor" in his U.S. Geological Survey Professional Paper 109, published in 1919.

Island site.[32] Dr. Anderson then passed this information along to Desbarats,[33] and Stefansson's attempt to prevent the duplication of geographic names was rejected.[34] It is conceivable also that Stefansson was attempting to prevent Dr. Anderson from receiving credit for naming the locality. His claim of having seen the bay earlier than Dr. Anderson while passing in the spring with a dog team deserves Dr. Anderson's pat response, "Buncombe," because the low-lying coastline and adjoining land would have been blanketed with snow at that season.

The Alaska Heads West

After delays caused by the slowness of the propeller replacement, fog, contrary winds and ice blocking the exit of the inner harbour, Dr. Anderson and the *Alaska* finally started west for Baillie Islands on September 6. From Baillie Islands he continued on to Herschel Island to secure additional coal and oil stored there, as well as supplies expected on the Hudson's Bay Company's supply ship, *Ruby*, from Vancouver. The *Ruby* had not arrived, however, by the time Dr. Anderson's men had loaded the *Alaska* (ice forced the *Ruby* to return south before reaching Herschel Island), so they returned to Baillie Islands, arriving there on September 15 in the midst of a fierce northwest gale. The storm continued for six days, during which time the tide rose several feet. A sudden rapid fall of the storm tide on September 20 left the *Alaska* hard aground. The crew was then forced to unload the *Alaska* in order to refloat it, which took them four days. Dr. Anderson then reluctantly decided that it was too late for the *Alaska* to continue east safely, because it would have taken three days to reload it, the hours of darkness were rapidly increasing, a considerable amount of heavy ice was blowing into the region from the north, and there was no safe harbour between Baillie Islands and Bernard Harbour. He therefore had the schooner prepared for the winter.

Dr. Anderson Sleds East from Baillie Islands

Two months passed before Dr. Anderson considered it safe to undertake the four-hundred-mile journey to Bernard Harbour with dogs and sled. He was familiar with that stretch of coast, having travelled along it three years before, so had few qualms about making the journey. Leaving Captain Sweeney and the engineer Blue to take care of the schooner with its more than adequate supply of food and fuel, he departed on November 20, accompanied by Castel, Sullivan, Ikey Bolt, seven dogs, one sled and supplies. They passed a few isolated Inuit trapping camps in Franklin Bay, Darnley Bay and near Pearce Point, but encountered no Inuit farther

---

[32] Chipman 1920 (September 24, 1920).

[33] Anderson 1920c (September 29, 1920).

[34] The name was transferred in the early 1920s to the larger mass of water in Dolphin and Union Strait sheltered by both Teddy Bear Island and Chantry Island, appearing first on a map accompanying Chipman and Cox's Expedition report of the geography of the Arctic Coast (O'Neill, 1924). The change might have been influenced by the construction of a Hudson's Bay Company trading post in 1916 on the same gravel peninsula as the Expedition's house but on its southeast side, facing Chantry Island and the larger bay to which the name "Bernard Harbour" is now applied, because the company's ships were too large to enter the inner harbour. The inner harbour behind Teddy Bear Island remains unnamed to this day.

east. On December 10 they came upon Chipman and O'Neill near Keat's Point, struggling in the dim light and cold to make reconnaissance topographical and geological observations from Bernard Harbour west to Pearce Point. As fog and falling snow were severely interfering with their work, the two surveyors decided to turn about and accompany Dr. Anderson's party back to Bernard Harbour. Together they arrived at the base camp on Christmas Day.

Establishing the Southern Party's Base Camp at Bernard Harbour

In September, just two days after the departure of the *Alaska* for Herschel Island, the five scientists at Bernard Harbour had set to work building their winter quarters. Chipman proved to be a well-organized leader. He and his companions used most of the spare lumber brought from Nome to build the staff house. It was oriented north-south, measured twelve feet by sixteen feet, with one window on each side, and was designed to serve as both office and bunkhouse for the men. Within four days they had framed the house, installed the shiplap siding on all four sides, hammered the roof boards in place, and piled flat strips of turf along the sides and back for insulation. Then they spread a layer of tar paper over the roof and piled a layer of turf on top of it. Next they lined the insides of the walls with white drill cloth. Unfortunately this failed to keep the cold out, and the walls were cold and damp each winter. The men then constructed bunks along the sides and back of the house—two double-tiered bunks and two single ones. The single bunks were for Dr. Anderson and Jenness, the former because of his status as executive head and the latter because of his frequent comings and goings. Jenness' bunk was also used for the occasional visitor when Jenness was away.

Using provision cases (wooden boxes containing peas, rice, pemmican, lard or hard bread) as walls, they created a twelve-foot by sixteen-foot extension onto the north end of the house, which became a makeshift shed when they roofed the walls with the remaining lumber and canvas. Later they lined the provision-box walls on the outside with blocks of snow. After Christmas the men set up a galley and a dining table, as well as an eight-foot by ten-foot tent with two bunks, inside this shed, leaving the main house with its six bunks for the scientists and tables for their map work and report writing. The cook and Castel slept in the shed tent. Palaiyak and Ikey Bolt slept in another tent, which had a snowhouse erected around it for shelter and warmth, and was located near the entrance to the shed. A coal stove in the main house kept the inside temperature between 60° F and 70° F during the day, and 40° F to 45° F at night. Johansen set up a tent nearby for his insect work, and Dr. Anderson put up one for the preparation work on his bird and mammal specimens. Chipman later erected a pole nearby with an anemometer on it for the collection of wind-speed data.

Two major problems immediately confronted the scientists: a shortage of fuel and maintaining a supply of fresh water. The schooners had not brought an adequate supply of coal to Bernard Harbour, which was the main reason why Dr. Anderson had hurried west on the *Alaska*. As the *Alaska* did not return that fall, the men had to regularly comb the shore for driftwood. Fortunately, there was a considerable amount of such wood available, which had drifted east over the years from some rivers as far to the west as the Mackenzie River. The men gathered and stacked it in piles, then before freeze-up loaded it for transport to the camp in the launch that Cox had used during the summer. At the main camp, the wood was cut into short lengths for use in the stove as needed. After freeze-up, early in October, when there was sufficient snow on the ground, one of the men would periodically take a sled and team of dogs along the shore and bring back a large load of the wood. In this way they accumulated sufficient driftwood to last them until Christmas. After Christmas the men had to travel much farther west along the coast to obtain the driftwood, for there was very little driftwood of any kind east of Chantry Island. Careful use of the small supply of coal they had brought with them from Herschel Island and the driftwood they scavenged along the shore for miles west of their camp enabled them to eke out their fuel until the summer of 1915, when the *Alaska* returned with a sufficient supply of coal to last the camp until the men completed their assignments and went south in 1916.

The first stove they installed in the house was unsatisfactory, so in the third week of October they removed the stove from the *North Star* and installed it in the house. It burned either wood or coal, and gave off roughly three times as much heat as their first stove. Chipman was cook on the day the stove was replaced and celebrated the improvement by baking bread and two pies!

As for fresh water, the closest source proved to be a lake a mile away. For the first few weeks and during the following two summers someone had to walk to the lake to fetch it. After the snow came, fresh snow could be melted for some of their water needs. Later, when the lake froze, a couple of the men would take a sled and team of dogs to the lake and cut blocks of ice, which the dogs pulled back to the camp on the sled. No easier arrangement ever materialized, and obtaining fresh water remained a constant problem for the two years they were there.

As the autumn progressed the amount of daylight diminished, and the need for lighting increased. The Southern Party had been supplied with a large and heavy piece of lighting equipment, but the men were unable to get it to work in the cold. Apart from that, they had three lanterns and a tin lamp that they had obtained from the geologist Leffingwell on Flaxman Island the year before. Good lighting was extremely important for Cox, Chipman and O'Neill, all of whom were expected to create maps of the areas they surveyed. The problem had not arisen the previous year at Collinson Point because there had been adequate lighting devices on board both the *Alaska* and the *Mary Sachs*, some of which could be removed and brought into the house.

An additional problem at Bernard Harbour that fall was the lack of a camp cook. Sullivan, whom Dr. Anderson had hired at Herschel Island, had gone west with the *Alaska* early in September, requiring the men at Bernard Harbour to fend for themselves. As the person in charge of the camp, Chipman at first assigned each of the scientists to the cooking duties on a daily rotation basis, later changing the daily assignments to weekly ones. Only Johansen grumbled over the arrangement. The two Inuit, Palaiyak and Siberia Mike, were not included in these assignments. The five scientists all took turns until Christmas when Sullivan returned with Dr. Anderson from the icebound *Alaska*.

Contacts with the Copper Inuit

Several Copper Inuit appeared and out of curiosity approached the scientists on the day the Southern Party arrived at Bernard Harbour. Thereafter one or more of them arrived and lingered around the camp almost daily for the next two years, always curious to watch what the scientists were doing, especially about the house they constructed, bewildered by the language the scientists spoke and their behaviour, and eager to touch or handle any and all of the scientists' possessions. Everything was so unfamiliar to them. Their name for the scientists' camp was "Igloopuk" (Cox, 1916).

Initially when they wanted to approach the camp they went through a sequence of upraised arm motions to signify they were friendly, expecting a similar gesture in return. Later such formalities were eliminated as they became better acquainted. A few of the visitors soon were guilty of absconding with one or other camp item they fancied. One man, for example, trotted off with some green canvas one day, appearing a few days later with a handsome pair of green trousers his wife had made for him. When charged with stealing the cloth he denied it, but after being told that he would no longer be allowed to visit them unless he returned the material, he sulked for a time, then started back to his camp, stopped, removed his green trousers and continued on home clad only in his parka and boots.

Until Dr. Anderson returned, the only men who could converse at all with the local Inuit were Palaiyak, Siberia Mike and Jenness. The local dialect created some difficulties until they became familiar with it. Jenness did most of the communicating for the Southern Party since he was expected to learn as much as possible about these little-known people, and he was in charge of trading activities. Dr. Anderson had acquired a working knowledge of the language[35] on his previous expedition in the region between 1908 and 1912.

Both local Copper Inuit and scientists had difficulty pronouncing their respective names. The following names for the white men at Bernard Harbour were given them by the Copper Inuit from the Dolphin and Union region. These would have been lost had it not been for Billy Natkusiak, who later told Stefansson about them, and he wrote them down.[36] Of course, spelling is one thing, pronunciation is another. The reader is left to figure out the latter.

Dr. Anderson	Annasinna
Chipman	Hipimanna (or Sipimanna)
Cox	Kaxsinna
Jenness	Yaninna[37]
Johansen	Juahinna (or Juöhinna)
O'Neill	Onina
Castel	Ä no (for Aarnout)
Cockney	Kagni

Four of the scientists (Chipman, Cox, Johansen and O'Neill) were unable to understand most of the conversations with their visitors, so generally went about their business, but usually

---

[35] The name for the language spoken by the Inuit across North America is *Inuktitut*. The name for the dialect spoken in the Coronation Gulf region by the Copper Inuit is *Inuinaqtun*.

[36] Stefansson, 1918 (March 7, 1916).

[37] Condon (1996, p. 58) reported years later that Jenness' Copper Inuit name was "Jennessi."

kept a wary eye on them to ensure that certain items did not disappear. Initially, items such as pots, pans, spoons or canvas disappeared with each visit. At some point Jenness conveyed to the visitors that all trading would cease if the petty thieving continued, and its frequency diminished thereafter. The scientists generally tolerated the presence of the curious visitors for Jenness' sake and appreciated the good sense of humour shown by most of them, but at times found them rather annoying. Discovering lice on their fur clothing or on themselves after such visits generally reduced their tolerance levels.

As soon as the camp was in reasonable order, Jenness turned his attention to studying and trading with the Copper Inuit, who were now visiting the camp regularly. Initially these activities were done at the base camp, but as Jenness' confidence with their language grew he started visiting them in their nearby camps, sometimes remaining overnight with one family or another. Later he went farther afield.

Late in October when Dolphin and Union Strait froze over, the Copper Inuit built snowhouses on the ice in the strait outside Bernard Harbour, near the Liston and Sutton Islands, in order to hunt seals that were plentiful there. Those islands were about fourteen miles north of the Expedition's camp. At first Jenness visited the Copper Inuit there only occasionally, but his visits became more frequent after he established a warm friendship with a middle-aged man named Ikpukhuak and his wife, Higilak, who had come across the strait from Victoria Island. Also known as Taqtu, Higilak had been a widow with two children when she married the childless Ikpukhuak some years previously. She also happened to be a shaman or medicine person. Throughout that first winter Jenness often stayed for several days with them in their snowhouse (igloo) at the islands to learn about the culture and folklore of their people, to observe their daily activities and seal-hunting methods, and to trade with the various people in the settlement. At other times Ikpuk (as Jenness called him) and Higilak would bring Higilak's two children—her twelve-year old daughter Kannayuk (known among the scientists as "Jennie", probably because Jenness "adopted" her as his little sister) and her older son Avrunna and his wife, Milukattuk—with them to the base camp.

Sometime during that first autumn, Jenness and Chipman retrieved from their large cache of provisions and equipment two identical Edison recording phonographs with about two hundred blank records. Jenness intended to record Copper Inuit songs with these machines at some future time. They then stored one of the phonographs and some records on the *North Star* so that they were out of the way. This proved to be an unfortunate mistake, however, for during the winter the schooner sprang a leak and filled with water, destroying both the phonograph and the blank records.

With camp facilities finally organized, Chipman planned to travel west to Pearce Point, ostensibly to make an initial topographical and geographical study of the intervening terrain in preparation for his coming spring's work in that region. He also wanted to find out if Captain Klengenberg, whose camp was near the cape, had any news of the *Alaska*, and hoped he might perhaps meet Dr. Anderson or some other person coming from that vessel. (He did.) Chipman was aware, however, that ice and snow conditions were unsafe for travelling along the coast before the latter half of November, by which time there was very little daylight for travelling, for the sun would cease to rise above the horizon at that latitude by the end of November.

Chipman and O'Neill Undertake Initial Field Work

After getting their caribou-skin boots and fur socks repaired by some local Copper Inuit women, Chipman and O'Neill headed west on November 18. They managed to travel only ten miles on the first day because of their heavily loaded sled. A blizzard kept them confined to their tent the next day, then they continued on to Stapylton Bay. On November 24 they crossed that wide bay and moved on to Cape Young. Travel conditions were good thereafter, although the temperature was well below zero Fahrenheit. They left caches near Clifton Point and Clinton Point for when they returned along that coast.

On December 4, just west of Deas Thompson Point, they encountered a second blizzard, which lasted for almost a week. Chipman estimated the winds reached fifty miles per hour at times, for he was barely able to stand upright when he ventured out to feed the dogs, fetch wood or bank snow against the walls of his tent. Being confined within a small tent for that length of time under such cold, windy and dark conditions would have certainly been an extremely unpleasant experience for the two men.

On December 10, shortly after he and O'Neill were finally able to get under way again, Chipman noticed four men and a dog team approaching from the west. It proved to be Dr. Anderson, Castel, Sullivan and Ikey Bolt, who had started east from the *Alaska* at Baillie Islands on November 20. Dr. Anderson brought news of the *Alaska*, of course, but also news that a schooner stocked with Hudson's Bay Company supplies had been sent to the mouth of the Horton River for the winter. The presence of this latter schooner was welcome news, for it would be a closer source for supplies than Bernard Harbour should Chipman and O'Neill be in need of any in the spring when they returned west to map around Darnley Bay.

Chipman and O'Neill now turned about and joined Dr. Anderson's party heading for Bernard Harbour. On their first night, Dr. Anderson introduced Chipman and O'Neill to what he said was "hooch"—pemmican boiled with flour or hard bread.[38] It had been so named by Shackleton's party in Antarctica in 1908, Dr. Anderson explained, and in Chipman's opinion was "... about as fine trail food as one could wish to eat."[39] The pemmican he referred to on that occasion was the American-made, Underwood brand, containing "... 53% meat, 35% grease (lard and tallow mixed), 3% sugar and 9% dried fruit mostly raisins."[40]

Travel conditions occasionally proved difficult, with open ocean water close to shore in a few places between Pearce Point and Clifton Point, leaving them only a narrow strip of frozen shore between the water and the cliffs along which they could drive their dogs. At Deas Thompson Point even that option was not available to them, and they were forced to detour inland behind the point. At other times they travelled over the sea-ice. Their progress

---

[38] This was certainly a different use of the word "hooch" than that familiar to the present author. *Webster's Dictionary* states the word derives from the Alaskan Indian word *hoochinoo*, which was a crude alcoholic beverage made by the Hoochinoo Indians. The word later came into use for liquor made or obtained surreptitiously during Prohibition years in both Canada and the United States (Webster, 1972, p. 674). That was obviously not the meaning of the word as Dr. Anderson used it.

[39] Chipman, 1916 (December 10, 1914).

[40] *Ibid.* This was the pemmican Stefansson had cabled Dr. Anderson not to bother having tested late in March 1913, because there was not time. Dr. Anderson had the tests made in spite of the cable.

was slow, largely because of the few hours of limited daylight at their disposal. When they travelled by limited moonlight, the rough coastal surface appeared featureless and forced them to take every step carefully lest they stumble. A blizzard then detained them for two days, but on December 22 they crossed Stapylton Bay and reached Cape Bexley the next day. There they found seven recently deserted snowhouses, with a trail coming from Stapylton Bay and heading north over the sea-ice. Chipman later told Jenness about these houses and trails, knowing that Jenness would probably wish to visit the Inuit who made them. Finally, weary and foot-sore, the parties of Dr. Anderson and Chipman reached the base camp shortly after noon on Christmas Day.

Chipman's reconnaissance journey to the west had provided very little topographical or geological information, but had given both men valuable experience in operating a dog team, camping and surviving under some of the Arctic's most severe weather conditions.

Christmas at Bernard Harbour, 1914

In their absence, the men at the base camp had decided not to bother cooking anything festive until at least Chipman and O'Neill returned. Besides, Johansen had been assigned to cook on Christmas Day, although his culinary abilities were notoriously limited. However, with the arrival of Dr. Anderson's party as well as Chipman's, Sullivan promptly resumed his culinary duties and every one made an effort for the evening meal, somehow producing as special treats some Christmas cake, candy, H & P biscuits, a bottle of an unidentified alcoholic beverage and Chipman's prize box of cigars. Mrs. Anderson had sent a Christmas parcel the previous spring, which they now opened and found to contain a package of twenty-five sheets of writing paper with matching envelopes for each of the scientists, on each of which was printed "Canadian Arctic Expedition." I have seen thank-you letters to Mrs. Anderson on this note paper from at least three of the men, written from Bernard Harbour and containing comments about her husband, among the R.M. Anderson papers at the Library and Archives Canada.

BERNARD HARBOUR, 1915

20

Darkness, snow and cold, stormy weather prevented the scientists other than Jenness from undertaking any scientific work during the first few weeks of the new year. Chipman recorded that the temperature at their Bernard Harbour base camp in January averaged -23° F compared with -18° F at Collinson Point the previous January, while the wind at both localities had averaged ten miles per hour. There was little sense trying to hunt caribou in the winter darkness. Furthermore, although the scientists had learned from Captain Joe Bernard and the local Copper Inuit that herds of caribou normally migrated southward in November across Dolphin and Union Strait from Victoria Island several miles east of Bernard Harbour, this winter the customary herds did not appear. Only four caribou had been shot for the Expedition by Christmas, although thirty seals had been shot at the edge of the ice in the outer harbour. Apparently the lateness of freeze-up of Dolphin and Union Strait had forced the caribou to cross Coronation Gulf farther east. Fortunately, Jenness was successful in obtaining by trade about five hundred pounds of caribou meat from the Copper Inuit living on Victoria Island and later some seal meat after they had moved to their sealing campsite on the ice in Dolphin and Union Strait. These meat acquisitions met the needs of the men at the base camp for many weeks.

During the dark cold days of January, the scientists kept busy reading, writing, attending to camp chores, fetching ice for drinking water or driftwood for fuel, and playing chess, cards or cribbage. Both the Northern and Southern Parties had been supplied with a good number of quality books, which had been donated by several publishers.[1] Many of the books were about Arctic subjects, and both Northern and Southern Parties had a complete set of the latest Encyclopaedia Britannica. Almost all of the books assigned to the Northern Party were lost with the sinking of the *Karluk*, while most of those belonging to the Southern Party were on the *Alaska* during the winter of 1914–1915 and therefore not available to the men at Bernard Harbour.

House confinement day after day in the cold darkness of January did not lead to the best of temperaments. It did not help, either, that there was ice on the inside of the walls around some of the bunks, dampening blankets, and forming smelly mould on the walls and books. Chipman grumbled in his diary that Dr. Anderson spent many hours repairing his fur clothing or boots when, in Chipman's view, he should have been devoting his time to his zoological and administrative work. It might not have occurred to the unmarried Chipman that Dr. Anderson could still be grieving over the death of his newborn son, about which he had learned only the previous spring. Burdened also with worries about the post-natal emotional

---

[1] The Macmillan Company in New York, for example, donated three hundred books (Brett, 1913).

condition of his wife (who had pleaded with him to return south immediately to be with her),[2] with thoughts about his ongoing feud with Stefansson, and with all of the administrative duties that kept him from having time to pursue his own zoological work, he may well have been depressed. His sewing tasks, about which Chipman complained, may therefore have been somewhat therapeutic in addition to being necessary.

Dr. Anderson's Attempted Journey to Fort Norman

The dull monotony of the dark Arctic winter was slightly relieved by the reappearance of the sun on January 15, after forty-five days below the horizon, and then by the departure on February 2 on a five-hundred-mile journey to Fort Norman of Dr. Anderson and Castel with a team of dogs pulling a toboggan. They intended to cross the west end of Coronation Gulf to the mouth of the Coppermine River and go overland from there using snowshoes via the Dismal Lakes and Great Bear Lake to Fort Norman on the Mackenzie River. Dr. Anderson expected his trip to take two months. It was a route with which he was somewhat familiar, for he had taken it a few years previously. He took with him, in addition to two months' supplies, government reports by each of the scientists on their work as well as all personal mail. He was determined to get the news to the Department of the Naval Service, the Geological Survey of Canada, and to newspapers and other interested organizations or persons, that his Southern Party was safely established at Bernard Harbour, that Wilkins had gone with some men in search of Stefansson's party as instructed, and that on the assumption Stefansson would be found, personnel were establishing a base camp on Banks Island for his further explorations.

With Dr. Anderson and Castel also went Jenness, Johansen, Palaiyak, dogs and two sleds. The latter three men went only as far as the first occurrence of trees (the tree-line) some sixty miles up the Coppermine River, however, where Johansen examined the spruce trees, which prospectors in 1912 had reported were dying. Jenness hoped the trip would enable him to meet a group of Copper Inuit who lived near the Coppermine River, because Stefansson had placed them in a separate tribe from the Inuit who lived around Bernard Harbour or those on Victoria Island. And all of the men hoped to shoot some caribou to bring back a supply of fresh meat to the base camp.

Deep snow between the Coppermine River and Great Bear Lake proved too difficult for Dr. Anderson's dogs, however, and he and Castel were forced to return to Bernard Harbour after getting almost to Great Bear Lake. He brought back the outgoing mail, having failed to encounter any white men or First Nations people (Indians) near Great Bear Lake to take it to Fort Norman. It was later taken west to the *Alaska* at Baillie Islands, from where it went in the summer to Herschel Island and thence south.

First Encounter with the Shaman Uloksak

Meanwhile, at the tree-line, Johansen quickly discovered that an infestation of bark beetles, of which he identified three species, was the probable cause of the death of many of the spruce

---

2 Mrs. R.M. Anderson, 1913 (November 7, 1913).

trees. After establishing that fact and collecting samples of the beetles for further study, he, Jenness and Palaiyak left Dr. Anderson's party, taking one of the two sleds and teams of dogs, returned to the mouth of the Coppermine River, and followed sled trails out on the ice in Coronation Gulf to a large snowhouse village. The band of Inuit in the Coppermine River region hunted seals on the ice in Coronation Gulf during the winter, then during the summer moved south to hunt caribou and catch fish near Great Bear Lake. They had occasionally encountered white men near that lake in the past, so were not unduly surprised when the three white strangers arrived.

Here on the ice of Coronation Gulf, Jenness met the hunter and shaman, Uloksak Meyok, a man about thirty years old. Jenness referred to him as the "rich man" because he had two wives and some unusual possessions, which he now proudly showed to his visitor. These included two small porcelain cups and saucers, a Roman Catholic breviary in Latin, an illustrated scripture book in French, a Winchester rifle and a Hollis double-barrelled fowling piece. Uloksak claimed that he had obtained these objects in trade at Great Bear Lake the previous summer. Elsewhere in the village, Uloksak told Jenness, someone had a priest's cassock and a crucifix. Such priestly objects in the hands of people with so little contact with the white men seemed extraordinarily unusual to Jenness.

Jenness' party remained at the village for three days, during which time he took photographs and head measurements, and examined the eyes, hair and skin of many of the villagers. Physical anthropologists in those days believed that head measurements provided a strong clue

Figure 36 CAE house after its completion, Bernard Harbour, with schooner *North Star* in the harbour on the right, October 5, 1914.

Photo: © Canadian Museum of Civilization, photo by Fritz Johansen, 42235 as to racial origin. Satisfied from his observations among these people that there were none of the so-called "Blond Eskimos" about whom Stefansson had written and spoken, and that these Inuit differed little or not at all from the ones he had encountered around the Expedition's base camp, Jenness and his companions then headed back to Bernard Harbour.

Soon after they reached it, a crowd of the Inuit from the snowhouse village they had just left arrived with twenty-two loaded sleds, curious to see where the white men lived and eager to trade with them. They promptly constructed snowhouses on the ice in the inner harbour, near the *North Star,* and then crowded around and inside the Expedition's house, some to simply look, others to trade. Jenness bought from them clothing of all kinds, bows and arrows, stone pots and lamps, and wolverine skins, giving in exchange such items as sewing needles, boxes of matches, lengths of calico cloth, knives of several sizes and rifle cartridges.[3] Two or three of these Inuit had obtained rifles in trade from Captain Joe Bernard in 1910 or 1911, and needed ammunition for them. Jenness also recorded two of their songs onto wax cylinders on his working Edison portable phonograph during their visit. The visitors stayed for several days, and then went to hunt seals at the Copper Inuit village on the ice near the Liston and Sutton Islands.

At one point while the Coppermine River Inuit were camped at Bernard Harbour, the shaman Uloksak Meyok held a seance to determine if Jenness was the person who had murdered a young Inuk man who had just died at the settlement near the Liston and Sutton Islands. The man in question had stolen a fox-trap from the base camp and his relatives claimed that Jenness, whom they thought was a shaman, had cast a spell upon him, causing his death. Jenness attended the seance and played an active role in it. Uloksak subsequently ruled the dead man had been murdered by some other white man far away. Jenness knew that Uloksak was out of ammunition for his rifle and could only obtain it from Jenness, because he was the Expedition's authorized trader. This made the outcome fairly certain, but Jenness' life might otherwise have been in great danger. In spite of his complete innocence in the case, if bereaved family members continued to think him guilty, revenge was a likely response.[4]

Chipman and O'Neill's Coastal Mapping from Darnley Bay to Bernard Harbour

Chipman, O'Neill and Ikey Bolt left the base camp on March 17, heading northwest for the southern end of Darnley Bay, which Sir John Richardson had left unmapped in 1848. Their task was to map the coast from there back to Bernard Harbour. Bad weather delayed their travel progress so greatly that they took twenty-six days to cover the two hundred miles to the camp of Patsy and Etna Klengenberg in the cove near Pearce Point. The youngsters' father, Captain Charles Klengenberg, was supposed to have left dog food there for the Expedition's use, but the camp was empty. Chipman did discover two notes, however, which stated that the

---

[3] A detailed list of Jenness' various trades with the Copper Inuit appears as Appendix 3 in Jenness, 1991, pp. 670–696.

[4] Jenness, 1928, pp. 85–89; Jenness, S.E., 1991, pp. 399–400.

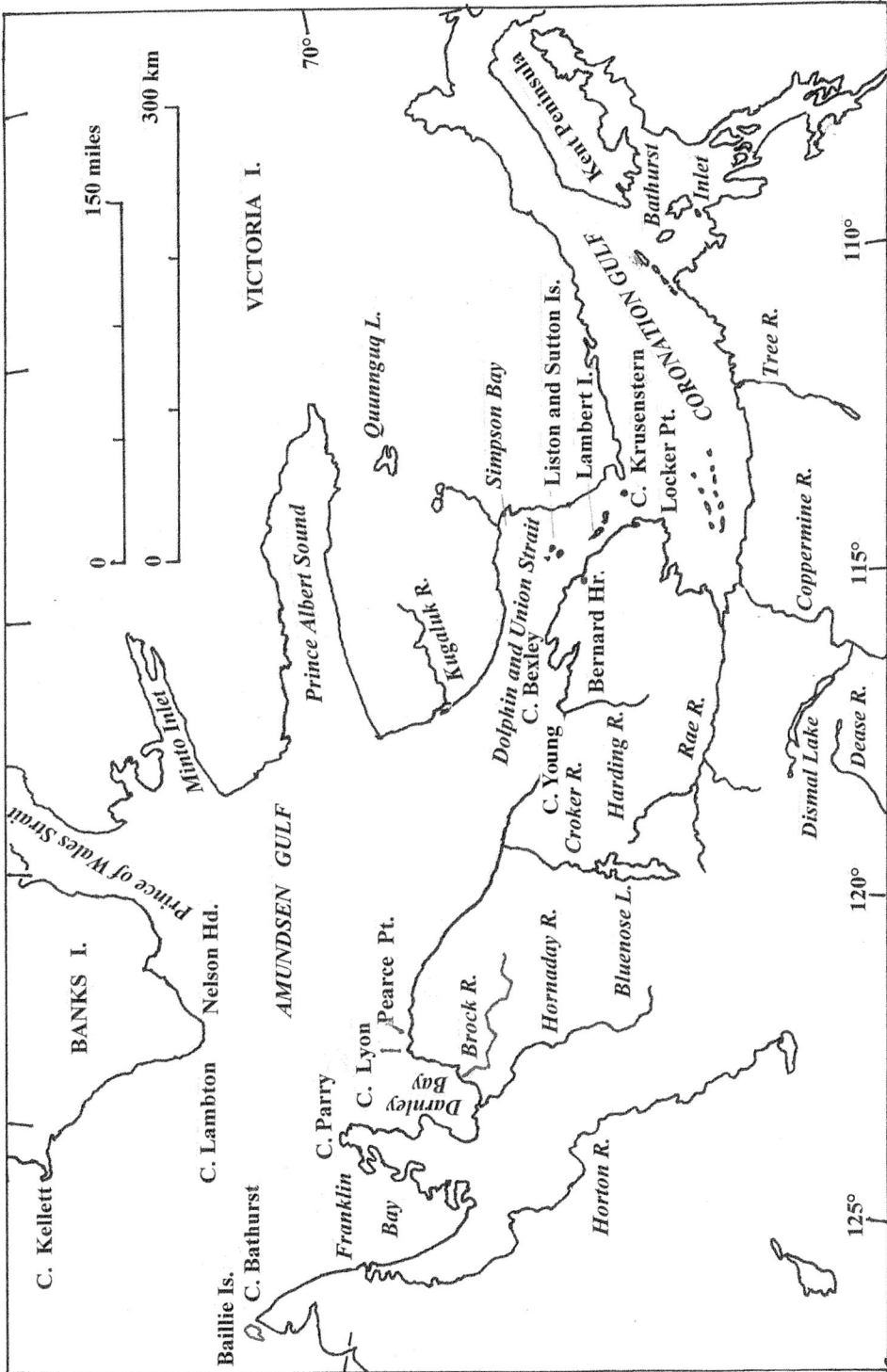

Map 15 Map showing the stretch of coast from Cape Parry to Kent Peninsula, which was mapped by the Southern Party, and the locations of many of the places mentioned.

Klengenbergs were at their camp in Darnley Bay. Chipman and his two companions then followed Klengenberg's trail across the peninsula to Darnley Bay, where the captain had a beached boat, but again they found neither persons nor dog food. Undaunted, they continued on until they came to the camp of an Inuk man, Wicksuak, who made them welcome and fed their dogs. Leaving O'Neill at Wicksuak's, from where he could study the local geology, Chipman and Ikey Bolt continued south along Darnley Bay in search of the Klengenbergs, without success.

At the southern end of Darnley Bay they found the schooner *Argo*[5] and the trading camp of Scotty McIntyre and Gallagher Arey from whom Chipman purchased two bags of flour. This quantity, he felt, would be sufficient to get his party back to Bernard Harbour, and so he and Ikey returned to Wicksuak's camp. From Wicksuak, Chipman obtained seal meat and blubber for his dogs and caribou meat for his men. In exchange for the meat and blubber and various services provided—O'Neill had been housed and fed for a week, Wicksuak had made a new ridgepole for Chipman's tent, and Mrs. Wicksuak had made soles for their water boots and undertaken other remedial sewing for them—Chipman gave Wicksuak a note payable from the supplies at the *Alaska* at Baillie Islands for a sack of flour, two pounds of tea, ten pounds of sugar, and enough calico for Mrs. Wicksauk to make *attigis* (parkas) for herself and her children.

Chipman's party then commenced their topographical and geological study of the mainland coast eastward from the southern end of Darnley Bay. This part of the bay had not previously been explored. Almost immediately they discovered two fairly large rivers flowing into Darnley Bay, to which Dr. Anderson later assigned the names Hornaday River and Brock River. The former had been known locally as the "Big River," the latter as the "Little River."[6] The three men started up the previously unmapped "Little River" on April 20, and charted its course and geology for more than fifteen miles. At that point their progress was halted by a steep-walled canyon and waterfalls, and they were forced to turn back.

Captain Klengenberg was at his boat camp when they returned to it. During their visit, O'Neill asked the captain if he would allow his fourteen-year-old son Patrick (known as Patsy) to join their Expedition for a year. The youth was already a competent hunter and dog-driver. If he was with them, both Chipman and O'Neill could do their geographical and geological field work while either Ikey or Patsy drove their sled and the other obtained fresh meat for them all. The captain agreed to the suggestion, eager for his son to have more contact with white men and recognizing that by being with the scientists Patsy could learn to speak, read and write English, as well as a variety of scientific matters.

Once Patsy was packed and ready, Chipman's party continued their coastal mapping eastward. By taking numerous time-consuming latitude and longitude determinations along the coast, Chipman obtained control points that enabled him to make many modifications

---

[5] This was the ship of which Dr. Anderson had spoken when he met Chipman before Christmas.

[6] Dr. William T. Hornaday was the Director of the New York Zoological Park and a well-known advocate for wildlife conservation in the United States and Canada. Dr. R. W. Brock was the Director of the Geological Survey of Canada, who urged the addition of the scientists and the Southern Party to the Canadian Arctic Expedition 1913–1916.

to the existing map of the region. The map he was correcting was the old British Admiralty Chart, which had originally been drawn in 1826 by Dr. John Richardson when he mapped the coast in a boat rowed by its crew from the Mackenzie River Delta to the Coppermine River. While Chipman surveyed the coastline, O'Neill examined the bedrock that was exposed along much of the way between Darnley Bay and Bernard Harbour. He found it to be mainly a light grey to buff-coloured dolomite (a sedimentary rock much like limestone but composed of both calcium and magnesium carbonate). As he found the dolomite was neither copper-bearing nor fossil-bearing, O'Neill concluded that it required no closer examination at that time.

Cox's Coastal Survey from Bernard Harbour to the Rae River

On March 24 at Bernard Harbour, with the parties of both Dr. Anderson and Chipman away from the base for many weeks, Cox undertook a ten-day geographical and geological examination of the coast southeast from the base camp in Dolphin and Union Strait to Cape Krusenstern and Locker Point at the west end of Coronation Gulf. Jenness accompanied him to assist him. Just west of Locker Point on March 30 they unexpectedly encountered Dr. Anderson and Castel returning from their failed effort to take the mail overland to Fort Norman. After a brief consultation, Cox cached most of his supplies for later use, and he and Jenness returned to Bernard Harbour with Dr. Anderson and Castel, as Jenness needed to prepare for a summer on Victoria Island with Ikpukhuak's family.

On April 16 Cox left Bernard Harbour for Locker Point to renew his mapping activities, this time accompanied by the cook, Sullivan. To his annoyance and inconvenience, the cache he had left near that point in March had been pilfered by some Copper Inuit. Commencing from where he had terminated his mapping while with Jenness, Cox continued his coastal survey to the mouth of the Rae River at the southwest end of Coronation Gulf. Then, instead of returning to Bernard Harbour the way he had come, he and Sullivan turned northwest and proceeded up the frozen Rae River for about seventy miles to its source. From there, they struggled across the height of land, helping their dogs to drag their sled over the snow-covered tundra and height of land for six days before following a frozen stream down to the shore of Stapylton Bay. Cox halted periodically along their route between Coronation Gulf and Stapylton Bay and waited for suitable weather conditions in order to record a solar reading for latitude with his sextant at noon and a time reading for longitude. These activities provided him with a series of accurate location points for plotting later on the map that he and Chipman would prepare. During their trek over the tundra, he and Sullivan encountered many caribou migrating north, so were able to obtain fresh meat whenever they needed it. After reaching Stapylton Bay they continued to the northwest, mapping along the coast to Cape Young, where they met Chipman's party working slowly east from Darnley Bay and connected their results. Chipman's party then went directly on to Bernard Harbour while Cox and Sullivan continued their coastal mapping from Stapylton Bay to Bernard Harbour, which they reached on May 25. When Cox completed that mapping project, he and Chipman had mapped the Arctic coast from Darnley Bay to the mouth of the Rae River, leaving still to be mapped the coast from Rae River east to Bathurst Inlet, and the coast around that inlet.

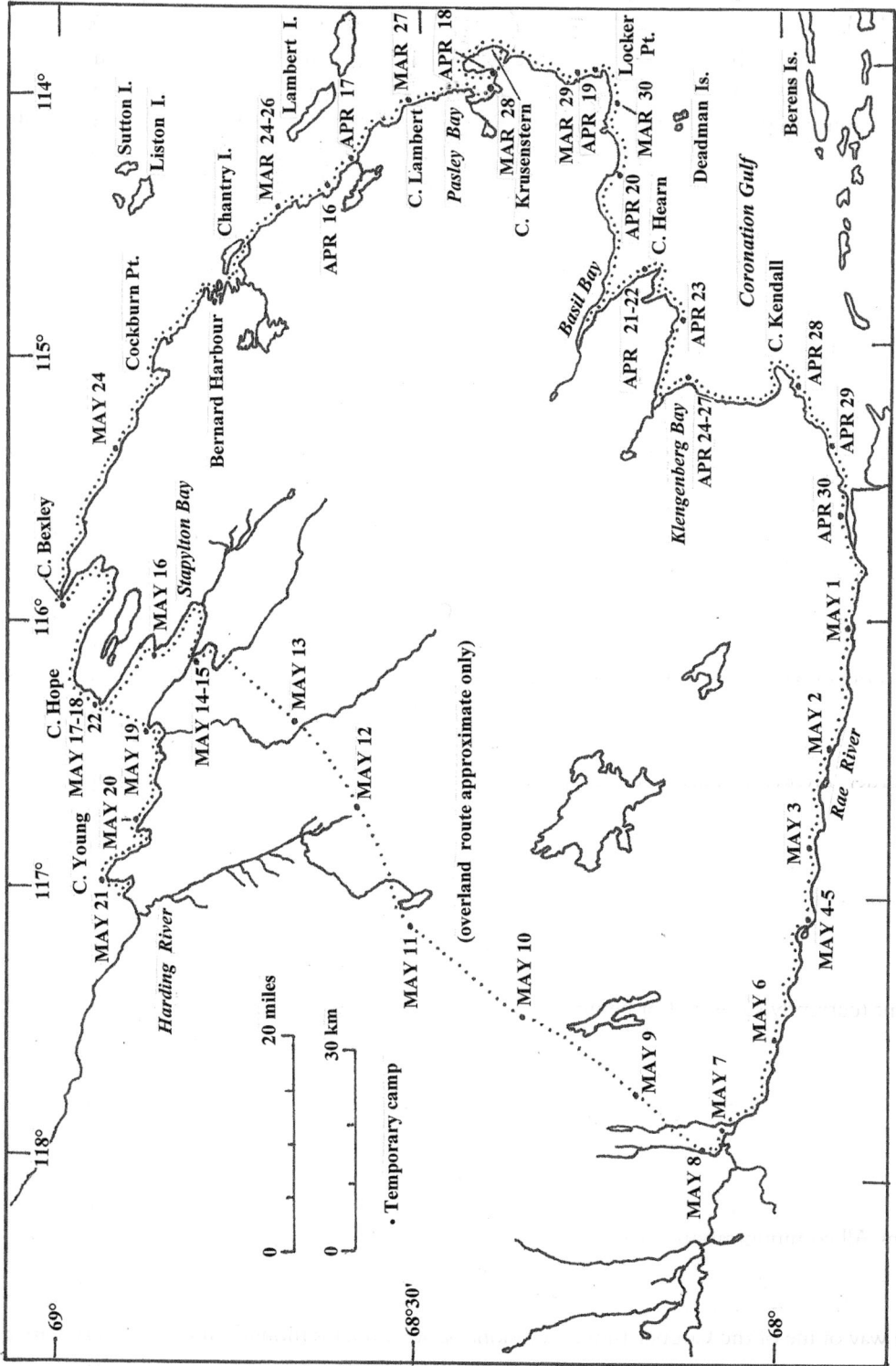

Map 16 Route mapped by Cox between Bernard Harbour, Rae River and Stapylton Bay, 1915.

Jenness' Summer with the Inuit on Victoria Island

By the late winter of 1915, Jenness had still found no real evidence of Stefansson's "Blond Eskimos" among the Inuit he had seen around Bernard Harbour, Simpson Bay on Victoria Island, or in the western part of Coronation Gulf. He decided therefore to seek them where Stefansson had originally seen them—in Prince Albert Sound—and persuaded Ikpukhuak and Higilak to "adopt" him as a family member for the summer of 1915 and take him there. They left Bernard Harbour on April 13, joined forces with a dozen or more other Copper Inuit at a snowhouse settlement on the ice of Dolphin and Union Strait, and for the next month hunted seals on the ice for their skins and oil. Then they moved to Victoria Island, cached most of their possessions at the coast in order to lighten their loads, and headed north across the Wollaston Peninsula.

For the first few weeks they lived largely on fish, which they caught through holes in the ice of many of the small lakes they encountered. Late in May Ikpukhuak suddenly informed Jenness without explanation that he had changed his plans and would not take him to Prince Albert Sound. However, they did go to Lake Quunnguq, more than halfway across the Wollaston Peninsula, and there they met half a dozen Inuit who had just come from Prince Albert Sound. Jenness had ample chance to examine the visitors closely during three days of feasting and dancing, and again he saw no characteristics to justify calling any of them "Blond Eskimos."

After the visitors headed back to Prince Albert Sound, Ikpukhuak and his people wandered to the southwest of Lake Quunnguq and cached their sleds for the summer at a small lake just west of Ammalurtuq Lake. From there they wandered on foot from lake to lake for the next several weeks in the immediate area, fishing through the ice for their food, since caribou were not to be found. About the end of July Jenness developed a severe, incapacitating, digestive disorder, involving sharp abdominal pains and dysentery. Higilak's shamanistic utterances provided no assistance, and his condition worsened. He was finally spared death by returning south to the coast with Ikpukhuak and obtaining some of the rice, sugar and pemmican he had cached there. They also brought back some seal oil they had cached at the coast, which provided them with fuel with which to cook their meals. Fog and rain had kept the bushes wet for more than a month, preventing the Inuit from lighting fires for cooking. During that time their diet consisted entirely of dried caribou meat and uncooked fish. Whatever his ailment (currently suspected of being caused by parasites in the raw fish he ate), Jenness ended up with a sensitive digestive system for the rest of his life, necessitating the avoidance of certain foods, especially those cooked with spices.

For the next two months he and Ikpukhuak's immediate family—other family members having gone their separate ways—roamed on foot to the west end of the Wollaston Peninsula and back, hunting caribou as they went. Each day Jenness discussed assorted subjects with Higilak or Ikpukhuak and after they camped would make extensive notes from these discussions. All communications between them were, of course, in their native language (Inuktitut), which Jenness had mastered fairly well by this time.

In this manner, week after week, Jenness accumulated anthropological information about the way of life of the Copper Inuit. Occasionally, however, his thoughts took him in different directions. During a brief halt near the south coast, early in August when the weather was

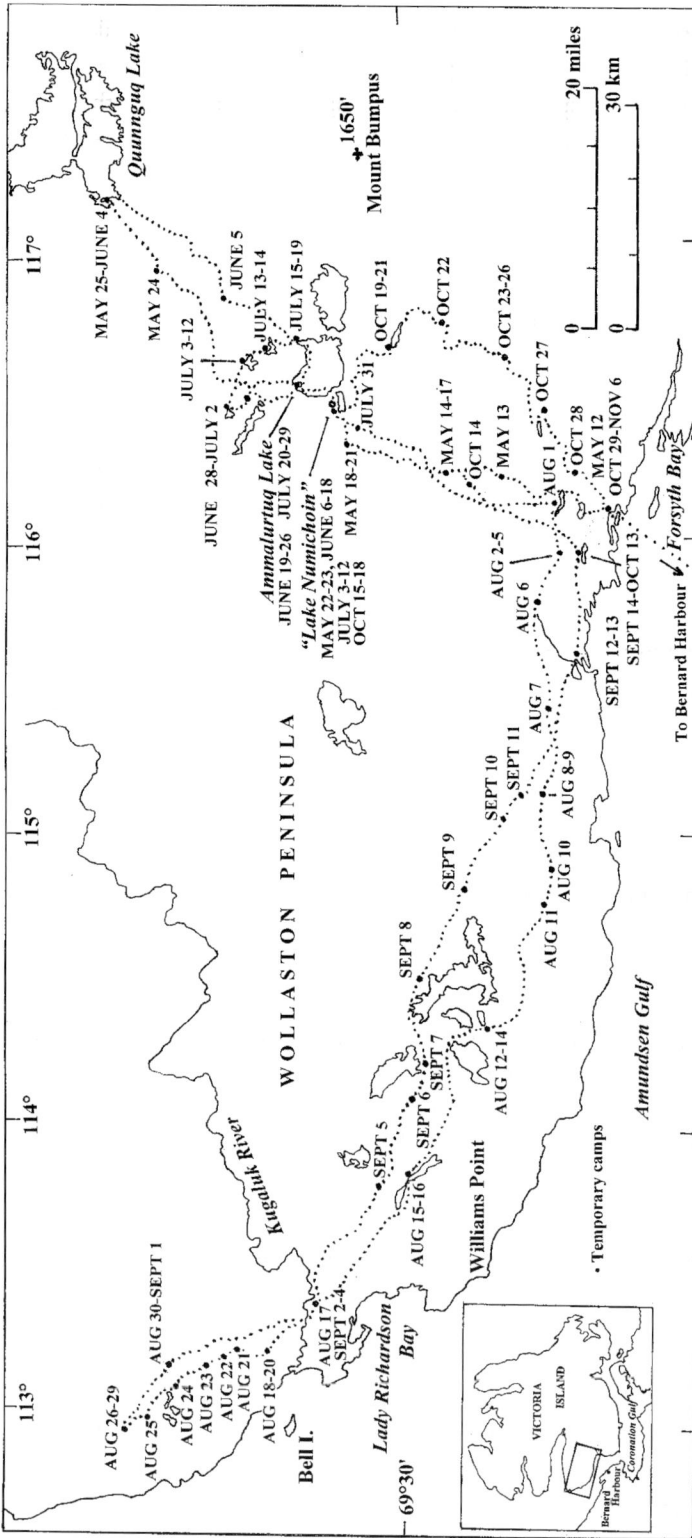

Map 17 Route taken by Jenness and his Copper Inuit "family" about the Wollaston Peninsula, southwest Victoria Island, May to November, 1915.

reasonably warm, "I gathered three or four flowers for Johansen—among them a wild rose—and had a glorious bathe in a deep pool …"[7] These plants would be of great interest and value to his colleague, Johansen, who was responsible for reporting on the botany of the entire region, and would not have the opportunity to collect on Victoria Island. Cleanliness in that region was another matter.

"A bath is a rare luxury in the Arctic," Jenness later remarked in *The People of the Twilight*,

> … where there is often no water to wash even the face or hands. My [Inuit] companions solved the problem by never washing. Contact with the snow in winter purified them, and in summer the deep tanning of their skin masked its uncleanliness … my bathing, being purely voluntary, certainly excited much interest, and when I stole away to the pool for the second bath, Ikpuck notified the whole party. Innocently devoid of clothes, I was disporting gaily in the water when the sound of voices made me look up. Men, women and children lined the bank above me, intently watching the exhibition. None of them knew how to swim, or even that it was possible, and they gasped with admiration when I dived out of sight and reappeared a few yards away. *Honi soit qui mal y pense.* Ikpuck, The Runner [Higilak's son Avrunna], and even the latter's wife [Milukattuk] stripped off their clothes immediately and joined me in the water, eager to receive a first lesson in the newly discovered art.[8]

Jenness and Ikpukhuak's family returned to their cache on the south coast by mid-September, and waited there for the fall freeze-up of Dolphin and Union Strait so that they could cross over to Bernard Harbour. Their more distant relations rejoined them over the next few weeks. By the latter half of October there was enough snow on the ground and the lakes were sufficiently frozen for them to trek inland to retrieve their cached sleds and then return to the coast. At the end of the first week in November Ikpukhuak decided that it was safe once again to travel on the sea-ice, and they returned across the strait to Bernard Harbour.

For returning Jenness safely, Ikpukhuak was rewarded with a rifle and ammunition and some provisions, as had been agreed upon in the spring. Jenness' seven months with these Copper Inuit was a scientific first. In addition to the large amount of information he acquired, his sojourn with them instilled in him a great admiration for their basic intelligence and wisdom as well as for their cheerful nature and their ability to survive under very harsh environmental conditions. The simple nomadic life of these people, however, prompted some writers of the day to describe them as "stone-age people," a term employed by anthropologists then for several little-known Native people in other parts of the world. On the other hand, they were simply demonstrating their ability to "live off the country," a capability Stefansson kept claiming he had discovered for the white man.

Getting the Mail Out

On April 21 Dr. Anderson and Palaiyak started west with the camp's winter mail, the same mail he had been unable to send south from Fort Norman in February. Near Deas Thompson

---

[7] S.E. Jenness, (Ed.), 1991, p. 492.
[8] Jenness, 1928, pp. 154–155.

Point they met Chipman, O'Neill, Ikey Bolt and Patsy Klengenberg, who were slowly mapping eastward from Cape Lyon. The Dr. then turned the mail over to Palaiyak and Ikey Bolt to take the rest of the way to the *Alaska* at Baillie Islands, from where it would be taken to Herschel Island and be sent south. Dr. Anderson also sent instructions to Captain Daniel Sweeney on the *Alaska* to proceed with the *Alaska* from Baillie Islands to Herschel Island when navigation commenced, and collect mail, provisions and a good deal of coal to bring back to Bernard Harbour. The mail did go south, and the *Alaska* reached Bernard Harbour with incoming mail, provisions and a large amount of coal four and a half months later, on September 5.

Johansen's Spring and Summer Activities

Johansen was the one scientist among the six whose working season was the shortest. In that northern climate, plants and insects flourished for only a few weeks each year. As a result, when they burst into maturity, he was intensely active. During the many months of the year when the temperature remained below zero, however, he was frequently at loose ends, although there were fish and other aquatic creatures he could study occasionally when they were caught through holes made in the ice in the bay or in a nearby lake. Dr. Anderson taught him how to type after they both returned from the Coppermine River trip, and he used his newly-acquired skill to prepare a report of his activities for the Department of the Naval Service. He also enjoyed taking the sled and dogs to gather driftwood from the coast or ice from the lake. His cooking abilities proved to be the worst of the group during the fall of 1914, but by the following spring, after the other men (including the cook Sullivan) headed off for their various activities, Johansen became a real housekeeper, cleaning the kitchenware and house, and fetching wood and water. He also kept a wary eye on visiting Inuit to ensure that equipment did not disappear. Occasionally he made minor trades with one or two of the Inuit. After the middle of May, he started to find insects and various plants to study, and from then until August kept extremely busy collecting and preserving both insects and plants in the region surrounding the base camp, searching wherever he could reach on foot.

Johansen had counted on using the *North Star* during the summer of 1915 to carry out some ocean dredging and sounding as part of his marine activities, but was unable to do so as events turned out. However, early in September, following the arrival of the *Alaska* from Baillie Islands and Herschel Island, he was able to take a number of soundings and carry out some dredging in Dolphin and Union Strait from the *Alaska* when he accompanied it and Captain Sweeney to Stapylton Bay on a driftwood-collecting mission.

Wilkins' Journey to Coronation Gulf

In Chapter 12, Wilkins had left Stefansson at the start of the latter's second ice trip northwest of Banks Island. Wilkins was about to set off on April 25 with the engineer Crawford and Billy Natkusiak from near Cape Kellett at the southwest corner of Banks Island on a five-hundred-mile sled trip to the Coronation Gulf region in search of the Southern Party. Stefansson had instructed him to find the Southern Party's base camp and bring back the schooner *North Star*, seven men, supplies and as many dogs as he could obtain. On his way back he was to pick up several Inuit women and their children who lived at Baillie Islands, in particular a woman named Minnie (better known as Guninana), the wife of Alingnak. Minnie

Figure 37 Wilkins, Crawford and Billy Natkusiak near Cape Bexley, west of Bernard Harbour, May 18, 1915.
Photo: Reproduced with the permission of Natural Resources Canada 2011, courtesy of the Geological Survey of Canada, J.R. Cox, 39679

knew a great many Inuit folk tales, which Stefansson wanted to write down. Stefansson also wanted Natkusiak to have the opportunity to find an Inuit wife so that there would be one more seamstress for the Northern Party. Later, when Wilkins on his return trip reached Herschel Island to collect the mail and supplies for the men on Banks Island, he was to leave Crawford there, terminating his service with the Expedition. Neither Stefansson nor Wilkins had been satisfied with Crawford's service and felt it better to get rid of him in this manner.

Stefansson had given Wilkins a letter stating that he was second-in-command of the Northern Party, with the power to act with all of the authority held by Stefansson himself. This letter was intended to enforce Wilkins' authority over any possible resistance from members of the Southern Party, in particular from Dr. Anderson, to the demands he brought from Stefansson. Wilkins and his two companions took with them the mail from Stefansson and other members of the Northern Party, which he was to leave with the police at Herschel Island when on his way back to Banks Island for dispatch south. Wilkins had with him his moving-picture camera and a Graphlex camera with which he hoped to obtain some good pictures of the Copper Inuit, for that was one of the main reasons why the Gaumont Company in London had lent him to the Expedition. At this point, however, Wilkins did not even know whether he was still in the company's employ—he was not, it turned out later.

Wilkins' party proceeded along the south coast of Banks Island for about twenty-five miles, then took a shortcut across the hilly southern part of the island. On May 3, after climbing up and down a series of ridges for several days, he and his men emerged on the east coast northeast of Cape Cardwell. From there they took an erratic course across the ice in Prince of Wales Strait to the north side of Minto Inlet, then turned south on May 9 to cross the mouth of that inlet.

Soon they encountered several Copper Inuit who were living in snowhouses on the ice while hunting seals for their food, furs for clothing and footwear, and seal oil for fuel. This was the first encounter with white men for a few of these Inuit, and after an initial period of arm-waving from a distance to signal that they were friendly and unarmed, they gathered around Wilkins and his companions with much curiosity. All acted most cordially. With Natkusiak translating to the best of his ability, Wilkins obtained from these people, through trade for knives: several fish, two dogs, some fur clothing and pieces of native copper. The copper, he was told, had come from the head of Minto Inlet. One or two of the Inuit remarked that they had met Stefansson and Natkusiak in Prince Albert Sound.[9]

Moving on again, Wilkins' party kept a short distance off shore where the sleds moved more easily, travelling around the west end of the large peninsula, which since 1972 has carried the name of the Expedition's ethnologist, Diamond Jenness. Then they crossed the mouth of Prince Albert Sound and headed slowly east along the south coast of the Wollaston Peninsula. Halting the sleds briefly when they reached Williams Point, Wilkins climbed to its summit in order to assess the ice conditions between that point and the mainland across Dolphin and Union Strait. Judging the ice in the strait to be safe for travelling, he and his men headed across it and reached the mainland a mile east of Cape Bexley on the afternoon of May 17.

Once on land they followed some sled tracks to the west for a short distance and came unexpectedly upon two members of the Southern Party, the assistant geographer Cox, who was just completing the geographic mapping of Stapylton Bay, and the cook James Sullivan. They all exchanged news, and then Cox gave Wilkins directions to the Southern Party's base camp—fifty miles southeastward along the coast. Wilkins and his party reached the camp on May 20, finding it occupied by only the naturalist Johansen and the sailor Castel.

Dr. Anderson and the other scientists were currently carrying out field work many miles to the west around Langley Bay and were not due to return to the base camp until June 1. With more than a week at his disposal, Wilkins decided to make a trip southeast to Coronation Gulf in order to photograph Copper Inuit living and hunting seals on the ice. Before leaving Bernard Harbour, however, he spent three days visiting a small encampment of Copper Inuit who were located a few miles southeast of the Expedition's house. There he met Uloksak Meyok, the young Copper Inuk to whom Jenness had given the casual title "the rich man" because of his having two wives and a collection of curious religious possessions. Wilkins perceived him as being the most successful hunter among his people and having more authority than the others in their camp. He was also the only polygamous Inuk Wilkins (or the ethnologist Jenness) encountered during their three years in the Arctic. Wilkins took the opportunity of his visit to film part of an Inuit caribou hunt and also scenes of them cooking, eating and dancing.[10]

On May 25 he and Natkusiak packed up and, with Uloksak as their guide, headed southeast for Coronation Gulf. Uloksak brought his own sled and supplies and also Koptana, the younger of his two wives. Together they followed the coast to Pasley Cove on the north side of Cape Krusenstern, then crossed a low portage to the shore of Coronation Gulf. From

---

[9] In 1911.

[10] Wilkins Photos Nos. 50901, 50903 and 50904, taken May 24, 1915 near Bernard Harbour, are in the Wilkins CAE Collection, LAC.

Figure 38 Snowhouse and Copper Inuit sleds on blocks of snow, Duke of York Archipelago, Coronation Gulf, February 24, 1915.
Photo: © Canadian Museum of Civilization, photo by Diamond Jenness, 37018

there, in thick fog, they continued across Coronation Gulf to the Berens Islands, on one of which they found a sandy beach with the tent camp of perhaps two dozen Copper Inuit. These people told him that they lived year-round in the region near the mouth of the Coppermine River, hunting caribou upriver in the summer and seals on the ice in the gulf in the winter. Some of them remembered Natkusiak from several years earlier when he and Stefansson had visited them briefly, so promptly welcomed and dined Wilkins' party. Fog greatly impeded Wilkins' efforts to film them during the next twenty-four hours, but in spite of it he obtained several very good still photographs.[11]

Meanwhile Natkusiak bartered for a wife. The father of Tupik, a young Inuk woman who had caught Natkusiak's eye, decided, however, that she was too young to be taken away from the village. He suggested instead that Natkusiak stay with them for a year, or alternatively, come back for her in a year. Unable to commit himself to either option, Natkusiak was next told about a slightly older woman, Kaudluak (or Kaullu), a widow in her twenties and the mother of a girl five or six years old. He was busy negotiating an arrangement with the widow's brother when Wilkins unexpectedly decided that it was time to return to Bernard Harbour. Natkusiak quickly completed the deal and, in exchange for a rifle and some ammunition, was permitted to take the widow for his wife, along with her daughter.[12]

---

[11] Wilkins Photos Nos. 50906–50921, taken May 28 and 29, 1915, on one of the Berens Islands, Coronation Gulf, are in the Wilkins CAE Collection, LAC.

[12] Kaudluak, bewildered and dazed by the suddenness of the news of her acquisition by Natkusiak, was not even permitted to collect her spare clothes and personal effects, as they were not included

The journey back to Bernard Harbour was disrupted by frequent efforts on the part of the unhappy widow to run away, but they all arrived safely on May 31 at the Copper Inuit encampment four miles southeast of the Southern Party's base camp. There Natkusiak left his unhappy bride and her daughter. A few days later her brother arrived from the Berens Islands, returned the gun to Natkusiak, and took his sister and her daughter back to Coronation Gulf. Natkusiak's trial marriage had failed.

Wilkins Stirs up Trouble

Dr. Anderson, Chipman, O'Neill and Patsy Klengenberg arrived at Bernard Harbour from the west a week earlier than anticipated. Cox and Sullivan showed up a day later. Johansen and Castel wasted little time in telling them of the visit by Wilkins and the demands he had brought from Stefansson. Most of the men promptly expressed their anger at this latest act of interference by Stefansson, much of it turned upon the messenger, Wilkins. On May 31, Cox recorded in his field notes:

> Wilkins holds that S[tefansson] as leader of the whole expedition has the right, according to orders, to conduct the expedition just as he pleases and now that the Government has been foolish enough to give him unlimited powers, we have but one alternative— either obey his orders no matter with what results to our work or—accept the responsibility of placing ourselves in the position of mutineers. We do not consider that it can be looked at in that light. The last orders that came in [from the government in Ottawa] stated very definitely that since the S[outhern] party had necessitated the greater part of the expenditure, nothing should be done to interfere with its success. The [*North*] *Star*—as the situation now stands—is essential to the success of this summer's and next fall and spring work.[13]

Cox, perhaps the calmest of the group, then added the following interesting assessment of Wilkins' position. "Wilkins, whose feelings, I think, really lie with the Southern Party, admirably carries out his unwished for ambassadorial job. He presents Stefanson's [*sic*] demands and case admirably, impartially and clearly."[14]

Dr. Anderson and all the other Southern Party scientists except Jenness were at the base camp the next day (June 1), when Wilkins returned from Coronation Gulf.[15] Following the customary welcoming greetings, much subdued in light of Wilkins' reason for being there, Wilkins presented Stefansson's letter to Dr. Anderson, with its list of demands for equipment, men and dogs. As the official photographer of the Canadian Arctic Expedition, he had been a friend and welcome member on both Northern and Southern Parties in 1914. Now, however, as Stefansson's emissary, making demands as if he were Stefansson himself, Wilkins found

---

in the transaction (Wilkins, 1916 [May 29, 1915]).

[13] Cox, 1916 (May 31, 1915).

[14] *Ibid.*

[15] Jenness was somewhere on Victoria Island with his adopted Inuit family.

himself regarded as a threatening adversary. Almost no one in the camp would speak to him, an uncomfortable situation when one recalls that they were housed in a building that measured only twelve by sixteen feet.

At the top of his list of wants, Stefansson had instructed Wilkins to bring back to Banks Island the schooner, *North Star*. Other items included a depth-measuring wire device, sleds and sled dogs, and several people, both white and Inuit. An unpleasant chill hovered over the camp as Dr. Anderson promptly refused to turn over most of the items that Wilkins demanded, especially the schooner, for the use of which the Southern Party already had summer plans. Its departure, he argued, would seriously impede the work of his own men, which he had official instructions not to allow. Drawing upon the authority granted him by Stefansson, Wilkins then threatened to take the schooner, dogs and other items by force, a genuinely brazen action on his part, in view of the number of men who opposed him.

The reason why the *North Star* was so necessary for the work of the Southern Party that summer was, of course, that it was the only vessel then available to them, for the *Alaska* was aground at Baillie Islands. Of this unplanned situation, understandably, neither Stefansson nor Wilkins had any knowledge when they parted in early April. Because of this transportation problem, when Wilkins made his threat, the geographer Chipman flatly declared that without the use of the *North Star* he could not continue his surveying work in Coronation Gulf and Bathurst Inlet and would return to Ottawa. Wilkins suggested compromises, but Dr. Anderson remained firm. As far as he was concerned, Stefansson had interfered with the operations of the Southern Party once too often.

Cox's Compromise

The impasse continued for several days. Then the cool-headed assistant topographer, Cox, offered a workable compromise. It called for some of the scientists to transport their lone *umiak* and outboard motor by dog sled east to Bathurst Inlet for use during the summer. Later, when navigation was possible, the *North Star* could bring equipment, supplies, the remaining scientists and a motor launch to a pre-selected site at the mouth of Bathurst Inlet, where they would join forces with the men who had sledded east. Wilkins could then take the schooner to Banks Island. After some consideration, both Dr. Anderson and Wilkins accepted this proposal, and a strained calm settled at last over the camp.

On June 10, O'Neill, Cox, Natkusiak and a Copper Inuk named Mupfi who was familiar with the region into which they were going, headed east for Coronation Gulf with the *umiak*, sleds and supplies. Cox had earlier planned to map the south coast of the gulf from the Rae River east to the mouth of Bathurst Inlet, but the change of plans now prevented this, and the task was left for Chipman to complete.

The men remaining behind at Bernard Harbour kept busy while they waited for the ice in the harbour and strait to disappear. Chipman undertook the detailed plane-table mapping of the harbour, using as his assistant the young Patsy Klengenberg to hold the stadia rod which allowed Chipman to make his readings. Dr. Anderson, with the assistance of Patsy when Chipman did not need him, cleaned the bird and mammal specimens he and others had collected and wrote descriptions of them, and Johansen roamed about the neighbourhood seeking plants, insects and fish for his collections. Wilkins worked with Castel and Crawford

Figure 39 Chipman and Patsy Klengenberg setting forth to map Bernard Harbour with plane table and stadia rods, July 18, 1915.

Photo: © Canadian Museum of Civilization, photo by George Hubert Wilkins, 51632

to raise the partly submerged *North Star* in the harbour so that it could be made ready for the summer's work. He also photographed birds and mammals for Dr. Anderson, and individual Copper Inuit and their clothing, artifacts and activities for Jenness.

Ice swept east from Amundsen Gulf into Coronation Gulf that summer, blocking the exit from the inner harbour and making navigation in Dolphin and Union Strait impossible during July. As a result, the *North Star*, with the scientists, dogs, provisions and equipment, was unable to leave Bernard Harbour until August 9. After a brief but precarious time among the fast-moving broken masses of ice where Dolphin and Union Strait met Coronation Gulf, the *North Star* motored safely across the gulf to the mouth of the Tree River. There the ship's crew left a cache of supplies for the use of the scientists when they were on their way back to Bernard Harbour from Bathurst Inlet in the fall, and then proceeded slowly east along the coast in search of the O'Neill-Cox party. From the directions written on a series of notes they found on beacons placed prominently along the coast, they finally located O'Neill, Cox and Natkusiak on August 12 at Cape Barrow, at the mouth of Bathurst Inlet. There the remaining provisions, launch, equipment and a few dogs were quickly unloaded, and Dr. Anderson and Chipman disembarked to commence their field work in Bathurst Inlet.

Figure 40 Dr. Anderson (on right) preparing specimens with assistance from Patsy Klengenberg, Bernard Harbour, June 10, 1915.
Photo: © Canadian Museum of Civilization, photo by George Hubert Wilkins, 50932

Figure 41 Copper Inuit fishing camp near mouth of Tree River, Coronation Gulf, July 1915.
Photo: Reproduced with the permission of Natural Resources Canada 2011, courtesy of the Geological Survey of Canada, J.R. Cox, 39721

Wilkins Takes the *North Star* to Banks Island

Wilkins, Castel and Crawford then took Natkusiak on board the *North Star*, and sailed immediately back to Bernard Harbour. They remained there only long enough to pick up the items Dr. Anderson had agreed to send Stefansson, then headed for Herschel Island to leave Crawford and the outgoing mail, and collect the incoming mail and supplies for people at the base camp near Cape Kellett.

Before reaching Herschel Island, however, Wilkins stopped at the tiny settlement at Baillie Islands on August 14 to pick up several Inuit, in accordance with Stefansson's instructions. While there, he was handed a recent note from Stefansson instructing him not to go to Herschel Island but instead to proceed directly to the main camp near Cape Kellett. The note explained that Stefansson had chartered Captain Louis Lane's schooner, *Polar Bear*, and gone to Herschel Island, and would get the needed supplies and mail. Wilkins was to wait briefly at Cape Kellett for Stefansson before proceeding north along the west coast of Banks Island to establish a new base camp on Prince Patrick Island.

In response to these unexpected instructions, Wilkins discharged Crawford at Baillie Islands instead of at Herschel Island, left the Northern Party's outgoing mail there and reluctantly sailed for Cape Kellett the next day. He reached it on August 17 and then waited a week for Stefansson as instructed.

A week later, with no sign of either Stefansson or the *Polar Bear*, Wilkins reluctantly started north with the now heavily laden *North Star* along the west coast of Banks Island, hoping to find a suitable site for a new northern base camp for Stefansson's explorations. Twenty-four hours later he passed Norway and Bernard islands, only to be halted a few miles farther north near Robillard Island by a massive stretch of year-old ice. Further progress appeared out of the question, so he steered the *North Star* into a small sheltered cove nearby on Banks Island and waited, hoping for a chance to continue northward.

That chance came a week later, when he noticed a narrow lead running northward toward Cape Prince Alfred. Hastily, he got his men and dogs on board, started the ship's engine and headed along the lead, only to be forced to turn back when the lead ended after a few miles. Turning back to the sheltered cove, he had his men unload the ship, pull it on shore, then create a fairly comfortable camp for both men and dogs. All of them then settled down to wait for the arrival of Stefansson. It was too late for Stefansson to use the *North Star* for exploration that summer.

Exit the *North Star*

Stefansson succeeded in adding new fuel to Dr. Anderson's animosity towards him that summer by insisting on taking the schooner *North Star* from the men of the Southern Party. His dictatorial demands had prevented its use for their summer work in 1915, thereby not only reducing the extent of the work they had hoped to carry out, but also causing them much inconvenience, hardship and personal danger to undertake what they did complete by 1916. Ironically, Stefansson then changed his plans—as he so often did—and did not use the trim little schooner again.

Stefansson's explanation to Wilkins was that he (Wilkins) was a year late getting the *North Star* to the Norway Island region. One can well imagine how Wilkins must have felt upon hearing that, after having devoted several months to travelling some five hundred miles by sled to Bernard

Harbour, arguing fiercely over the schooner, arousing the hostility of almost all of the members of the Southern Party, and then sailing the schooner well over five hundred miles to its ultimate destination on the west coast of Banks Island, only to be told it was no longer needed.

Summer at Bernard Harbour

After the departure of the *North Star* for Bathurst Inlet on August 9, Johansen was left in charge of the camp at Bernard Harbour until the other scientists returned many weeks hence. Also left at the camp were the cook, Sullivan and Patsy Klengenberg. Sullivan had asked to go with the *North Star* so that he could quit the Expedition when it reached Herschel Island on its return from Cape Bathurst, but was refused permission to do so by Dr. Anderson.

Johansen spent the month of August studying the freshwater life in the lakes near the camp and some marine life, in particular, char and sculpins, which he caught with a fish net. Though sometimes accompanied by Patsy in these activities, he was limited to working near the shore with the canoe and generally preferred to do his collecting alone. Patsy spent much of his time hunting caribou.

The *Alaska* reached Bernard Harbour on September 5 bringing mail, supplies, Ikey Bolt, Siberia Mike and his wife, "Sis," and several other Inuit hired by Stefansson a few weeks earlier to serve as hunters and seamstresses.[16] Palaiyak, who had helped Ikey take the mail west to the *Alaska* and then had gone with it to Herschel Island, had remained at Herschel Island and later that summer rejoined Stefansson's Northern Party.

Cox and O'Neill at Tree River

Meanwhile, after sledging across Coronation Gulf from Bernard Harbour, Cox, O'Neill and Natkusiak established a temporary camp on June 16 on a rocky point near the mouth of the still-frozen Tree River, between the Coppermine River and Bathurst Inlet. There they decided to stay until the ice along the coast melted sufficiently to allow them to

Figure 42 Mingeouk, five-year-old daughter of Copper Inuk Mupfi and his wife, Kilauluk, measures herself alongside a lake trout weighing about 35 pounds, which was caught at the mouth of the Tree River. Mupfi was an assistant to Cox and O'Neill, Tree River, Coronation Gulf, July 1, 1915. Photo: © Canadian Museum of Civilization, photo John J. O'Neill, 38554

---

[16] Ambrose Agnavigak, his wife Unalina, Adam Ovoiyuaq, and Eunice Añayu Annarihopopiak, the wife of Captain Daniel Sweeney.

continue eastward to Cape Barrow by *umiak*. The summer of 1915 was late and cold, forcing them to remain at Tree River until the end of July. One of their first activities was to place a fish net in a small lead of open water near the mouth of the river, hoping thereby to obtain enough char for themselves and their dogs for the present, and to dry and cache a considerable number for dog food for their trip back to Bernard Harbour in the fall. They may thus have been the first white men to fish for char on the river. Today, the Tree River is world famous for its Arctic char and lake trout fishing, with sports fishermen flying in from far and wide. One of the best known of these was former U.S. President George H.W. Bush, who fished there with his grandson Jeb in August 1997.[17]

Cox obtained a time sight for longitude on July 1 and a noon solar reading for latitude the following day. Both scientists also collected an assortment of plants for Johansen while they were camped there. A day's hike in the granitic hills east of the Tree River revealed to Cox that the highest elevation was 570 feet above sea level. O'Neill then found Pleistocene-age marine shells in clay deposits at an elevation of 500 feet, marking a former level to which the ocean had covered the land a few thousand years earlier following the retreat of the continental ice mass from the region.

On July 7, after most of the snow had disappeared from the countryside, Cox, O'Neill and Natkusiak started a six-day exploration of the Tree River valley, commencing with a six-mile ride upriver in their outboard-motor-powered *umiak*. Rapids and two waterfalls at that point forced them to abandon the *umiak* and backpack their food and equipment two miles to the second and higher waterfall, near which they camped. Hills flanking the valley rose to elevations ranging from one thousand feet west of the river to two thousand feet to its east. Among these hills, the many lakes were still frozen, even though it was mid-July. A variety of wildflowers—yellow buttercups and poppies, blue Arctic lupins, and carmine dwarf rhododendron—added much colour to the spring green of the leafing willows along the river, together providing an altogether eye-appealing landscape. However, the mosquitoes were so thick that they almost hid the scenery. Cox put it this way:

"Northern Canada is cursed, not by its interminable distances, its vast stretches of swamp and muskeg, its isolation or the bitter cold of its winters; all of these one can, with a certain amount of determination triumph over … It is the astounding atmosphere of mosquitoes that envelops the whole face of the country in the summer time that is the real curse."[18]

They therefore travelled at night, when it was cooler and they were less troubled by the mosquitoes.

After proceeding upriver for about twenty-five miles, Cox and O'Neill concluded that they had obtained sufficient information about the geography and geology of the region for their needs and returned to the coast. During their absence, Mupfi by pre-arrangement had brought their *umiak* back to the mouth of the river. He and several other friendly Inuit had then netted and dried a large quantity of fish for dog food for the two scientists. The Arctic char had suddenly started "running" during their absence. Over the next few days, Cox, O'Neill and Natkusiak caught and dried about six hundred pounds of fish for their use in the autumn

---

[17] Bush, 1997.

[18] Topographer J.R. Cox, quoted (without being identified) on p. 26B, in Chipman and Cox, 1924.

when they returned from Bathurst Inlet, caching it meanwhile on the island at the mouth of the large bay (Port Epworth) into which the Tree River empties.

After their return to the coast, O'Neill spent a day examining small pockets of talc–chlorite schist enclosed in granite half a mile east of the mouth of the Tree River. These small deposits of schist supplied the material that the local Copper Inuit used regularly to carve their cooking utensils. The Inuit told O'Neill (with Natkusiak translating) that there were smaller deposits on the river flowing into Grays Bay farther east and also near Cape Barrow.

Before Cox, O'Neill and Natkusiak started east from Tree River in their *umiak*, their Copper Inuit helper Mupfi, with their approval, took most of their sled dogs with his dogs and moved his family inland for the summer to hunt caribou. In the fall they all returned to the coast and rejoined the scientists when they returned from Bathurst Inlet en route to Bernard Harbour.

Cox and O'Neill Map the Coast from Tree River to Cape Barrow

On July 17, the temperature reached 78° F, sufficiently warm that Cox managed to bathe in the harbour near the mouth of Tree River. He observed in his field notes that the sky was clear but smoky from forest fires far to the south, where trees were abundant. The run of fish up the river ceased towards the end of July almost as suddenly as it had commenced four weeks earlier. Meanwhile, stormy weather and ice in Coronation Gulf prevented Cox's party from starting east in their outboard-motor-powered *umiak* for Cape Barrow until July 30. Even

Figure 43 View of Southern Party's lonely base camp at Bernard Harbour, with the house largely hidden behind the white tent and the *North Star* leaving through the narrow entrance to the inner harbour, July 20, 1915. A sandspit from Teddy Bear Island almost blocks that entrance.

Photo: © Canadian Museum of Civilization, photo by Dr. R.M. Anderson, 38866 on that date, ice formed during the night, and they were forced to pick their way carefully among the floating pieces of ice to avoid serious damage to the *umiak* as they progressed slowly along the rocky coast. At a few prominent sites they stopped to leave a stone cairn or wooden monument with a message for Dr. Anderson, telling of their plans in anticipation of his arrival soon. On the evening of August 2 they left one of their messages on Hepburn Island, then rounded Cape Barrow, immediately encountering water roughened by a north-east breeze.

O'Neill soon determined that the northeasterly-trending islands along the south side of Coronation Gulf were formed of a single rock type known as diabase. This diabase has since been recognized as layers (sills) injected in a molten state millions of years ago along beds of ancient (Late Precambrian) sandstones, all gently dipping northerly. They form part of a large collection of similar-layered rocks that give some of the Coppermine River valley its charac-teristic hill-and-valley topography. He also determined that older granitic rocks were exposed along the coast east of Tree River as far as Cape Barrow, with the granitic rocks and the younger sediments being in an almost straight, north-south, faulted contact for over thirty miles south from the mouth of the Tree River.

About a mile south of Cape Barrow in Bathurst Inlet they found a well sheltered place to camp and decided to wait there for the arrival of the *North Star*, Dr. Anderson and Chipman. They returned to the cape a few days later and left a note at the base of a pole of driftwood standing upright as a beacon, on which was written the location of their camp. By August 5, Bathurst Inlet appeared to be clear of ice, and four days later Coronation Gulf appeared likewise. But where was the *North Star*? Was it still submerged at Bernard Harbour? Had the men been unable to get its engine working after its submergence in the harbour? Had it encountered trouble crossing Coronation Gulf? Heavy seas and rain occurred on the night of August 10, accompanied by one peal of thunder—a rare meteorological event in the Arctic.

Reunion at Cape Barrow

Chipman suddenly appeared in the opening of their tent on the morning of August 12, after he had walked across the cape from a small sheltered harbour on its west side, where the *North Star* had just tied up. His appearance brought tremendous relief to Cox and O'Neill, but not to Natkusiak, who at that time had no desire to return north with Wilkins. All four men walked back across the cape and helped unload the *North Star*. Then Natkusiak reluctantly stowed his gear on board, not wanting to return to the Northern Party without a wife, and Wilkins headed the schooner west for Banks Island. The four scientists left behind—Chipman, Cox, Dr. Anderson and O'Neill—then discussed their plans for the days ahead and decided to cache some of their supplies at the cape.

A Problem with Transportation

From Cape Barrow, they started their survey into Bathurst Inlet on August 16, using a launch, an *umiak* and a canoe. They regretted not having the *North Star* for this work, for rough water frequently limited their activities thereafter because of the small size of their three water-craft. Having no alternative, however, they slowly and carefully worked their way south along the

Figure 44 *North Star* unloading supplies and equipment in small harbour at Cape Barrow, August 12, 1915.
Photo: © Canadian Museum of Civilization, photo by Kenneth G. Chipman, 43277

Figure 45 Cox, determining latitude with sextant, in small harbour at Cape Barrow; Dr. Anderson examining the condition of the canoe behind him, *umiak* with outboard motor beyond, Coronation Gulf, August 14, 1915.
Photo: © Canadian Museum of Civilization, photo by Kenneth G. Chipman, 43279

west coast of the deep inlet, with Chipman and Cox plotting the outline of the coast and noting the nature of the terrain, while O'Neill identified the rock types and looked for evidence of copper mineralization.

For a brief period they operated as two separate parties. Dr. Anderson and O'Neill used the *umiak* and Evinrude outboard engine, while Chipman and Cox operated the launch. However, the Evinrude engine ceased to work after a few days. As it was not practical for them all to proceed in the launch with their three dogs, it became apparent that one person would have to camp and stay with the dogs, provisions, *umiak* and the defunct Evinrude engine, while the other three continued their survey around the coast in the launch.

Mapping Bathurst Inlet

Chipman bravely volunteered to remain alone with the three sled dogs for the next several weeks on a sheltered beach near Kater Point, not knowing when or even if the others would return. Dr. Anderson, Cox and O'Neill then continued southward on August 25 along the west coast of Bathurst Inlet, towing the canoe behind their launch as they progressed into Arctic Sound, then up the Hood River to its first cascade. There they camped on August 27, then explored inland briefly to obtain a sense of the topography, geology, and animal and bird life of the region. A fat young caribou bull replenished their meat supply while they were up the river. Returning to the coast, they continued around the bottom of Arctic Sound. There Dr. Anderson shot a healthy barren ground bear for the National Museum of Canada's collection of Arctic mammals. He had heard there were muskoxen in that region and hoped to obtain a specimen, but did not see a single one. Then while Cox plotted the shape of the

Figure 46   View of "Snug Harbour" near Kater Point, Bathurst Inlet, where Chipman and three sled dogs were alone for several weeks. His round tent is near the water's edge, September 17, 1915.

Photo : Reproduced with the permission of Natural Resources Canada, 2011, courtesy of the Geological Survey of Canada, J.R. Cox, 39666.

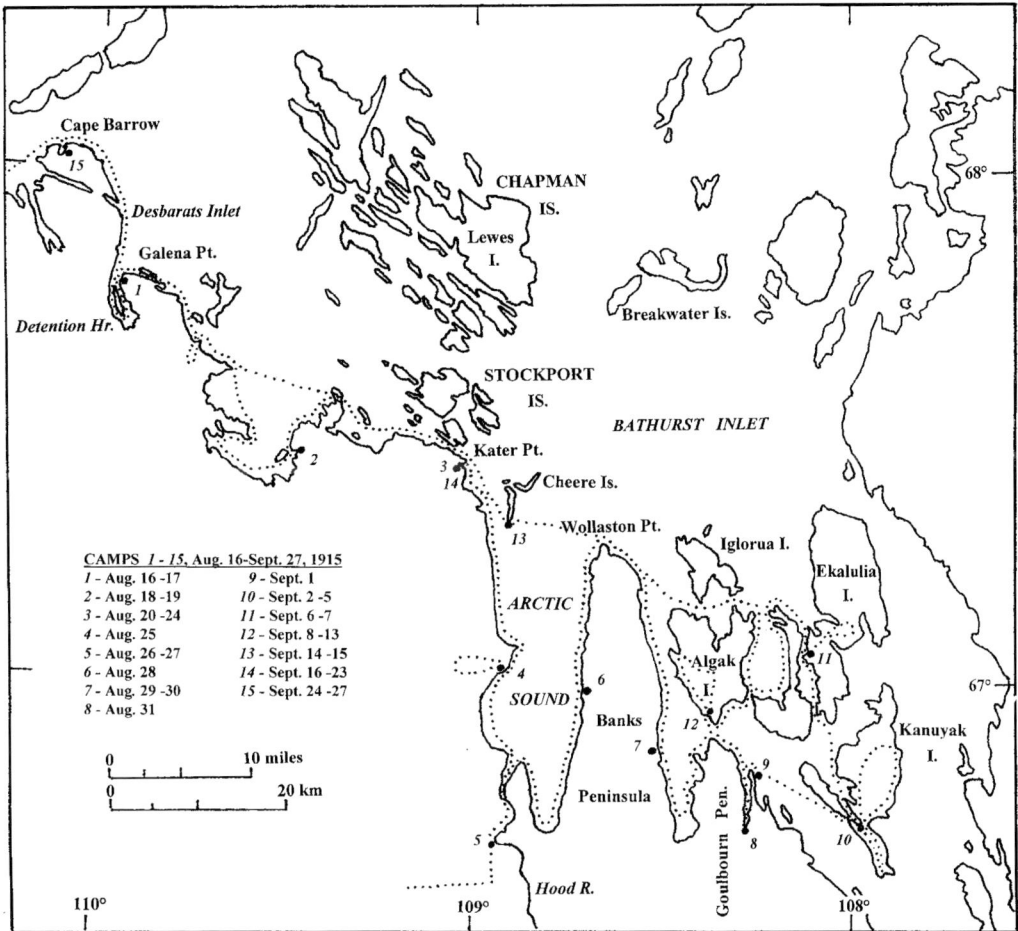

Map 18 Route followed by Dr. Anderson, Cox and O'Neill, mapping by motor boat in Bathurst Inlet, August 16 to September 27, 1915.

coastline and made some notes, O'Neill discovered small amygdules[19] of native copper in the rocks on the Banks Peninsula, which flanked the east side of Arctic Sound.

The only previous mapping of Bathurst Inlet had been done by Sir John Franklin and his men in 1821, using large birchbark canoes his men had somehow laboriously carried from Great Slave Lake overland to the Coppermine River. Thus it is understandable why Cox had to make many modifications to the outlines of the coast and islands in Bathurst Inlet as they were shown on the old British Admiralty Chart.

However, time restrictions and the need to focus upon the area containing copper-bearing rocks prevented Cox and his two companions from making a complete survey to the south end of Bathurst Inlet. Consequently, they crossed from Banks Peninsula to Kanuyak Island

---

[19] Amygdules are small mineral masses deposited in cavities in a solidified igneous rock.

and from there to Ekullialuk Island, one of a group of islands that Franklin had thought was only one island, and named "Barry Island."

It was September 8 when Dr. Anderson and his companions reached that locality, their farthest from Bernard Harbour. With the winter freeze-up fast approaching, they decided it was too risky to continue any longer before heading back to pick up Chipman at Kater Point. Their decision was a timely one, for sub-freezing temperatures, driving winds, snow squalls and trouble with the launch's engine then prevented them from attempting the long trip north to Kater Point for almost a week. Their arrival at Kater Point on September 16 was timely, too, for Chipman's supply of food had grown desperately low.

Return to Tree River

Strong northwesterly winds, heavy snowfalls and freezing temperatures then held them at Kater Point for a week. This latest delay raised serious concerns among them about their being able to reach Tree River, many miles to the west, where they had left their sleds and most of their dogs.

Snow was now accumulating in gullies to depths of four or five feet, and the air temperature seldom rose above the freezing level. The four men played bridge in their tent to while away the time until the winds and high seas diminished. By September 20 a six-foot snowdrift almost concealed their *umiak*, and they had increasing difficulty in obtaining heather for their fuel needs. Finally September 24 dawned sunny and calm. Hastily they dug out the *umiak* and canoe, loaded their supplies, equipment and dogs into the launch, and headed for Cape Barrow, towing the *umiak* and the canoe. Cox had to stop the launch almost every half-hour to oil the engine, but after nearly eight hours, they arrived safely, though well chilled, at the small sheltered cove on the west side of Cape Barrow where the *North Star* had left Dr. Anderson and Chipman early in August.

Winds and high seas again detained them, giving Cox and O'Neill ample time to undertake repairs on the launch's engine. The morning of September 28 dawned sunny and calm, but a thin layer of ice covered the harbour. Cox cleared a way through to the harbour entrance with the launch, but the ice cut through the wood along its sides, causing it to leak and delaying their departure while they made repairs. Despite these added problems, however, they managed to set off shortly before noon, following the shore closely to the west. On September 29 they progressed only as far as the eastern side of Grays Bay before increasing winds forced them to take shelter. The winds continued to hinder their progress the next day, but they pushed onwards, stopping only briefly to examine a twelve-foot stone monument erected years before by the Hudson's Bay Company explorers Peter Dease and Thomas Simpson in 1839. They reached the large bay (Port Epworth) at the mouth of the Tree River the following evening.

Greatly relieved to be back at the mouth of the Tree River, close to the caches of food for both men and dogs, which they had left there in August, the four scientists discussed their situation and decided that it was too late in the season to risk continuing in the launch to Bernard Harbour, more than one hundred miles to the northwest. They therefore moved their camp on October 3 to a more suitable site on a willow flat on the east side of the bay, where they hauled the launch onto the beach. They made this move just in time, for the temperature dropped to 15° F on the night of October 5, and the harbour froze over.

Figure 47 Cox, O'Neill and Chipman at the base of a stone monument erected in 1839 by explorers Dease and Simpson, about 15 kilometres east of Tree River, Coronation Gulf, September 30, 1915.
Photo: © Canadian Museum of Civilization, photo by Dr. R.M. Anderson, 38750

During the next three weeks Mupfi's family and several other Copper Inuit families set up their tents nearby. They had taken adequate care of the scientists' seven dogs over the summer, had shot a number of caribou and presented some of the meat to the scientists. The men's large cache of fish on the island in the mouth of the harbour was intact, so all fared fairly well for the next while.

By October 27 Dr. Anderson concluded that it would be safe to travel on the ice in the gulf and headed his men, sleds and dogs across Coronation Gulf, leaving the launch, *umiak* and canoe at Tree River. They reached Bernard Harbour safely and without incident on November 9.

In spite of having to use small open water-craft, the four scientists, within two months, had succeeded in making a preliminary survey of the south coast of Coronation Gulf from Tree River to the southeast side of Bathurst Inlet, made numerous corrections to the available map of Bathurst Inlet, collected many mammal, bird and plant specimens for the National Museum in Ottawa, mapped the distribution of several rock types, delineated a band of native-copper-bearing rocks in Bathurst Inlet, and determined the contact between what has since proven to be an economically important band of mineralized older granitic rocks east of the Tree River and younger sedimentary rocks to the west of that river. Considering the conditions under which they had to operate, theirs was a truly commendable accomplishment. They had not, however, been able to complete their intended work. Had they been able to use the *North Star*, they probably could have finished their survey of the Bathurst Inlet, for they could have carried out field work on many days when the water was too rough for them to venture on it safely with their launch.

Unexpected Guests and News of the War

When Dr. Anderson, Chipman, Cox and O'Neill reached Bernard Harbour on November 9, they found not only Johansen, who had been left in charge of the camp, Jenness, just back from seven months on Victoria Island, cook Sullivan, and Patsy Klengenberg, but also several strangers. These included Corporal W.V. Bruce of the Royal North-West Mounted Police, Church of England missionary Rev. Herbert Girling and engineer J.E. Hoff. The corporal had come east on the *Alaska* from Herschel Island in search of information about two French Roman Catholic missionaries who had been reported missing from Great Bear Lake since 1913. Rev. Girling hoped to establish a mission at Bernard Harbour, but his schooner *Atkoon* had been blown ashore at Clifton Point many miles to the west. Hoff was the engineer Stefansson had hired to replace Daniel Blue, the engineer of the *Alaska*, who had died of scurvy and pneumonia during the previous winter and was buried at Baillie Islands.

There were also several unfamiliar Inuit whom Stefansson had hired in August at Herschel Island to assist the Southern Party. They were: Ambrose Agnavigak, his wife Unalina, and her small daughter, Annie Fitzgerald (by Unalina's previous marriage to Sergeant F.J. Fitzgerald of the RNWMP);[20] Adam Uvoiyuaq, who was Palaiyak's brother; Eunice Añayu Annarihopopiak, also known as "Red Calico," whom Captain Sweeney had married at Herschel Island during the summer of 1915; and "Sis" or "Ciss," the Iñupiaq wife of Siberia Mike, with her son Georgie and a younger (unnamed) child. There was also news that Palaiyak had rejoined Stefansson's Northern Party during the summer.

In the mail that was brought to Bernard Harbour on the *Alaska* came the first news of World War I in Europe. There were also government letters with clear instructions that all members of the Southern Party were to leave the Arctic in the summer of 1916 and come south to Ottawa.

Visit by Coppermine River Inuit

A few dozen Copper Inuit arrived at Bernard Harbour about the middle of November from the Coppermine River region and built snowhouses along the beach near the base camp. Several weeks of singing and dancing followed, which the scientists found interesting and enjoyable, for the visitors seemed to be such carefree people. During this period the visiting Inuit women sewed new fur garments for everyone before they all moved out to the seal-hunting areas around the Liston and Sutton Islands. The sewing of the caribou skins had to be done before they moved out on the sea-ice, for they had a strict taboo against sewing the skins of caribou and other land animals after they hunted seals. That was why they had built their snowhouses on the shore this time instead of on the harbour ice. The previous March, when many of the same Inuit came to the base camp, they were not sewing clothes, so they camped on the ice in the bay.

---

[20] Sergeant Francis Joseph Fitzgerald (1869–1911) was in charge of the famous "Lost Patrol," four members of the RNWMP who got lost in the winter of 1911 and died attempting to take the mail over the mountains from Fort McPherson to Dawson.

A Snowhouse for Tide Studies

In December the scientists constructed a snowhouse on the ice on Dolphin and Union Strait alongside the small gravel island—Teddy Bear Island—at the entrance to the inner harbour. For a week they kept a record of the height of the tide every half-hour around the clock, thereby establishing the maximum local rise of tide as two and one-half feet. The men took shifts recording the data, often stumbling about in their efforts to find their way in the darkness to the snowhouse or returning to the main house.

Caribou Migration

The fall migration southward of the barren-ground caribou from the western part of Victoria Island crossed Dolphin and Union Strait east of Bernard Harbour during November 1915. Expedition members shot about thirty of them, and the skins and bones from several of them became specimens for the National Museum in Ottawa. Added to the ten carcasses Jenness had brought back with him from Victoria Island, this proved to be an adequate supply of meat for the camp during the first part of the winter.

The dark days of November and December, when the sun failed to come above the horizon, were quiet times at Bernard Harbour. Only Jenness kept active, at first visiting the Inuit in their snowhouses on the beach near the base camp, then later following them to their seal-hunting grounds near the Liston and Sutton Islands. Back and forth he travelled between the base camp and the snowhouse village near the islands, where he gathered ever-increasing amounts of information about their activities and beliefs. Meanwhile, the other scientists worked indoors over their field notes and planned their spring work. All of them knew that they had to complete their projects in the region by the coming summer, when they were required to head for Nome and points south.

Christmas 1915

With everyone present and healthy, the scientists made a real effort to celebrate Christmas Day. Jenness brought his two young "adopted daughters," Jennie and Alunak,[21] from the sealing camp on December 23 to liven up the occasion with their youthful enthusiasm and joyful laughter. On Christmas day itself the two girls giggled and exclaimed in delight over the gifts of sewing needles, calico and other useful items they found in cloth bags someone had specially created for them. After all the presents were distributed, several of the men organized a series

---

[21] Jennie Kannayuk was the twelve- or thirteen-year old daughter of Higilak and step-daughter of Ikpukhuak. Kannayuk is the Inuit word for the sculpin, an ugly, spiny bottom-dwelling seafish with a big head and wide mouth, not the prettiest name for a lively young girl. However, according to Mary Carpenter Frost (personal communication, January 2009), such a name makes a great deal of sense to the Inuit, for they frequently select a child's name that gives an opposite impression to reality. Thus, in the case of Kannayuk, to ensure that their delightful baby girl did not grow up to be vain, her parents would have given her, almost in jest, a name of something unattractive. Alunak was the daughter of Wikkiak, an Inuk from southwest Victoria Island, and his wife, Unahak, and was also about thirteen years old.

of outdoor races for everyone present—jumping matches, tugs of war and the like. Most of these were won by the Copper Inuit entrants.

The cook Sullivan exceeded himself in providing an elaborate dinner later that day, which even included the traditional British Christmas desserts—plum pudding and mince pies—together with assorted candies.

Some newly arrived Inuit visitors from south of Bathurst Inlet then performed a dance for the Expedition members, using one of the cook's frying pans for a drum. Jenness later made four wax-cylinder recordings by having the visitors sing into the large horn on his recording machine, which provided him with songs that differed markedly in sound and dialect from the songs he had recorded by Ikpukhuak and many other Coronation Gulf Inuit. Quite possibly on that occasion Jennie Kannayuk sang one or more of the songs she recorded for him, which are preserved on the wax cylinders in the collections of the Canadian Museum of Civilization. Her songs are a special delight to hear because she sometimes interrupted her singing by breaking into unexpected, joyful giggling and laughter. During the next few weeks Jenness recorded many other songs of the Copper Inuit who visited the base camp from the sealing camp in Dolphin and Union Strait. These recordings form a priceless record of part of the Copper Inuit's past and largely forgotten culture.[22]

---

[22] In July 1989 I brought to Coppermine (now called Kugluktuk) cassette tape recordings of the songs that my father recorded while at Bernard Harbour in 1915 and 1916, and presented them to Rosemarie Meyok, the granddaughter of Jennie Kannayuk. These tapes had been specially recorded for me through the kindness of Dennis Fletcher, then the audio technician at the Canadian Museum of Civilization. On them Fletcher clearly announced before each song the names of the singer and the identifying recording number as given in volume XIV of the Reports of the Canadian Arctic Expedition (*Songs of the Copper Eskimos*) by D. Jenness and Helen H. Roberts (1925). I also presented Rosemarie with a copy of volume XIV, which contains the music for all of the Copper Eskimo songs. The tapes aroused so much local interest at the time that within twenty-four hours, additional copies were circulating around the settlement. Some of the elders recognized the voices of their long-departed friends or relatives, and rejoiced to hear their traditional songs again, many of which had been forgotten for years.

FAREWELL TO THE ARCTIC, 1916

21

In response to the written instructions they had received in the government mail from Ottawa in the summer of 1915, which stipulated that they were to leave the Arctic in the summer of 1916 as originally planned, the men at Bernard Harbour spent the sunless period from mid-November 1915 to mid-January 1916 writing up their field notes, evaluating the work they had accomplished, and planning the completion of those tasks still unfinished.

Another Failed Trip to Fort Norman

The sun made its first appearance in mid-January. On January 26, Dr. Anderson set off with Corporal Bruce, Adam Uvoiyuaq and a dog team to take the mail overland to Fort Norman by way of the Coppermine River and Great Bear Lake. A support party consisting of the cook Sullivan, Ikey Bolt and Ambrose Agnavigak accompanied them with a dog sled of supplies to Bloody Fall on the Coppermine River. There they left a cache of provisions for Dr. Anderson's return trip and then hunted caribou to supplement the dwindling meat supply at the base camp. Unfavourable weather and much snow greatly delayed Dr. Anderson's party, which by February 10 had scarcely passed Bloody Fall. By that time much of their dog food was consumed. Dr. Anderson then realized that if he continued at such a slow pace he would not get back to Bernard Harbour in time to complete his field work in Bathurst Inlet. He therefore terminated his trip to Fort Norman, concentrated briefly on securing caribou meat and then returned to Bernard Harbour, arriving back at the base camp on February 27.

Late in January, the Copper Inuit brothers, Kohoktok and Mupfi, both employed for some months by Dr. Anderson, went with their families to Tree River with a sledload of supplies from Bernard Harbour, which they cached on the island at the mouth of the river for the use of the men completing the survey of Bathurst Inlet. They then returned to Bernard Harbour, bringing with them several boxes of geological and zoological specimens left there by Dr. Anderson's men the previous November.

Jenness' Visit with the Bathurst Inlet Inuit

On February 16, eleven days before Dr. Anderson returned from his failed attempt to get the outgoing mail to Fort Norman for dispatch south, Jenness left Bernard Harbour, accompanied by Reverend Herbert Girling, Patsy Klengenberg, Higilak's married son Avrunna, and the shaman Uloksak Meyok and Haqungaq, the newest and youngest of his three wives, and headed east to visit the Inuit around Bathurst Inlet. Stefansson, during his brief trips to the Coronation Gulf region in 1910 and 1911, had separated the Eskimos he met into a number of small tribal divisions, to which he had applied various names, according to the districts they lived in, such as Puivlirmiut, Akulliangmiut, Noahognirmiut and Kogluktomiut.[1] By February 1916 Jenness had studied the people in all of the so-called "tribes" mentioned by Stefansson except those in the Bathurst Inlet region. It was those people he now intended to visit at their sealing camp somewhere near Cape Barrow. Rev. Girling wished to meet the people he hoped soon to convert to Christianity, and was taking advantage of Jenness' knowledge of the Native language and his rapport with the people to assist him. Uloksak wished to visit relatives to the east and was willing to carry some of Jenness' supplies on his sled in exchange for sharing them. When they camped briefly with some Coppermine River Inuit near Locker Point, Jenness noticed that a man named Noqallaq in the camp possessed a bone and ebony crucifix and a cassock with a hood. He later reported this observation, and others like it, to Corporal Bruce on the latter's return to Bernard Harbour, which was of considerable aid to the police in their investigations of the two missing French Oblate priests, Fathers Jean-Baptiste Rouvière and Guillaume Le Roux.

Jenness and his party continued the next day across Coronation Gulf. On February 22 they stopped at a small settlement of seal-hunting Inuit about ten miles north of the mouth of Tree River. The people there proved to be so congenial that Jenness decided to spend a few days with them, trading, collecting oral information about their daily activities, taking head measurements and repairing his sled, which had been slightly damaged from the heaviness of its load. This unexpected delay annoyed Uloksak, who soon departed to the east with his wife. Jenness and his companions left for Cape Barrow five days later.

Following a trail seaward from the northwest corner of Bathurst Inlet, Jenness and his companions soon came upon a snowhouse village, whose inhabitants were the very Bathurst Inlet people he wanted to visit. Some of them had never seen a white man before. From these people he learned that Uloksak had just started back to Bernard Harbour with ten families from their village. This meant that Jenness would have to find his own way back to Bernard Harbour.

A blizzard kept him in this settlement for several days, during which time he measured several heads and did a little trading. As these people had had no luck in their recent seal-hunting efforts and were nearly starved, Jenness gladly shared his food with them. This act, however, reduced his own resources to a three-day supply, so as soon as the weather improved he and his companions hastened to the site of the Expedition's supplies cached the previous autumn at the mouth of the Tree River. En route Patsy shot a caribou, which was a welcome acquisition for both men and dogs. At Tree River Jenness took some of Dr. Anderson's cached supplies and then started across Coronation Gulf for the base camp. After struggling with bitterly cold winds and drifting snow, he and his two companions reached Bernard Harbour on March 19, to find they had arrived ahead of the ten families of Inuit from Bathurst Inlet being guided by Uloksak. When that group arrived a short time later, Jenness persuaded some of them to sing into his recording machine to add to his collection of Inuit songs.

---

[1] These were the Inuit who lived in the districts Stefansson had labelled Puivlik, Akulliakattak, Noahognik, and Kogloktok, respectively. The last group occupied part of the Coppermine River valley.

Dr. Anderson's Croker River Trip

During Jenness' absence, Dr. Anderson, Chipman and Siberia Mike headed west from Bernard Harbour on March 6 to map the Croker River, the largest river between Darnley Bay and Coronation Gulf. At that time nothing was known about it except the location of its mouth. The two hundred miles of terrain between Stapylton Bay, east of which lay the domain of the Copper Inuit, and the bottom of Darnley Bay were uninhabited, apart from occasional temporary camps of traders or trappers like Scotty McIntyre and the Klengenbergs. Dr. Anderson wanted to see if the scarcity of humans was related to the animal population, or lack of it, in that region. More particularly, he wondered if it was inhabited by musk-oxen, for neither he nor any of his Southern Party had yet seen one. The Croker River, named by Franklin's companion Sir John Richardson in 1826, had a broad delta onto which it emerged from some steep-walled dolomitic hills about five miles inland.

Today the river's source is known to be Bluenose Lake,[2] a large elongated body of fresh water nearly fifty miles inland. Dr. Anderson and Chipman found that the Croker River flowed mostly northward but with short sharp bends, between almost vertical canyon walls, which grew higher the farther upriver they progressed. The river's width was between one hundred and one hundred and fifty feet for several miles inland. Some twelve miles upriver, a large creek flowed into it from the southeast, which

Figure 48 Dr. Anderson and Siberia Mike sledding in the canyon of Croker River, March 19, 1916.
Photo: © Canadian Museum of Civilization, photo by Kenneth G. Chipman, 43312

offered the first opportunity for Dr. Anderson and Chipman to climb to the top of the canyon to examine the surrounding countryside. Once they reached the top of the canyon they found themselves on a smooth, rolling upland, which dipped gently northward towards Amundsen Gulf. Chipman's aneroid barometer indicated that the canyon at that locality was about three hundred feet deep.

About twenty-five miles upriver Dr. Anderson's party emerged from the narrow gorge on to an upland with small snow-covered hills. One conspicuous hill rose well above the surrounding plain, seven miles distant. Sir John Richardson had seen this hill from Amundsen Gulf in the summer of 1826, when he was more than fifteen miles away, and named it "Mount Davy." On March 21, 1916, the first day of spring and a sunny day with a temperature ranging

---

[2] The existence of Bluenose Lake was only discovered in 1948 after the Royal Canadian Air Force had conducted systematic aerial photographic flights in that part of the Arctic.

between –46° F and –9° F, Chipman walked several miles to this conspicuous hill, photographed it,[3] described its shape as being "like an inverted cup,"[4] and then climbed to its summit, perhaps the first human to do so. The hill proved to be composed entirely of glacial gravels and was surrounded as far as the eye could see by many smaller and lower gravel hills, although higher hills lay farther to the south.[5] From its summit, Chipman was able to sketch the valley of the Croker River southward for an additional fifteen miles or so, but did not see the thirty-three-mile-long lake that is its source. Nor did he see any bedrock within the gravel hills.

Vegetation was unusually sparse on this upland, with no ground willows and few clumps of grass. The only signs of life were a few tracks of Arctic hares and one Arctic fox. Dr. Anderson was not surprised, therefore, that he saw neither caribou nor muskoxen, for none could survive in such a barren region. He then concluded that the scarcity of caribou west of Cape Bexley explained the absence of Inuit in that region. Chipman had by then mapped the course of the river for nearly forty miles, so they felt that they had accomplished their mission and returned to the coast, arriving at Bernard Harbour on April 2.

O'Neill and the Copper Resources in Bathurst Inlet

Roughly two weeks before Dr. Anderson and Chipman arrived back at the base camp, O'Neill, Cox, Ikey Bolt and Kohoktok started for Bathurst Inlet with two heavily laden sleds and provisions for about ten weeks. With Kohoktok was his wife, Munnigorina, and their baby son, Itayuk. The temperature was between –36° and – 41° F. O'Neill sought to complete his survey of the copper-bearing area in that inlet, and Cox wanted to finish his geographical mapping of the islands that he had not been able to examine the previous summer. En route to Cape Krusenstern they met Jenness, Rev. Girling and Patsy Klengenberg returning from their visit to the Bathurst Inlet Inuit. Jenness reported having had a cold, unpleasant trip but being pleased with the results. After a brief exchange of news, O'Neill and his companions continued on their way across Coronation Gulf to the mouth of Tree River. From there, they proceeded east along the coast to Hepburn Island, arriving on March 30 at the cache at Cape Barrow they had left the previous fall.

Comments in O'Neill's diary suggest that he found travelling with a small baby to be a nuisance because the entire party had to halt frequently while the mother, Munnigorina, removed her naked child from the hood of her *attigi* where she carried it, and nursed it or cleaned it up.

"Maniguan [Munnigorina] takes the kid out with just a short jacket & lets it suckle, exposed to 36° below zero for more than 5 minutes at a time (yesterday). This happens at

---

[3] K.G. Chipman, Photo Nos. 43315 and 43316, GSC.

[4] Chipman, 1916 (March 21, 1916). He gave the height of the hill as two hundred feet.

[5] "Mount Davy is a unique and spectacular case of a mega-kame … a candidate for the world's biggest kame" (St-Onge and McMartin, 1995, p. 13). It is a well dissected circular mound with its summit more than 300 feet above the surrounding plain (according to these authors), and lies about 15 miles east of the Croker River and 18 miles from the coast, near the margin of a wide belt of northwest-trending drumlins, which form the Inman River drumlin field. This kame is composed of stratified sand and gravel and was deposited from a subglacial stream as a fan or delta at the margin of the melting continental glacier about 10,000 years ago. Little was known about continental glaciation in 1916.

intervals all day. A square of heavy deerskins for diaper. It dried by beating in soft snow, then carried in leg of boot."[6]

Five days later he added, "Had to take 20 min. stop 'to feed kid.' "[7] Somehow the naked child survived the sub-zero temperatures during the few moments these activities regularly required.

For the next several weeks O'Neill and Cox carried out their mapping assignments, working under challenging weather conditions—snow, fog and sub-zero temperatures. In spite of those deterrents, however, they managed to examine and map the outline and geology of more than two hundred islands that form the Chapman Islands and the Stockport Islands in the northwestern part of Bathurst Inlet. O'Neill concluded that almost all of the rocks on these islands were thick volcanic flows of fine-grained amygdaloidal basalt, which he interpreted as being geologically related to similar-looking rocks up the Coppermine River and along the south coast of Coronation Gulf west of Tree River. Small pockets of native copper occurred throughout the volcanic flows.

While O'Neill examined the rocks for copper on both the shores and the interiors of the many islands, Cox trekked around the islands with the sled and dogs, sketching their shapes and taking occasional solar readings for latitude and longitude determinations when the skies were clear. Ikey and Kohoktok built snowhouses wherever they camped until May. Thereafter snow conditions were no longer suitable, and they set up their tents. Ikey, being from Alaska, had little experience at building snowhouses and seemed to have difficulty learning. "Ikey still putting block on block without overlap; he will never learn," grumbled O'Neill in mid March.[8] They supplemented their provisions with meat from an occasional caribou they shot on the Stockport Islands during the period they worked there, and to increase the efficiency of their work, they used Primus stoves for their cooking, thereby avoiding wasting time searching for driftwood.

O'Neill's party then moved from the Stockport Islands to Kater Point and followed the coast south. On May 1 they crossed the mouth of Arctic Sound and continued past Cape Wollaston to the Barry Islands, where they made camp on the west side of Iglorua Island. There they were unexpectedly joined by five sleds of Inuit from Victoria Island. Because of the scarcity of seals that spring, these Inuit had followed O'Neill's trail south from the Lewes Islands near the mouth of Bathurst Inlet, where they had camped together briefly. Now they were heading for a fishing site they knew about, which was south of Cape Wollaston.

After mapping Iglorua Island, O'Neill and Cox moved their camp to the north side of Algak Island, near to the place where Ikey and Kohoktok had just killed several caribou. Cox surveyed the northwestern part of Algak Island on May 4; then bad weather halted further work for a day. On May 6 they moved east to Ekalulia Island, which Cox mapped and O'Neill examined for copper. Later, while en route to Algak Island to make camp, they sighted and shot several caribou and also found some much-needed driftwood. They completed their

---

[6] O'Neill, 1916 (March 18, 1916).

[7] *Op. cit.* (March 23, 1916).

[8] *Op. cit.* (April 30, 1916).

examination of Algak Island on May 9, while Ikey and Kohoktok moved their camp to the northwest side of Kanuyak Island,[9] then spent the next day replenishing their meat supply. On May 12, O'Neill, Ikey and Kohoktok moved the camp west to the Goulbourn Peninsula while Cox waited in vain for clear weather to obtain solar readings for latitude and longitude on Kanuyak Island. Alone he completed his survey of Kanuyak Island and then proceeded west across the strait to rejoin O'Neill, Ikey and Kohoktok. O'Neill, meanwhile, had determined that the copper-bearing rocks did not extend farther south than the islands he had examined. Once he had that information, they had at last completed their field work.

Now began the race against time to return the hundreds of miles around the coast to Bernard Harbour, before travelling with dog sleds on the sea-ice became unsafe. Fog, snow and gale winds prevented them from starting until May 16. That day they reached the east side of Banks Peninsula and camped alongside two tents of Inuit, which included one couple who had never seen a white man before. Cox obtained spring-fur boots, five trout and a lashing for his sled in trade from these people, observing that they were "fine people—quiet and pleasant to deal with."[10] The following day O'Neill and Cox moved their camp to Cape Wollaston, where they unexpectedly met Dr. Anderson and Adam Uvoiyuaq.

Dr. Anderson, Chipman and Bathurst Inlet

Dr. Anderson had left Bernard Harbour for Bathurst Inlet by dog team on April 29, hoping to obtain one or two muskoxen specimens near Arctic Sound for the National Museum in Ottawa. Then he planned to join the O'Neill party. In addition to Adam, he was accompanied by the engineer Hoff and Siberia Mike. The four men reached Tree River on May 4, where they spent several days digging out the cached supplies, *umiak*, canoe and launch from the drifted snow. As the hull of the launch was badly cut up from its use in ice the previous September, Dr. Anderson decided to abandon it. Accordingly, he had Hoff and Mike remove its Gray engine, load it on their sled along with the *umiak*'s defunct Evinrude engine, and take them back to Bernard Harbour. O'Neill and Cox would bring back the *umiak* and the canoe after they completed their field work in Bathurst Inlet.

Chipman arrived at Tree River on May 7, after mapping the south coast of Coronation Gulf east from the Rae River, where Cox's coastal mapping had ended. There he encountered Dr. Anderson's party. Chipman was accompanied by Ambrose Agnavigak and his wife and daughter, Annie Fitzgerald, and by D'Arcy Arden, a former government topographer and friend of Chipman. Arden had taken on the life of a hunter and guide, and was living in a cabin at Great Bear Lake. He had just guided two RNWMP officers (Inspector C.D. La Nauze and Constable D.E.F. Wight) and their Inuvialuit interpreter, Ilavinirk, down the Coppermine River from Great Bear Lake, where by sheer chance he had met his friend Chipman just east of the mouth of that river.

After reporting to Dr. Anderson on the results of his surveying from Rae River to Tree River, Chipman asked for and obtained the Dr.'s permission to leave the Expedition once he

---

[9] The Inuktitut word "Kanuyak" refers to the presence of copper in the rocks (Chipman and Cox, 1924, p. 41B; Jenness, 1995, p. 97).

[10] Cox, 1916 (May 16, 1916).

completed his mapping to Cape Barrow. In his eagerness to return to Ottawa and to his fiancée, Marjorie Pennock, Chipman was contemplating backpacking with Arden from the Coppermine River overland to Fort Norman (the same journey that had twice proven too difficult for Dr. Anderson during the winter months) and proceeding from there south to Edmonton and east to Ottawa. He was second-in-command of the Southern Party, so that it was essential that he secure Dr. Anderson's approval on such a change of plans.

Dr. Anderson and Adam sledded east from Tree River on May 9, passing Cape Barrow on May 14 and following the west coast of Bathurst Inlet south. Anxious to locate O'Neill and Cox, they crossed the mouth of Arctic Sound to Cape Wollaston, where they met their two colleagues on May 16. O'Neill and Cox reported that they had just completed the surveying and examination of the more than two hundred islands in Bathurst Inlet. Dr. Anderson then continued a short distance along the coast to a small Inuit settlement, where he obtained the Native names for the local islands and other geographical features around the inlet, specimens of the tomcod they were catching and some general information. He and Adam rejoined the O'Neill-Cox party the next day, and they collectively decided that it was time to start back to Bernard Harbour.

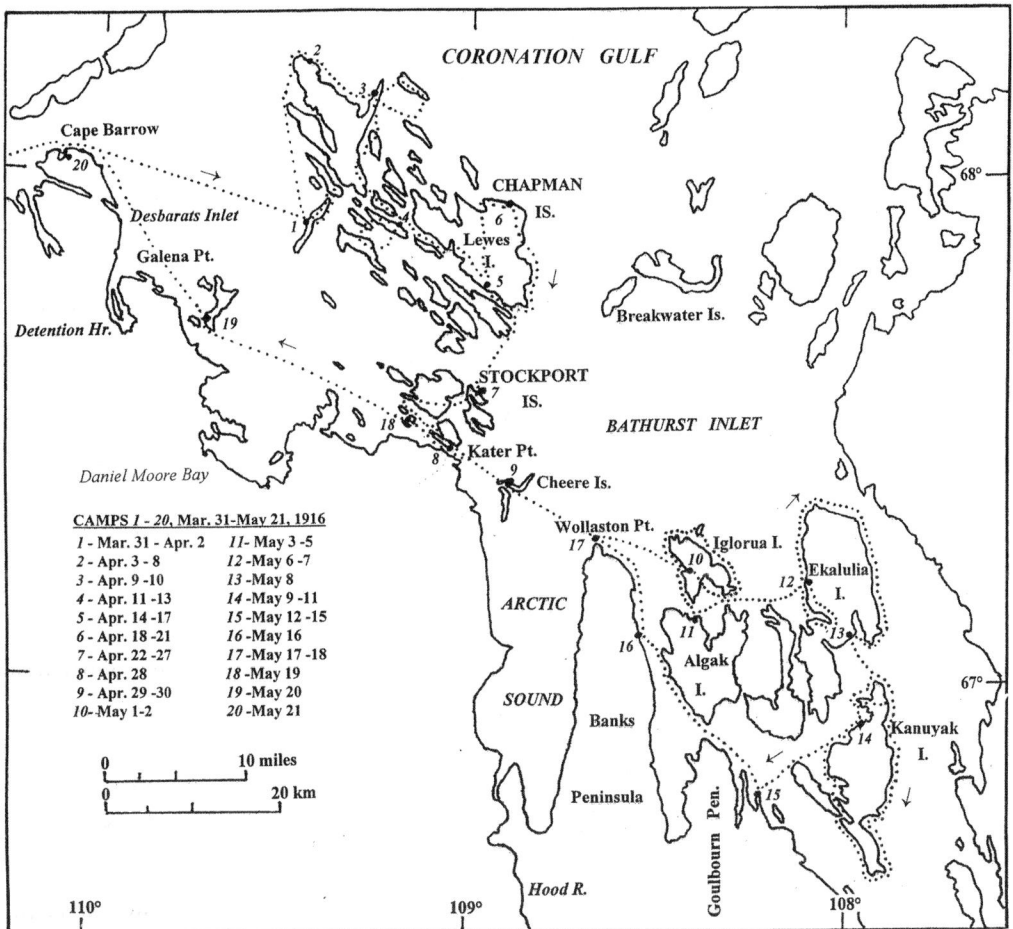

Map 19 Route followed by Cox and O'Neill mapping with sleds and dogs in Bathurst Inlet, March 31 to May 21, 1916.

Both parties started for Cape Barrow on May 19, but Dr. Anderson suddenly decided to turn off and head into Arctic Sound. He hoped to contact some Inuit there who had visited Bernard Harbour during the winter, because he had arranged with them to obtain for him some muskoxen skins as well as other mammal skins. Several miles down the sound Adam shot two bull caribou. While he was skinning the caribou, two Copper Inuit men appeared from the nearby hills. They told Dr. Anderson they had seen neither muskoxen nor Inuit in the area. Greatly disappointed by this news, Dr. Anderson turned about and headed his sled for Kater Point, catching up to O'Neill and Cox at the entrance to Detention Harbour, just south of Cape Barrow. The next day they continued to Cape Barrow, encountering frequent pools of water up to two feet deep along their route. At the cape they found Chipman.

Chipman had now completed the coastal mapping from Darnley Bay east to Stapylton Bay, and from the Rae River east to Cape Barrow. Cox had meanwhile mapped from Stapylton Bay to Rae River, so between them they had successfully completed their assigned task of mapping the more than six hundred miles of Arctic coast between Darnley Bay and Bathurst Inlet. Their field work was finished.

Thereafter the three small parties (those of Dr. Anderson, O'Neill and Chipman) travelled at night, when it was cooler. On May 24 they reached their cache at the mouth of the Tree River. They spent three days there, making arrangements with local Copper Inuit they knew to bring some of their cached supplies from Tree River to Bernard Harbour. Continuing west, they reached the mouth of the Coppermine River on May 31.

A Gathering at the Mouth of the Coppermine River

About one hundred and twenty- five Copper Inuit were gathered at the mouth of the Coppermine River when Dr. Anderson's men arrived on May 31, making preparations for their spring trek inland to the tree-line, where they would hunt caribou for the summer. With them were RNWMP Constable Wight and his interpreter and guide, Ilavinirk. From Constable Wight Dr. Anderson now learned that shortly after the constable's arrival with Inspector La Nauze at Coronation Gulf in the spring, the Inspector had established that the two missing French priests had been murdered in 1913 and had arrested two Copper Inuit men on suspicion of being the murderers. One of the suspects was a man named Sinnisiak from Victoria Island, the other a man named Uloksak Avingak from the Coppermine River region. These two suspects and the Inspector were waiting at Bernard Harbour to go west on the *Alaska* in the summer.

Cox then learned from Constable Wight and D'Arcy Arden that Inspector La Nauze had been obtaining furs and specimens by trading with the Inuit, using Expedition supplies he had "… apparently commandeered from Sweeney. He traded White's [*sic* Wight's] rifle & wanted to trade Illavinik's [*sic*] too & send them back to [Fort] Norman 600 miles without rifles!"[11] Cox was rather troubled by the Inspector's actions.

The shaman Uloksak Meyok, whom Dr. Anderson had employed earlier as a hunter and guide, was also among the Inuit assembled at the mouth of the river. Dr. Anderson now learned that this man had obtained fairly legitimately the supply of white-men's goods he had shown

---

[11] Cox, 1916 (June 1, 1916).

them, having been friendly with the two French priests in the summer of 1911 or 1912 at Lac Rouvière, where they had their cabin (a place the Inuit called Immaernirk). When he learned that the two priests had been killed, this Uloksak and some others had gone straight to the priests' cabin and taken some of the possessions cached there in the belief that it was all right to do so.

Uloksak Meyok was now preparing to lead Constable Wight and Ilavinirk to the place where the two missing priests were thought to have been murdered, "… on the west side of the [Coppermine] river, about at the first trees…,"[12] so that the constable could obtain photographic evidence and look for their bones. Thereafter Constable Wight intended to continue backpacking with Ilavinirk overland to Great Bear Lake, then to proceed to Fort Norman and the police headquarters at Herschel Island. Dr. Anderson was pleased to see Ilavinirk again, for the latter was the Mackenzie River Delta Inuk who had assisted Dr. Anderson for many months on the latter's previous (1908–1912) Arctic expedition.

Figure 49 Ihumatak, a young Copper Inuit girl, at the mouth of the Coppermine River, wearing a long black robe that had belonged to one of the two Oblate priests murdered on the Coppermine River in 1913, June 1, 1916. Photo: © Canadian Museum of Civilization, photo by John R. Cox, 39728

Chipman Heads Overland to Fort Norman

The following day, June 1, Chipman bid his companions a temporary farewell, for he expected to see them all again in Ottawa in a few months. Then with Arden, Uloksak Meyok and his three wives, and other Copper Inuit, as well as several pack dogs, he started the long trek over the snow-covered ground to Great Bear Lake. Describing the scene in his diary that day, Chipman wrote "All the crowd for up the Coppermine pull out toward evening. Looked like an army on the move. Our outfit was the last to pull out leaving at 11 p.m."[13] When they reached Great Bear Lake he and Arden proceeded across the still-frozen lake and down the Great Bear River to the Mackenzie River and Fort Norman. From Fort Norman, Chipman travelled upriver by steamer early in the summer to Fort McMurray and thence by train to Edmonton and Ottawa. He reached Ottawa about the end of August, just a few weeks before Dr. Anderson and the other men from Bernard Harbour arrived from Nome and Seattle.

---

[12] Anderson, 1916a (May 31, 1916).
[13] Chipman, 1916 (June 1, 1916).

Figure 50 Group of Copper Inuit packing inland from the mouth of the Coppermine River towards Great Bear Lake, June 2, 1916.
Photo: © Canadian Museum of Civilization, photo by Kenneth G. Chipman, 43343

Dr. Anderson and His Men Return to Bernard Harbour

Within an hour of Chipman's departure, Dr. Anderson's party was heading around the southwest coast of Coronation Gulf en route to Bernard Harbour, judging it to be too late in the season to cross on the ice in the gulf safely. They camped three days later at Locker Point, and on June 6 they finally reached the base camp.

Johansen's Exploration Trip

Meanwhile at Bernard Harbour, the naturalist Johansen had grown increasingly restless as the winter progressed. While all the other scientists had travelled hundreds of miles from the camp, exploring, hunting and living new experiences, he had spent nearly two years at Bernard Harbour, and had not been able to explore any terrain except that which was within easy walking distance of the base. He had sought to accompany Chipman and Dr. Anderson early in March on their trip up the Croker River, but they had preferred not to have him along. As an alternative scheme, Dr. Anderson suggested Johansen make an exploratory sled trip of his own to Victoria Island.

Accordingly, on March 6, the same day that Dr. Anderson, Chipman and Siberia Mike headed west for the Croker River, Johansen and Adam Ovoiyuaq set forth to explore part of the south coast of Victoria Island as far east as the Mackenzie River, stating they would return about April 1. They crossed the strait by way of the Liston and Sutton Islands, then visited Lady Franklin Point and the Miles Islands, and continued east past the Richardson Islands to Cape Murray. Here and there along the way Johansen collected botanical, zoological and geological specimens.

His failure to return by April 8 was of sufficient concern to Jenness that he thought it advisable to go in search of Johansen. He knew from first-hand experience that Johansen was not a good traveller and that he had set out with a poor team of dogs. Accordingly, Jenness and Patsy Klengenberg departed on April 10 for the Liston and Sutton Islands, then turned east to Victoria Island, where they picked up Johansen's trail going north and followed it to the point at the mouth of Forsyth Bay. There they found a small stone cairn with a note in it from Johansen, written two days earlier, stating that he was heading for Bernard Harbour. It was one of several notes Johansen later said he had left along his route, evidently patterning his actions after those of early explorers whom he sought to emulate.

Johansen greeted Jenness and Patsy Klengenberg on their return to Bernard Harbour on April 13. He had returned to the base camp the day after Jenness and Patsy left and did not understand why anyone had been worried about him. He felt that his trip had been reasonably successful. Among the items he had collected were some corals from the rocks on one of the Liston and Sutton Islands. These proved to be of Silurian age (about 400 million to 435 million years old), which provided an indication of the geological age for the dolomitic limestones that underlay miles of coast and interior in that region and had previously been thought to be non-fossiliferous and therefore impossible to date. Determining the ages of rocks is fundamental to arranging their relationships to surrounding rocks, which in turn helps provide guidance to the occurrence of economic mineral deposits. Thus Johansen's seemingly

Figure 51 Copper Inuit spearing salmon in the fishing creek four miles southeast of the CAE base at Bernard Harbour, June 29, 1916.

Photo: © Canadian Museum of Civilization, photo by George Hubert Wilkins, 51185

inconsequential find of some fossils in an exposure of rocks may have acquired much more practical value in the future development of the region than anything else he collected or accomplished on his short journey.

RNWMP and Others at Bernard Harbour

Several newcomers were at Bernard Harbour when Dr. Anderson, O'Neill, Cox and their Inuit companions returned on June 6. Wilkins and Palaiyak had arrived from the Northern Party. Wilkins had received a rather chilly reception when he appeared because of his actions of the previous summer. This time, however, he brought no new demands or problems, being merely the conveyor of news of Stefansson's latest achievements and plans. Wilkins explained that he wanted to leave the Arctic with the Southern Party and intended to devote his remaining free time to photographing Expedition personnel and local Inuit at the camp, including the two Copper Inuit prisoners.

Another newcomer was RNWMP Inspector C.D. La Nauze, who had been sent the previous summer on the "Great Bear Lake patrol" to investigate the disappearance of two French Oblate priests, missing since 1913. He had been accompanied by the Inuit interpreter, Ilavinirk. After wintering at Fort Confidence on Great Bear Lake, the Inspector had been guided

Figure 52 Ikpukhuak and Higilak in their finest dancing clothes, at Bernard Harbour, July 11, 1916.
Photo: © Canadian Museum of Civilization, photo by G.H. Wilkins for D. Jenness, 36913

overland to Coronation Gulf by D'Arcy Arden. Within a month he had arrested on suspicion of murder two Copper Inuit men—Sinnisiak and Uloksak Avingak. By the time Dr. Anderson returned to Bernard Harbour from Bathurst Inlet, the two prisoners were quietly carrying out tasks around the camp, such as fetching water and gathering firewood, while everyone awaited the start of the navigation season and the departure west of the Southern Party. Somehow they were all accommodated at the crowded base camp.

Corporal Bruce had carried out several police investigations since his arrival at Bernard Harbour from Herschel Island on the *Alaska* the previous September. He had compiled quite a lot of information about the two missing French priests, both from the scientists and from various Copper Inuit, and had recovered an assortment of personal objects that had belonged to the missing priests.

During the preceding weeks Cox and O'Neill also had recovered items that were of interest to the police. These were a few pages from notebooks and one or two other items they had obtained from some Inuit in Bathurst Inlet, items which had belonged to two young adventurers, American H.V. Radford and Canadian G.R. Street. These two men in 1911 headed for Bathurst Inlet from Hudson Bay, intending to continue west to Fort McPherson, but were killed in 1912 by the Inuit in Bathurst Inlet. Their deaths were subsequently reported to the RNWMP, who sent Inspector F.H. French on an investigative patrol[14] about the same time that Inspector La Nauze, Corporal Bruce and Constable Wight were investigating the case of the two missing priests.[15] The Radford and Street deaths were ultimately interpreted as the result of an unfortunate argument between them and the Inuit they encountered in Bathurst Inlet, and no charges were filed.

The Fates of the Two Copper Inuit Prisoners

Sinnisiak and Uloksak Avingak were subsequently taken on the *Alaska* to Herschel Island and then escorted south by Inspector La Nauze to Edmonton, where in 1917 they were tried for murder. They were found not guilty. The federal government prosecutors were dissatisfied with that judgment, however, for the Canadian government wanted to make an example of the two men to demonstrate to the Inuit across the Canadian Arctic that the white man's laws were the laws of the land. A follow-up trial was therefore held a few weeks later at Calgary. On this occasion the two were found guilty of murdering the two priests and were sentenced to death. The trials aroused nation-wide interest at the time, and the sentences of the two Inuit were subsequently reduced to life imprisonment in the North, in part as a result of the intervention of Right Reverend Gabriel Breynat, the Roman Catholic Bishop of the Mackenzie vicarage.

After two years of detainment at Fort Resolution on Great Slave Lake, both men were permitted to return to their people in the summer of 1919. Both subsequently worked with the police as helpers and guides for a time thereafter, having learned a little English during their imprisonment.[16]

---

[14] French, 1919.

[15] La Nauze, 1917.

[16] Moyles, 1979; Jenkins, 2005.

Confusion over the identity of the prisoner Uloksak arose in a report of one of the trials and persisted in many subsequent accounts of the murders and trials for the next three-quarters of a century. Simply explained, there were two Copper Inuit with the name Uloksak (or a variant of that spelling) who were associated with the Expedition at Bernard Harbour. The man accused of murder was credited with being a good hunter, a shaman and a polygamist, all of which was incorrect. These were, in fact, the attributes of the other Uloksak, the man Jenness referred to as "the rich man." The man found guilty of murder was Uloksak Avingak, whereas the polygamous shaman, hunter and "rich man" was Uloksak Meyok.[17] To avoid perpetrating that former confusion, I have deliberately included the second names of both men.

Cache Left at Bernard Harbour

Before its departure from Bernard Harbour, the Southern Party cached a large quantity of unused supplies there, which were intended for the use of the Northern Party should they have need of it. The supplies included 2,100 pounds of rolled oats (much of which Stefansson had ordered in 1915), 500 pounds of pemmican, 240 pounds of lard and 200 pounds of sugar. These supplies were left in the temporary custody of Jenness' friends Ikpukhuak and his wife, Higilak. The Expedition's canoe and paddles were given to Ikpukhuak in appreciation for his reliable service. Later that summer Rev. Girling and his two companions (W.H.B. (Billy) Hoare and G. Eldon Merritt) arrived to occupy the Expedition's building and to establish their Church of England mission there, and they took charge of the supplies. As no member of the

Figure 53 Jennie Kannayuk and Kila Arnauyuk in their finest attire, at Bernard Harbour. Partly loaded schooner *Alaska* on left, almost ready to leave for Nome, July 11, 1916.

Photo: © Canadian Museum of Civilization, photo by Dr. R.M. Anderson, 38997

[17] S.E. Jenness, 1992.

Figure 54 Southern Party preparing to leave Bernard Harbour on CAE schooner *Alaska* for Nome, Alaska, July 13, 1916.
Photo: © Canadian Museum of Civilization, photo by George Hubert Wilkins, 51269

Northern Party ever came near Bernard Harbour thereafter (with the possible exception of Harold Noice between 1917 and 1921), the supplies were quickly put to good use, in due course, by Rev. Girling and his colleagues.

The Southern Party Heads for Nome and Ottawa

On July 13, 1916, the sixty-five-foot schooner *Alaska,* fully loaded with twenty-six people, twenty-two dogs, equipment and supplies, steamed out of Bernard Harbour for the last time, captained by Daniel Sweeney, and headed west for Nome. At Pearce Point Dr. Anderson agreed to take on board the family and dogs of Captain Charles Klengenberg, taking them west to the Baillie Islands and towing their small boat. The Klengenberg family, including Patsy Klengenberg and Ikey Bolt with them, disembarked at the Baillie Islands. Ikey had decided in that short time that he wanted to marry Klengenberg's oldest unmarried daughter, Etna (or Edna). Another young Inuvialuk man, Mungalina, also left the Expedition while the *Alaska* was at the Baillie Islands.

The Klengenberg family moved to the mouth of the Coppermine River a few weeks later, established a trading post at the tree-line on the Coppermine River, and within a few years became widely known as traders in the Coronation Gulf region. Their success was in no small part due to the training in English received by both Patsy Klengenberg and Ikey Bolt from the scientists on the Southern Party while at Bernard Harbour.

The *Alaska* reached Herschel Island on July 28, and Inspector La Nauze and Corporal Bruce departed from the ship with their two Inuit prisoners, as did Ambrose Agnavigak and his wife, Unalina, and daughter Annie, Silas Palaiyak and his brother Adam Ovoiyuaq[18], and Siberia Mike and his wife and son, Georgie Mike.

On July 31 Constable Alexander Lamont turned over to Dr. Anderson an envelope containing $240 in $5 bills from the Department of the Naval Service in Ottawa. Dr. Anderson promptly questioned the constable about the amount, for he had requested $2,500 from Ottawa in August 1914. He was told that in August of 1915 Stefansson had opened the envelope and removed all of the money except the $240, then returned the envelope to the police without comment. These funds were for Dr. Anderson to pay the wages of his Inuit employees when the Southern Party left the Arctic in 1916. Lacking the funds he needed now to pay them, he was forced to supply his departing Inuit assistants with residual provisions and bank drafts written on the Department of the Naval Service. Dr. Anderson telegraphed for additional funds when he reached Nome in order that his scientific colleagues and the *Alaska's* remaining

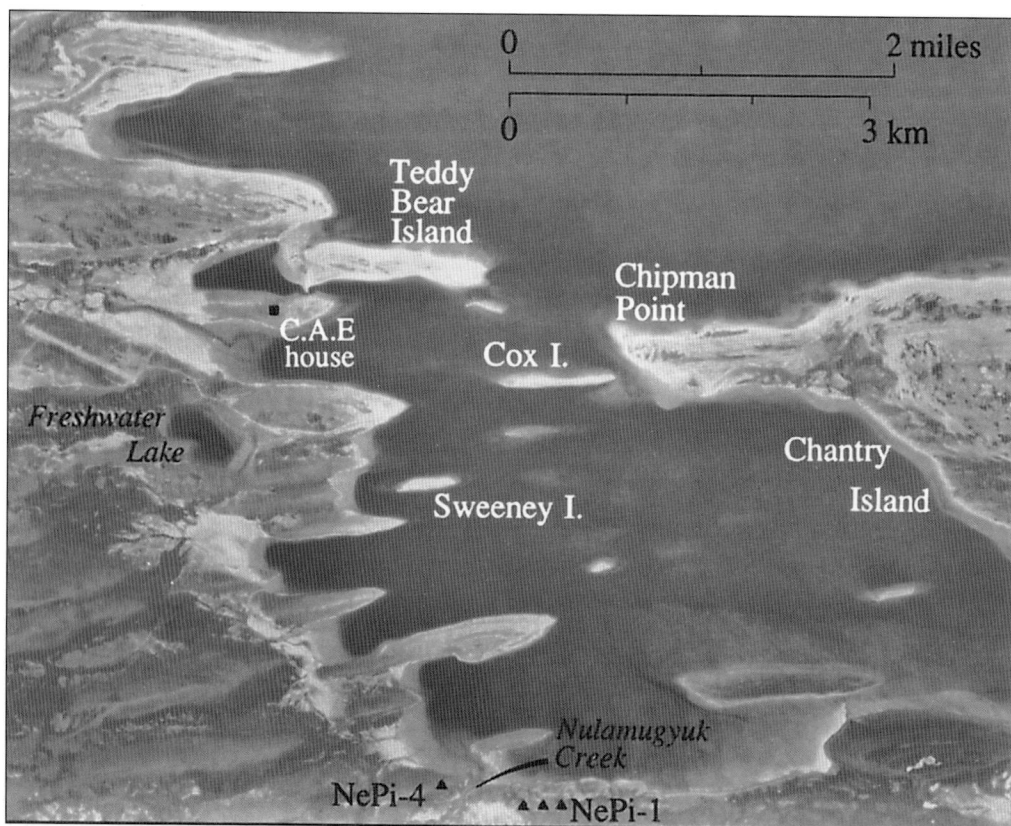

Figure 55 Aerial view of Bernard Harbour showing the site of the CAE house (■), the small lake from which fresh water was obtained, and the location of two Dorset culture tent-ring sites (▲) straddling the mouth of the fishing creek. Teddy Bear Island almost blocks the entrance to the inner harbour where the Expedition ships wintered.
RCAF Airphoto No. A19211-42 with names added.

---

[18] "Adam promptly contacted pneumonia and died a week later on August 17, 1916 (Bruce, 1917). A photograph of his grave marker on Herschel Island appears in Bockstoce (2000, p. 37), showing his name spelled Ovayoak."

Figure 56 Captain Charles Klengenberg's family at Baillie Islands, standing in front of their scow-schooner, *Laura Waugh*, which the Captain called the *Homely Hippopotamus* and which took him with his family to the Coppermine River a month later. (l. to r.): Etna, Jorgen, Kenmek, Bob, Patsy (with cap), Andrew, Charles and Lena. July 26, 1916.

Photo: Reproduced with the permission of Natural Resources Canada, 2011, courtesy of the Geological Survey of Canada, J.R. Cox, 39692

Figure 57 Siberia Mike (in plaid overalls), Ikey Bolt (with hat and jacket) and Mungalina in white parka, CAE Southern Party Inuit members from Siberia, Alaska, and the Mackenzie River respectively, at Baillie Islands, July 26, 1916.

Photo: © Canadian Museum of Civilization, photo by George Hubert Wilkins, 51321

Figure 58 Cox taking final solar readings at the Canada-Alaska boundary monument at the north coast, August 4, 1916. Photo: © Canadian Museum of Civilization, photo by Dr. R.M. Anderson, 39337

crew members could have funds to get home to Ottawa. He then reported the "disappearance" of his money shortly after he reached Ottawa.[19]

While at Herschel Island Dr. Anderson's men stored various residual Southern Party provisions pending their shipment to Cape Kellett by the first available ship for Stefansson's Northern Party. On August 3, the *Alaska* continued west, stopping briefly at Kamarkak, an Inuit settlement thirty miles west of Herschel Island, to let off Captain Sweeney's Iñupiaq wife, Eunice (Añayu), and her infant boy, Daniel Jr. The latter had been born on the *Alaska* on July 16 during its voyage from Bernard Harbour to Herschel Island.[20] Mother and child wished to return to her relatives, who lived nearby.

A little while later the *Alaska* stopped at the U.S.–Canada boundary to give Cox the opportunity to obtain solar readings for longitude and latitude and to correct the Southern Party's chronometers at the boundary monument erected there. Once Cox obtained his readings the *Alaska* continued west, reaching Barrow on August 8, after minor difficulty with ice. Dr. Anderson left most of the Southern Party's dogs, the *umiak* and the sleds at Barrow, thereby reducing the clutter on the deck of the heavily loaded vessel in anticipation of a rough voyage to Nome.

The weary men finally reached Nome on August 14, 1916, after being forced to take shelter from a gale near Cape Prince of Wales for a few hours on August 11. They unloaded the *Alaska*, then had it dragged up on the beach, from where it was later sold. The men meanwhile got cleaned up and changed to city clothes,[21] and on August 29 left Nome and

---

[19] Anderson, 1916b (November 10, 1916).

[20] Anderson, 1916a (July 16, 1916). The child died a few months later (Anderson, 1917 [May 1, 1917]).

[21] Jenness submitted a receipt from a store in Nome for $99.75 for a suit of clothes and a suitcase to replace the items he had lost on the *Karluk* (RG42, Vol. 468, File 84-2-5, Subfile 9, LAC). Judging from the fit of his suit-coat, as seen in Fig. 60, the store did not have a suit small enough for his frame.

Figure 59 CAE schooner *Alaska* for sale on beach at Nome, Alaska, August 27, 1916.
Photo: © Canadian Museum of Civilization, photo by Dr. R.M. Anderson, 39729

the Arctic for Seattle on the coastal steamer *Northwestern*. Johansen, ever the odd fellow, disembarked at Prince Rupert, and took the train to Ottawa from there. The others continued on to Seattle, where Mrs. Anderson and assorted members of the press greeted them upon their arrival on September 11. They were interviewed and feasted in Seattle, then departed for Ottawa, which they reached later in September, to report on their three years of northern activities, adjust to city living again and settle down to writing detailed accounts of their work.

Dr. Anderson stayed briefly in Seattle to arrange for the shipment to Ottawa (by way of Vancouver) of all the specimens and equipment brought south by the Expedition. He and his wife then proceeded to Victoria, where he finished up other Expedition matters. With that completed, he and his wife enjoyed a brief but much-deserved holiday before they finally headed for Ottawa.

His arrival in Ottawa brought to a close the successful field activities of the Southern Party Dr. Anderson had managed over the past three years. Two more years would pass before Stefansson brought his Northern Party's field activities to a close. Then followed several years of administrative problems and aggravating delays that accompanied the efforts of both the Deputy Minister of the Department of the Naval Service and the Director of the Geological Survey of Canada to get the official narratives of both parties published as well as the many scientific reports resulting from the scientists' field activities. Dr. Anderson played a major role in expediting the publication of most of those reports.

Figure 60 Five members of the Southern Party and unidentified friends on the coastal steamer *Northwestern*, bound from Nome to Seattle, August 1916. (Back row, centre): Wilkins, O'Neill and Cox; (front row): Jenness and Dr. Anderson. Photo: Courtesy of S.E. Jenness, photographer unknown

Figure 61 Two Copper Inuit prisoners, RNWMP, interpreters, witness and judicial officials at murder trial in Edmonton, August 1917. Front row (l. to r.): Ilavinirk (interpreter), Koeha (witness), Uloksak Avingak (prisoner), Sinnisiak (prisoner), Patsy Klengenberg (interpreter). Back row (l. to r.): C.C. McCaul (prosecuting attorney), RNWMP Inspector C.D. La Nauze (arresting officer), James Wallbridge (defence attorney) and RNWMP Constable D.E.F. Wight (assistant to Inspector La Nauze). Photo: Library and Archives Canada, e010859171

PART 5
AFTERMATH

PUBLICATIONS AND OTHER POST-EXPEDITION MATTERS[1] 22

The Canadian government wisely recognized the need to make the scientific findings of its Expedition available to the public within a reasonable period of time. Accordingly, shortly after Dr. Anderson and the members of his Southern Party reached Ottawa, the Deputy Ministers of the two departments mainly responsible for the Expedition—G.J. Desbarats of the Department of the Naval Service and Dr. R.G. McConnell of the Department of Mines—agreed in principle to the publication of a series of scientific reports resulting from the findings of the Expedition. They then established an interdepartmental committee to formulate the basic policies for the publication program and to oversee the publishing activities that would follow.

Committees

The initial interdepartmental committee was given the name "Canadian Arctic Expedition Committee 1913–1916." Its tasks were to establish the policies to be followed in publishing the reports on the Expedition findings, and to decide how many reports would be published, the size and type of paper to be used in the printing, the number of copies of each report to be printed, and related printing matters.

Appointed to chair this committee was Dr. Edward E. Prince, the Dominion Commissioner of Fisheries at the time and an employee of the Department of the Naval Service. The two Deputy Ministers then asked four other government scientists to assist Dr. Prince: Dr. Charles Gordon Hewitt, the Dominion Entomologist in the Department of Agriculture; Dr. A.B. Macallum, the Chairman of the Commission for Scientific and Industrial Research; Mr. James M. Macoun, the Chief of the Biological Division at the Victoria Memorial Museum from 1913–1920; and Dr. R.M. Anderson, zoologist, who was the recently returned leader of the Expedition's Southern Party and understandably the most knowledgeable among them about the Expedition. Each committee member was to be responsible for editing reports in his own section, while Dr. Anderson was appointed general editor. The committee members subsequently held a series of meetings, generally in Dr. Anderson's office at the Victoria Memorial Museum, and early in 1918 submitted to the two Deputy Ministers a plan recommending the following actions:

- the publication of a series of about forty-eight biological reports on the various specimens collected by the scientific members of the Expedition;

---

[1] The information in Chapter 22 has been gathered from many letters written by Dr. Anderson between 1919 and 1926, copies of which are included in CMNAC/1996-077, Series A—R.M. Anderson Collection, Archives, CMN. Only the more significant of these are cited.

Figure 62 Victoria Memorial Museum Building, home of the Geological Survey of Canada and National Museum of Canada, on Metcalfe Street in Ottawa, in 1927. Dr. Anderson's office was on the fourth floor in the east wing (left of the centre of the building), Jenness' office was on the third floor (window hidden behind the telephone pole).
Photo: © Canadian Museum of Civilization, 69308

- the printing of between 3,000 and 4,000 copies of each report as "separates," 250 of which should go to the author or authors who prepared the report in appreciation for his or their endeavours;

- the printing of an additional 1,000 copies of each report on superior paper for assembly later into approximately ten bound volumes, each of which would carry the title "Scientific Results of the Canadian Arctic Expedition 1913–1916" and be available for distribution to libraries and scientific institutions.[2]

As events unfolded, the ten volumes initially planned covered almost exclusively the biological materials collected by Dr. Anderson and Johansen of the Southern Party. No plans were put forward at that early stage for reports on the geographical, geological and anthropological work of Chipman, Cox, O'Neill and Jenness. Both Cox and Jenness had joined the Canadian Army and gone oversees early in the summer of 1917 to help fight in World War I, so that consideration of what reports they might prepare was set aside until their return.

The two Deputy Ministers also established a second interdepartmental committee—the "Arctic Biological Committee." It consisted of the same five individuals as the policy-making

---

[2] Canadian Arctic Expedition Committee, 1913–1916, 1918.

committee, but its functions differed. It was the action committee, established to assume responsibility for the biological *specimens* (insects, plants, birds, mammals, fish—both marine and non-marine—and any other zoological forms) collected by the members of the Expedition. More specifically, this committee's tasks were:

- to select the specialists in universities in Canada, the United States or Great Britain to study and report on each of the various types of specimens;

- to supervise generally the work of those specialists and edit the resulting reports;

- to recommend publication of each forthcoming report, once it was considered acceptable, to the appropriate government department (Naval Service or Mines);

- to recommend means of distribution of surplus biological specimens not needed by either government department; and

- to recommend to whom complimentary copies of the printed reports ought to be sent.

Dr. Anderson was notified of his appointment to this second committee in January 1917, barely three months after his return from the Arctic. Judging by the large mass of correspondence that he wrote and received over the next decade in connection with this program, he was an extremely active committee member.

Disagreements quickly arose between members of the two government departments over publication details, as they had on other aspects of the Expedition from its inception. These were mainly between members of the publishing committee, all of whom were scientists, and G.J. Desbarats, the Deputy Minister of the Naval Service, who was an engineer and administrator. A staunch supporter of Stefansson's plans and subsequent actions, initially through admiration, later through face-saving necessity, Desbarats fervently believed that his department's assumption of so much of the Expedition's costs justified him having a major decision-making role regarding certain publication details. It helped his cause, too, that the new Deputy Minister of Mines, Dr. R.G. McConnell, showed little interest in the Expedition's publication matters.[3]

CAE Reports

The first few soft-covered Expedition reports were published in July 1919, and provided details on certain types of insects collected on the Expedition. Some months earlier—and following at least one visit to Ottawa by Stefansson—Desbarats informed the Arctic Biological Committee that he wanted three changes made to the cover text they had proposed for all reports. The committee unanimously rejected all three changes, giving reasons for their objections. Desbarats

---

[3] Geologist Dr. R.G. McConnell was Deputy Minister of Mines from 1914 to 1920, replacing Dr. R.W. Brock following the latter's sudden departure in the summer of 1914 to become the first Dean of Applied Science at the fledgling University of British Columbia. Brock had been keenly interested in and strongly supportive of the Expedition.

yielded to their objections on two, but overruled them on the third. Thus came about the changing of the dates in the main heading of each report from 1913–1916 to 1913–1918[4] to include the two extra years Stefansson remained in the Arctic. The committee promptly appealed this action to McConnell, the Deputy Minister of Mines, but he refused to participate in their disagreement with Desbarats, his counterpart in the Department of the Naval Service. In consequence, with the exception of the first two or three reports published, all reports thereafter appeared with the five-year date spread on their title pages. However, Desbarats allowed some of the reports to carry the sub-heading, "Southern Party 1913–16" on their title pages when they clearly contained the work of Southern Party members only. Ironically, the entire series of publications dealt almost exclusively with the activities and collections of the Southern Party, which left the Arctic in 1916.

The Department of the Naval Service agreed to process and pay for the publication of six of the first ten volumes—Volume 1 and Volumes 6 to 10—while the Department of Mines (on behalf of the Geological Survey and the National Museum) was to publish the remaining four—Volumes 2 to 5. Most of these ten volumes were expected to contain two or more reports.

Sometime in 1919, after Cox and Jenness had returned from the war in Europe, the publications committee approved plans for the publication of five additional volumes in the Expedition's series. Volume 11 would contain a Part A by O'Neill on the geology of the region examined by the Southern Party, and a Part B by Chipman and Cox on the geography of that region. The other four volumes (12 to 15) would be prepared almost exclusively by Jenness[5] and deal with assorted anthropological aspects on the Copper Inuit, including their songs, folklore, language, string-figure games, physical features and culture.

In that same year, 1919, the name of the operations committee was changed from "Arctic Biological Committee" to "CAE Editorial Committee" to reflect the expansion of the series from ten to fifteen volumes and their topics to include geological, geographical and anthropological subject matter. Dr. Anderson was appointed general editor, with the heavy responsibility of overseeing the processing and publication of all reports in the series. It was a monumental and thankless task, involving hundreds of hours of his time, and left him little time for anything else. He thus had no time to write the narrative account of his Southern Party (Volume 1, Part B) in spite of repeated urging from the Deputy Minister of the Naval Service, or to study, describe and discuss the numerous mammal and bird specimens he had collected, which were to form Volume 2, Parts A and B, respectively.

Jenness had done some organizational work with his Arctic specimens and notes before leaving to serve with the Canadian armed forces in Europe in 1917, and renewed his study of them in the fall of 1919. With diligent effort he completed and published Volume 12A in 1922 (his classic account of the life of the Copper Inuit), Volume 12B in 1923 (the physical characteristics of the Copper Inuit), Volume 13A (Inuit myths from Alaska, the Mackenzie Delta and Coronation Gulf) and 13B (Inuit string figures) in 1924, Volume 14 (Songs of the

---

4 Desbarats, 1919 (February 26, 1919).

5 John Cameron wrote Part C of Volume 12 (Cameron and Ritchie, 1923) and Helen H. Roberts of New York transcribed the music to paper from the original wax recording cylinders for D. Jenness' Volume 14 (Roberts and Jenness, 1925).

Copper Inuit) in 1925 and Volume 15A (Inuit language) in 1926. Early in 1926 he was appointed Chief of Anthropology, and his new administrative duties required most of his attention thereafter. However, sometime in the 1930s he completed Volume 15B (which provided grammatical support for the Eskimo vocabulary in Volume 15A). Somehow funds were found and this report was published in 1944. Ultimately he published an additional and final volume in the series (Volume 16) in 1946, long after the interdepartmental committees had been disbanded. It described and discussed the Copper Inuit material he had brought south (clothing, household items, tools, weapons, kayaks and sundry other items), clearly information that deserved publication. The Department of Mines (which became the Department of Mines and Resources in the mid-1930s, including the Geological Survey) processed and financed the last three reports that were published.

Following the defeat in December 1921 of Arthur Meighen's Conservative government by William Lyon Mackenzie King's Liberal party, the Department of the Naval Service was suddenly merged with the Department of Militia and Defence and the Air Force Board in the summer of 1922, to form a new Department of National Defence. Desbarats became the new department's first Deputy Minister, bringing his strong administrative influence on the Expedition's publication program to a speedy end. Fortunately, Frank McVeigh, the Deputy Minister in the newly created Department of Marine and Fisheries, which assumed charge of many of the responsibilities of the former Naval Service Department, including the affairs of the Canadian Arctic Expedition, was sympathetic for a few years to the former department's publishing commitments. In 1925, however, McVeigh sent word to the editorial committee that funds for publishing the Expedition's reports would cease on March 31, 1926, when the 1925–1926 fiscal year terminated. It is certainly to the credit of Dr. Anderson, the interdepartmental publications committees and all the scientists involved, that so many of the reports were prepared and published within the seven-year period, 1919–1926. When finally completed in 1946, the Expedition series consisted of sixty-four individual scientific reports bound in thirteen volumes (Volumes 3 to 5 and 7 to 16). Volumes 1 and 2 were never written.[6] Volume 6 was written, but not published.

The one thousand copies of each volume that were specially bound had a nominal price of one dollar. Individual separates were initially priced at ten cents apiece for the shorter ones, fifty cents for the longer ones (which were mostly anthropological items). The Department of Mines had mailing lists numbering in the several hundreds of institutions and specialists to which separates were shipped when printed.[7]

Only one report of the sixty-four published in the Expedition's series attained considerable popularity—Jenness' *Life of the Copper Eskimos* (Volume 12A), but even its popularity was nowhere near that attained by Stefansson's book, *The Friendly Arctic*, published a few months earlier. Most of the remaining sixty-three reports were too specialized to receive any widespread acclaim or interest, but they added valuable original information to the scientific literature about the Arctic. And although copies of each report were mailed to dozens of scientists and libraries around the world, many copies of some reports slowly gathered dust in museum storage and were finally disposed of as the twentieth century came to a close.

---

[6] A complete list of all of the Expedition reports is given in Appendix 4.
[7] Anderson, 1924b (July 29, 1924).

Why no Volumes 1 and 2?

Volumes 1 and 2 are conspicuously absent from the Expedition's report series. Each of these two volumes was to consist of two parts, but the four parts were never written. Odd explanations lie behind their omission.

Volume 1 was expected to be the star publication within the entire series. Plans called for it to contain as its Part A a general introduction that set forth the purposes of the entire Expedition and the preparations made to get it safely launched, followed by the narrative of the Northern Party, all written by Stefansson. Part B was to be the narrative of the Southern Party, written by Dr. Anderson.

Deputy Minister Desbarats took upon himself the responsibility of coaxing the needed manuscripts from each author. Unfortunately Stefansson was busy enriching his reputation and his pockets at the time by publishing and lecturing across the U.S. and Canada about the "friendly" Arctic. Additionally, as Dr. Anderson commented, he "… was not in the habit of writing scientific reports."[8] The obvious reasons, of course, were that such reports were very time-consuming to prepare and added nothing to his financial situation. Furthermore, by early 1921 he was completing his popular narrative of the Expedition, *The Friendly Arctic*, and planning his next Arctic project, and so was always on the move and difficult to contact.

Like Stefansson, Dr. Anderson was also busy, but remained close to his office coordinating and editing the dozens of Expedition scientific reports and seeing them through to publication. As Dr. Anderson was an employee of the Department of Mines and not of the Department of the Naval Service, however, Desbarats lacked the authority to demand his immediate action on Part B of the first volume. Nevertheless, Desbarats did insist, undoubtedly with pressure from Stefansson, that Dr. Anderson prepare a report of the work of his Southern Party for the Annual Report of the Department of the Naval Service, and this 37-page account was published in 1918 in that obscure publication in sufficient time for Stefansson's use.[9]

In November 1921 Stefansson's popular account of the Canadian Arctic Expedition, his 784-page book entitled *The Friendly Arctic*, reached the bookstores and rapidly rose to the top of the *New York Times* list of best-selling books in the U.S. In this book he took many departures from his general narrative to discuss a wide assortment of topics unique to the Arctic or his interests, such as seal- or caribou-hunting techniques, survival procedures, snowhouse construction and native folklore, all aimed at convincing his readers that life in the Arctic was neither harsh nor dangerous when one used one's wits. However, informed readers such as Dr. Anderson and *Karluk*-survivor McKinlay, both of whom were also experienced in Arctic conditions, had no difficulty in finding many errors of fact in Stefansson's book and disagreed virtually one hundred per cent with both the book's title and its central thesis, that white men could "live off the country" almost anywhere in the Arctic.

In his book, Stefansson covered in detail most of the activities of the Northern Party, provided an account of the *Karluk* tragedy based upon the recollections of John Hadley, one of its survivors, and added a summary of the work done by the Southern Party, based almost entirely on the 1918 report Dr. Anderson had prepared for the Annual Report of the

---

[8]   Levere, 1993, p. 418.
[9]   Anderson, 1918.

Department of the Naval Service. The contents of his book make it obvious why he had no interest in writing Part A of Volume 1 in the Expedition series. His popular book already provided a detailed narrative of the Expedition—what else could he write about? Additionally, his book was so full of "lies," according to both Dr. Anderson and McKinlay, that in order to write an accurate scholarly version for the government he would have to produce quite a different account. In his inimitable manner, however, Stefansson avoided telling the chairman of the Arctic Biological Committee, Dr. Prince (when the latter sought some idea of Stefansson's progress on Volume 1, Part A in 1919), that he would not be writing the report for the government on his Northern Party. Instead he replied that his schedule was much too busy at the moment, but he hoped to be able to undertake the writing sometime in the near future. And of course that "future" never came.

Behind these two incidents lies a story about "gamesmanship." "Explorers playing the Waiting Game," read a headline in the *Ottawa Journal* of April 24, 1923, with the subheading "Stefansson and Anderson Each Seek Last Word."[10] The article refers to the delays in the publication of their respective reports owing to differences that arose between the two men while in the North, and hints that if their differences could not be settled soon, the series of government reports might be completed without their contributions. However, Stefansson and Dr. Anderson each had his own reason for delaying his respective part of Volume 1, reasons as different as the personalities of the two men. Stefansson wanted to see what Dr. Anderson wrote so he could rebut any statements in it with which he disagreed. Indeed, presumably with that in mind he had urged Deputy Minister Desbarats many months earlier to press Dr. Anderson to complete his narrative of the Southern Party. Dr. Anderson was wary of Stefansson's ways by then, having been "stung" by the latter around the time the 1913–1916 Expedition got under way.

In 1912, Dr. Anderson had sent Stefansson an account of his 1908–1912 activities in the Arctic, along with his Arctic photographs, on the understanding that the two men would share equal authorship and royalties in a book about their joint "Stefansson-Anderson Arctic Expedition 1908–1912." Instead, Stefansson incorporated Dr. Anderson's account and the photographs the latter had sent him into *My Life with the Eskimo*, a book published under his own name. He then assigned limited acknowledgment to Dr. Anderson's contribution and only one-tenth of the royalties, after pocketing a $2,000 advance.[11]

Having been treated in this manner once by Stefansson, Dr. Anderson had no intention of allowing him to pull the same sort of trick on him a second time. Thus he was not willing to write his Part B of Volume 1 of the Canadian Arctic Expedition Reports Series before Stefansson wrote his Part A, being certain that his manuscript would be promptly passed along to Stefansson by Desbarats after it was received, resulting in key subjects being usurped and discussed by Stefansson. Dr. Anderson said nothing of this to Desbarats, of course, but mentioned it later in letters to friends.

---

[10] Anonymous, 1923a (*Ottawa Journal*, April 24, 1923).

[11] A revised and abridged edition of Stefansson's 1913 book *My Life with the Eskimo* was published in 1924 from which Dr. Anderson's account of his work was omitted (although not his photographs) and for which Dr. Anderson then received no royalties whatsoever.

Somehow this impasse between the two men got leaked to the *Ottawa Journal*, which revealed that "Neither Stefansson nor Anderson proposes that the other shall have the last word; and, as a result, each has been waiting for the other to publish his official report, in order that any statements therein with which the 'survivor' does not agree may be controverted."[12] The aspect that most interested the newspapers at the time was the feud between them.

Dr. Anderson did make a determined start on his detailed account of the activities of the Southern Party, producing an eleven-page topic outline listing thirty sections or chapters,[13] and nearly two dozen pages of introductory text.[14] But that was apparently as far as he got.

Different reasons account for the absence of both parts of Volume 2. Part A was to have been *The Mammals of Western Arctic America*, written by Dr. Anderson. Before he could start writing that report, however, he needed to spend a great deal of time examining the mammal specimens he had collected in the Arctic, in addition to other Western Arctic specimens already in the Museum's collections and perhaps elsewhere. Unfortunately, that time never became available to him. From 1917 until 1926, when funds for publication of the Expedition reports were finally stopped, he was too busy serving as secretary for the two editorial committees, corresponding with dozens of scientists in Canada, the U.S. and Europe, in connection with obtaining reports on the identifications of thousands of the Expedition's biological specimens, editing the reports they submitted, and finally herding their manuscripts through various stages of the printing process, to have any time for his own research. Furthermore, he had been promoted to Chief of the Biological Division in 1920, with all the duties and headaches that position entailed.

The problem with Part B of Volume 2 was different. It was to have been entitled *The Birds of Western Arctic America* and to have been prepared jointly by Dr. Anderson and his associate, Percy A. Taverner. Dr. Anderson held a PhD in zoology from the University of Iowa State, with ornithology as a major, and had some years earlier published a fine book on "The Birds of Iowa." He also had seven years of Arctic experience and specimen collecting. He had personally collected and skinned most of the Arctic bird specimens from the 1913–1918 Expedition and naturally felt very possessive about them. He also regarded himself as the Museum's authority on birds, especially on Arctic birds. Taverner had no university education whatsoever and no Arctic experience at that time, but had acquired a considerable knowledge about birds through self-training and lengthy association with bird experts in southwestern Ontario and adjoining U.S. states. He had been hired as the Museum's ornithologist two years before Dr. Anderson came on the scene, and had developed "territorial rights." In addition to his exceptional knowledge of birds, he was talented with his hands, creative, and a competent artist. Taverner undoubtedly felt some insecurity after the arrival of his more educated co-worker, who was in due course promoted to be his divisional chief, and Dr. Anderson resented Taverner being in charge of the birds. It was not realistic, therefore, to expect these two men to work together, studying and writing about the bird specimens Dr. Anderson had collected.

Nearly twenty years later Taverner justified his position on staff by publishing his authoritative book, *Birds of Canada*, which deservedly brought him wide acclaim. He had drawn most

---

[12] Anonymous, 1923a (*Ottawa Journal,* April 24, 1923).

[13] Anderson, 1919a.

[14] Anderson, 1919b.

of the black–and–white drawings in his book himself, and arranged for many of the coloured illustrations to be painted by an artist friend, Allan Brooks.

Two Unpublished Expedition Reports

Two other important Expedition reports were never published, for which manuscripts or notes still exist. The first was a manuscript entitled "Fishes," by the naturalist and marine biologist Fritz Johansen, which was scheduled to become Volume 6, Part A in the CAE Reports Series. It was to have provided details on the characteristics and distribution of many varieties of Arctic fish, a subject of considerable economic and social interest then as well as now.

Johansen was a lonely and somewhat eccentric man, very much concerned with making an international reputation for himself and just possibly envisioning himself as some day attracting public attention as Stefansson managed to do so well. He was a virtual slave to his biological research, with little concern for time, but often frittered away his time doing things quite unrelated to his intended work. As a person he could be quite entertaining, but in the working environment he somehow managed to irritate almost everyone he had to work with. In the Arctic he spent much of his three years roaming here and there by himself, for he did not mix well with the other scientists. He had been hired by the Department of the Naval Service[15] and was paid by them, but received his work instructions from the Geological Survey of Canada.[16] Back in Ottawa after the Expedition ended, he was constantly behind in his work. Dr. Anderson and other members of the committees overseeing the preparation of the Expedition's reports had to nudge him frequently to prepare the several reports he had agreed to write. Additionally, the quality of his writing in English was below average, perhaps because it was not his native tongue. Yet three of his reports were ultimately published between 1922 and 1924 (Volume 5, Part C; Volume 7, Parts G and N).

Johansen's report on the Arctic fishes, however, was to be his last and most significant manuscript, but months went by without any sign of it. When it finally reached the editorial committee, it was much too long, poorly written and in need of thorough revision. Here Johansen and the appropriate specialist on the committee, Dr. A.G. Huntsman,[17] came to an impasse after the latter endeavoured to revise the manuscript to render it acceptable for publication. Johansen would not agree to any of the changes in his manuscript and refused to let it be published under his name with the modifications called for by both Dr. Huntsman and the general editor, Dr. Anderson. The unacceptable manuscript, despite its fundamental importance, became a victim of the cessation of government funding for the Expedition's publications in 1926, and today reclines silently in the Archives of the library of the Canadian Museum of Nature, drawing only occasional examination by researchers.[18] Part of it, however, was included many years later in a Museum publication by V. Walters.[19]

---

[15] Dr. Anderson was actively involved in the preliminary negotiations to hire Johansen.

[16] This was because he was attached to Dr. Anderson's Southern Party, which was directed by the Geological Survey.

[17] Dr. Huntsman replaced Dr. Hewitt after the latter's sudden death in February 1920.

[18] Johansen, 1926. "Fishes of Arctic America."

[19] Walters, 1953.

A different story lies behind the non-publication of Diamond Jenness' ninth Expedition manuscript, tentatively entitled "Contributions to the Archaeology of Western Arctic America." As mentioned in Chapter 19, Jenness had collected in the early summer of 1914 some three thousand archaeological specimens from three long-abandoned Alaskan native villages on and near Barter Island, northern Alaska. Being outside Canadian territory, these specimens necessarily had a lower priority than his Copper Inuit material. However, he had done some preliminary research on them, fully intending to complete his study and write a report on the sites and material, when he was unexpectedly appointed Chief of the Division of Anthropology in 1926. His new duties immediately burdened him with a considerable administrative workload, and he never thereafter found the time to complete his northern Alaskan studies. At the time he had written more than one hundred pages of preliminary notes for his archaeological report, which today still reside in the archives of the Canadian Museum of Civilization.[20]

Jenness' Barter Island ethnographic collection was finally studied in the 1960s and later reported on at length in 1987 in a private report by the American archaeologist, E.S. Hall Jr.[21] In addition to providing detailed descriptions of Jenness' artifacts and discussions of their implication and significance, Hall obtained radiocarbon dates on some of Jenness' Barter Island material, which dated the sites at about 1500 AD, confirming their "Thule" connection.

Five other Expedition reports were originally planned but never materialized. Either no specialists were found in 1918 to study the kinds of specimens needing examination or researchers assigned to carry out specific studies had failed to do so. In one instance, the researcher died before completing his task. The omission of reports on them may not have been a serious loss to science, however, for the biological species they involved were either of minor significance or the sizes of the Expedition's collections were hardly sufficient to justify publication.

Appendix 4 lists all of the reports (published and unpublished) stemming from the field work of the Canadian Arctic Expedition, 1913–1918, together with their dates of publication. The reports deal almost exclusively with specimens collected by, and studies made by or on behalf of members of the Southern Party.

Stefansson's Interference with the Publications

Stefansson occasionally came to Ottawa between 1918 and 1922 to deliver public lectures and to attend meetings of the Royal Commission on Reindeer and Musk-Ox. He had repeatedly urged the Canadian government to establish such a Commission to consider developing a reindeer industry in the North like that in Alaska, as well as to domesticate the muskoxen. The Commission was established in May 1919, and he had been appointed one of its directors.[22] While in Ottawa he generally visited G.J. Desbarats, and once or twice also Dr. C. Camsell (who succeeded Dr. McConnell as Deputy Minister of Mines in 1920 upon the latter's retirement)

---

[20] Jenness, D. 1914b.

[21] Hall, 1987. For brief accounts of my father's Barter Island activities, in addition to what is set forth in Chapter 19, see also S.E. Jenness, 1990, and S.E. Jenness (Ed.), 1991.

[22] Among the expert witnesses who appeared before this Royal Commission were Dr. R.M. Anderson, K.G. Chipman, Fritz Johansen and D. Jenness (Ashlee, 2008, p. 190), all formerly with the Southern Party of the Canadian Arctic Expedition.

to obtain news about both the North and the progress of the government publications resulting from his Expedition. During more than one of those visits he insisted that Desbarats have changes made to parts of certain of his Expedition's publications.

I mention six examples here that I have learned about from Dr. Anderson's correspondence, but there may have been more. The changes Stefansson demanded were of two kinds: (1) changes to the wording on the title and cover page of each publication and to amplify the importance of the roles played by the Department of the Naval Service and of Stefansson; and (2) changes to the authors' interpretations, where they differed from those of Stefansson. The modifications or delays asked for (or even insisted upon) by Stefansson were viewed by Dr. Anderson as unacceptable interference, because Stefansson "… had done absolutely nothing towards helping with any reports …"[23] and "… had brought back practically nothing in 1918 that could go in a scientific report …"[24]

Stefansson insisted that action be taken on the following matters:

(1) Change the dates assigned to the Expedition from its original 1913–1916 to 1913–1918, to include the extra two years Stefansson stayed in the Arctic beyond the Expedition's original scheduled period. Desbarats overruled the unanimous objections of the publications committee and insisted that this change be made.

(2) Include prominently on the title page the heading "Department of the Naval Service. Report of the Canadian Arctic Expedition under Vilhjalmur Stefansson, 1913–1918." When word of this reached Dr. Anderson in 1919, he promptly expressed his certainty that if Stefansson's name was going to be inserted in this manner in every report, most of the scientists would refuse to write their reports.[25] Most affected by such a change would have been Jenness, who ultimately authored five of the thirteen CAE published volumes. Stefansson's request was rejected.

(3) Jenness was to change certain statements, or his report (Vol. 12, Part A, *Life of the Copper Eskimos*) was to be withheld. Jenness was understandably indignant when he learned of Stefansson's demands. While writing this report in 1919, Jenness, out of the respect he held at that time for Stefansson as an ethnologist, had kept Stefansson informed of the manuscript's progress during its preparation. He completed the manuscript early in 1920 and submitted it to the Geological Survey's editorial unit, and in due course it was approved for publication.[26] However, the Geological Survey had a backlog of geological manuscripts awaiting publication at that time, so the Department of Mines (on behalf of the Geological Survey) arranged to

---

[23] Anderson, 1921d (March 28, 1921).

[24] *Ibid.*

[25] *Ibid.* Of the scientists on the Expedition, Johansen, Mamen, McKinlay, Murray and Wilkins were employed by the Department of the Naval Service. The others—Dr. Anderson, Beuchat, Chipman, Cox, Jenness, Malloch and O'Neill—were in the employ of the Geological Survey.

[26] In the summer of 1920, Stefansson asked Jenness to take his (V.S.'s) ethnographic notes from his three Arctic trips, and edit and publish them. Stefansson could not do it himself, according to Dr. Anderson, because "… he does not care to do anything more in ethnology. He announced that in 1914." Jenness recognized there was "… some value [in Stefansson's notes] if weeded out properly, and they would be lost if not handled by some proper person … (Anderson, 1921a [March 5, 1921]). Jenness chose instead to write his own reports on the Copper Inuit. Stefansson's Arctic ethnographic notes were finally edited and published by anthropologist Gisli Pálsson (2001).

have the Department of the Naval Service publish Jenness' manuscript, and the manuscript was forwarded to that department's Deputy Minister Desbarats. In June 1920 Jenness received a letter from Stefansson, which contained a list of criticisms of Jenness' manuscript. Desbarats had evidently shown it to Stefansson or provided him with a copy, for Jenness had not sent Stefansson a copy of his final manuscript. In response to Stefansson's comments, Jenness willingly made a few simple changes, but refused to make others, which called for him to alter his fundamental interpretations about the Copper Inuit to agree with those held by Stefansson. As a result, the publication of his manuscript was delayed by Desbarats until some sort of agreement was reached. The report was finally published early in 1922 after a delay of nearly two years. That was also several months after the publication of Stefansson's *The Friendly Arctic*.

(4) Jenness was to delete certain passages in his manuscript (Vol. 12, Part B, *Physical Characteristics of the Copper Eskimos*). In April 1923, while Stefansson was in Ottawa on a brief visit, Dr. Edward E. Prince, the Chairman of the Canadian Arctic Expedition Publication Committee, showed him a paragraph in the page proofs of Jenness' Part B of Volume 12 on the Copper Inuit. After perusing the proofs, Stefansson informed Dr. Prince that he strongly objected to some passages, as they disagreed with his views, and wanted them deleted, adding that he intended to see Dr. Camsell about the matter. Dr. Camsell was out of town at the time, however, so Stefansson wrote to him about it, including the following offensive slur on Jenness' integrity:

> I have not the slightest objection to having Mr. Jenness controvert my opinions in any truthful and legitimate way, whether through argument or the testimony of others, but I do distinctly object to being misquoted in reports of my own expedition and to having these reports contain such deliberate attempts to mislead the public.[27]

Dr. Prince may have been somewhat intimidated by Stefansson's reaction, for he promptly wrote Dr. Camsell informing him of the meeting with Stefansson and of the latter's strong reaction to a part of Jenness' manuscript. Dr. Prince's solution to the problem was if the "offending paragraphs were eliminated, all trouble would cease."[28]

In response to Dr. Prince's letter of April 19, Dr. Camsell, as well as his assistant, L.L. Bolton, examined the page proofs and the so-called "offending passages" of Jenness' report and undoubtedly discussed their contents with Jenness. Dr. Camsell then replied to Dr. Prince, "I have looked through the report and cannot see that any alteration should be made, nor can I see that there are any statements which Mr. Stefansson, as a scientist, can take exception to."[29] His letter then continues that Mr. Jenness had presented his evidence in detail and drawn his conclusions in proper scientific fashion, with no personal reference to Mr. Stefansson. As well, in its preparation Jenness had received assistance from both Dr. Boas of Columbia University and Dr. Hooton of Harvard University. Additionally, the manuscript had been closely examined by Dr. Sapir, who found nothing in it that was not strictly scientific or that could give an offence.

---

[27]   Stefansson, 1923a (April 18, 1923).
[28]   Prince, 1923 (April 19, 1923).
[29]   Camsell, 1923b (April 26, 1923).

Some while later Stefansson visited Dr. Camsell, urging him to insist that parts of Jenness' manuscript be modified to ensure that its interpretation on the Copper Inuit agreed with his own. Instead of agreeing to do so, however, Dr. Camsell raked Stefansson over the coals for using the word "mutiny" in his book, *The Friendly Arctic,* in describing the behaviour of Dr. Anderson and some of his colleagues at Collinson Point, Alaska, in 1914. Stefansson admitted he had been wrong to use the word and suggested that his press agent was to blame for his having done so.[30] Captain Bartlett recognized this trait of Stefansson, of not accepting the blame for his actions, describing him as "a great buck passer."[31]

It would appear that Stefansson also discussed his concern with the Deputy Minister of the Naval Service, for a few days later Dr. Anderson, as editor of the publication series, was made aware of Stefansson's interference and wrote Dr. Camsell that Stefansson had induced Desbarats "… to suggest suppression or emasculation of some sections, because, as Mr. Stefansson alleged, it did not look right to have anything printed in the scientific reports of the expedition which disagreed with the views of the commander."[32]

For Stefansson, public image was of vital importance, as Dr. Anderson's quotation reveals, because he made his living by public lectures, writing and getting newspaper publicity. But for the dedicated scientist Jenness, Stefansson's actions amounted to intolerable and unscientific interference.

Although Jenness, in his report, had not mentioned the controversial topic "the Blond Eskimos," I believe that is what the fuss was mainly about. The outcome, I understand, was that Jenness had to insert or modify a footnote (probably the one on page 46B in his report, which mentions that neither he nor the Rev. H. Girling had seen any sign of the "blond" Eskimos Stefansson had reported to be located north of Dolphin and Union Strait). Yet he was permitted to retain the following important conclusion: "The theory of any European admixture among the Copper Eskimos may be rejected without further consideration."[33] This effectively provided a professional *coup de grâce* to one of Stefansson's most appealing and hence rewarding public topics, the possible genetic connection of the "blond Eskimos" among the Copper Inuit with a band of Norse settlers who had disappeared from western Greenland late in the fifteenth century.

This report of Jenness was published a month later, but being a specialized government report received limited distribution and did not attract the attention of the news media. As a result, Stefansson's popular hypothesis on the origin of some of the Copper Inuit (the "Blond Eskimos") continued to surface from time to time for many more years. The negative results recently obtained from DNA studies on Copper Inuit descendants in Cambridge Bay, Victoria Island,[34] should finally put an end to Stefansson's once-popular hypothesis.

---

[30] Anderson, 1923e (June 13, 1923).

[31] Bartlett, 1922 (February 6, 1922).

[32] Anderson, 1923b (April 24, 1923).

[33] Jenness, 1923, p. B47.

[34] This conclusion was mentioned also in Chapter 1. Recent archaeological studies by Pat Sutherland, however, have identified yarn spun from Arctic hare fur from four Dorset sites in northern Labrador, Baffin Island and northern Ellesmere Island, yarn that is similar to Norse yarn dating to roughly 1250 AD in a Western Greenland site (Pringle, 2009). The Dorset people did not know

(5) Stefansson sought a more leading role in the publication of the Expedition reports. For some reason or other, he felt slighted because members of the Southern Party were writing about their own scientific findings. Citing the geologist O'Neill's geological report as an example, Stefansson complained to Prime Minister Borden by way of Desbarats in July 1919,[35] that he, as the Expedition commander, should be writing about the scientific results of the Expedition. Desbarats backed Stefansson's view when he was questioned by Borden, and added that the Geological Survey had been disinclined to recognize Stefansson's authority throughout the entire course of the Expedition.

Desbarats then proposed that Stefansson be made editor of the Expedition's publication committee. Inquiry into this possibility revealed that the members of the Southern Party would be unlikely to prepare *any* reports if Stefansson was appointed editor. Thus the Prime Minister became acquainted with the fact that not only were the various members of the Expedition's Southern Party at odds with Stefansson, but the two government departments were not seeing eye to eye either. Fortunately, the suggestion was set aside, and Stefansson was not given the opportunity to edit any of the scientific reports.

(6) Stefansson pressured Desbarats to hurry up the preparation of several maps showing the routes Stefansson took during his various ice trips.[36] He needed these maps, which the Canadian government had agreed to prepare, for his popular book *The Friendly Arctic*, although it was not a government publication. Their preparation by the Geodetic Survey of Canada was delayed, however, because Stefansson was slow to supply the geographic names he wanted to appear on the maps. Then the information he submitted about many of the new names he had proposed was inadequate to get the approval of the Canadian Board of Geographic Names. Even more delays resulted from the amount of correspondence that followed, seeking clarifications between the Geodetic Survey, the Canadian Board of Geographic Names, Desbarats and Stefansson.

Stefansson finally had to come to Ottawa to discuss the specific problems with the Geographic Board to get his geographic names approved, and with the Geodetic Survey to expedite the maps. He had been largely to blame for the delays in the preparation of the maps to that time, but thereafter he was impatient about further delays. In an effort to expedite matters, he evidently persuaded a school-map publisher in Chicago, the Denoyer-Geppert Company, to write Desbarats for permission to use Stefansson's new geographic names on an Arctic map they intended to publish, although these had not yet been approved by the Canadian Board of Geographic Names. Several of the maps in Stefansson's *The Friendly Arctic* that carry the names "Department of the Naval Service" and "Geodetic Survey of Canada" attest to the dedicated efforts of the many persons involved at that time, but reveal nothing of the feelings some of them may have held towards the irritating problems and delays Stefansson had caused.

---

how to create yarn. This yarn thus offers strong evidence of early Norse presence in the Canadian Arctic about the same time but considerably farther east than where Stefansson suspected it.

[35] Stefansson, 1919a (July 10, 1919).

[36] Details on this topic are found in RG42, Vol. 471, File 84-2-7, LAC.

Dr. Anderson's Editorial Role

From 1917 to 1926 Dr. Anderson served as secretary for the interdepartmental publications and editorial committees, and as general editor of the report series. No better choice for those arduous tasks could have been made. In the secretarial capacity he played a key role in negotiating and coordinating with the several dozen scientific specialists in Canada, the U.S., Great Britain and France who agreed to study the many types of specimens brought back by the Expedition from the Arctic.[37] He subsequently devoted thousands of hours corresponding with scientists and authors to ensure that the specimens assigned them were studied and written about within the tight schedules established by the various committees, and then an equal or perhaps greater number of hours editing the submitted manuscripts and seeing them through publication as reports. Undoubtedly time was also spent in seeking the return of specimens from some of the specialists. Thankfully, Dr. Anderson saved his voluminous collection of correspondence about these activities, and they exist today in the Archives of the Canadian Museum of Nature in Gatineau, Quebec.[38] It is truly a paper monument to the dedication he first showed as executive head of the Southern Party in the Arctic. By his own estimation, "This editorial work occupied much of his time from 1917 to about 1939." This estimate was followed by the handwritten comment "No help from Mr. V.S. in any respect."[39]

Unauthorized Publications

In Victoria in June 1913, shortly before the departure of his Expedition for the North, Stefansson told its members that they were not permitted to publish any account of the Expedition until a year after its completion. As the Expedition was originally scheduled to remain in the North for three years, that publication ban extended until late in 1917. As events unfolded, however, Stefansson remained in the North for two additional years, theoretically extending the publication ban until late in 1919. Stefansson understandably refrained from mentioning that this publication prohibition was to protect the exclusive newspaper contracts he had made with the *London Chronicle, The New York Times* and *The Globe* in Toronto, for which he expected to be handsomely paid. These contracts, plus book royalties and lecture fees following the completion of the Expedition were how he expected to be compensated for his time and work in the North, since at his own request he received no salary from the Canadian government while leading the Expedition.[40]

---

[37] The Canadian government's policy at the time, and hence that of the publications committee, was to have the specimens studied and reported on by Canadian scientists knowledgeable with the material they were sent. However, as such specialists in Canada were few and far between, some specialists were sought in the U.S., Great Britain and France.

[38] They form a collection of eighty-three boxes of papers designated CMNAC/1996-077, Series A—R.M. Anderson Collection, Archives, CMN. I am ever grateful to Dr. David R. Gray for having drawn my attention to the existence of these valuable documents in 1986.

[39] This statement is written in Dr. Anderson's handwriting on the bottom of page 197 in one of his copies of Stefansson's *The Friendly Arctic* (Call No. G670 1908 s73 1922) in the Rare Book Room of the library, CMN.

[40] Stefansson gave two reasons why he preferred to serve without government pay: "(1) I have a preference or what may be called a personal prejudice in favor of doing such work as this without

His publishing arrangements and restrictions might have worked satisfactorily had the *Karluk* not become separated from the rest of the Expedition and then ended up on the bottom of the Chukchi Sea. Four survivors of that disaster whose service with the Expedition ended unexpectedly after the first year—McKinlay, Maurer and Chafe from the *Karluk*, and Stefansson's secretary, McConnell, whose contract Stefansson did not renew in June 1914—published one or more articles about the Expedition after they left the Arctic in 1914. McConnell, for example, published half a dozen articles about the Expedition in a Nome newspaper in 1914, a detailed article about the rescue of the *Karluk* survivors ("Got Karluk's Men as Hope was Dim") in the *New York Times* of September 15, 1914, and two more the following year in *Harper's Magazine*.[41] McKinlay published an article in the *Manchester Guardian*[42] in 1914 about his experiences on the Expedition as well as four articles in a Scottish weekly newspaper, *The People's Journal*.[43] Maurer published four articles in *World Magazine* in 1914;[44] and Chafe published one article in the *Geographical Review* in 1918.[45] Of these four men, only Chafe received no remuneration for his literary effort, having chosen to publish in a scientific journal. However, McKinlay donated his payments to wartime charities.[46]

A fifth man, Johansen, while in the Arctic with the Southern Party, quietly wrote and submitted a series of twelve articles in Danish about the Expedition in the form of lengthy letters to the Danish newspaper *Politiken*, which were published in Copenhagen between July 1913 and July 1917.[47] Perhaps he understood Stefansson's ban to apply only to English-language publications.

---

pay—in fact, were I wealthy, I should be glad to pay for the mere privilege of engaging in the one work that is thoroughly congenial to me; (2) I had made contracts with the Macmillan Company, The United Newspapers for the sale to them of the official narrative of the Expedition for a price which I considered a fair return for my services to the Expedition as well as sufficient to enable me to pay my own expenses for a year or two till I could get my scientific results off my hands and be ready for other work. This sum—1250 pounds—was less than the highest salary paid by the government on the scale of pay under which the Expedition sailed from Victoria" (Stefansson, 1916). Naturally, no mention was made of the freedom of action which that arrangement allowed Stefansson, a condition Desbarats came to deeply regret when Stefansson did not obey instructions to return to Ottawa, failed to submit complete financial accountings of disbursements, and spent vastly more money than the government could easily justify.

[41] McConnell, B.M., 1914a,b,c; 1915a,b.

[42] The article in the *Manchester Guardian* (McKinlay, 1914e) was written "within two hours of landing at Liverpool on my way home in 1914, without any reference to diary notes; and was done at Dr. Bruce's urgent request" (McKinlay, 1925). Dr. W.S. Bruce was Director of the Scottish Oceanographical Laboratory, a man greatly admired by McKinlay. Both McKinlay and Dr. Anderson suspected that Hadley and Stefansson used McKinlay's accounts of the *Karluk* from the *Manchester Guardian* and *The People's Journal* as the basis for the Hadley "point of view" on the latter's *Karluk* story, as it was presented in Stefansson's *The Friendly Arctic*.

[43] McKinlay, W.L., 1914a,b,c,d.

[44] Maurer, F.W., 1915a,b,c,d.

[45] Chafe, E.F., 1918.

[46] McKinlay, 1925 (May 13, 1925).

[47] Johansen, F., 1913a,b,c; 1914a,b,c,d; 1916; 1917a,b,c,d. English translations of these twelve articles are with the Johansen Papers, MG30 B165, LAC.

Captain Bartlett of the *Karluk* also entered the publishing arena, producing in 1916 his highly readable book, *The Last Voyage of the Karluk, Flagship of Vilhjalmur Stefansson's Canadian Arctic Expedition of 1913–1916*, which included details of his heroic journey to eastern Siberia in search of ships and men to rescue the *Karluk's* survivors from Wrangel Island.

Stefansson later complained that his newspaper contracts were broken and he lost a considerable amount of money because these men published stories about the Expedition when they did.

Expedition Costs

The Canadian Arctic Expedition, through both mishap and Stefansson's uncontrolled mishandling of government funds, proved extremely costly, both in lives lost and in dollars spent. "With the exception of the Franklin expedition, no other expedition in the Canadian Arctic during the past two centuries has had such an appalling record."[48] Seventeen men on Stefansson's Expedition lost their lives, but he evidently did not consider this loss as excessive.[49] Questions about the size of his expenditures during wartime arose both in the Canadian Parliament and in the newspapers, to the embarrassment of both Prime Minister Borden and the Minister of the Department of the Naval Service, J.D. Hazen.[50] As the senior government department backing the Expedition, the Department of the Naval Service was held responsible for many of the Expedition's failings. The expenditures and losses of life were even more embarrassing to that department's Deputy Minister George J. Desbarats, the government official ultimately required to explain the many extravagant cheques that Stefansson distributed around the Arctic in rebuilding and operating his new Northern Party, for the spending of thousands of dollars in cash Stefansson could not account for, for the loss of the *Karluk* and many lives, and for Stefansson's ignoring of official instructions from Ottawa to wind up his Arctic activities and return south.

As Stefansson originally planned his Expedition, the amount of money needed to pay for a few scientists and sailors, and the purchase of a ship, equipment, and supplies so that he could study the "Blond Eskimos" on Victoria Island and explore for unknown land in the Beaufort Sea for three years was in the order of $50,000. This cannot be regarded as a realistic amount, however, for Stefansson had not taken the time to make detailed plans and cost estimates when he proposed that figure. The additional needs of the scientific personnel who were added to the Expedition's plans on the insistence of the Canadian government (especially by the Geological Survey of Canada) when it agreed to sponsor Stefansson's Arctic Expedition, increased the original estimated cost of the Expedition by about fifty percent, that is, to about $75,000.

By the time the last cheques were paid, however, the cost of the Expedition to the Canadian government between April 1912 and March 1920, according to the Auditor General's sessional reports, was $559,972,[51] more than ten times Stefansson's original estimate. And that

---

[48] Anonymous, 1923b.

[49] Stefansson, 1921b, p. 73.

[50] John Douglas Hazen was Minister of Marine and Fisheries as well as Minister of the Naval Service from 1911 to 1917, then was appointed Chief Justice of the Appeal Division, Supreme Court of New Brunswick.

[51] "Summary of amounts expended for CAE 1913–18," MG30 B40, vol. 10, LAC.

was after deducting the money from the sale of the *Alaska* and the Southern Party's unused supplies ($4,675), the *Polar Bear* ($5,000), and the fox skins Stefansson sold ($4,235). The share of the costs allotted to the Department of the Naval Service was $516,332, to the Geological Survey, $43,640. And since the salaries of the scientists were not indicated in the government's tabulation and a few residual bills were still outstanding in 1920, the total cost would have been closer to $600,000.[52]

The cost of the Expedition was far greater than mere dollars, however. Seventeen men—scientists, sailors and one Inuk—all Expedition members, lost their lives.

At this point, I will mention a statement having to do with Stefansson's perception of the value of human life, one he evidently made on more than one occasion following his return from the Arctic. I vividly recall having seen it several years ago in one of his many writings, both published or in letters, but have been unable to unearth the exact source of the statement despite numerous searches. When asked to comment on the large number of men lost on his Expedition, his reply was that on his Expedition relatively few men had died to enable him to acquire thousands of square miles of land for Canada, while in World War I in Europe (which was taking place at the same time as his Arctic activities), thousands of men had died to achieve but a few feet.[53] This reply may well be basically true, but I find it extraordinarily insensitive.

Expedition Accomplishments

It is reasonable now, so many years later, to ask "What did the Canadian Arctic Expedition actually accomplish?" and "Was it worth the price paid for it?" The short answers are "a great deal" and "yes," as the following list covering both Northern and Southern Parties reveals. The contribution of Stefansson's Northern Party was almost entirely the result of his exploration, having lost its scientific members with the *Karluk*. The Expedition's scientific contribution was almost entirely that done by the members of the Southern Party, who accomplished an amazing amount in the two years they operated in Canadian territory, given the conditions under which they were required to operate.

Northern Party

(1) Stefansson and members of his reconstructed Northern Party discovered five of the last six large unknown islands in the North American Arctic, and claimed them for the Dominion of Canada. These were: Brock, Borden, Meighen and Lougheed Islands, with Stefansson's "Borden Island" later proving to be two large islands.[54]

---

[52] There were evidently some additional expenses resulting from late submission of accounts, but I have not seen an official statement showing the ultimate total expenditures by the Expedition.

[53] A variant of this story appears in *The Friendly Arctic* (p. 73), where Stefansson wrote "... I could never see how anyone can extol the sacrifice of a million lives for political progress who condemns the sacrifice of a dozen lives for scientific progress." McKinlay (1976, p. 34) mentioned this latter quotation, adding that as the ice pack swept the *Karluk* ever farther from Stefansson, the men on the ship felt "... not so much like soldiers sacrificing ourselves to a great cause, as lambs left to the slaughter."

[54] This fact was first revealed by aerial photographs taken between 1946 and 1948. The name Borden Island was retained for the more northern island of the two. The name "Mackenzie King Island"

The study of aerial photographs in 1948 revealed the existence of the sixth and last large unknown Arctic island in Canadian-claimed territory. In 1951, fully three decades after the publication of Stefansson's *The Friendly Arctic,* this last bit of unknown territory was named Stefansson Island.[55] Stefansson's man Storkerson had seen its west coast in 1917 and had referred to it as "Leffingwell Island" at that time, after the American geologist with whom he had been associated while on the Mikkelsen-Leffingwell Expedition of 1906–1907, but it was not completely mapped nor truly recognized as separate from Victoria Island at that time. Ironically, Leffingwell became one of Stefansson's avowed critics, so its present name seems more appropriate.

(2) Stefansson and his men discovered several small islands as well, only a few of which he named. These included Bernard Island[56] and Jenness Island,[57] the former off the west coast of Banks Island, the latter between Brock Island and Borden Island farther north. Both are too small to appear on most northern maps, as was also a small unnamed island immediately north of Meighen Island. Stefansson also claimed the presence of several small islands close to Lougheed Island, but the existence and names of these (and other geographical claims) were hotly disputed for some years afterwards.[58] In 1917, Storkerson named Elvina Island, Kilian Island and Mikkelsen Islands, off the northeastern coast of Victoria Island. Storkerson, in actual fact, was the first person to see most of the islands that Stefansson received the credit for having discovered.

(3) Stefansson established the outer edge of the Continental Shelf at several locations in the Beaufort Sea, ruling out the possibility of any continent in that region north of Alaska and the islands of the western Canadian Archipelago. There was, therefore, no "Crocker Land," for which both Stefansson's and MacMillan's expeditions were searching.[59] There was also no land north of Alaska as had been predicted in 1911 by R.A. Harris of the U.S. Coast and Geodetic Survey from assorted hydrographic evidence.[60]

---

was given to the larger southern island, honouring the Canadian Prime Minister when the existence of a strait dividing Stefansson's original Borden Island was discovered.

[55] D. Rowley, 1952; S.E. Jenness, 1997b.

[56] Stefansson's name, Bernard Island, set off a mild administrative storm after he commented to Deputy Minister Desbarats in a letter dated June 1, 1921 that Captain Peter Bernard, after whom Stefansson was naming the island, always spelled his name Beneard. Shortly afterwards, when the Secretary of the Geographic Board of Canada learned of this spelling discrepancy he indignantly wrote Desbarats that he considered "the spelling Bernard for Bernard I. as indefensible if it commemorates "a man who spelled his name Beneard' " (Douglas, 1921a). A week later, perhaps following an irate telephone call from Desbarats, he sent a second letter apologetically requesting that the sentence with that comment be "cancelled" from his previous letter (Douglas, 1921b).

[57] Jenness Island is a gravel and sand island with a maximum elevation of less than fifty feet, and a maximum length of less than a mile, lying just off the southwest tip of Borden Island in the High Arctic. It would be interesting to know what prompted Stefansson to name such an insignificant geographical feature after my father.

[58] Stefansson, 1921a; McDiarmid, 1923; Anonymous, 1924, 1925. See also correspondence between the Geographic Board and Desbarats between June and September, 1921, RG42, Vol. 471, File 84-2-7, LAC.

[59] MacMillan beat Stefansson by a few months to the recognition that there was no such continent.

[60] Harris, R.A., 1911, pp. 7 et seq., 30 et seq.

(4) Stefansson added dozens of depth soundings and considerable data on oceanic currents in previously unexplored parts of the Arctic Ocean north of Alaska. However, there proved to be problems with the geographical accuracy of his data. A member of the Geodetic Survey in Ottawa struggled for a year to establish reliable locations for Stefansson's depth sites and island locations, with only partial success.

(5) Stefansson's activities confirmed that a small number of white explorers or scientists could survive for several months with care, luck and sufficient ammunition on the Arctic ice by shooting and eating seals, and on land by shooting and eating caribou or muskoxen. Stefansson was obsessed with demonstrating that white men could "live off the country," giving the impression that it was his own idea, but John Rae and even Sir John Richardson before Rae had demonstrated it long before Stefansson did. Dr. Anderson mentioned this in letters to his friends on more than one occasion, but sometimes added, based on his own experience, that "... a white man cannot be expected to do much scientific work under similar conditions."[61]

(6) Under Stefansson's direction, Storkerson successfully demonstrated that man could use a large Arctic sea ice mass as a base from which to gather data on polar drift in the Beaufort Sea, and his findings disproved some conceptions about polar current at that time.

(7) Stefansson and Wilkins redefined parts of the western coastlines of Prince Patrick Island and Banks Island, respectively, and Storkerson mapped many miles of the northern coast of Victoria Island. Wilkins collected some tidal records near Cape Kellett on Banks Island, the first such information obtained north of the Canadian mainland.

(8) Several members of Stefansson's original Northern Party took ocean-depth measurements, geologist Malloch took solar readings for location determinations, and oceanographer Murray studied several kinds of marine life he collected by dredging while the *Karluk* drifted within its entrapping ice mass. Unfortunately, all of their derived information and observations were lost when the *Karluk* sank, and the men subsequently perished.

(9) Stefansson's explorations and flag-planting on previously unknown High Arctic islands provided Canada with a good basis for claiming sovereignty over the islands north of its mainland.

Southern Party

(1) On Dr. Anderson's Southern Party, geographers Chipman and Cox made many corrections to the available map (a British Admiralty Chart dating from the 1850s) of the mainland coast of Northern Canada from the Alaska-Yukon boundary to the delta of the Mackenzie River, and from the south end of Darnley Bay east to Bathurst Inlet. They also mapped parts of the Firth, Brock, Croker, Rae, Tree and Hood Rivers, and much of the delta of the Mackenzie River, obtaining a number of channel soundings in that delta. And when not otherwise occupied, they surveyed the harbours at Collinson Point, Alaska, and at Bernard Harbour, northern Canada, when they sheltered their ships between 1913 and 1916.

(2) Geologist O'Neill mapped the rocks exposed in the lower parts of the Firth, Mackenzie, Tree and Hood Rivers, the geology of Herschel Island, and the geology along the coast from the southern end of Darnley Bay east to Bathurst Inlet. While doing so, he collected representative

---

[61] Anderson, 1909 (November 15, 1909).

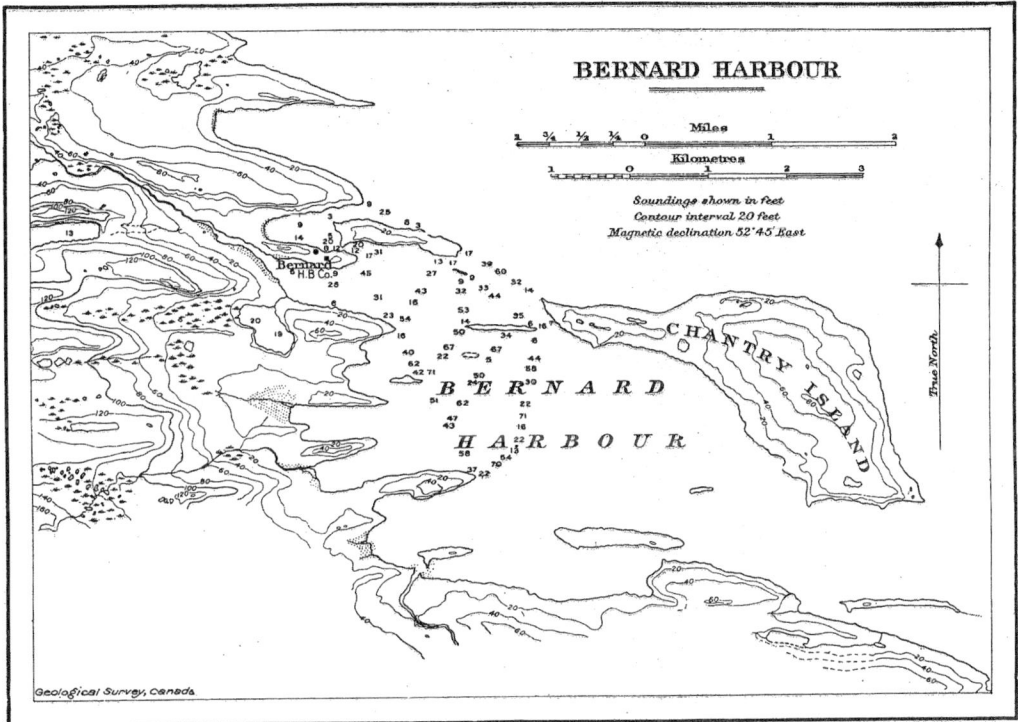

Map 20 Topographic map of Bernard Harbour prepared by Chipman and Cox, with the assistance of Patsy Klengenberg (from Chipman and Cox, 1924). The location of the CAE base camp is marked by a small black dot (•) immediately north of the location of the Hudson's Bay Company post, which was erected a month after the departure of the Southern Party in 1916. The post is marked by a small black square (■) and the name "Bernard Harbour."

specimens of the rocks, minerals and fossils he observed for the Geological Survey of Canada.[62] He also established the boundaries of a wide band of Precambrian basalt rocks containing disseminated native copper in Bathurst Inlet, and concluded that such rocks were too low-grade at the time to be of commercial interest.

(3) Naturalist and marine biologist Johansen made extensive collections for the Department of the Naval Service of insects, plants, fish and other aquatic life around Collinson Point, Demarcation Point, Herschel Island and Bernard Harbour. He also discovered the fossils that permitted the geological dating of the widespread dolomitic limestones west of Coronation Gulf as of Silurian age.

(4) Ethnologist Jenness studied the physical features, language and daily activities of the Copper Inuit, and collected their songs, myths, skin clothing, weapons, cooking ware, sewing tools and other artifacts, bringing south for the National Museum of Canada a fine collection of more than three thousand items. He also shipped back some three thousand archaeological objects from Barter Island, northern Alaska, which is the largest extant collection

---

[62] Most of O'Neill's rock and mineral specimens appear to have disappeared (probably through discard) in intervening years, a common fate of specimens collected by geologists once they have been studied and their geological reports have been published.

of Thule-culture material from northern Alaska. Both collections are now stored by the Canadian Museum of Civilization.

(5) Zoologist Dr. Anderson successfully coordinated the activities of his Southern Party, which completed its intended tasks within its assigned three-year time period, and sent regular and detailed reports of his Party's progress to the Deputy Ministers of both government departments—Naval Service and Mines. He also collected for the National Museum of Canada 616 specimens of 73 species of Arctic birds, and 422 specimens of 22 species of mammals from various places across the Arctic between northern Alaska and Bathurst Inlet.

(6) Members of the Southern Party collected data on meteorological conditions and tidal variations at Collinson Point in northern Alaska, and at Bernard Harbour northwest of Coronation Gulf.

In sum, the Canadian Arctic Expedition discovered new land, accumulated valuable scientific information on the Beaufort Sea and on the Inuit across the Arctic, and brought south hundreds of specimens of Arctic birds, animals, insects, aquatic and plant life, and rocks and minerals from both northern Alaska and northern Canada. It also obtained a large number of Thule artifacts from northern Alaska and Copper Inuit material from the Coronation Gulf region, and determined that the copper deposits in the region were uneconomical at that time. That was a considerable accomplishment for three years in the Arctic and reflects well on the dedication and industriousness of the scientists involved.

Considerable publicity was given to the exploration achievements of Stefansson's reconstituted Northern Party following his return south in 1918, as he had hoped, much of it through the popularity of and controversy over his book, *The Friendly Arctic,* as well as his various lectures and writings. His scientific achievements—oceanic depth recordings, current data, and biological and geographical observations, obtained rather haphazardly rather than systematically—received much less attention.

The Arctic achievements of the Southern Party, on the other hand, were considerable but received little publicity. This is not surprising, however, for publicity and personal acclaim were not the goals of any of its members, with the possible exception of Johansen. Their assignments involved methodically completing an assortment of scientific tasks in the Arctic, and preparing and submitting reports for publication after their return south. They successfully accomplished virtually all of these tasks, and their work provided a sound foundation for later scientific studies, most of which took place more than thirty years later. All told, the collections and information obtained by the Expedition more than justified its considerable cost to the Canadian government.

A change of Canada's government in 1926 and other factors soon resulted in the fading of interest in the Expedition from the public's mind. A renewal of interest in the Expedition has arisen in the past three decades, however, largely through the publication of several books on parts of the Expedition or about individuals connected with it, the recent appearance of an excellent detailed government website about the Expedition,[63] featuring the careful research of Dr. David R. Gray, and a much greater interest today, in general, about Arctic development and sovereignty.

---

[63] See http://www.civilization.ca/cmc/exhibitions/hist/cae/splashe.shtml

Unheralded Contributors

Most accounts of the Canadian Arctic Expedition have focused upon those persons who were perceived to have been the main contributors—the Expedition's two leaders, the scientific personnel, the ships' crews and captains, and on the grim experiences of the personnel who survived the last days of the *Karluk*, but otherwise had little opportunity to contribute. There has, for example, been little mention of the contributions of the various Native people to the Expedition's success or of any other behind-the-scenes individuals.

Inuit

Inuit men and women from northern Alaska, the Mackenzie River Delta region, Herschel Island, Baillie Islands and even the Coronation Gulf region made valuable contributions to the day-to-day operations of the Expedition, yet their contribution remains largely unrecognized to this day.

The Inuit men looked after and drove dogs and sleds, moved supplies, set up and took down tents, built snowhouses, hunted, fished, cooked, occasionally served as guides and undertook other tasks for various members of both Stefansson's Northern Party and Dr. Anderson's Southern Party.

With the ill-fated *Karluk* in 1913–1914 were two Iñupiat men from the Barrow region, Kuraluk and Kataktovik. The former helped sustain many of the *Karluk* survivors by his successful hunting on Wrangel Island over the better part of eight months, while the latter helped Captain Bartlett reach Native settlements in the Bering Strait and raise the alarm that resulted in the rescue of the *Karluk* survivors on Wrangel Island.

Stefansson's reconstituted Northern Party included the following Iñupiat and Inuvialuit men: Billy Natkusiak, Alingnak, Emiu ("Split-the-Wind"), Illun, Palaiyak, Pikalu, Pipsuk, Ulipsinna, and Fred Wolki. Of these the Alaskan Billy Natkusiak worked for the Expedition (both the Northern Party and the Southern Party) for the longest period of time, from 1914 to 1917. When his services were no longer needed by Stefansson and the Northern Party in 1917, he settled and hunted foxes on Banks Island for a few years, becoming widely known in later years as Billy Banksland. Palaiyak, likewise, worked for both parties.

The Inuit men who worked with Dr. Anderson's Southern Party between 1913 and 1916 were: Fred Adluak, Jimmy Asetsaq, Jerry Payuraq, Aiyakuk, Ipanna and Ikey (Angutisiak) Bolt, some of whom were from Point Hope; Siberia Mike, originally from northern Siberia; Ambrose Agnavigak, Mungalina, Silas Palaiyak and his brother Adam Uvoiyuaq from the Mackenzie River Delta area or western Arctic; and Patsy Klengenberg, from the Amundsen Gulf region east of the Mackenzie Delta, but originally from Alaska.[64] Of these, Ikey Bolt served the Expedition for the longest period of time, from 1913 to 1916. The Southern Party also occasionally employed as hunters and guides the following Copper Inuit, men not previously accustomed to working with white men: Ikpukhuak, and his stepson Avrunna, Uloksak

---

[64] Dr. Anderson wrote of Patsy, "Captain C. Klingenberg's *[sic]* young son Patsy was with the Expedition from May 1915 to July 1916, and proved an exceptionally valuable member of the party, as interpreter, hunter and assistant preparator of scientific specimens" (Anderson, 1916c [December 15, 1916]).

Meyok, and Mupfi (sometimes spelled Maffa) and his brother Kohoktok. Ikpukhuak and his step-son were from southwestern Victoria Island, Uloksak Meyok was from the Coppermine River area, and Mupfi and his brother were from the Tree River region. The two Inuit who were charged with murder, Sinnisiak and Uloksak Avingak, also worked for nearly a year under direction around the Bernard Harbour camp while awaiting transport south for trial; Sinnisiak was from southwest Victoria Island, Uloksak Avingak from near the mouth of the Coppermine River.

Inuit women also played a major role on both parties, chiefly as seamstresses and cooks. A few of them were the wives of sailors on one or other of the Expedition ships; some also had young children. Most of them worked for Stefansson's Northern Party. These included: Violet Mamayauk Gonzales (wife of Captain Gonzales), Guninana (wife of Alingnak), Pannigabluk (wife of Stefansson), Pukalook (first wife of Illun), Kutok (second wife of Illun), Pusimmik (wife of Pikalu), Annie Seymour (wife of Bill Seymour), Uiniq Elvina Klengenberg Storkerson (wife of Storker Storkerson), Jennie Thomsen (wife of Karl "Charles" Thomsen) and Uttaktuak Lopez (wife of Peter Lopez), all of whom came from the Western Arctic. Keruk (also known as "Auntie," wife of Kuraluk), hired at Cape Smyth, was an invaluable member among the *Karluk* survivors on Wrangel Island.

Inuit women with Dr. Anderson's Southern Party were: Jennie Thomsen at Collinson Point in 1913–1914; "Sis" (or "Ciss," wife of Siberia Mike),[65] Higilak (wife of Ikpukhuak), Munnigorina (or Manigyorina, wife of Kohoktok), Añayu Eunice Sweeney (wife of Captain Dan Sweeney), and Unalina (wife of Ambrose Agnavigak), at Bernard Harbour in 1915–1916. All were from the Western Arctic except Higilak, who was from the Dolphin and Union Strait region, and Munnigorina, who was from Coronation Gulf. Two or three other unidentified Copper Inuit women undertook brief sewing tasks for the scientists at Bernard Harbour in the first winter, 1914–1915.

Day in and day out, the Inuit women sat patiently outdoors or in tents, igloos or makeshift cabins, preparing the skins of either caribou and seals (according to the season and location—on land or ice), then skilfully sewing clothing and footwear with those skins for the men on the Expedition as well as for themselves and the children. Without this clothing and footwear, the members of the Expedition would have accomplished considerably less exploration and field work, and several of them might possibly have perished.

The Western Arctic Iñupiat and Inuvialuit were generally accustomed to being paid for their employment with the white man at that time, but the Southern Party's activities marked the introduction of the white-man's "pay for work" concept to the Copper Inuit. Payment then normally took the form of objects of trade—rifles, ammunition, knives, matches, cloth, needles, pots and pans, and the like, for the paper money in common use farther south would have been of no use to them.

---

[65] Sis (or Ciss) died from burns received a month after she left the *Alaska* at Herschel Island in August 1916. Her husband, Siberia Mike, had obtained work as helper and interpreter for the RNWMP at Herschel Island (Bruce, 1917).

Deputy Ministers

Four deputy ministers were involved with the activities of the Canadian Arctic Expedition, three of them sequentially in the Department of Mines. The roles they played in the story of the Expedition merit mention.

George J. Desbarats, Deputy Minister of the Department of the Naval Service, deserves a great deal of credit for the tremendous amount of time and effort he put into trying to make the Canadian Arctic Expedition a success. In this he faced many major physical obstacles, commencing with the shortness of time he was given to oversee its organization and coordination with Dr. R.W. Brock, his counterpart in the Department of Mines in 1913.[66] Desbarats had been appointed Deputy Minister of Marine and Fisheries in 1908 and of Naval Service in 1910, a position he retained until 1922, when the Department of Naval Service was amalgamated with the Department of Militia and Defence and the Air Board, and in 1924 he became the first Deputy Minister of the resultant Department of National Defence. During the years 1913–1922 when he was involved with the activities of the Canadian Arctic Expedition, "… Desbarats' authority and influence considerably exceeded those of most Deputy Ministers."[67] Desbarats showed almost unquestioning faith in Stefansson in the early days of the Expedition, and strong loyalty later on defending the latter's actions when embarrassing questions came from Parliamentary members, or from members of the Geological Survey, the press or the public.

After Dr. Anderson's return to Ottawa in 1916, he and Desbarats frequently held and expressed conflicting views in matters concerning the publishing of the Expedition reports,

Desbarats   McConnell   Camsell   Brock

Figure 63 The four Deputy Ministers who administered most of the affairs of the Canadian Arctic Expedition 1913–1918. Left to right: G.J. Desbarats, Naval Services (1913–1922), R.G. McConnell, Geological Survey (1914–1920), Dr. C. Camsell, Geological Survey (1920–1946), Dr. R.W. Brock, Geological Survey (1913–1914).

Photo: Desbarats, Library and Archives Canada, e010858588; McConnell, Camsell, Brock, reproduced with the permission of Natural Resources Canada, 2011, courtesy of the Geological Survey of Canada, 68776, 91823, 201772

---

[66] Dr. Brock served as both Director of the Geological Survey and Acting Deputy Minister of Mines from December 1908 until January 1914, at which time he became Deputy Minister in addition to his directorship (Zaslow, 1975, p. 264, 307).

[67] Tucker, 1952, p. 151.

initially with regard to their format, later in connection with their content, especially in connection with two of Jenness' ethnographic reports. Desbarats' seemingly dogmatic views and demands irritated Dr. Anderson greatly at the time, but with the passage of time were seen to have been reasonably fair. Some thirty years later, reminiscing on his struggles with Desbarats, Dr. Anderson made the following observation in a letter to Captain Bob Bartlett:

> I came to the opinion that the old boy was [Desbarats] a very good type of government executive and acted as squarely as possible when dealing with circumstances which were thrust upon him and to a larger extent beyond his control. We were perhaps inclined to consider him a partisan of old "V.S.," but probably a good deal of that was due to the defensive smoke screen of misquotations put out by the great "Arctic authority"… He outfitted the expedition well at the start and followed it up to the last in spite of war conditions, and about the only fault was that his orders and instructions were too vague, and when "V.S." got too far away and out of his control, and was getting away with it after his return, about all he could do was to "cover up" for his own superior officers, and try to be a peace-maker for the whole business.[68]

Desbarats' task was certainly not an easy one.

Dr. Reginald W. Brock, the Director of the Geological Survey of Canada in 1913 and a year later also Deputy Minister of the Department of Mines, was the person whom Stefansson approached first in February 1913 for supplementary funding for his new Arctic Expedition. Brock had granted Stefansson a small amount of funds in 1908 for his previous Arctic expedition, which Stefansson had repaid with a collection of Inuit clothing and artifacts four years later. It was also Brock who recognized the merit of having a few scientists from the Geological Survey of Canada added to the Expedition. And it was Brock who took Stefansson to see Prime Minister Robert Borden, where the agreement was reached for the Canadian government to assume full responsibility for the entire Expedition. Brock was an enthusiastic supporter of the Expedition during its early stages.

Dr. R.G. McConnell assumed Brock's duties as Deputy Minister after Brock resigned unexpectedly in the summer of 1914 to become Dean of the Faculty of Applied Sciences at the University of British Columbia. Brock's duties as Director of the Geological Survey were then attended to by William McInnes as "Directing Geologist." The Great War in Europe and McConnell's own geological and administrative interests governed his actions while he was Deputy Minister, leaving him little time for the activities of the Canadian Arctic Expedition. His apparent disinterest in the Expedition thus placed more of the responsibilities for it on Desbarats' shoulders.

Dr. Charles Camsell became Deputy Minister of Mines upon the retirement of Dr. McConnell in 1920, and was promptly transferred to Ottawa from the Geological Survey's Vancouver office, of which he had been in charge. He remained in his new capacity until his retirement in 1946. He soon showed a greater interest than his predecessor in the activities of

---

[68] Anderson, 1944 (May 10, 1944).

the Canadian Arctic Expedition, several members of which were already busily preparing reports of their completed field work. In due course he also provided a sympathetic ear to the difficulties Dr. Anderson had been encountering with Desbarats and Stefansson, and in 1922 to the request from four members of that Expedition to charges publicly declared against them by Stefansson. (For more on this subject see the following chapter.)

Others

Behind the scenes at the Geological Survey in 1913 two men with the ear of their director played an influential role on the Expedition during its formative period: **W.H. Boyd**, Chief of the Topographic Division, and **O.E. LeRoy**, a fast-rising geologist under Dr. Brock and "a likely candidate for the directorship had he lived."[69] These two men helped bring about the division of the Expedition into two distinct parties during its formative period—an exploration section and a scientific section—through their recognition that the Naval Service Department had made a grave error in placing Stefansson in command of the Expedition and their desire for the Geological Survey to have more control over its own men. They realized that their Geological Survey colleagues could not carry out worthwhile field investigations under the vague plans then envisioned by Stefansson. They even urged Dr. Brock to withdraw Chipman, Cox, Malloch and O'Neill from the Expedition,[70] owing in part to Stefansson's lack of executive ability and his irresponsibility.[71] Their input and encouragement to the men in the field helped to an immeasurable but unheralded extent in the success of the Southern Party.

One other person, in spite of no official link to the Expedition, nevertheless played an important behind-the-scenes role in its success. That was **Mae Belle Anderson**, the wife of Dr. Anderson, who deserves recognition for her unpaid and largely unknown involvement. In 1913 she accompanied her husband, to whom she had been married for less than a year, all the way to Nome at the start of the Expedition, taking some historically valuable photographs while at both Victoria and Nome. At Nome she agreed, in response to specific requests from some of the Expedition's scientists, to "do what she could for them" by keeping the scientists' loved ones informed about any news from the Expedition after she returned to her family home in Iowa. And keep in touch she did. From Iowa for the next three years, she faithfully sent news of the Expedition and words of encouragement regularly to one wife (Mrs. Murray),[72] several mothers (Mrs. Beuchat, Mrs. McKinlay, Mrs. Mamen and Mrs. Malloch), and one sweetheart (Miss Marjorie Pennock, Chipman's fiancée). Her news sources were articles in the various American newspapers she perused regularly, letters from her husband, and also from the Director of the Geological Survey, from whom she also obtained information and news periodically during the first year in response to her letters of inquiry.

---

[69] Zaslow, 1975, p. 312. Captain LeRoy was killed in action at the battle of Passchendaele, in 1917.

[70] LeRoy, O.E., to R.W. Brock. 1913?

[71] LeRoy, O.E., to R.G. McConnell. 1914?

[72] Johansen, too, was married when he joined the Expedition, but may actually have been separated from his wife, who was living in Denmark with their young daughter. To my knowledge, Mrs. Anderson did not contact Mrs. Johansen with news of the Expedition, but Johansen probably did.

In this way she tried to comfort the relatives, especially those whose husbands or sons were on the *Karluk*.[73]

Curiously, she also exchanged several personal letters with Stefansson during the first year the Expedition was in the Arctic, urging him among other things to encourage her husband to write popular articles as he, Stefansson, did so frequently. His letters in response to hers[74] provide an interesting insight into the troubled state of his mind after he lost his flagship *Karluk* and the personnel of his Northern Party, and merit examination by anyone writing about him. The correspondence ended after Mrs. Anderson learned of her husband's defence of his Southern Party men from the demands of Stefansson at Collinson Point in 1914.

More significantly and certainly more enduring, in addition to her morale boosting endeavours through her correspondence, Mrs. Anderson acted as an unofficial archivist for the Expedition, and saved vast numbers of newspaper clippings and articles dealing with the Expedition. This material, now housed in the R.M. Anderson Collection at Library and Archives Canada, provides a great additional source of valuable information on the Expedition.

Further, through her husband after the completion of the Expedition, she successfully urged the Canadian government to provide a pension to the widow of oceanographer James Murray and the widowed mother of Henry Beuchat.[75] In response to her efforts the government supplied a compassionate allowance of $1,000, which although small at least represented a form of recognition for the two men's efforts. She also played a leading role in the creation and installation of a memorial plaque to those men who lost their lives while working for the Canadian Arctic Expedition.

On the negative side, however, Mrs. Anderson played a significant role for many years after the end of the CAE, along with her husband, in attacking Stefansson's actions in the Arctic as well as his character, in many of her letters to friends and acquaintances. She was understandably trying to defend Dr. Anderson's reputation from the unpleasant charges Stefansson had levelled against him in *The Friendly Arctic*, but that is another story, yet to be told.

The Memorial Plaque

A few years after the completion of the Expedition, Mrs. R.M. Anderson wrote Prime Minister William Lyon Mackenzie King, urging him to have a bronze memorial plaque designed and put on display, possibly in the Parliament Buildings, to honour those members of the Canadian Arctic Expedition who had died while on duty in the Arctic. As a result of her suggestion, a handsome plaque bearing the names of sixteen deceased men was placed on display in 1926 in the foyer of the relatively small Public Archives Building beside the Royal Canadian Mint on Sussex Drive, Ottawa. And there it remained for viewing for many years.

---

[73] Wrangel Island survivor William L. McKinlay wrote Mrs. Anderson after World War I to say that his parents, and especially his mother, "… can never forget the wonderful source of comfort you provided to them in their days of anxiety about me" (McKinlay to Mrs. Anderson, 1922a).

[74] Stefansson's letters to Mrs. Anderson are among the latter's papers in the R.M. Anderson Collection, MG30 B40, LAC.

[75] Auditor General Report, 1920. Stefansson also arranged that The Explorers Club in New York contribute to a fund for the destitute Madame Beuchat (Stefansson, 1964, p. 217).

The plaque contained the following names of the men who died, listed alphabetically: Alexander Anderson, Ship's Officer; Charles Barker, Ship's Officer; Peter Bernard, Ship's Master; Henri Beuchat, Anthropologist; Daniel Wallace Blue, Ship's Engineer; John Brady, Seaman; George Breddy, Ship's Fireman; Edmund Lawrence Golightly, Seaman; John Jones, Ship's Engineer; Alistair Forbes Mackay, Surgeon; George Stewart Malloch, Geologist; Bjarne Mamen, Topographer; Thomas Stanley Morris, Seaman; James Murray, Oceanographer; André Norem, Steward; and Charles Thompson, Seaman.

Historian Levere commented years later on the Expedition's considerable loss of life, "Plenty of critics, then and now, have argued that the sacrifice was less for Canada and for science than for Stefansson's irresponsibility as a leader."[76]

There are sixteen names on the plaque commemorating the Expedition's death toll, but the number is now known to have been seventeen. The missing name is that of the Alaskan Iñupiaq Pipsuk, who transferred to the *Polar Bear* from the *Challenge* in the summer of 1917, and drowned at Barter Island on July 22, 1918, after his kayak capsized while he was tending his fish net. Stefansson did not mention Pipsuk or his fatal accident in *The Friendly Arctic* and may not have known about him or his death, for he was recuperating in the hospital at Fort Yukon when the tragedy occurred. Dr. Anderson mentioned the loss of seventeen Expedition men in a letter written in 1921,[77] so would appear to have known about Pipsuk's fate.

The last name on the plaque, Charles Thompson, was a sailor on the *Mary Sachs* and husband of the Iñupiaq woman, Jennie. Thompson claimed to be Norwegian, so that his name should have ended in "son."[78] Yet Stefansson spelled it Thomsen in *The Friendly Arctic,* and

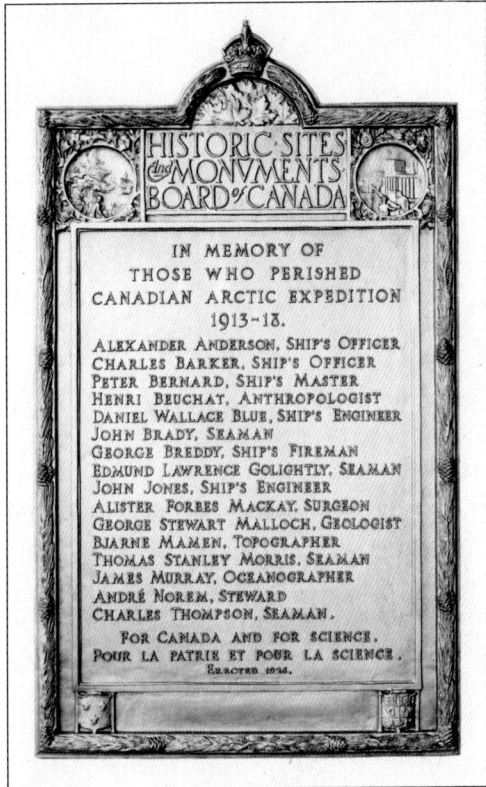

Figure 64 Plaque commemorating the men who died while with the Canadian Arctic Expedition. It was installed in 1926 in the foyer of the Public Archives Building (which was then on Sussex Drive alongside the Mint in Ottawa). The plaque went missing when the Archives moved to its new building on Wellington Street in 1967.

Photo: © Canadian Museum of Civilization, IMG 2010-0243-0001-Dm

---

[76] Levere, 1993, p. 402, footnote 104.
[77] Anderson, 1921f (November 21, 1921).
[78] According to Dr. Anderson, "Names ending in '-sen' are generally Danish, while Norwegian and Swedish names end(ing) in '-son, …' " (Anderson, 1925). Thomsen claimed he was Norwegian, yet Stefansson showed his signature was spelled "Thomsen" (Stefansson, 1921b, p. 649). Dr. Anderson

showed that spelling in the man's signature on the last note from both Bernard and Thomsen in *The Friendly Arctic* (p. 649). And the river on Banks Island which Stefansson named after the man is likewise spelled Thomsen. I have followed Stefansson's spelling in this book, as did Ashlee in her recent book on Storkerson.[79]

Unfortunately, the plaque disappeared in the 1960s when the Public Archives were moved to their present location on Wellington Street, and it has never reappeared.

Giving the Expedition Its Due Credit

The Canadian Arctic Expedition 1913–1918, in spite of its tragic losses of life and shocking over-expenditures, was the most successful northern accomplishment by the Canadian government in the early years of the twentieth century. In addition to its discoveries of new land, it added much to Canada's knowledge of the Arctic territory it claimed as its own, and supplied an abundance of biological, mineralogical and ethnological material for the National Museum of Canada. The public accolades showered upon Stefansson (primarily in the United States) for his Expedition's many achievements were largely deserved. His Expedition was the vision of a remarkable Arctic explorer, one who managed to survive four years of isolation and enormous dangers on the Arctic ice and on uninhabited land regions with no physical injury to himself other than a sprained foot. His northern interests sparked Canada's need to wave its flag, which was then the British Union Jack, in the Arctic and to assert to the world its sovereignty over the islands and waters north of the North American mainland and west of Greenland. Concerns over foreign exploration among those islands and the likely consequences thereof were what induced Canada's Prime Minister Borden to sponsor the Expedition in the first place.

Once the excitement and glamour of Stefansson's Arctic island discoveries died down, however, what remained of enduring value were the many biological and ethnographical collections made by Dr. Anderson and his Southern Party members, and the thirteen volumes of reports published almost exclusively on the findings and collections made by the same persons.

Concerns over Arctic Sovereignty

Claiming sovereignty of the Arctic lands north of the Canadian mainland by "showing the flag" was the Canadian government's prime reason for offering to finance Stefansson's Expedition in 1913. At that time, exploration in the Arctic still attracted considerable public attention, although the transmission of exciting news items still might take months to reach the public. Adventurous men from the United States, Great Britain, Norway, Sweden, Denmark, Greenland and Germany conducted expeditions into that vast region at one time or another, largely without asking the permission of the Canadian government on the tacit assumption that the region was open territory.

---

spelled it the anglicized way, "Thompson," in his field notebook—see e.g. Anderson (1916a), August 11, 1914—but "Thomson" in a letter to the Deputy Minister of the Naval Service (1921b).
[79] Ashlee, 2008.

Today, almost one hundred years after Stefansson's Expedition left the Arctic, the matter of Arctic sovereignty is still an ongoing concern of the Canadian government. The early expeditions were primarily looking to discover new lands, but increasingly the later expeditions sought scientific information and resource assessment. A few foreign territorial claims were made in the past and required political settlement. Norway, for example, claimed Axel Heiberg Land, Ellef Ringnes Land and Amund Ringnes Land in 1902, following their discovery by members of Norway's Otto Sverdrup Expedition. These areas did not become Canadian territory until Canada reached a financial settlement with Norway in 1930.[80] None of these "lands," now known to be islands, was inhabited by people in the early 1900s. In recent years Denmark and Canada have disputed the ownership of small, rocky and uninhabited Hans Island, which lies between Ellesmere Island and Greenland. No settlement has yet been reached. Even more recently the Russian government has claimed ownership rights over vast undersea areas of the Arctic, locally in regions that Canada may want to dispute.

The Canadian government has plans to increase northern patrols by land and sea to help assert its sovereignty over the islands and surrounding waters, and also to build one new, powerful ice-breaker. But even if these plans prove to be more than mere political gestures, what could it do if Russia suddenly decided to send part of its naval fleet from Anadyr to Murmansk by way of our Northwest Passage?

The rapid diminution of Arctic sea-ice north of the Canadian mainland, about which one now reads almost daily in our newspapers, or hears by means of our televisions or assorted satellite-linked electronic devices, will likely lead to a greatly increased number of ocean-going vessels wanting to cross North America from the Atlantic Ocean to the Pacific Ocean or vice versa by way of Canada's Northwest Passage, either on the water's surface or beneath it. Nearly a century after Stefansson and his assistants on the Canadian Arctic Expedition freely roamed over the ice in the frozen North in search of new land to claim in the name of Canada, Canada is far from prepared to deal with the new access to the region the warmer climate will permit.

In 1913 the matter of Arctic sovereignty greatly influenced the sponsoring of the Canadian Arctic Expedition lest other countries, particularly the United States, discovered lands north of the Canadian continental mainland and laid claim to them. Today Arctic sovereignty exists with even more urgency than it had then, but the emphasis has changed. Satellite cameras and manned space ships have given us proof that all lands on planet Earth are now known and can be watched (excepting those still concealed beneath the Greenland and Antarctic ice caps).

Today, perceived threats to our political ownership of the North remain. Concerns have arisen over rival claims to potential offshore petroleum deposits, over how to control the probable increase in trans-Arctic shipping activities that may result from the recently recognized climatic warming, and about possible military threats from foreign countries. Additionally, the needs and demands of the Inuit have changed over the years and challenge the minds of government officials charged with overseeing Arctic affairs today. As a consequence, sovereignty concerns of the past have not dissipated, they have simply undergone a marked change in form.

---

[80] Kenney, 2004, pp. 125–128.

THE STEFANSSON–
DR. ANDERSON FEUD

23

No book on the Canadian Arctic Expedition 1913–1918 would be complete without mentioning the unfortunate feud that arose between the two Expedition leaders and developed into a lifelong animosity on the part of one of them.[1] Many others took one side or the other, and the resulting bitterness lasted for years.

Early Encounters

Stefansson and Dr. Anderson first met in 1903 when both were students at the State University of Iowa in Iowa City.[2] They were apparently only casual acquaintances then.[3] Both men graduated that same year with the degree of Bachelor of Philosophy.[4] Dr. Anderson continued his studies at the same university, obtaining his PhD in zoology in 1906. He then accepted the position of assistant commandant at Blees Military Academy, Macon, Missouri, one he soon found not to his liking, for it offered him little opportunity to teach and none to pursue his interests in zoological research.[5]

From 1903 to 1906, Stefansson pursued graduate studies at Harvard University. He took courses in religion during his first year after receiving a scholarship from that department; then

---

[1] Some time after his return from the Arctic in 1918 Stefansson sought Dr. R.W. Brock's assistance to help him make up with Dr. Anderson "… and let bygones be bygones." Dr. Brock apparently believed that Stefansson sincerely desired the restoration of their friendship at the time and viewed the feud between them as having "taken up far too much of their time" and having "been given too much publicity already." Dr. Brock also believed that if he had seen Stefansson's manuscript for *The Friendly Arctic* before it was published, which Stefansson tried to have him do, he could have suggested alterations that "would have been satisfactory to all concerned." (Brock, 1923 [April 24, 1923]).

[2] Established in Iowa City in February 1847, only 59 days after Iowa became a state, this public institution was named the State University of Iowa. And that was its name in 1903, when both Dr. Anderson and Stefansson graduated from it. Today it is known as the University of Iowa.

[3] Diubaldo, 1978, p. 39.

[4] The names of both Dr. Anderson and Stefansson are shown as recipients of Bachelor of Philosophy degrees on the official program of the Forty-Third Annual Commencement of the State University of Iowa's College of Liberal Arts. (A copy of this program is in CMNAC/1996-077, Series A—R.M. Anderson Collection Box 65, Folder 19, Archives, CMN.) Dr. Anderson wrote in ink on this commencement program that Stefansson claimed in his book, *Hunters of the Great North* (1922, p.7), that his degree from Iowa was a Bachelor of Arts, adding the comment "Very forgetful!" Stefansson made the same claim in his autobiography *Discovery* (1964, p. 38).

[5] Diubaldo, 1978, p.39.

switched departments when offered a fellowship in anthropology. He remained there for almost two years; then was advised to leave without graduating.[6]

Stefansson spent part of the next two years among the Inuit along the Arctic coast near the mouth of the Mackenzie River, hired as anthropologist on the Anglo-American Polar Expedition 1906–1907 by one of its co-leaders, Ernest de Koven Leffingwell.[7] He spent very little time with the members of the expedition, however. In 1907 while at Herschel Island he met Captain Christian Klinkenberg Jorgensen,[8] a Danish sailor, hunter and trapper, whose account of some Inuit on Victoria Island with light hair and bluish eyes greatly aroused Stefansson's curiosity, and Stefansson resolved to visit and study them.

In 1908 Stefansson persuaded the American Museum of Natural History in New York to provide him with a small amount of money for a one-man anthropological expedition to the central Arctic to study the unusual Inuit about whom the Danish captain had spoken. Dr. Anderson learned of Stefansson's plans and wrote inquiring about going north with him.[9] Recognizing that the addition of Dr. Anderson would provide zoological professionalism to his expedition, Stefansson advised him to offer his services to the museum without salary, just as he himself had done, pointing out the amount of public attention that such a trip would arouse. Stefansson also hinted at the possibility of employment with the museum afterwards. Meanwhile, Stefansson convinced the Director of the American Museum of the value of having a professional zoologist collecting specimens of Arctic birds and animals for the museum, of which they had few at the time. Sensing a great opportunity, the American Museum of Natural History agreed to have the two men join forces, Stefansson to study the Inuit, Dr. Anderson to study and collect bird and animal specimens for the museum.[10] Both men shared one common trait; both sought public recognition through their activities when they embarked on what was originally planned as a one-year undertaking, then soon stretched to two years,

---

[6] Dr. Anderson mentioned Stefansson's hasty departure from Harvard University and subsequent interest in working for the Canadian government in Ottawa in a letter to an American friend. "Near the end of the school year 1906, he was 'allowed to resign his fellowship' [at Harvard], to leave without expulsion, on the charge of using his connection with the faculty to obtain examination questions and sell them to students. That was brought out by investigation from here [Ottawa], and explains why he never took a position as ethnologist in this country" (Anderson, 1922b). Dr. Anderson learned of this as a result of an investigation made for the Geological Survey of Canada sometime shortly before 1920 when Stefansson "… was considering a job here" (Anderson, 1921a).

[7] Stefansson, 1964, p. 59.

[8] Ashlee, 2008, p. 47. Sometime soon after he decided to remain in the Arctic Klinkenberg dropped his family name Jorgensen, and anglicized his middle name to Klengenberg, by which his descendants in the Arctic are known today. He soon afterwards became widely known as "Captain" Klengenberg or simply "Charlie" Klengenberg. Stefansson and Dr. Anderson, however, generally used the original form of his name, "Klinkenberg," in both their written and spoken references to the man. Late in his life Stefansson added the comment … "his real name was Klengenberg" (Stefansson, 1964, p. 73), perhaps in recognition of the fact that the descendants had adopted that spelling.

[9] Stefansson, 1964, p. 101

[10] Diubaldo, 1978, pp. 39–42.

and in due course to four. Stefansson's course utilized popular lectures and publications, and eye-catching newspaper submissions and interviews. Dr. Anderson chose the more conservative path of scientific publishing when he had time.

The Seeds of Conflict

Thus came about the Stefansson-Anderson Arctic Expedition 1908–1912, the first of the two Arctic expeditions involving these two men. Between 1908 and 1912, almost totally dependent upon meagre funds primarily provided by the American Museum and the Geological Survey of Canada, with some assistance from the Meteorological Service of Canada, the Royal Ontario Museum and one or two private sources, Stefansson travelled the Arctic coast between northern Alaska and Coronation Gulf on foot and by dog team, making extensive ethnographic collections, while Dr. Anderson collected large numbers of bird and mammal specimens between the Mackenzie River Delta and Coronation Gulf. During much of that time Stefansson was ably assisted by the Alaskan Iñupiaq, Billy Natkusiak, while Dr. Anderson was aided by the Mackenzie Delta hunter, Ilavinirk. Both Stefansson and Dr. Anderson were also occasionally accompanied by one or more other Inuit hunters and the Iñupiaq widow and seamstress, Pannigabluk.

Some friction arose between the two men on occasions during their 1908–1912 expedition, as when Stefansson berated Dr. Anderson over his unwillingness to fully appreciate the Inuit lifestyle,[11] but the two men had generally gotten along fairly well. This was largely because they were together for only about four months of the four years,[12] and apparently also because Dr. Anderson chose to hold his tongue and temper. Looking back on that expedition a decade later, however, Dr. Anderson showed some early misgivings about Stefansson's behaviour and also some hostility towards the man:

> He was an agreeable cuss in many ways, and treated me all right as long as he could use me. I made excuses for his delinquencies and thought that on the whole he was fairly decent until after I was mixed up in the expedition in 1913 … Stefansson has absolutely no moral sense, and considers anything justifiable to accomplish an end he has in view. He admitted that to me several years ago, early in the expedition, when he calmly admitted having told many lies to me and made many promises, in order to keep me in line long enough to get the expedition organized. After that he treated all promises as scraps of paper.[13]

The American Museum of Natural History and the Geological Survey of Canada subsequently benefitted richly from the collections brought back by the two men. Stefansson quickly published a popular account[14] of his Arctic studies during the 1908–1912 period, and then a detailed ethnographic report.[15]

---

[11] Diubaldo, 1978, p. 42.
[12] Anderson, 1928b (December 13, 1928).
[13] Anderson, 1922b (February 22, 1922).
[14] Stefansson, 1913.
[15] Stefansson, 1919m.

The Feud Gathers Strength

During the winter of 1912–1913 Stefansson planned his next Arctic expedition and obtained partial sponsorship from the National Geographic Society on condition that Dr. Anderson was part of the expedition. The latter had married his Iowa fiancée, Mae Belle Allstrand, upon returning from the Arctic in the fall of 1912, however, and had little interest in further field work until he had written up the results of his Arctic investigations and collections for the American Museum of Natural History. Moreover, he hoped to be hired full-time by that museum, although that employment was not a sure thing. Once Stefansson had accepted the offer of full sponsorship by the Canadian government, he was faced with having to persuade a reluctant Dr. Anderson to join his Expedition as part of the arrangement. Towards that goal, he then cleverly persuaded Dr. Brock, Director of the Geological Survey of Canada, to offer full employment to Dr. Anderson as a zoologist with Canada's relatively new Victoria Memorial Museum, on condition that he accompany Stefansson on his new Arctic Expedition. He thus succeeded in getting Dr. Anderson to become his second-in-command and to take charge of the scientific southern part of Stefansson's Expedition. Both men later regretted the arrangement.

That spring saw a series of incidents initiated by Stefansson, which increased Dr. Anderson's resentments. These initially included Stefansson's rushing off to Europe in the spring of 1913, leaving all sorts of time-consuming tasks for Dr. Anderson to attend to. In the absence of Stefansson, Dr. Anderson, although busily struggling to complete his own assignments for the American Museum of Natural History, spent much time attending to correspondence that properly required Stefansson's attention.

Later, as we have seen, a series of serious differences of view erupted between the two men, commencing with the disclosure in Victoria of Stefansson's newly imposed news restrictions and the heated argument over Stefansson's contracts with major newspapers, which Dr. Anderson cited as the start of their real feud. Similar inflammatory meetings followed, although on differing grounds, at Nome in July 1913, and at Collinson Point early in March 1914, reflecting the increasing bitterness among the government scientists for the way Stefansson kept interfering with their activities. The world, however, heard little about these personnel problems.

Publicly Dr. Anderson traced the start of his trouble with Stefansson to the outset of their second Expedition in 1913. The *Ottawa Citizen* of January 14, 1922, quoted him as stating:

> The friction … started from the time that Mr. Stefansson informed members of the expedition [in Victoria] that he had sold out the newspaper rights of both parties. Mr. Stefansson explained at the time that no member of the expedition could write a line to his family except through Ottawa, and that no member of the expedition could send a telegram back to his relatives. This was obviously done to protect Mr. Stefansson's newspaper rights.

> The feeling spread among the men that a supposedly great scientific expedition was really, at the bottom, a newspaper and magazine exploiting scheme. This matter of furnishing news to the newspapers remained a cause of trouble through the whole expedition. It

might not have been so with another commander in chief, but Mr. Stefansson ever had his eye on his news-reading public. He admitted to me that what he had done was ethically wrong, but said he felt justified in using any means to get the expedition started.[16]

From 1913 onwards the feud between the two men festered, with periodic outbursts, for the next three years while both were in the Arctic, despite the infrequency of their contact. Stefansson had not been pleased when senior government officials modified his original plans by insisting upon adding several men from the Geological Survey of Canada to undertake separate scientific studies on the Arctic mainland under the leadership of Dr. Anderson. This action resulted in the creation of a dual-purpose Expedition with direction from two dissimilar government departments and two dissimilar leaders. Additionally, with his here-today, there-tomorrow mode of operation and his frequently changed plans of action and activities, Stefansson could not or did not understand the seemingly regimented working style of Dr. Anderson and the loyalties of the government scientists. He revealed his early resentment to their inclusion by criticizing them soundly in a letter in 1915 to Desbarats in Ottawa.

> I am unalterably ... of the opinion that men of the Government services are as a class unsuited for expeditions such as this one was intended to be as I planned it. The trouble is what I should call a moral one ... These men feel the Govt owes them a living ... These men are receiving higher pay than ever paid before on a polar expedition but make it clear their presence is a favor to the Govt and expedition ... They make their service conditional on being supplied with unnecessary items of food – e.g., butter. [17]

Perhaps the key words in the quotation above are "... for expeditions such as this one *was intended to be as I planned it*." After three controversial, heated and exasperating meetings with Stefansson—in Victoria, Nome and at Collinson Point—all of which had to do basically with the scientists' seeking a better understanding of Stefansson's plans and trying to confirm the reality of the dual form of authority on the Expedition, the Geological Survey men held little respect thereafter for Stefansson's leadership or his frequent interference with their work.

Following Dr. Anderson's return to Ottawa in 1916, he initiated a program of letter-writing to newsmen, missionaries, scientists, the heads of some professional societies and even museum executives, informing them of various actions by Stefansson that he had found objectionable over the years, and making little attempt to conceal his mounting bitterness against his rival. Many of the details he included in those letters ought not to have been mentioned, perhaps, but they were, and he kept several copies of them scattered among his many files as ammunition for his quietly mounted verbal assault on Stefansson.

In due course word of some of these letters reached Stefansson, who initially chose to ignore Dr. Anderson's verbal attacks on him. As time went on, however, and he learned of more such letters, he decided to retaliate. He had already complained about Dr. Anderson's actions (which he regarded as insubordination) in several letters written from the Arctic to Desbarats

---

[16] Anonymous, 1922a.
[17] Stefansson, 1915 (February 13, 1915).

as well as to other senior government officials. In the summer of 1919 he struck back more pervasively by complaining to both Desbarats and Prime Minister Borden about disagreements he had with the Geological Survey and the publication committee. Desbarats' suggestion that Stefansson be appointed editor for the Expedition's publications series was quickly rejected by both Dr. Anderson and the publications committee, so Stefansson decided to make a direct, all-out assault, on Dr. Anderson in the book he was writing about the Expedition.

Storm over *The Friendly Arctic*

Stefansson's book about the Canadian Arctic Expedition, *The Friendly Arctic*, all 784 pages of it, was published in New York late in 1921. Its charges of attempted mutiny, alleged conspiracy, insubordination, disobedience[18] and other objectionable actions by the men of the Southern Party, and Dr. Anderson, in particular, almost immediately caused a human explosion at the Geological Survey in Ottawa. Such charges were considered extremely serious, and when they were further publicized by means of Stefansson's lectures across the United States and Canada, in frequent newspaper interviews and through the wide distribution of his book, Dr. Anderson felt he could remain silent no longer. He had largely ignored Stefansson's earlier complaints about him, he claimed, but because these charges were levelled publicly against some of his Southern Party colleagues as well, he felt compelled, as their former leader, to respond in kind. Furthermore, he recognized that Stefansson's charges also reflected upon the honour of the Geological Survey. Since their return from the Arctic in 1916 the six scientists from the Southern Party (Dr. Anderson, Chipman, Cox, Jenness, Johansen and O'Neill) had remained largely silent publicly about the various conflicts they had had with Stefansson while in the Arctic. Apart from Dr. Anderson, they preferred, instead, to set aside the unpleasantness of the past and busy themselves with their Arctic collections and data. Besides, Stefansson was seldom in the news, and the war in Europe and the recovery from it after it finally ended in 1918 occupied people's minds. Furthermore, few if any of the Southern Party members had any interest whatsoever in contributing to Stefansson's insatiable desire for publicity, even if it was negative publicity. But foremost, all were aware that government employees were not supposed to voice controversial matters in public without first obtaining administrative approval.

Meanwhile, Cox and Jenness had joined the Canadian Army and gone off in 1917 to fight in France, returning in 1919. A year later, in 1920, Cox and O'Neill resigned from the Geological Survey and left Ottawa to work in India. Cox remained out of the country for several years, but O'Neill returned to Canada in 1921 and commenced a life-long teaching career at McGill University.

Stefansson's book about the Canadian Arctic Expedition received almost immediate acclaim and wide distribution throughout the United States.[19] Indeed, it was (in today's

---

[18]   Stefansson, 1921b, pp. 120, 123. See also Anonymous, 1922e, p. 32.

[19]   *The Friendly Arctic* had not been Stefansson's original title for his book. It was suggested by Gilbert Grosvenor of the National Geographic Society (Stefansson, 1922). Little did he think at the time how much argument the word "Friendly" would stir up over the next century.

terminology) a "best-seller." And the publicity it received drew large numbers of people to the many lectures Stefansson subsequently delivered across the U.S. and Canada, during which he spoke glowingly about the wonders of the North.

On page 123 in his book, Stefansson deliberately charged Dr. Anderson and some of his scientist colleagues with "threatened mutiny" at Collinson Point for defying his orders to supply Stefansson's first ice trip with sleds, dogs, equipment and other supplies. Stefansson was commander of an Expedition that was under the jurisdiction of the Department of the Naval Service and, in naval circles, refusal to obey orders was considered mutiny. But the scientists in the Southern Party, with the exception of the naturalist Johansen, worked for the Department of Mines and took their orders from that department. Even Johansen took his instructions from the Mines Department, although he had been hired by the Department of the Naval Service. Most of them, in consequence, felt no loyalty to the Department of the Naval Service despite its senior rank, and hence felt no obligation to respond to Stefansson's instructions in 1914. Dr. Anderson's reason for refusing to go along with some of Stefansson's demands then, and again the following year at Bernard Harbour, was that by doing so he would have seriously disabled the activities of the Southern Party. In addition, Dr. Anderson objected strongly to Stefansson's attempts to override the orders that Dr. Anderson had received from the Geological Survey. This was exactly the kind of unpleasantness with Stefansson that Dr. Anderson had anticipated before the Expedition headed North, unpleasantness that he had tried to avoid by asking Stefansson, at both Victoria and at Nome, to give him a clearly written statement of their respective areas of authority. These were statements that Stefansson always managed to avoid putting in writing.

During the first few weeks after getting Stefansson's book, Dr. Anderson pencilled many marginal comments in his two copies of the book,[20] voiced his views of Stefansson and his book both strongly and widely in private letters to friends, but said nothing to the press. Being a dedicated scientist, he was greatly troubled by the many inaccuracies he found in *The Friendly Arctic*. "It is over 700 pages," he wrote to an acquaintance, "and I could pick out and nail a lie on nearly [every] page."[21] While perhaps a trifle exaggerated, this statement certainly reflected his feelings at the time. In a letter to a military friend several weeks after the book appeared, he wrote:

> ... I would not have minded the general account so much, it is really good readable fiction, but he went out of his way to make a large number of wholly imaginary, not to

---

[20] Dr. Anderson had at least two copies of *The Friendly Arctic*, one bearing the date 1921 on its title page, the other with the date 1922 on its title page. Dr. Anderson inserted different comments in each copy. His 1922 copy is enclosed in a green box with the call number G670 1908 S73 1922. Both copies are housed in the Rare Book Room in the library of the Canadian Museum of Nature, Gatineau (Aylmer), Quebec.

[21] Anderson, 1921g (December 3, 1921). W.L. McKinlay, the sole scientist to survive the *Karluk* disaster, shared the same sentiments as Dr. Anderson about Stefansson's book. In 1980 during a visit I had with him in Glasgow, he commented angrily on the number of "lies" in the book. Both McKinlay and Dr. Anderson evidently retained their feelings of hostility towards Stefansson throughout their lives.

say maliciously and criminally libellous statements in regard to the scientific staff of the expedition, and as I was the ringleader, it was up to me, in justice to the other men, to take the first crack at Stefansson as a liar and an imposter.[22]

Assault by Newspaper

Soon after the appearance of Stefansson's book, three of the members of the former Southern Party—Dr. Anderson, Chipman and Jenness—met at the Victoria Memorial Museum in Ottawa to discuss the contents of the book, its various charges against them, and what they could do about it. By mail they sought the views of a fourth member, O'Neill, in Montréal.[23] All were sufficiently irritated by Stefansson's accusations that they decided to mount a public assault upon him and his book through the newspapers. They hoped by this public declaration to induce the Canadian government to investigate, or at least to inquire, into Stefansson's charges in defence of their reputations as well as that of the Geological Survey.

Both Dr. Anderson and Stefansson knew that government officials were not likely to be greatly concerned about the injustices done to the members of the Expedition, for Stefansson had convinced them that such things happened regularly on expeditions. But the government might be "… willing to investigate scandals"[24] that could prove embarrassing to senior government officials or members of Parliament.

The first volley came from Dr. Anderson, who gave an interview to a member of the Canadian Press on January 13, 1922, commenting on the charges against him and his colleagues that had been published in Stefansson's *The Friendly Arctic*. This interview was duly reported the next day in many newspapers across Canada. *The Citizen*, one of the two daily newspapers in Ottawa at the time, carried the interview under the heading "Arctic voyage of Stefansson may be probed."[25] *The Ottawa Evening Journal* published a shorter version on the same day under the heading "Karluk journey was big scheme to exploit news." A much longer version was sent at the same time to the Associated Press in the United States. In his interview Dr. Anderson stated that Stefansson had unjustly charged the scientific staff of the Expedition with "attempted mutiny" and "insubordination" while at the same time being insubordinate himself by attempting to change the activities of the Expedition without authority from Ottawa, and ignoring government instructions to terminate the Expedition and return to civilization. He also voiced other criticisms of Stefansson's actions while in the Arctic, some of which had floated around Ottawa as rumours since 1916.

The Canadian Press representative also interviewed the ethnologist Jenness at the Museum, who provided the following written statement:

---

[22] Anderson, 1922b (February 22, 1922).

[23] The fifth and sixth members of the former Southern Party were either unavailable (Cox was out of the country) or held different views on the subject (Johansen). McKinlay was the seventh member, who gave his moral support to Dr. Anderson, but did not participate in the newspaper campaign.

[24] Anderson, 1922a (January 24, 1922).

[25] Anonymous, 1922a (*Ottawa Citizen*, January 14, 1922).

The southern party of the Canadian Arctic expedition was altogether separate from the northern party, and Dr. Anderson, the commander of the southern party, was responsible, not to Mr. Stefansson, but directly to the Canadian Government. This point was raised at the very outset of the expedition, even before it left Victoria. One of the main reasons for the meeting in Nome, of which Mr. Stefansson gives such an absurd account [in his book], was to have reaffirmed, once for all, the independence of the two parties, and Dr. Anderson's direct responsibility to the Canadian Government, and this Mr. Stefansson did with his own mouth in the presence of nearly all the staff.

After the loss of the Karluk Mr. Stefansson found himself without the resources he required for the formation of a new Northern party and sought to draw on the supplies and equipment of the Southern party. For this he had absolutely no authority, his jurisdiction being limited to the Northern party only. Dr. Anderson rightly refused to allow his party to be crippled for Mr. Stefansson's new enterprise, and in this he was supported by every scientific member on his staff. Later, the Canadian Government officially endorsed his action.

All talk of a mutiny then in the Southern party of the expedition is false. Dr. Anderson's firm attitude against Mr. Stefansson's unjustified aggression had the approval of every member of his staff as well as of the Canadian Government.[26]

Jenness had obtained the permission of the Deputy Minister, Dr. Camsell, to publish this statement, which appeared in *The Ottawa Evening Journal* on January 14, 1922, under the banner "Ethnologist slams Mr. Stefansson also." However, the *Journal* statement did not include two additional sentences indicating the strong loyalty felt towards Dr. Anderson at the time. These sentences read:

"The members of the Southern party have nothing but praise for the wise leadership of Dr. Anderson, and for the support he gave them on every occasion in carrying out their scientific work. It was his courage and ability that made the work of the Southern party so conspicuous a success."[27]

The public comments by Dr. Anderson and by Jenness in the two Ottawa papers on Saturday, January 14, were followed two days later[28] by a lengthy statement from the Southern Party's geologist O'Neill. His comments, directed more at Stefansson and the Expedition than at the charges in his book, appeared on Monday, January 16, in *The Ottawa Evening Journal* under the heading "Says Stefansson lived in luxury." *The Citizen* on that date published, under the banner "Explorer issues denial but other members persist," a brief response from Stefansson,

---

[26] Jenness, 1922c.

[27] A typed copy of Jenness' statement is in CMNAC/1996-077, Series A—R.M. Anderson Collection, Box 66, Folder 13, Archives, CMN. It includes the final two sentences presented here, which are missing from the newspaper account. A note mentioning that omission is typed across the bottom of the document.

[28] The intervening day was a Sunday. Ottawa had no Sunday newspapers at that time.

together with O'Neill's comments and the statement made by Jenness, which had appeared two days earlier in *The Ottawa Evening Journal*.

Chipman, the Southern Party's second-in-command, a man I knew to be serious, responsible and cautious, was also drawn into the fray. He, too, prepared a statement for the press, had it approved by Dr. Camsell for publication, then decided to add some additional comments and asked Dr. Camsell to return it to him. That delay on the weekend resulted in his statement not reaching the newspapers because, on Monday, January 16, when it would have been posted to the newspaper, Dr. Camsell sent the following memo to his assistant deputy minister, L.L. Bolton:

"Advise Anderson, Chipman and Jenness that the Minister instructs that no more interviews are to be given the Press <u>re</u> Stefansson at present, until the whole matter can be laid before the regular Minister, Hon. C. Stewart, who will decide what action is to be taken on behalf of our men."[29]

O'Neill's name is notably missing from Dr. Camsell's memo because he was no longer a member of the Geological Survey, and thus beyond Dr. Camsell's administrative jurisdiction. Thus silenced, the three aggrieved government men revamped their assault approach, in close communication with O'Neill.

Curiously, either by sheer chance or deliberate intent, a long and glowing review of Stefansson's book, *The Friendly Arctic,* appeared in *The Citizen* on January 16 and 17, over the initials E.W.H., whose actual identity remains unknown. Judging from the tone of the review and the attention it gave to the Stefansson-Anderson feud, it would appear to have been a deliberate and clever public response, possibly written at least in part by Stefansson himself, to Dr. Anderson's assault of January 14.

On January 24, O'Neill wrote Dr. Camsell informing him of an important editorial on the subject in the *Montreal Standard* on January 21. The scientists, he added, simply wanted a square deal (presumably by way of a government inquiry or investigation) in response to the charges being spread by Stefansson in lectures and press releases all over the United States and Canada, and in his book, *The Friendly Arctic*.

During January and February the Minister of Mines was busily touring the countryside seeking a seat by means of which he could obtain re-election.[30] As a result, a request dated February 25, 1922 from the four Southern Party members (Dr. Anderson, Chipman, Jenness and O'Neill) for a government inquiry of the charges against them in Stefansson's book was

---

[29] Camsell, 1922 (January 15, 1922). As the Minister of Mines, the Honourable Mr. C. Stewart, was out of town at the time, the instructions to Camsell had evidently come from another cabinet minister.

[30] The Thirteenth Parliament had been dissolved early in October 1921. Charles Stewart, a farmer from Alberta, was appointed Minister of Mines as well as Minister of the Interior on December 29, 1921, and under the rules of the day, had to face a by-election to qualify for his appointments. The Fourteenth Parliament did not commence until March 8, 1922, so Mr. Stewart, a Liberal, was in the West in January seeking unsuccessfully to locate a safe constituency in which to obtain a seat in the House of Commons (Johnson, 1968, pp. 551 and 728). He finally won a seat in Argenteuil, Quebec, in a by-election March 2, a few days before Parliament opened. Thus he had other things on his mind when he returned to Ottawa than asking the government to investigate the charges levelled by Stefansson against Dr. Anderson and other members of the Geological Survey.

stalled on the Deputy Minister's desk. In the days that followed Dr. Camsell listened either individually or collectively to the concerns of Dr. Anderson, Chipman and Jenness about Stefansson's charges, and was especially sympathetic when they spoke of the unfairness of having to remain silent because they were government employees. He also expressed his understanding of their reasons for requesting such an inquiry and assured them he would take up the matter with his Minister upon the latter's return to Ottawa. Days passed without further news, the explanation being the Minister was terribly busy following his return from winning his by-election.

Suggested Compromise

Seeking an alternative to the government inquiry requested by the maligned foursome, Dr. Camsell suggested that they consider responding to Stefansson's charges through the major English geographic journals rather than the newspapers, thereby more or less giving them permission to take their assault into new territory. It was quickly pointed out, however, that such an approach to the three main geographic journals was inappropriate, because one of them did not publish discussions, and Stefansson had friends in senior positions at the other two who could be counted on to suppress their communiqués or to provide Stefansson the opportunity to respond and thereby further promote his publicity. After some additional discussion, the four scientists agreed to have Jenness prepare a "Letter to the Editor" for the major U.S. scientific journal *Science*. As an ethnologist he would respond to Stefansson's book *The Friendly Arctic* from a professional point of view, drawing attention to a number of significant misstatements or errors in the book. In this fashion they could defend their reputations professionally as well as strike at Stefansson's pride (and pocketbook) by drawing public attention to the book's imperfections.

Jenness' carefully worded "Letter to the Editor" appeared in the July 7, 1922 issue of *Science* after being approved first for content by his Expedition colleagues and then for publication by Dr. Camsell. It was responded to eight months later in the March 20, 1923 issue of the same journal by two letters, one from Stefansson himself, the other signed by two of his supporters, B.M. McConnell and Harold Noice.[31] The latter letter struck sharply and insinuatingly at Jenness' literary methods in a manner that looked suspiciously like the sophisticated literary creativity of Stefansson himself. Both of Stefansson's young supporters, it is worth adding, soon afterwards fell out of favour with him as a result of seeking their own fame at his expense.

Scientists Seek Government Inquiry into Stefansson's Charges

Meanwhile, on March 6, 1922, Dr. Camsell finally forwarded the scientists' petition for a government inquiry to his Minister, the Honourable Mr. Charles Stewart. Two days later the *Ottawa Morning Journal* reported that the petition of the Southern Party's scientists had been sent to the Minister of Mines, asking for the appointment of a commission by the government: "to investigate the organization, conduct and events of the expedition, and to determine the

---

[31] McConnell and Noice, 1923, pp. 368–373.

truth or falsity of charges made by Mr. Stefansson ... Mr. Stefansson has slandered the members of the scientific staff of the expedition and, indirectly, the Geological Survey of the Department of Mines."[32]

Stefansson had also: "... made statements which reflect upon the undersigned and are such serious reflections upon their personal honor that they are determined in justice to their personal and professional reputations to clear themselves of the unjustifiable charges made against them."[33]

When Stefansson learned of this new action in Ottawa, he also asked Dr. Camsell to call upon the Canadian government to convene a commission to investigate *his* charges. The scientists sought justice and the defence of reputations—theirs and that of the Geological Survey—whereas Stefansson sought more publicity.

On March 16, while in Ottawa on one of his frequent visits there, Stefansson openly admitted to a reporter from the *Ottawa Journal* during an interview that he had deliberately made the offending charges against Dr. Anderson and his colleagues in *The Friendly Arctic* that led them to ask for a government inquiry into his charges.[34] He was then asked, "Did you anticipate when you were writing your book that this situation [the anger over the charges made by Stefansson] would result?" "I did. I knew it would," he replied, "... I learned from my ... friends that stories were being circulated here. I had desired to be magnanimous, but it became necessary to take cognizance of them [the newspaper attacks on him]."[35]

The naturalist Johansen, who had been with the Southern Party in the Arctic, spent some time with Stefansson while he was in Ottawa on this occasion, and Stefansson remarked to him that "... he put the things in the book because it would interest the public." During his visit to Ottawa Stefansson managed to speak in his defence with both the Honourable Mr. Stewart, the Minister of Mines, and the Honourable Mr. Thomas Crerar, the leader of the Progressive Party from 1921 to 1925, who admired Stefansson's work.[36]

Finally, on the evening of March 16, in response to their petition, the Minister of Mines, the Honourable Mr. Charles Stewart met with Dr. Anderson, Chipman, Jenness and

---

[32] Camsell, 1922 (January 15, 1922).

[33] Anonymous, 1922c (March 8, 1922).

[34] In May 1963, my father received a book entitled *Stefansson—Ambassador of the North* as a gift from its author, D.M. LeBourdais (1963). On its title page appears the author's inked inscription "To Diamond Jenness who thirty years ago advised me not to publish some parts of this book, with best wishes," followed by the author's signature. When I chanced upon the book some while later and asked him the significance of its inscription, my father replied in a low and somewhat gruff voice, quite unusual for him, that the book discussed some unpleasant events that had been buried years ago and were best left that way. Sensing that my question had struck an unexpected raw nerve, I did not press him for details. LeBourdais' book does, however, draw attention to the fact that the conflict between the two men is very much a part of the Expedition story, and that topics that proved too distasteful and painful to discuss for many years have, with the passage of time, become a part of the historical record.

[35] Anonymous, 1922d (March 17, 1922).

[36] Anderson, 1922e and 1922g (March 20, 1922, and April 11, 1922).

Dr. Camsell.[37] The fourth petitioner, O'Neill, was unable to come from Montréal that day. Captain Robert Bartlett had also hoped to attend the meeting in support of their cause, but was unable to get permission and leave from his U.S. government employer to come from New York. Dr. Anderson later reported that his threesome:

> … talked over the charges we had laid against Mr. Stefansson's book and told him [the Minister] why we were asking for an inquiry into the events of the expedition. He admitted that as civil servants we had not had a fair deal, being obliged to be silent while Mr. Stefansson had the privilege of going up and down the country disparaging our work and slandering and libelling us personally. He promised to bring our request before the Government.[38]

The petitioners and Dr. Camsell felt that the interview had been a satisfactory one. Mr. Stewart admitted at the time, however, that he knew practically nothing about the case or even about the Expedition.[39]

The Government's Response

Mr. Stewart may then have discussed the matter with one or two other Ministers, but it is doubtful whether the subject went much farther than that. In any case, he shortly afterwards instructed his Deputy Minister, Dr. Camsell, to inform the petitioners that no action would be taken on their requests for a government inquiry. This Dr. Camsell did, but also mentioned the Minister's decision in a note he sent to the scientific journal *Science*, which appeared as a "Letter to the Editor" in the June 8, 1923 issue of that scholarly journal. There it would likely be seen by some other scientists but few politicians or the general public, and although his note was primarily a response to Stefansson's earlier criticisms of the Geological Survey men in the same journal, it also expressed the formal displeasure of his Minister over the entire "Collinson Point mutiny affair." And it included the following statement with the intention of producing closure on the subject: "An inquiry was asked for not by one of the parties to the controversy alone, but by Mr. Stefansson as well. The minister declined to grant the request of either party for the reason that no good could come of such an inquiry and much harm might be done."[40]

The newly re-elected Minister's refusal to seek an official government inquiry ended the public commentaries by Dr. Anderson, Chipman and Jenness. Professor O'Neill, however,

---

37 The Honourable Charles A. Stewart had succeeded in getting elected to the House of Commons in a by-election for Argenteuil, Quebec on March 2, 1922 (Johnson, 1968. p. 551). He promptly went west, where he had a farm, and returned to Ottawa on March 15.

38 Anderson, 1922f (March 25, 1922).

39 Anderson, 1922c (March 17, 1922).

40 Camsell, 1923, pp. 665–666. In addition, the Southern Party scientists were officially instructed to refrain from any further public discussion of the affair through the news media. Camsell had written to Dr. Brock about the feud, seeking his view on the matter (Camsell, 1923a). Brock responded that both Stefansson and Dr. Anderson had been right on some points of this conflict and wrong on others, and no good would result from continuing this controversy (Brock, 1923).

operating outside the control of the government, was able to continue public reference to the feud for a while longer. In addition, Dr. Anderson referred to it in some of his personal letters for many years thereafter.

Mr. Stewart made the only decision open to him at the time. He had to avoid at all costs the embarrassment such an official inquiry would cause to his former Prime Minister, Sir Robert Borden, and Borden's successor, Arthur Meighen, both of whom had publicly praised Stefansson and his actions as Expedition leader. He also needed to spare certain senior officials, especially those in the Department of the Naval Service, from having the public learn more about Stefansson's gross abuse of government funds during the war years, about that department's inability to control Stefansson's activities in the North, and about Stefansson's deliberate inattention to their instructions to return south in 1916. Additionally, he would certainly not have wanted the public to learn about the friction between his Mines Department and that of the Department of the Naval Service.

Life Goes On

One last blast from the Stefansson camp occurred several months after Dr. Anderson, Chipman and Jenness were instructed to cease their newspaper communiqués and a month after Jenness' comments on Stefansson's *The Friendly Arctic* appeared in the journal *Science*. In August of 1922, both Dr. Anderson and Jenness received a letter from Andrew Haydon, a member of an Ottawa law firm, which stated that his firm had received a complaint from Mr. V. Stefansson of their actions "… in publishing, broadcast, in the Public Press and otherwise, false and malicious statements and reports regarding his [Stefansson's] control and management of the Arctic Expedition." Mr. Haydon's letter further notified them that "… unless you forthwith publish a statement of retraction with an apology for the wrong done to Mr. Stefansson," he would be forced "… to take such proceedings as will vindicate the rights of my client."[41] On the advice of the Director of the Victoria Memorial Museum, both Dr. Anderson and Jenness ignored the lawyer's letters, and nothing further happened.

In 1928, several months after the funding for the Expedition's publications program had been terminated, Dr. Anderson wrote a friend that the Minister of Mines (Mr. Charles Stewart) had finally issued "… a statement in mild parliamentary language, that we [the scientists of the Southern Party] were gentlemen and scholars who had done our duty and that Stefansson was unjustified in his attacks, but naturally that did not have much circulation outside of Ottawa…"[42]

Thereafter the feud between Stefansson and Dr. Anderson subsided, although it continued to smoulder for decades.[43] Stefansson easily found other matters to occupy his mind and time, seldom mentioning the subject again. Dr. Anderson, however, continued to vent his bitter feelings occasionally about Stefansson in his letters to acquaintances for the rest of his life.

---

[41] Jenness, 1922 (August 19, 1922). Dr. Anderson received a similar letter.

[42] Anderson, 1928b (December 13, 1928).

[43] I first became aware of the long-lived feud in 1980 when I attended a lecture on Stefansson by Professor R.J. Diubaldo in Ottawa and heard several reactive questions following the lecture. It was apparent to me then that there was still heat in the dwindling ashes of the feud.

In the early 1920s Stefansson fell out of grace in Canada with both the Canadian government and its general population, and seldom returned to the country of his birth thereafter. His lecture tours across Canada, which had previously been both popular and financially rewarding, were largely cancelled or simply not scheduled. His well publicized feud with Dr. Anderson and other members of the Geological Survey of Canada played a role in this downfall, but the chief causes were two entirely separate activities that Stefansson undertook after his return from his 1913–1918 Expedition. These were (1) his attempt in 1921 to settle a small group of people on Wrangel Island in order to claim ownership of that island for Canada (or Great Britain); and (2) his arrangements between 1920 and 1922 with the Hudson's Bay Company to establish a large number of reindeer from Norway on land in eastern Baffin Island that he had leased from the Canadian government. Both activities failed miserably, with the deaths of four of the five persons he sent to Wrangel Island[44] and many of the reindeer taken to Baffin Island,[45] and resulted in considerable embarrassment for both the Canadian government and the Hudson's Bay Company. Thereafter Stefansson received little interest or friendship from either organization and led no more Arctic expeditions. His feud with Dr. Anderson also eliminated virtually any respect for him that might have been held prior to 1920 by members of the Geological Survey of Canada.

Many of the views of life in the Arctic that Stefansson described so well in his book, *The Friendly Arctic,* are now known to be untrue. Nevertheless, his perception of how easy it was for individuals to survive in that vast but little-known region remained the general view of the Canadian and American public until after World War II. Commencing in the late 1940s, aerial photography, government and private mapping, marine investigations, and the influx of persons from the south (educators, health and religious personnel, exploration and mining personnel, traders, trades people and others) provided vast amounts of new information about the Arctic, which necessitated modifications to many of his visions of the Arctic.

---

[44] Niven, 2003.

[45] Ashlee, 2008.

APPENDIX 1

Personnel on the Canadian Arctic Expedition, 1913–1918

Name	Origin	Party	Time Served	Comments
Leaders				
Vilhjalmur Stefansson	U.S./Can.	Both	1913–1918	Leader of Expedition
Dr. Rudolph M. Anderson	U.S.	South	1913–1916	Executive head, S.Party
Scientists				
Henri Beuchat, ethnologist	France	South	1913–1914	Perished near Herald I.
Kenneth G. Chipman, geographer	Canada	South	1913–1916	2nd in command, S.Party
John R. Cox, asst. geographer	Canada	South	1913–1916	
Diamond Jenness, ethnologist	N. Zealand	South	1913–1916	
Fritz Johansen, marine biologist	Denmark	South	1913–1916	
George S. Malloch, geologist	Canada	North	1913–1914	Perished on Wrangel I.
Bjarne Mamen, asst. to geologist	Norway	North	1913–1914	Perished on Wrangel I.
William L. McKinlay, magnetician	Scotland	South	1913–1914	Rescued from Wrangel I.
James Murray, oceanographer	Scotland	North	1913–1914	Perished near Herald I.
John J. O'Neill, geologist	Canada	South	1913–1916	
Other Professionals				
Dr. Alistair Forbes Mackay, surgeon	Scotland	North	1913–1914	Perished near Herald I.
Burt M. McConnell, secretary	U.S.	North	1913–1914	1-year contract
George H. Wilkins, photographer	Australia	Both	1913–1916	
Non-Professionals (ship's crew, Inuit and others)				
General				
Ole Andreasen	Norway	North	1914; 1917	On Stefansson's first ice trip
Peder Pedersen	Denmark	South	1914	Worked on Mackenzie Delta
Storker Storkerson	Norway	North	1914–1918	Led ice trip in 1918
Karluk crew				
Robert A. Bartlett, captain	Nfld.	North	1913–1914	Alerted rescue efforts
C.T. Pedersen, captain	U.S.	North	1913	Quit in Victoria
James F. Allen, 1st mate	U.S.	North	1913	Discharged at Victoria
Sandy Anderson, 1st mate	Scotland	North	1913–1914	Perished on Herald I.
Thomas Anderson, 1st engineer	?	North	1913	Discharged at Nome
Charles Barker, 2nd mate	Canada	North	1913–1914	Perished on Herald I.
John Brady, seaman	Canada	North	1913–1914	Perished on Herald I.
George Breddy, fireman	Canada	North	1913–1914	Shot to death, Wrangel I.
Ernest F. Chafe, messroom boy	Nfld.	North	1913–1914	Rescued from Wrangel I.

Name	Origin	Party	Time Served	Comments
Edmund Golightly, a.k.a. "Archie King", seaman	England	North	1913–1914	Perished on Herald I.
Fred Maurer, fireman	U.S.	North	1913–1914	Rescued from Wrangel I.
T. Stanley Morris, seaman	Canada	North	1913–1914	Perished near Herald I.
John Munro, 1st engineer	Scotland	North	1913–1914	Rescued from Wrangel I.
J. Ridley, fireman	?	North	1913	Discharged at Esquimalt
Robert Templeman, steward	?	North	1913–1914	Rescued from Wrangel I.
Hugh "Clam" Williams, seaman	?	North	1913–1914	Rescued from Wrangel I.
Robert J. Williamson, 2nd engineer	?	North	1913–1914	Rescued from Wrangel I.
T. Wiseman, fireman	?	North	1913	Discharged at Nome
Alaska crew				
Otto Nahmens, captain	U.S.	South	1913–1914	Quit at Herschel I.
Daniel Sweeney, captain	U.S.	South	1914–1916	Obtained from *Belvedere*
Charles Brooks, steward	?	South	1913–1914	Quit at Herschel I.
Daniel Blue, engineer	Scotland	South	1913–1915	Died at Baillie Is.
J.E. Hoff, engineer	?	South	1915–1916	Hired at Herschel I.
Louis Olsen, seaman	Denmark	South	1913–1914	Quit at Herschel I.
Andre Norem, cook	Norway	South	1913–1914	Suicide at Collinson Pt.
James Sullivan, steward	England	South	1914–1916	Hired at Herschel I.
Mary Sachs crew				
Peter Bernard, captain	Canada/U.S.	South	1913–1915	Perished on Banks I.
W.J. "Levi" Baur, cook	Switzerland	North	1914–1917	Obtained from *Belvedere*
James R. Crawford, engineer	?	South	1913–1914	
Karl "Charlie" Thomsen, seaman	Norway	Both	1913–1915	Perished on Banks I.
North Star crew				
Aarnout Castel, seaman	Holland	Both	1914–1918	With Stefansson after 1915
Polar Bear crew				
Henry Gonzales, captain	Portugal	North	1915–1917	Dismissed at Baillie Is.
"Charlie" Karsten Andersen, seaman	Denmark	North	1915–1918	
Otto Binder, seaman	U.S.	North	1917–1918	Obtained from *Challenge*
Peter Donohue, seaman	U.S.	North	1917–1918	Obtained from *Challenge*
Manuel Fernandez, seaman	?	North	1918	Two months at Barter I.
Jim Fiji (a.k.a. James Asasela), seaman	Hawaii	North	1915–1917	
Adelbert G. Gumaer, seaman	?	North	1917–1918	On Storkerson's ice trip
John Hadley, 1st mate/captain	England	North	1915–1918	Sailed *Polar Bear* to Nome John
John J. Jones, 2nd engineer	U.S.	North	1915–1916	Died of heart attack
Herman Kilian, chief engineer	U.S.	North	1915–1918	Brother of Martin Kilian
Martin Kilian, seaman	U.S.	North	1915–1918	Brother of Herman Kilian
E. Lorne Knight, seaman	U.S.	North	1915–1918	
Karl August "Charlie" Lewin*	Sweden	North	1917–1918	Helped prepare last ice trip

Name	Origin	Party	Time Served	Comments
Peter Lopez, seaman	Portugal	North	1915–1917	Husband of Uttaktuak
August Masik, 2nd mate	Russia	North	1917–1918	From *Challenge*
Harold Noice, seaman	U.S.	North	1915–1917	Resigned at Baillie Is.
Walter Rasmussen, acting 2nd mate	?	North	1918	Two months
William Seymour, 2nd /1st mate	Australia	North	1915–1917	Carpenter, husband of Annie
Anthony Shannon, seaman	U.S.	North	1917–1918	Obtained from *Challenge*
Inuit (men)				
Adluak, Fred	Alaska	South	1913–1914	Seaman on *Mary Sachs*
Agnavigak, Ambrose	Canada	South	1915–1916	Husband of Unalina
Aiyakuk	Alaska	South	1914	Helped Jenness at Barter I.
Alingnak	Canada	North	1915–1916	Husband of Guninana
Asetsaq, Jimmy	U.S.	North	1913	Quit at Colville River
Avrunna	Canada	South	1916	Higilak's son
Dick	?	South	1914	Assisted Cox briefly
Emiu (Split-the-Wind)	U.S.	North	1915–1918	With Stefansson
Hopson, Alfred	Alaska	South	1913–1914	Young interpreter for Jenness
Ikey (Angutisiak) Bolt	Alaska	South	1913–1916	Hired at Barrow
Ikpukhuak	Canada	South	1916	Custodian of supplies
Illun	U.S.	North	1915–1917	Husband of Pukalook & Kutok
Ipanna	Alaska	South	1914	Stepson of Aiyakuk, Barter I.
Kataktovik	U.S.	North	1913–1914	Went with Bartlett to Siberia
Klengenberg, Patsy	U.S.	South	1915–1916	Son of Chas. Klengenberg
Kohoktok	Canada	South	1916	Brother of Mupfi
Kuraluk	U.S.	North	1913–1914	Husband of Keruk
Memoganna, Roxy	Canada	South	1914	Worked for Cox
Mungalina	Canada	South	1915–1916	Assistant cook
Mupfi	Canada	South	1916	From Tree River region
Natkusiak, Billy	U.S.	North	1914–1917	Bought *North Star* in 1917
Ovoiyuaq, Adam	Canada	South	1915–1916	Brother of Silas Palaiyak
Palaiyak, Silas	Canada	Both	1914–1916	Worked with both parties
Pausanna	U.S.	North	1918	With Storkerson at Barter I.
Payuraq, Jerry	U.S.	South	1913	Quit at Barrow
Pikalu,*Walter	U.S.	North	1915–1917	Husband of Pusimmik
Pipsuk*	Alaska	North	1917	Drowned at Barter island
Siberia Mike	Siberia	South	1914–1916	Asst. engineer on *Alaska*
Sinnisiak	Canada		1916	Charged with murder
Takluk, John	U.S.	North	1918	Barter Island, Alaska
Ulipsinna,* hunter	Canada	North	1917	On *Polar Bear*
Uloksak Avingak	Canada	South	1916	Charged with murder
Uloksak Meyok	Canada	South	1915–1916	Shaman, hunter
Wolki, Fred	Canada	North	1917–1918	On ice trip, 1918

Name	Origin	Party	Time Served	Comments
Inuit (women)				
Amaganna	Canada?	North	1916–1917	Seamstress at Cape Kellett
Arnauyuk, Kila	Canada	South	1915–1916	Ikpukhuak's adopted daughter
Añayu Annarihopopiak, Eunice	U.S.	South	1915–1916	Sweeney's wife (Red Calico)
Guninana	Canada	North	1915–1917	Seamstress, wife of Alingnak
Higilak (a.k.a. Taqtu)	Canada	South	1914–1916	Wife of Ikpukhuak
Ikiuna, Topsy	Canada	North	1915–1917	Wife of Billy Natkusiak
Iyituarryuk	Canada?	North	1916–1917	Seamstress at Cape Kellett
Kannayuk, Jennie	Canada	South	1915–1916	Daughter of Higilak
Keruk	U.S.	North	1913–1914	Wife of Kuraluk
Kutok	U.S.	North	1915–1917	Seamstress, 2nd wife of Illun
Mamayauk,* Mamie	Canada	North	1915–1917	Seamstress, wife of Ilavinirk
Mamayauk,* Violet	Canada	North	1915–1917	Seamstress, wife of Gonzales
Munnigorina	Canada	South	1915–1916	Wife of Kohoktok
Pannigabluk, Fanny	U.S.	North	1914–1917	Seamstress, mother of Alex
Pukalook*	U.S.	North	1915	Seamstress, 1st wife of Illun
Pusimmik (or Puchimuk), Bessie	U.S.	North	1915–1917	Seamstress, wife of Pikalu
Seymour, Ikugana Anna	U.S.	North	1915–1917	Seamstress, wife of Seymour
"Sis" (or "Ciss")	?	South	1915–1916	Wife of Siberia Mike
Storkerson, Uiniq Elvina Klengenberg	U.S.	North	1914–1917	Seamstress, wife of Storkerson
Thomsen, Jennie	U.S.	North	1913–1917	Seamstress, wife of Thomsen
Unalina (sister of Palaiyak)	Canada	South	1915–1916	Seamstress, wife of Ambrose
Uttaktuak	Canada	North	1915–1917	Wife of Lopez
Inuit/Iñupiat (children)				
Aida* (born 1915)	U.S.	North	1914–1917	Daughter of Storkerson & wife
Alashuk* (Alex) (born 1910)	Canada	North	1913–1918	Son of Pannigabluk and Stef.
Annie, age 3 (1913)	U.S.	North	1913–1917	Daughter of Thomsen & Jennie
Annie Fitzgerald*	Canada	South	1915–1916	Daughter of Unalina
Bessie* (born 1918)	U.S.	North	1918	Daughter of Storkerson
Daniel Jr. (born 1916)	U.S.	South	1916	Son of Sweeney & Añaiyu
Georgie Mike, age 4 (1916)	U.S.	South	1915–1916	Son of Siberia Mike and "Sis"
Helen, age 8 (1913)	U.S.	North	1913–1914	Daughter of Kuraluk & Keruk
Itayuk (or Ivallu) (born 1916)	Canada	South	1916	Son of Kohoktok & Munnigorina
Martina,* age 4 (1914)	U.S.	North	1914–1917	Daughter of Storkerson & wife
Mingeouk, age ca. 5 (1915)	Canada	South	1915–1916	Daughter of Mupfi & Kilauluk
Mugpi, age 3 (1913)	U.S.	North	1913–1914	Daughter of Kuraluk & Keruk
?? (born 1915)	U.S.	North	1915–1917	Son of Thomsen & Jennie
Others (mostly for brief employment)				
Ayak, Sydney	U.S.	North	1917–1918	Hired for final ice trip
Ayaki	U.S.	North	1917–1918	Dog driver; brother of Ayak

Name	Origin	Party	Time Served	Comments
Bishop, A.	?	North	1918	Hired at Crow River
Daniel, Fred	?	North	1918	Hired at Crow River
Dick	Canada	South	1914	Worked for Cox on Firth River
Ijronna	?	North	1917	Hired near Fort McPherson
Inukok (or Inokook)	U.S.	North	1918	Hired by Stefansson
Mike	?	North	1915	Freighting on Banks Island
Mike ("Maikis")	?	North	1917	Hired near Fort McPherson
Moses, Peter	?	North	1918	Hired at Old Crow
Naipaktuna	?	North	1918	Guide on trip to Yukon
Saryoak	?	North	1918	Helper on trip to Yukon
Statak	Canada	North	1917–1918	Son of Pisuktuak; age 15
Stewart, Abraham	Canada	North	1917–1918	Young helper to Stefansson
Taliak	U.S.	North	1914	On Stefansson's first ice trip

Notes

1. Names followed by an asterisk (*) were provided by Jette Ashlee and David R. Gray.
2. The alias "Archie King," for reasons unknown, was used by a young Englishman whose real name was Edmund Lawrence Golightly.
3. The wives of Kohoktok, Mupfi and Uloksak Meyok (Munnigorina, Kilauluk, and Kukilukkak, Koptana and Haqungaq respectively) generally accompanied their husbands, but were rarely if ever employed by the Expedition.
4. Ikiuna, also known as Topsy Ikiuna, was the adopted daughter of Alingnak and Guninana. She was born in 1903.
5. Stefansson (1921b, pp. 679, 681) hired a Loucheaux Indian boy in December 1917, who subsequently helped care for him during his three months of illness at Herschel Island, but did not identify the boy in The *Friendly Arctic*. From his diary (Stefansson, December 23, 1918), we learn that he was 18-year-old Abraham Stewart (Stefansson spelled it Stuart), son of Kenneth Stewart, a trader living a few miles north of Fort McPherson. Abraham was on Stefansson's payroll from December 27, 1917 to April 21, 1918 (RG42, Vol. 471, File 84-2-5, Sub-file 56, LAC).
6. Siberia Mike was a native from Siberia. With wife "Sis" and son Georgie Mike, he was hired by Stefansson at Herschel Island to work briefly with J.J. O'Neill on the Firth River in 1914. He and his wife and son joined the Southern Party at Bernard Harbour in September 1915 until 1916, after being rehired at Herschel Island by Stefansson in August. Stefansson hired him again at Herschel Island to go up the Mackenzie River with him (Stefansson, December 19, 1917). A young sailor from the *Polar Bear* named Mike was with the Northern Party at Cape Kellett, Banks Island, in October of 1915 (Jenness, 2004, p. 289, 390, note 7). No further information is known about him.
7. Mamie Mamayauk, the wife of Ilavinirk, joined Stefansson's Northern Party as a seamstress at Barter Island in late 1917 or early 1918. She had spent the winter of 1915–1916 at Great Bear Lake while her husband, Ilavinirk, had assisted RNWMP Inspector C.D. La Nauze in solving the mystery of the disappearance in the Coronation Gulf region of two French Oblate priests. Ilavinirk had assisted Dr. R.M. Anderson during the Stefansson-Anderson Expedition of 1908–1912.
8. Annie Fitzgerald was the daughter of RNWMP Sergeant F.J. Fitzgerald and Unalina. Following the Sergeant's tragic death in 1911, Unalina married Ambrose Agnavigak, who adopted Annie. Annie later married Jasper Andreasen, one of two sons of Ole Andreasen (CMC website).
9. Dick was an Inuk from Herschel Island, who was hired with his dogs and sled to work with J.R. Cox in mapping the Canadian part of the Firth River in April 1914.
10. Karl August Lewin was a sailor born in Kalmar, Sweden, whom Stefansson found living and trapping at Demarcation Point, Alaska, and hired in November 1917 to assist in the preparations at Barter Island for his 1918 ice trip. Lewin left after working for only a few weeks (Ashlee, 2008, p. 166).
11. Stefansson and Storkerson hired quite a number of local white, Inuit and First Nations people in 1917 and 1918 in connection with the planned 1918 ice trip over the Beaufort Sea. Information on such persons is sparse, so that the above list may not be quite complete.
12. The Canadian Museum of Civilization offers a well-organized account of the Canadian Arctic Expedition in text and pictures on one of its website at (http://www.civilization.ca/cmc/exhibitions/hist/cae/splashe.shtml). It is entitled "Northern People, Northern Knowledge—The Story of the Canadian Arctic Expedition 1913–1918." Created by Dr. David R. Gray, the website is well worth visiting.

APPENDIX 2

Geographical Names and the Canadian Arctic Expedition, 1913–1918[1]

Fifty-three geographical features in the Canadian Arctic today carry the names of thirteen of the fourteen scientific members and fifteen of the more than fifty non-scientific members on Vilhjalmur Stefansson's Canadian Arctic Expedition 1913–1918.

The lone Expedition scientist after whom no geographic feature in the Arctic is named is Dr. John J. O'Neill, geologist. In 1966 the Canadian Hydrographic Service proposed to name a promontory on southwestern Victoria Island "Cape O'Neill" in his honour, but later withdrew its proposal. In 1995, I proposed the name "O'Neill Island" to the Canadian Permanent Committee on Geographical Names for one of the many unnamed Stockport Islands that he examined for copper on the west side of Bathurst Inlet in 1915, but the name was not adopted.

Several of the Expedition's scientists named coastal features in the central Arctic. For example, the zoologist Dr. Anderson named Bernard Harbour, Brock River, Hornaday River and Ekalulia Island (in Bathurst Inlet), all of which he visited. The two geographers, Chipman and Cox, named Agiak Headland, Napaaktoktok River and Anialik River (in Coronation Gulf east of Kugluktuk), and Algak Island and Iglorua Island (in Bathurst Inlet), all of which they visited. The anthropologist Jenness named the Asiak River (east of Kugluktuk), Kugaluk River, Singialuk Peninsula, and Naoyat Cliff (on the southwest coast of Victoria Island), and Ekailuk River (near Cambridge Bay), most of which he visited. And Stefansson, of course, named many geographical features farther north, such as the Thomsen River, Storkerson Bay, Castel Bay and Sachs Harbour (not the settlement) on Banks Island, Bernard Island (off the west coast of Banks Island), and islands he discovered farther north, including Brock Island, Borden Island, Meighen Island, Mackenzie King Island, Lougheed Island and tiny Jenness Island.

Storkerson saw and named Elvina Island, close to Stefansson Island, for his Iñupiaq wife.

Several geographical locations were also named after the Expedition's ships—*Karluk*, *Alaska*, *Mary Sachs*, *North Star* and *Challenge*.

The following list demonstrates the influence of the Canadian Arctic Expedition in the western part of the Canadian Arctic.

---

[1] This listing is updated from S.E. Jenness (1987).

Name	Geographical Feature	N. Latitude	W. Longitude
Scientists			
Dr. R.M. Anderson (mammalogist)	Anderson Creek	67E 15'	117E 55'
	Anderson Headland	66E 22'	71E 12'
Henri Beuchat (anthropologist)	Beuchat Lake	70E 03'	127E 18'
	Cape Beuchat	77E 31'	113E 10'
Kenneth G. Chipman (geographer)	Chipman Point	68E 47'	114E 43'
John R. Cox (geographer)	Cox Island	68E 48'	114E 44'
	Cox Lake	67F 5'	116C 35'
Diamond Jenness (ethnologist)	Jenness Island	78E 17'	113E 55'
	Jenness River	67E 48'	81E 53'
	Diamond Jenness Peninsula	71E 00'	116E 00'
Fritz Johansen (naturalist)	Johansen Bay	68E 34'	111E 05'
Dr. Alistair Forbes Mackay (surgeon)	Cape Mackay	78E 20' 3"	113E 17' 12"
William L. McKinlay (magnetician)	McKinlay Lake	70E 25'	127E 37'
George S. Malloch (geologist)	Cape Malloch	78E 46'	110E 43'
	Malloch Dome	78E 12'	101E 15'
	Malloch River	78E 07'	101E 12'
	Malloch Hill	70E 01'	126E 57'
Bjarne Mamen (topographer)	Cape Mamen	77E 37'	110E 03'
	Mamen Lake	70E 30'	127E 59'
James Murray	Cape Murray	77E 58' 2"	115E 6' 11"
Vilhjalmur Stefansson (anthropologist, explorer)	Stefansson Creek	68E 49'	125E 17'
	Stefansson Island	73E 20'	105E 45'
	Stefansson Lake	68E 55'	124E 59'
	Stefansson Point	80E 09'	99E 39'
George Hubert Wilkins (photographer)	Wilkins Bay	73E 37'	124E 10'
	Wilkins Strait	78E 10'	112E 00'
	Wilkins Point	68E 47' 40"	93E 37' 50"
Non-Scientists			
Ole Andreasen (sailor)	Cape Andreasen	77E 21'	118E 46'
	Andreasen Head	70E 48'	96E 35'
Robert A. Bartlett (captain of the *Karluk*)	Bartlett Bay	79E 10'	74E 45'
	Bartlett Point	68E 56'	79E 25'
Peter Bernard (captain of the *Mary Sachs*)	Bernard Island	73E 36'	124E 14'
	Bernard River	73E 34'	124E 05'
Aarnout Castel (sailor)	Castel Bay	74E 12'	119E 35'
	Castel Butte	77E 40'	111E 23'
Jim Asasela Fiji (sailor)	Fiji Island	70E 10' 30"	125E 03'
	Jim Fiji Harbour	71E 10'	125E 05'

Name	Geographical Feature	N. Latitude	W. Longitude
John Hadley (sailor)	Hadley Bay	72E 30'	108E 12'
Martin Kilian (sailor)	Kilian Island	73E 35'	107E 53'
	Kilian Lake	72E 10'	111E 35'
E. Lorne Knight (sailor)	Knight Harbour	73E 31'	115E 18'
Otto Masik (sailor)	Masik Pass	71E 34'	122E 08'
	Masik River	71E 32'	123E 48'
Billy Natkusiak (hunter)	Natkusiak Peninsula	72E 45'	109E 45'
Harold Noice (sailor)	Noice Peninsula	78E 25'	104E 00'
	Noice Point	80E 08'	99E 04'
Elvina Storkerson (seamstress and wife of Storker Storkerson)	Elvina Island	73E 21'	107E 29'
Storker Storkerson (sailor)	Storkerson Bay	72E 56'	124E 50'
	Storkerson Lake	72E 49'	122E 49'
	Storkerson Peninsula	72E 30'	108E 30'
	Storkerson River	72E 56'	124E 29'
Daniel Sweeney (captain of the *Alaska*)	Sweeney Island	68E 46'	114E 45'
Charles Thomsen (sailor)	Thomsen River	74E 08'	119E 45'

Bernard Harbour, in Dolphin and Union Strait northwest of Coronation Gulf, is not included on this list, because it was named after Captain Joseph Bernard, who was not a member of the Canadian Arctic Expedition 1913–1918.

A half dozen more Expedition names have also been attached to geographic features around Barter Island on the northern Alaskan coast. As they are not in Canadian territory they have not been included in the table above. They are as follows:

Challenge Entrance   Mary Sachs Entrance

Challenge Island   Stefansson Sound

Karluk Island   Sweeney Point

That the schooner *Challenge* has two sites named for it is something of a surprise, and may stem from incidents involving the vessel other than during its brief association with the Expedition in the summer of 1917.

APPENDIX 3

The Collections of the Canadian Arctic Expedition

The government-sponsored Canadian Arctic Expedition 1913–1918 returned south with large ethnographical, archaeological, mineralogical and photographic collections as well as collections of birds, mammals, plants, insects, fish and invertebrates. In addition, most of the scientists and a few of the crew on the Expedition kept diaries or field notebooks, which here and there contain information about Expedition activities. Most of these diaries and notebooks ultimately were deposited in the Public Archives in Ottawa, an organization now known as Library and Archives Canada (LAC). Information on the location of other diaries is included at the end of the list of diaries.

The various biological, anthropological, geological and photographic collections were duly disseminated for study and retention among several government departments and universities, the Department of the Naval Service and the Department of Mines (which included the Geological Survey of Canada and the Victoria Memorial Museum) receiving the largest quantity. Over succeeding years, the names and internal organizations of most of the government departments involved have changed, complicating the task of ascertaining today whether or not some of the collections still exist and where they are located.

Nowhere to my knowledge was a list ever prepared to indicate the locations of these various collections, hence the uncertainty of some of the information on the following pages. The list below reveals the current locations of the collections I have examined or know about. Many specimens within some of the collections appear to have been mislaid or discarded years ago.

1. CAE Diaries

Diaries (or field notebooks) of the following twenty-five members of the Canadian Arctic Expedition 1913–1918 are housed in the Library and Archives Canada in Ottawa. Their archival identification numbers are included here.

Andersen, Karsten (seaman, Northern Party)—MG30 B7
Anderson, R.M. (head, Southern Party)—MG30 B40
Baur, W.J. (cook, Northern Party)—MG30 B8
Bernard, Peter (ship's captain, Northern Party)—MG30 B9
Castel, Aarnout (seaman, Northern Party)—MG30 B10
Chipman, K.G. (geographer, Southern Party)—MG30 B66
Cox, J.R. (geographer, Southern Party)—MG30 B123
Gumaer, Adelbert (seaman, Northern Party)—MG30 B1
Hadley, John (seaman, Northern Party)—MG30 B2
Jenness, Diamond (ethnologist, Southern Party)—MG30 B89
Johansen, Fritz (naturalist, Southern Party)—MG30 B165
Kilian, Herman (seaman, Northern Party)—MG30 B11
Kilian, Martin (seaman, Northern Party)—MG30 B12

Knight, E. Lorne (seaman, Northern Party)—MG30 B14

Lewin, Karl "Charlie" A. (temporary labourer, Northern Party)—MG30 B15

Mamen, Bjarne (topographer, Northern Party)—MG30 B20

McConnell, B.M. (Stefansson's secretary)—MG30 B24

McKinlay, W.L. (magnetician, on *Karluk*)—MG30 B25

Masik, August (seaman, Northern Party)—MG30 B3

Munro, John (chief engineer, *Karluk*)—with the W.L. McKinlay collection, MG30 B25

Noice, Harold (seaman, Northern Party)—MG30 B16

O'Neill, J.J. (geologist, Southern Party)—MG30 B171

Stefansson, Vilhjalmur (leader of Expedition)—MG30 B81

Storkerson, Storker (seaman, Northern Party)—MG30 B17

Williamson, Robert J. (2nd engineer on *Karluk*)—MG30 B44

Notes

1. The original diary of **John Hadley**, which covers only the period he was on Wrangel Island, is with McKinlay's diary and papers in the National Library of Scotland in Edinburgh. McKinlay personally informed me (S.E. J.) in 1980 that he had obtained it from Hadley when they left that island. The Hadley diary in the LAC in Ottawa under MG30 B2 is, according to McKinlay, not a direct copy but a totally different version created by Stefansson and Hadley between 1915 and 1917.

2. The original diary and field notes of naturalist **Fritz Johansen** are in the Arktisk Institut in Charlottenlund, Denmark. The diary is written in Old Danish. A photocopy of the original diary, together with an English translation, is with the Johansen papers at the LAC in Ottawa under MG30 B165.

3. The original diary of the magnetician **William L. McKinlay** is with the William Laird McKinlay Papers (DEM 357) at the National Library of Scotland in Edinburgh, where it is accompanied by a manuscript containing a more detailed account of his experiences on the Canadian Arctic Expedition than appears in his book *Karluk: The Great Untold Story of Arctic Exploration.* "Fairly extensive parts [of his original handwritten diary] are almost indecipherable, indeed I should say wholly so, to any but myself," was part of the reason McKinlay refused to turn it over to the Canadian government (McKinlay, 1925 [May 13, 1925]).

   The diary in the LAC in Ottawa is a typed "summary of events" supplied to the Canadian government by McKinlay over several years. Part 1 (covering July 25, 1913 to March 21, 1914) was sent to G.J. Desbarats, Department of the Naval Service, in 1915 (McKinlay, 1915 [July 31, 1915]); Part 2 (covering March 22, 1914 to June 25, 1914) was sent to F. McVeigh, Department of Marine and Fisheries in 1926 (Marine and Fisheries had assumed part of the Naval Service Department's responsibilities in 1922) (McKinlay, 1926 [May 27, 1926]); and Part 3 (June 26, 1914 to September 6, 1914) was sent to Ottawa in 1927.

4. The original diary of the photographer **George Hubert Wilkins** is with the Wilkins papers, Box 1, Folders 2–12, in the Stefansson Collection at the Rauner Special Collections Library, Dartmouth College, Hannover, New Hampshire.

5. The diary of (a) *Alaska* engineer **Daniel Blue** from July 21, 1914 to April 17, 1915, is in CMNAC/1996-077, Series A—R.M. Anderson Collection, Box 60, Folder 5, Archives, National Museum of Nature (CMN), Gatineau, Quebec. (b) A copy of part of *Alaska* Captain **Otto Nahmens** diary from July 10, 1913 to September 20, 1913, is in the same Dr. Anderson collection, Box 68, Folder 18.

6. The original diary of **Captain Robert Bartlett** is with the Robert A. Bartlett Papers (M8), Special Collections & Archives, Bowdoin College Library, Brunswick, Maine. A copy is in RG42, Vol. 475, Folder 84-2-37, LAC.

7. The papers of **Fred Maurer**, fireman on the *Karluk*, were retained by his family in Ohio, being most recently with his niece, Marian Reiss (now deceased).

8. **Karl August (Charles) Lewin** was employed at Barter Island to work for Storkerson in 1917–1918, but was unfit to accompany his ice trip. His diary during that short period is in MG30 B15, LAC.

2. CAE Photographs[1]

Ref.: Jenness, S.E. (compiler), 2003. Photographs and films of the Canadian Arctic Expedition 1913–1918—One of Canada's little known historic treasures. An annotated catalogue with brief biographies. 228 pp. Private publication. Copies with: Photo Archives Section, LAC; Archives, CMN; Archives, CMC; and Photo Archives, GSC.

Known black-and-white photographs taken by members of the Canadian Arctic Expedition 1913–1918 number about 3,800. The vast majority of negatives for these, mostly from roll film but including some glass-plate negatives, are housed among three different Canadian government agencies: Library and Archives Canada (LAC): the Canadian Museum of Civilization (CMC); and the Geological Survey of Canada (GSC). Included are photographs taken by Dr. Anderson, Chipman, Cox, Hadley, Jenness, Johansen, Maurer, McConnell, McKinlay, O'Neill, Stefansson, Storkerson and Wilkins. The locations of Maurer's and McConnell's original negatives are not known to me (S.E.J.), but Library and Archives Canada has prints of their photographs. McKinlay's original negatives are with his diary in Edinburgh, Scotland (see Note 3). Noice also took photographs, some of which may be among those in his book (Noice, 1924), but the location of his negatives is unknown to me. Captain Bartlett's book includes photographs that are not in the collections in Ottawa. They may be his own photographs, housed with his papers in Bowdoin College Library, Brunswick, Maine.

Most of the CAE negatives held by the Canadian Museum of Civilization are copies made from the originals before they were handed over to the Public Archives of Canada (now Library and Archives Canada) many years ago.

Numerous prints made for Stefansson from Wilkins' negatives, after they were received from London by the Victoria Memorial Museum, are today catalogued and housed with the Stefansson Collection in the Rauner Special Library at Dartmouth College.

Late in 2007 a private collector in Southern England offered to sell, on the American Internet site known as E-Bay, more than 150 Canadian Arctic Expedition photographs taken by G.H. Wilkins, the CAE photographer. A few were bought by collectors in Europe but the

---

[1] Jenness, S.E. (compiler), 2003.

majority were bought by two men in North America, one in New Jersey, the other in northern Canada. A lot of these photographs are not in the possession of the Canadian government. They include many of the nearly 200 by Wilkins that I listed as "location unknown" in my catalogue. Wilkins frequently took two or three photographs of the same scene or subject, with only the best one being catalogued. Many of these recently "discovered" negatives appear to be the discarded versions of Wilkins' numbered negatives in the Canadian Archives.

3. CAE Mammals and Birds

Specimens of Arctic mammals and birds collected by Dr. R.M. Anderson, Jenness, Wilkins and possibly others on the Expedition are housed now at the Canadian Museum of Nature in Gatineau, Quebec.

4. CAE Insects, Plants, Fish, and Other Marine and Freshwater Vertebrate and Invertebrate Specimens

Arctic insects, plants, fish, and other marine and freshwater vertebrate and invertebrate specimens collected on the Expedition mainly by Fritz Johansen are housed at the Canadian Museum of Nature, Natural History Building, in Gatineau, Quebec. A few Expedition insects are also housed at the Department of Agriculture in Ottawa.

5. Eskimo Clothing, Weapons, Utensils, Kayaks, Archaeological Specimens, Toys, Games, etc.

Anthropological specimens collected principally by Diamond Jenness, with some additional specimens collected by Stefansson, Wilkins and Hadley, are housed at the Canadian Museum of Civilization in Gatineau, Quebec.

6. Rocks, Minerals, and Fossils

Specimens of rocks and minerals, including native copper, collected primarily by geologist J. J. O'Neill, as well as Jurassic fossils from the Firth River and post-glacial marine shells from the Mackenzie River Delta, Darnley Bay, Coppermine River and one or two other Arctic coast localities, were housed for years with the Geological Survey of Canada. A few Silurian corals from Sutton Island collected by Fritz Johansen were housed in the same locality. In 2010 the fossil specimens were transferred to the Canadian Museum of Civilization for safekeeping. Dr. O'Neill's rock and mineral collections are no longer in the Geological Survey's collection and may have gone to McGill University when he moved there in the 1920s, and later been discarded. The university's Redpath Museum at 859 Sherbrooke Street West, Montreal, currently holds fifteen specimens of native copper from the Coppermine River area (exact sources not specified), which were donated by Dr. O'Neill.

7. Books

Many of the books taken north by the Expedition and assigned to the Southern Party were brought south and retained by Dr. R.M. Anderson, and are now included in the regular book holdings of the library of the Canadian Museum of Nature, Natural History Building, in Gatineau, Quebec. Some of these bear special Expedition bookmarks.

APPENDIX 4
Reports of the Canadian Arctic Expedition, 1913–1918

(Names of Canadian Arctic Expedition members are in bold-face type; almost all were Southern Party members. The prices for the reports when issued ranged from 10 cents to 50 cents, chiefly on the basis of the length.)

VOLUME 1. GENERAL INTRODUCTION, NARRATIVE, ETC.

Vol. 1   Part A. Northern Party, 1913–1918, by **Vilhjalmur Stefansson**   (not written)

Part B. Southern Party, 1913–1916, by **Rudolph M. Anderson**   (not written)

VOLUME 2. MAMMALS AND BIRDS

Vol. 2.   Part A. Mammals of Western Arctic America, by **Rudolph M. Anderson**

(not written)

Part B. Birds of Western Arctic America, by **R.M. Anderson** and P.A. Taverner

(not written)

VOLUME 3. INSECTS

Vol. 3   Introduction, by C. Gordon Hewitt   (Published December 10, 1920)

Part A. Collembola, by Justus W. Folsom   (Published July 10, 1919)

Part B. Neuropteroid insects, by Nathan Banks   (Published July 11, 1919)

Part C. Diptera   (Published July 14, 1919)

Crane-flies, by Charles P. Alexander

Mosquitoes, by Harrison G. Dyar

Diptera (excluding Tipulidae and

Culicidae), by J.R. Malloch

Part D. Mallophaga and Anoplura   (Published September 12, 1919)

Mallophaga, by A.W. Baker

Anoplura, by G.F. Ferris and

G.H.F. Nuttall

Part E. Coleoptera   (Published December 12, 1919)

Forest insects, including Ipidae,

Cerambycidae, and Buprestidae,

by J.M. Swaine

Carabidae and Silphidae, by H.C. Fall

Coccinellidae, Elateridae,

Chrysomelidae and Rhynchophora

(excluding Ipidoe)

Dytiscidae, by J.D. Sherman, Jr.

Part F. Hemiptera, by Edward P. Van Duzee   (Published July 11, 1919)

Part G. Hymenoptera and Plant Galls   (Published November 3, 1919)
         Sawflies (Tenthredinoidea),
         by Alex D. MacGillivray
         Parasitic Hymenoptera, by Charles T. Brues
         Wasps and Bees, by F.W.L. Sladen
         Plant Galls, by E. Porter Felt
Part H. Spiders, Mites and Myriapods   (Published July 14, 1919)
         Spiders, by J.H. Emerton
         Mites, by Nathan Banks
         Myriapods, by Ralph V. Chamberlin
Part I. Lepidoptera, by Arthur Gibson   (Published January 10, 1920)
Part J. Orthoptera, by E.M. Walker   (Published September 4, 1920)
Part K. Insect life on the Western Arctic Coast
         of America, by **F. Johansen**   (Published November 7, 1921)
Part L. General Index   (Published December 1922)

VOLUME 4. BOTANY

Vol. 4. Part A. Freshwater Algae and Freshwater Diatoms,
                  by Charles W. Lowe   (Published February 20, 1923)
         Part B. Marine Algae, by F.S. Collins,
                  Mme. Paul Lemoine and
                  Marschall A. Hume   (Published November 24, 1927)
         Part C. Fungi, by John Dearness   (Published June 1, 1923)
         Part D. Lichens, by G.K. Merrill   (Published July 16, 1924)
         Part E. Mosses, by R.S. Williams   (Published February 8, 1921)
         Part F. Marine Diatoms, by Alabert Mann (Published November 12, 1922)

VOLUME 5. BOTANY

Vol. 5. Part A. Vascular plants, by James M. Macoun
                  and Theo. Hohm

                                                 (Published October 14, 1921)
         Part B. Contributions to the Morphology,
                  Synonymy, and General Distribution of
                  Arctic Plants, by Theo. Holm   (Published February 10, 1922)
         Part C. General notes on Arctic Vegetation,
                  by **Fritz Johansen**   (Published October 7, 1924)

VOLUME 6. FISHES, TUNICATES, ETC.

Vol. 6. Part A. Fishes, by **F. Johansen**[1]   (Not published)
         Part B. Ascidians, etc., by A.G. Huntsman   (Published November 29, 1922)

---

[1] Part of Fritz Johansen's incomplete 1926 manuscript "The fishes of Arctic America" (CMNAC 95-043, Archives, CMN) is included in Walters, V., 1953, "The fishes collected by the Canadian

VOLUME 7. CRUSTACEA

Vol. 7. Part A. Decapod Crustaceans, by Mary J. Rathbun (Published August 18, 1919)

Part B. Schizopod Crustaceans,
by Waldo L. Schmitt (Published September 22, 1919)

Part C. Cumacea, by W.T. Calman (Published October 15, 1920)

Part D. Isopoda, by P.L. Boone (Published November 10, 1920)

Part E. Amphipoda, by Clarence R. Shoemaker (Published September 7, 1920)

Part F. Pycnogonida, by Leon J. Cole (Published January 3, 1921)

Part G. Euphyllopoda, by **F. Johansen** (Published May 10, 1923)

Part H. Cladocera, by Chancey Juday (Published June 23, 1920)

Part I. Ostracoda, by G.O. Sars (Published July 28, 1926)

Part J. Freshwater Copepoda, by C. Dwight Marsh (Published April 21, 1920)

Part K. Marine Copepoda, by A. Willey (Published June 25, 1920)

Part L. Parasitic Copepoda, by Charles B. Wilson (Published August 6, 1920)

Part M. Cirripedia, by H.A. Pilsbry (Not published)

Part N. The Crustacean Life of
some Arctic Lagoons, Lakes and Ponds,
by **F. Johansen** (Published December 30, 1922)

VOLUME 8. MOLLUSKS, ECHINODERMS, COELENTERATES, ETC.

Vol. 8. Part A. Mollusks, Recent and Pleistocene,
by William Healey Dall (Published September 24, 1919)
(Supplementary report to Part A) (Published March 27, 1924)

Part B. Cephalopoda and Pteropoda (Published August 6, 1925)
Cephalopoda, by S. Stillman Berry
Pteropoda, by William Healey Dall

Part C. Echinoderms, by Austin H. Clark (Published April 6, 1920)

Part D. Bryzoa, by R.C. Osburn (Published February 20, 1923)

Part E. Rotatoria, by H.K. Harring (Published December 21, 1921)

Part F. Chaetognatha, by A.G. Huntsman (Not published)

Part G. Actinozoa and Alcyonaria, by A.E. Verrill (Published April 28, 1922)

Part H. Medusae and Ctenophora, by H.B. Bigelow (Published June 30, 1920)

Part I. Hydroids, by C. McLean Fraser (Published August 24, 1922)

Part J. Porifera, by A. Dendy and L.M. Frederick (Published July 5, 1924)

VOLUME 9. ANNELIDS, PARASITIC WORMS, PROTOZOANS, ETC.

Vol. 9. Part A. Oligochaeta (Published September 29, 1919)
Lumbriculidae, by Frank Smith
Enchytraeidae, by Paul S. Welch

Part B. Polychaeta, by Ralph V. Chamberlin (Published November 16, 1920)

Arctic Expedition 1913-1918, with additional notes on the ichthyofauna of western Arctic Canada," Bulletin of the National Museum of Canada, no. 128, pp. 257–274.

Part C.	Hirudinae, by J.P. Moore	(Published February 4, 1921)
Part D.	Gephyrea, by Ralph V. Chamberlin	(Published June 20, 1920)
Part E.	Acanthocephala, by H.J. Van Cleave	(Published April 7, 1920)
Part F.	Nematoda, by N.A. Cobb	(Not published)
Part G–H.	Trematoda and Cestoda, by A.R. Cooper	(Published February 4, 1921)
Part I.	Turbellaria, by A. Hassell	(Not published)
Part J.	Polychaeta (Supplementary), by J.H. Ashworth	(Published September 29, 1924)
Part K.	Nemertini, by Ralph V. Chamberlin	(Not published)
Part L.	Sporozoa, by J.V. Mavor	(Not published)
Part M.	Foraminifera, by J.A. Cushman	(Published February 6, 1920)

VOLUME 10. PLANKTON, HYDROGRAPHY, TIDES, ETC.

Vol. 10. Part C. Tidal Observations and Results,
by W. Bell Dawson   (Published October 1, 1920)

VOLUME 11. GEOLOGY AND GEOGRAPHY

Vol. 11. Part A. The Geology of the Arctic Coast of Canada,
West of the Kent Peninsula, by **J.J. O'Neill**   (Published July 8, 1924)

Part B. Maps and Geographical Notes,
by **Kenneth G. Chipman** and **John R. Cox**   (Published July 8, 1924)

VOLUME 12. THE COPPER ESKIMOS

Vol. 12. Part A. The Life of the Copper Eskimos,
by **D. Jenness**   (Published January 12, 1922)

Part B. Physical Characteristics of
the Copper Eskimos, by **D. Jenness**   (Published May 23, 1923)

Part C. The Osteology of the Western and
Central Eskimos, by John Cameron   (Published June 23, 1923)

VOLUME 13. ESKIMO FOLK-LORE

Vol. 13. Part A. Eskimo Myths and Traditions from Alaska,
the Mackenzie Delta, and Coronation Gulf,
by **D. Jenness**   (Published November 15, 1924)

Part B. String Figures of the Eskimos,
by **D. Jenness**   (Published August 8, 1924)

VOLUME 14. ESKIMO SONGS

Vol. 14. Songs of the Copper Eskimos,
by Helen H. Roberts and **D. Jenness**   (Published December 8, 1925)

VOLUME 15. ESKIMO LANGUAGE AND TECHNOLOGY

Vol. 15. Eskimo Language

Part A.   Comparative Grammar and Vocabulary of
the Eskimo Dialects of Point Barrow,
the Mackenzie Delta, and Coronation Gulf,
by **D. Jenness**   (Published December 5, 1926)

Part B.   Grammatical Notes on Some Western Eskimo Dialects,
by **D. Jenness**   (Published 1944)

VOLUME 16. MATERIAL CULTURE OF THE COPPER ESKIMOS

by **D. Jenness**   (Published 1946)

VOLUME 17. CONTRIBUTIONS TO THE ARCHAEOLOGY
OF WESTERN ARCTIC AMERICA

by **D. Jenness**   (Not published)

REFERENCES

CMC = Canadian Museum of Civilization
CMN = Canadian Museum of Nature
DCL = Dartmouth College Library
GSC = Geological Survey of Canada
LAC = Library and Archives Canada

Published sources

"A friend in Canada." 1922. Letter to the Editor: "Arctic explorer's friend replies to Stefansson charge." *St. Louis Star*, March 12, 1922. CMNAC/1996-077, Series A—R.M. Anderson Collection, Box 65, Folder 24, Archives, CMN.

Anderson, R.M. 1915. Canadian Arctic Expedition, 1913–1914. In Summary Report of the Geological Survey Department of Mines for the Calendar Year 1914. pp. 163–166.

————. 1918. Canadian Arctic Expedition of 1913–1916. Report of the Department of the Naval Service for the Fiscal Year Ending March 31, 1917. pp. 11–47.

Anonymous. 1913. "Believe mutiny drove Stafansson [*sic*] from ship." *San Francisco News*, December 22, 1913. (Copy in V. Stefansson Collection, Wilkins Papers, Box VI, Folder 17. Rauner Special Collections Library, Dartmouth College, Hanover, New Hampshire.)

————. 1914. "Eight are saved of Karluk's men." *New York Times*, September 14, 1914.

————. 1922a. "Arctic voyage of Stefansson may be probed." *Ottawa Citizen*, January 14, 1922.

————. 1922b. "Youngest Arctic explorer to live among floes." *Honolulu Star Bulletin*, February 24, 1922. CMNAC/1996-077, Series A—R.M. Anderson Collection, Box 69, Folder 22, Archives, CMN.

————. 1922c. "Ask for probe into the charge of Stefansson: Four scientists of Southern Party send Petition to Minister of Mines." *Ottawa Morning Journal*, March 8, 1922.

————. 1922d. "Stefansson knew charges would arise. Says he made statements in book deliberately." *Ottawa Journal*, March 17, 1922. (A copy is glued inside the front cover of Dr. Anderson's 1922 copy of *The Friendly Arctic* housed in the Rare Book Room, Library, CMN.)

————. 1922e. "Stefansson's charges are refuted." *Toronto Star Weekly*, August 5, 1922. p. 32.

————. 1923a. "Explorers playing the waiting game." *Ottawa Journal*, April 24, 1923.

————. 1923b. "As to Stefansson." Editorial. *Ottawa Journal*, May 1, 1923. CMNAC/1996-077, Series A—R.M. Anderson Collection, Box 65, Folder 24, Archives, CMN.

————. 1923c. "How Stefansson had citizenship in 2 countries." *Ottawa Journal*, May 1, 1923. CMNAC/1996-077, Series A—R.M. Anderson Collection, Box 65, Folder 24, Archives, CMN.

————. 1923d. "First white man to set foot on island." *Ottawa Citizen*, May 1, 1923. CMNAC/1996-077, Series A—R.M. Anderson Collection, Box 66, Folder 30, Archives, CMN.

————. 1924. "The geographical work of the Canadian Arctic Expedition." *Geographical Journal*, vol. 63, no. 6 (June), pp. 508–525.

————. 1925. "The geographical work of the Canadian Arctic Expedition." *Geographical Journal*, vol. 65, no. 4 (April), pp. 340–342.

Ashlee, Jette Elsebeth. 2008. *An Arctic Epic of Family and Fortune: The Theories of Vilhjalmur Stefansson and Their Influence in Practice on Storker Storkerson and His Family*. Philadelphia, Pa.: Xlibris Corporation. 408 pp.

Auditor General Report. 1920. Ottawa: Government of Canada. p. 69.

Bartlett, R.A. 1914. "Crew of Karluk braved death as bergs sank ship." *Public Ledger*. Philadelphia, Pa. June 1, 1914. (Copy in MG30 B40, Vol. 10, File 17, LAC.)

Bartlett, R.A. and Hale, R.T. 1916. *The Last Voyage of the Karluk—Flagship of Vilhjalmur Stefansson's Canadian Arctic Expedition of 1913–1916.* Toronto: McClelland, Goodchild and Stewart. 329 pp.

Bockstoce, J.R. 1977. *Steam Whaling in the Western Arctic.* New Bedford, Mass.: Old Dartmouth Historical Society. 127 pp.

————. 2000. "Arctic discoveries—Images from voyages of four decades in the North." Seattle, Wash.: University of Washington Press. 122 pp.

Bush, G.W. 1997. "Tight lines." *Deh Cho Drum* (a newspaper). Fort Simpson. Sept. 4, 1997.

Cameron, John, and Ritchie, S.G. 1923. "Osteology and dentition of the western and central Eskimos." Report of the Canadian Arctic Expedition 1913–1918. "The Copper Eskimos." Vol. 12, Part C. 67 pp.

Camsell, C. 1923. "The friendly Arctic." (Reply to letter by V. Stefansson (1923) on the same subject.) *Science*, vol. LVII, no. 1484 (June 8, 1923), pp. 665–666.

Chafe, E.F. 1918. "The voyage of the Karluk and its tragic ending." *Geographical Journal*, vol. LI, no. 5, pp. 307–316.

Chipman, K.G. and Cox, J.R. 1924. Geographical notes on the Arctic Coast of Canada. Part B in Geology and geography. Report of the Canadian Arctic Expedition 1913–1918. Vol. 11, pp. 1B–57B.

Condon, R.G. 1996. *The Northern Copper Inuit—A History.* Norman, Okla.: University of Oklahoma Press. 216 pp.

Cooke, A. 1980. "A gift outright: The exploration of the Canadian Arctic Islands after 1880," pp. 51–60 in *A Century of Canadian Arctic Islands, 1880–1980.* Morris Zaslow (Ed.). Ottawa: The Royal Society of Canada. 358 pp.

Crich, G.E. 1990. *In Search of Heroes.* Air Ronge, Sask.: Northwinds. 283 pp.

Diubaldo, R.J. 1978. *Stefansson and the Canadian Arctic.* Montréal: McGill-Queen's University Press. 274 pp.

Finnie, R.S. 1978. "Stefansson's unsolved mystery." *North* (Nov.–Dec.), pp. 2–7.

French, Inspector F.H. 1919. Report of the Bathurst Inlet Patrol, Royal Northwest Mounted Police, 1917–1918, 93 pp. In Report of the Royal Northwest Mounted Police for the Year ending September 30, 1918.

Geological Survey of Canada. 1915. Annual Report for the Year Ending March 31, 1914. Ottawa: King's Printer.

————. 1916. Summary Report of the Geological Survey, Department of Mines, for the Calendar Year 1915. Ottawa: King's Printer.

Godfrey, W. Earl. 1966. "Birds of Canada." National Museum of Canada. Bulletin No. 203, Biological Series 73. 428 pp.

Gray, David R. 1979. CGS *Alaska. The Bulletin: Quarterly Journal of the Maritime Museum of British Columbia*, no. 43, pp. 19–22; no. 44, pp. 2–4, 12.

————. 1987. *The Muskoxen of Polar Bear Pass.* Markham, Ont.: Fitzhenry & Whiteside. 191 pp.

Greely, A.W. 1912. "The origin of Stefansson's Blond Eskimo." *The National Geographic Magazine,* vol. CXXVI, no. DCCLI (December), pp. 1225–1238. (A copy of the article is in CMNAC/1996-077, Series A—R.M. Anderson Collection, Box 65, Folder 19, Archives, CMN.)

Guthrie, R.D. 2004. Letter to the Editor: "Radiocarbon evidence of mid-Holocene mammoths stranded on an Alaskan Bering Sea Island." *Nature*, vol. 429, pp. 746–749.

Hall, Edwin S. Jr. 1987. *A Land Full of People, a Long Time Ago; an Analysis of Three Archaeology Sites in the Vicinity of Kaktovik, Northern Alaska.* Brockport, N.Y.: Edwin Hall and Associates. Technical Memorandum No. 24. 360 pp.

Hansard (House of Commons Debates) for June 10, 1925, vol. LX, no. 84, pp. 4266–4267. (A copy is in CMNAC/1996-077, Series A—R.M. Anderson Collection, Box 66, Folder 21, Archives, CMN.)

Harris, R.A. 1911. *Arctic Tides.* Washington: U.S. Coast and Geodetic Survey.

Horwood, H. 1977. *Bartlett, the Great Canadian Explorer.* Garden City, N.Y.: Doubleday & Co. 194 pp.

Hunt, W.R. 1986. *Stef—A Biography of Vilhjalmur Stefansson, Canadian Arctic Explorer.* Vancouver: University of British Columbia Press. 317 pp.

Jenkins, McKay. 2005. *Bloody Falls of the Coppermine—Madness, Murder, and the Collision of Cultures in the Arctic, 1913.* New York: Random House. 298 pp.

Jenness, Diamond. 1922a. "The Copper Eskimos, Part A: The life of the Copper Eskimos." Report of the Canadian Arctic Expedition 1913–1918. Vol. 12, 277 pp.

————. 1922b. "The friendly Arctic." Comments on a review of V. Stefansson's book by Professor Richard Pearl. In *Science*, vol. 56, no. 1436 (July 7, 1922), pp. 8–12.

————. 1922c. "Declares Stefansson Acted Aggressively." *Toronto Star*, January 14, 1922.

————. 1923. "The Copper Eskimos, Part B: Physical characteristics of the Copper Eskimos." Report of the Canadian Arctic Expedition 1913–1918. Vol. 12. 89 pp.

————. 1928. *The People of the Twilight.* Toronto: Macmillan and Company. 247 pp.

————. 1932. "The Indians of Canada." National Museum of Canada. Bulletin No. 65. Anthropology Series No. 15. 445 pp.

————. 1957. *Dawn in Arctic Alaska.* Minneapolis: University of Minnesota Press. 222 pp.

Jenness, Diamond, and Stuart E. Jenness. 2008. "Through darkening spectacles: The memoirs of Diamond Jenness." Canadian Museum of Civilization. Mercury Series, History Paper 55. 407 pp.

Jenness, S.E. 1987. "Geographical names and the Canadian Arctic Expedition, 1913–1918." *Canoma*, vol. 23, no. 1, pp. 29–32.

————. 1990. "Diamond Jenness's archaeological investigations on Barter Island, Alaska." *Polar Record*, vol. 26, no. 197, pp. 91–102.

————. 1992. Letter to the Editor [re. Uloksak]. *Arctic*, vol. 45, no. 2, pp. 208–209.

————. 1996. "Conflict and adversities—The Southern Party of the Canadian Arctic Expedition 1913–1916." *The Beaver*, vol. 76, no. 4, pp. 34–41.

————. 2004. *The Making of an Explorer: George Hubert Wilkins and the Canadian Arctic Expedition, 1913–1916.* Montréal: McGill-Queen's University Press. 417 pp.

Jenness, S.E. (Ed.). 1991. *Arctic Odyssey—The Arctic Diary of Diamond Jenness, 1913–1916.* Gatineau, Que.: Canadian Museum of Civilization. 859 pp.

Johansen, Fritz. (12 articles published in the Danish newspaper *Politiken* in Danish between 1913 and 1917; English translations are in the Johansen papers, MG30 B165, LAC):

————. 1913a. "Victoria, 10 June, 1913." *Politiken*, July 7, 1913.

————. 1913b. "Nome, 20 July, 1913." *Politiken*, August 30, 1913.

————. 1913c. "Point Barrow, 24 July, 1913." *Politiken*, December 17, 1913.

————. 1914a. "Outside the Colville River, Alaska, 1 September, 1913." *Politiken*, March 9, 1914

————. 1914b. "Outside the Thetis Islands, Alaska, 9 September, 1913." *Politiken*, March 11, 1914.

————. 1914c. "Cape Collinson, Alaska, 13 October, 1913." *Politiken*, January 19, 1914.

————. 1914d. "Collinson Point, Alaska, 9 December, 1913." *Politiken*, June 10, 1914.

————. 1916. "Demarcation Point, Alaska, May 1914." *Politiken*, December 10, 1916.

————. 1917a. "Bernard Harbour, N.W.T., Christmas 1914." *Politiken*, January 18, 1917.

————. 1917b. "Bernard Harbour, N.W.T., New Year, 1915." *Politiken*, June 11, 1917.

————. 1917c. "Bernard Harbour, December 1915." *Politiken*, July 21, 1917.

————. 1917d. "Bernard Harbour, N.W.T., April 1916." *Politiken*, July 26, 1917.

Johnson, J.K. (Ed.). 1968. The Canadian Directory of Parliament 1867–1967. Ottawa: Public Archives of Canada. 731 pp.

Kenney, Gerard. 2004. *Ships of Wood and Men of Iron: A Norwegian-Canadian Saga of Exploration in the High Arctic.* Canadian Plains Research Center. Regina, Sask.: University of Regina. 135 pp.

La Nauze, Inspector C.D. 1917. Reports Regarding the Great Bear Lake Patrol. Appendix O, pp. 190–253. In Report of the Royal Northwest Mounted Police for the year ending September 30, 1916.

LeBourdais, D.M. 1931. *Northward on the New Frontier.* Ottawa: Graphic Publishers Limited. 311 pp.

————. 1963. *Stefansson, Ambassador of the North.* Montréal: Harvest House. 204 pp.

Leffingwell, E. de K. 1919. "The Canning River region, Northern Alaska." United States Geological Survey Professional Paper 109. 251 pp.

Levere, Trevor. 1993. *Science and the Canadian Arctic—A Century of Exploration 1818–1918.* Cambridge: Cambridge University Press. 438 pp.

MacBeth, Madge. 1923. "Daring Captain Joe and his Teddy Bear." *Toronto Star Weekly*, April 7, 1923.

MacMillan, Donald B. 1915a. "In search of a new land (Part 1)." *Harper's Magazine*, vol. cxxxi, no. dcclxxxv, pp. 651–665. (Copy in CMNAC/1996-077, Series A—R.M. Anderson Collection, Box 65, Folder 19, Archives, CMN.)

————. 1915b. "In search of a new land (Part 2)." *Harper's Magazine*, vol. cxxxi, no. dcclxxxv, pp. 921–930. (Copy in CMNAC/1996-077, Series A—R.M. Anderson Collection, Box 65, Folder 19, Archives, CMN.)

————. 1918. *Four Years in the White North.* New York: Harper & Brothers. 426 pp.

Manning, Tom. 1956. "Narrative of a Second Defence Research Board Expedition to Banks Island." *Arctic*, vol. 9, nos. 1–2, pp. 3–77.

Markham, Sir Clements. 1909. *The Life of Admiral Sir Leopold McLintock.* London: John Murray. 370 pp.

Maurer, Fred W. (Four articles published in 1915; a typed copy of each is in CMNAC/1996-077, Series A—R.M. Anderson Collection, Box 68, Folder 11, Archives, CMN):

————. 1915a. "Part 1. The drift." *World Magazine*, June 6, 1915.

————. 1915b. "Part 2. The trail." *World Magazine*, June 13, 1915.

————. 1915c. "Part 3. Wrangel Island." *World Magazine*, June 20, 1915.

————. 1915d. "Part 4. Rescue." *World Magazine*, June 27, 1915.

McConnell, Burt M. 1914a. "Newsy letter about Stefansson party." *The Nome Daily Nugget*, August 31, 1914.

————. 1914b. "McConnell writes about expedition." *The Nome Daily Nugget,* September 14, 1914.

————. 1914c. "Got Karluk's Men as Hope was Dim." *New York Times*, September 15, 1914.

————. 1915a. "The rescue of the *Karluk* survivors." *Harper's Magazine*, February 1915, pp. 349–360.

————. 1915b. "Over the ice with Stefansson." *Harper's Magazine*, April 1915, pp. 672–685.

McConnell, Burt M., and Noice, Harold. 1923. Reply to "Letter to Editor" by D. Jenness (1922b). *Science*, vol. LVII, no. 1474 (March 24, 1923), pp. 368–373.

McDiarmid, F.A. 1923. "Geographical determinations of the Canadian Arctic Expedition." *Geographical Journal*, vol. 62, no. 10, pp. 293–302.

McGhee, R. 1978. *Canadian Arctic Prehistory.* Toronto and New York: Van Nostrand Reinhold Ltd. 128 pp.

McInnes, Tom (Ed.). 1932. *Klengenberg of the Arctic: an Autobiography.* London: Jonathan Cape. 360 pp.

McKinlay, W.L. 1914. (Four articles published in the Scottish paper *The People's Journal*, in 1914. A copy of each is in the R.M. Anderson Papers, MG30 B40, Vol. 11, File 1, LAC.)

————. 1914a. "Scotchman marooned in the Arctic—Tells of amazing adventures in the icy north: Only survivor of eleven scientists." *The People's Journal*, Saturday, November 28, 1914.

————. 1914b. "Adrift on an Arctic ice floe—How members of Stefansson's Expedition perished—Gramophone plays as Karluk goes down." *The People's Journal*, Saturday, December 5, 1914.

————. 1914c. "Tragedies of the frozen north—Providential escapes from awful life—Stefansson's men fight for life." *The People's Journal*, Saturday, December 12, 1914.

————. 1914d. "In the grip of the Polar ice field—Disease and death wage war on Arctic explorers." *The People's Journal*, Saturday, December 19, 1914.

————. 1914e. "The last of the Arctic ship Karluk." *Manchester Guardian*, November 23, 1914. MG30 B40, Vol. 11, File 1, LAC.

————. 1976. *Karluk—The Great Untold Story of Arctic Exploration*. London: Weidenfeld & Nicolson. 170 pp.

Moss, M.L., and Bowers, P.M. 2007. "Migratory bird harvest in northwestern Alaska: A zooarchaeological analysis of Ipiuatak and Thule occupations from the Deering Archaeological District." *Arctic Anthropology*, vol. 44, no. 1, pp. 37–50.

Moyles, R.G. 1979. *British Law and Arctic Men*. Saskatoon, Sask.: Western Producer Prairie Books. 93 pp.

Nasht, S. 2005. *The Last Explorer: Hubert Wilkins, Australia's Unknown Hero*. Sydney, Australia: Hodder. 346 pp.

Niven, Jennifer. 2000. *The Ice Master—The Doomed 1913 Voyage of the Karluk*. New York: Hyperion. 402 pp.

————. 2003. *Ada Blackjack—A True Story of Survival in the Arctic*. New York: Hyperion. 433 pp.

Noice, Harold. 1922. "The blond Eskimos." *American Anthropologist*, vol. 24, no. 2, pp. 228–231.

————. 1924. *With Stefansson in the Arctic*. New York: Dodd, Mead & Company. 270 pp.

O'Neill, J.J. 1924. "The geology of the Arctic Coast of Canada west of the Kent Peninsula." Part A in Geology and geography. Report of the Canadian Arctic Expedition 1913–1918. Vol. 11, pp. 1A–107A, Ottawa: King's Printer.

Ottawa Citizen. 1930. "Karluk belonged to famous cat family." Uncle Ray's Mail Bag. *Ottawa Evening Citizen*, April 12, 1930.

Ottawa Evening Journal. 1922. [Interview with D. Jenness]. January 14, 1922, p. 24.

Pálsson, Gísli. 2003. *Travelling Passions—The Hidden Life of Vilhjalmur Stefansson*. Winnipeg, Man.: University of Manitoba Press. 374 pp.

Pálsson, Gísli (Ed.). 2001. *Writing on Ice: The Ethnographic Notebooks of Vilhjalmur Stefansson*. Hanover, N.H., and London: University Press of New England. 393 pp.

Pearl, R. 1922. Book review: "The friendly Arctic," by V. Stefansson. *Science*, vol. LV, no. 1421, March 24, 1922, pp. 320–321.

Peary, R.E. 1907. *Nearest the Pole: A Narrative of the Polar Expedition of the Peary Arctic Club in the S.S. Roosevelt, 1905–1906*. New York: Doubleday, Page & Co. 410 pp.

Pharand, D. 1980. "Quel sera l'avenir du passage du Nord-Ouest?" *North?Nord*, Part I, summer/été, pp. 2–9, Part II, fall/automne, pp. 2–9.

Pringle, Heather. 2009. "Strands of evidence." *Canadian Geographic*, vol. 129, no. 2, pp. 44–48, 50, 52, 54, 56.

Roberts, Helen H., and Jenness, Diamond. 1925. "Songs of the Copper Eskimos." Report of the Canadian Arctic Expedition 1913–1918. Vol. 14, 506 pp.

Rodahl, K. 1949. "The toxicity of polar bear liver." *Science,* vol. 164, no. 4169, pp. 530–531.

Rodahl, K. and Moore, T. 1943. "The vitamin A content of bear and seal liver." *Biochemical Journal*, vol. 37, p. 166.

Rowley, Diana. 1952. "Stefansson Island." *The Arctic Circular*, vol. 5, no. 5, pp. 45–53.

Sperry, J.R. 1987. Letter to the Editor: " 'Eskimo'—a term of dignity or derogation?" *Arctic*, vol. 40, p. 364.

St-Onge, Denis A., and McMartin, Isabel. 1995. "Quaternary geology of the Inman River area, Northwest Territories." Geological Survey of Canada Bulletin 441. 59 pp.

Steckley, John L. 2008. *White Lies about the Inuit*. Peterborough, Ont.: Broadview Press. 168 pp.

Stefansson, V. 1913. *My Life with the Eskimo*. New York: Macmillan Company. 538 pp.

————. 1919a–f. "Solving the problem of the Arctic" (six articles with a common title). *Harper's Magazine*. (A copy of each part is in CMNAC/1996-077, Series A—R.M. Anderson Collection, Box 65, Folder 19, Archives, CMN.)

————. 1919a. Part I. "A record of five years' explorations," vol. 138, no. 827, pp. 577–590.

————. 1919b. Part II. "Ways and means of life on the ice," vol. 138, no. 828, pp. 721–735.

————. 1919c. Part III. "Drifting to Banks Island—The arrival of the Mary Sachs," vol. 139, no. 829, pp. 36–47.

————. 1919d. Part IV. "Hunting caribou and building snow houses," vol. 139, no. 830, pp. 193–203.

————. 1919e. Part V. "Our first discovery of New Land," vol. 139, no. 831, pp. 386–398.

————. 1919f. Part VI. "Conclusion—Further discoveries of New Land," vol. 139, no. 833, pp. 709–720.

————. 1919g–l. "Solving the problem of the Arctic" (six articles with a common title). *Maclean's Magazine*. (A copy of each part is in CMNAC/1996-077, Series A—R.M. Anderson Collection, Box 68, Folder 7, Archives, CMN. These are the same six articles as published in *Harper's Magazine*, 1919a–f, with minor changes; Part I states "Maclean's Magazine has secured the inclusive rights to publish Mr. Stefansson's story in Canada.")

————. 1919g. Part I. "A record of five years' exploration for the Canadian Government," vol. 32, no. 4, pp. 15–16, 87.

————. 1919h. Part II. "Subsisting in the North," vol. 32, no. 5 (May), pp. 16–19, 95, 97.

————. 1919i. Part III. "Drifting to Banks Island on the Ice," vol. 32, no. 6 (June), pp. 22–25.

————. 1919j. Part IV. "Wintering in the North," vol. 32, no. 7 (July), pp. 21–22, 77, 78.

————. 1919k. Part V. "We discover New Lands," vol. 32, no. 8 (August), p. 26–28, 50–52.

————. 1919l. Part VI. "Further discoveries of New Land," vol. 32, no. 10 (October), p. 35–37, 75–76.

————. 1919m. Stefánsson-Anderson Arctic Expedition. Anthropological Papers of the American Museum of Natural History, vol. XIV, New York. Reprinted in 1978 by AMS Press, New York. 475 pp.

————. 1921a. "The Canadian Arctic Expedition of 1913 to 1918." *Geographical Journal*, vol. 58, October, pp. 283–305.

————. 1921b. *The Friendly Arctic*. New York: Macmillan Company. 784 pp.

————. 1922. *Hunters of the Great North*. New York: Harcourt, Brace and Company. 301 pp.

————. 1923a. "The friendly Arctic," reply to "Letter to Editor" by D. Jenness (1922b). *Science*, vol. LVII, no. 1474, March 30, 1923, pp. 368–369.

————. 1924. *My Life with the Eskimo*. London: George Harrap. 308 pp. (revised and abridged version).

————. 1944. *Arctic Manual*. New York: Macmillan Company. 556 pp.

————. 1964. *Discovery—The Autobiography of Vilhjalmur Stefansson*. New York: McGraw-Hill Book Company. 411 pp. (See especially Chapters 15 to 23, pp. 133–213.)

Stefansson, Vilhjalmur, with the collaboration of John Irvine Knight. 1925. *The Adventure of Wrangel Island—Based on the Diary of Errol Lorne Knight*. New York: Macmillan Company. 424 pp.

Story, G.M., Kirwin, W.J., and Widdowson, J.D.A. (Eds.). 1982. *Dictionary of Newfoundland English*. Toronto: University of Toronto Press. 625 pp.

Storkerson, S.T. 1920. (Two articles on Storkerson's unusual ice-drift experiences. A copy of each is in CMNAC/1996-077, Series A—R.M. Anderson Collection, Box 68, Folder 7, Archives, CMN.)

————. 1920a. "Eight months adrift in the Arctic—The story of a remarkable exploration." *Maclean's Magazine*, vol. XXXIII, no. 5, March 15, 1920, pp. 9–11.

————. 1920b. "Eight months adrift in the Arctic—The complete story of some remarkable discoveries." *Maclean's Magazine*, vol. XXXIII, no. 6, April 1, 1920, pp. 12–13, 63–64.

Taverner, Percy A. 1940. *Birds of Canada*. Toronto: The Musson Book Company. 446 pp. (This is a new and revised edition of the 1937 original edition.)

Thomas, Lowell. 1961. *Sir Hubert Wilkins, a Biography*. New York and Toronto: McGraw-Hill Book Company. 294 pp.

Tucker, G.N. 1952. *The Naval Service of Canada—Its Official History*, Vol. 1: "Origin and early years." Ottawa: King's Printer. 436 pp.

Van Stone, James W. 1994. "The Noice Collection of Copper Inuit material culture." *Fieldiana Anthropology*, N.S. no. 20, Feb. 28, 1994, pp. 1–71.

Vartanyan, S.L., Arslanov, Kh. A., Tertychnaya, T.V., and Chernov, S.B. 1995. "Radiocarbon dating evidence for mammoths on Wrangel Island, Arctic Ocean, until 2000 BC." *Radiocarbon*, vol. 37, no. 1, pp. 1–6.

Walters, V. 1953. "The fishes collected by the Canadian Arctic Expedition, 1913–1918, with additional notes on the ichthyofauna of western Arctic Canada." Bulletin of the National Museum of Canada, no. 128, pp. 257–274.

Webster, N. 1972. *Webster's New World Dictionary of the American Language* (Second College Edition). New York: New World Publishing Company. 1,692 pp.

————. 1980. *Webster's New Geographical Dictionary*. Springfield, Mass.: G. & C. Merriam Co., Publishers. 1,370 pp.

White, James. 1924. "Letter to the Editor." *Geographical Journal*, vol. 63, no. 6, pp. 508–525.

Zaslow, M. 1975. *Reading the Rocks—The Story of the Geological Survey of Canada 1842–1972*. Macmillan Company of Canada Limited, Toronto, in association with the Department of Energy, Mines and Resources and Information Canada, Ottawa, 599 pp.

Zaslow, M. (Ed.). 1980. *A Century of Canadian Arctic Islands, 1880–1980*. 23rd Symposium. Ottawa: The Royal Society of Canada. 358 pp.

Archival sources

Anderson, R.M. 1909. Letter dated November 15, 1909 to Dr. Herman C. Bumpus, Director, American Museum of Natural History. MG30 B40, vol. 12, LAC.

————. 1914a. Letter dated January 4, 1914 to G.J. Desbarats. RG42, Vol. 476, 84-2-29, vol. 13, LAC.

————. 1914b. Letter dated January 19, 1914 to Mrs. R.M. Anderson. R.M. Anderson Correspondence, MG30 B40, Vol. 2, Folder 1, LAC.

————. 1914c. Letter dated August 15, 1914 to Chief Accountant, Department of the Naval Service. RG42, Vol. 478, File 84-2033, vol. I-2, LAC.

————. 1915. Letter dated February 13, 1915 to G.J. Desbarats, Ottawa. RG42, Vol. 476, 84-2-29, vol. 13, LAC.

————. 1916a. Canadian Arctic Expedition Records, 1913–1916, vols. I –VIII. CMNAC /1996-077, Series A—R.M. Anderson Collection, Box 59, Folders 1–4, Archives, CMN.

————. 1916b. Letter dated November 10, 1916 to I.J. Beausoleil, Chief Accountant, Department of the Naval Service. RG42, Vol. 471, File 82-405, LAC.

————. 1916c. Report to Deputy Minister of the Naval Service, Dec. 15, 1916. CMNAC/1996-077, Series A—R.M. Anderson Collection, Box 69, Folder 19, Archives, CMN.

————. 1916d. Memorandum, undated, to Department of the Naval Service, "Stores of the Canadian Arctic Expedition, Southern Party, returned to Nome, Alaska, 1916." CMNAC/1996-077, Series A—R.M. Anderson Collection, Box 68, Folder 36, Archives, CMN.

————. 1917. Letter dated May 1, 1917 to Daniel Sweeney. CMNAC/1996-077, Series A—R.M. Anderson Collection, Box 69, Folder 43, Archives, CMN.

————. 1918. Letter dated November 6, 1918 to Isaiah Bowman. Stefansson Collection, Rauner Special Library, Stefansson MS 5598, Folder IV-22, DCL.

————. 1919a. "History of operations of Canadian Arctic Expedition, 1913–1916, particularly of the Southern Party," 19 pp. (typescript). CMNAC/1996-077, Series A—R.M. Anderson Collection, Box 66, Folder 8, Archives, CMN.

————. 1919b. "The Canadian Arctic Expedition, 1913–1918," 19 pp. (incomplete typescript). CMNAC/1996-077, Series A—R.M. Anderson Collection, Box 66, Folder 8, Archives, CMN.

————. 1920a. Letter dated January 7, 1919 [sic, 1920a] to W.V. Bruce, Brandon, Manitoba. CMNAC/1996-077, Series A—R.M. Anderson Collection, Box 67, Folder 22, Archives, CMN.

————. 1920b. Letter dated July 20, 1920 to D.B. MacMillan. CMNAC/1996-077, Series A—R.M. Anderson Collection, Box 68, Folder 8, Archives, CMN.

————. 1920c. Memorandum dated September 29, 1920 to G.J. Desbarats. CMNAC/1996-077, Series A—R.M. Anderson Collection, Box 68, Folder 36, Archives, CMN.

————. 1921a. Letter dated March 5, 1921 to C.W. Townsend, Boston. CMNAC/1996-077, Box 70, Folder 45, Archives, CMN.

————. 1921b. Letter dated March 21, 1921 to Deputy Minister, Department of the Naval Service [Desbarats]. CMNAC/1996-077, Series A—R.M. Anderson Collection, Box 68, Folder 36, Archives, CMN.

————. 1921c. Letter dated March 26, 1921 to J.E. Law, University of California, Berkeley. CMNAC/1996-077, Series A—R.M. Anderson Collection, Box 68, Folder 6, Archives, CMN.

————. 1921d. Letter dated March 28, 1921 to W.B. Bell, U.S. Department of Agriculture, Washington, D.C. CMNAC/1996-077, Series A—R.M. Anderson Collection, Box 67, Folder 22, Archives. CMN.

————. 1921e. Letter dated November 19, 1921 to General A.W. Greely, Cambridge, Massachusetts. CMNAC/1996-077, Series A—R.M. Anderson Collection, Box 66, Folder 3, Archives, CMN.

————. 1921f. Letter dated November 21, 1921 to General A.W. Greely, Cambridge, Massachusetts. CMNAC/1996-077, Series A—R.M. Anderson Collection, Box 66, Folder 3, Archives, CMN.

————. 1921g. Letter dated December 3, 1921 to J.J. Underwood, Washington, D.C. CMNAC/1996-077, Series A—R.M. Anderson Collection, Box 70, Folder 16, Archives, CMN.

————. 1921h. Letter dated December 31, 1921, to J.J. Underwood, Washington, D.C. CMNAC/1996-077, Series A—R.M Anderson Collection, Box 70, Folder 16, Archives, CMN.

————. 1922a. Letter dated January 24, 1922 to E. de K. Leffingwell. CMNAC/1996-077, Series A—R.M. Anderson, Collection, Box 68, Folder 1, Archives, CMN.

————. 1922b. Letter dated February 22, 1922 to Major H.E. Colley. CMNAC/1996-077, Series A—R.M. Anderson Collection, Box. 67, Folder 33, Archives, CMN.

————. 1922c. Letter dated March 17, 1922 to Professor J.J. O'Neill. CMNAC/1996-077, Series A—R.M. Anderson Collection, Box 69, Folder 33, Archives. CMN.

————. 1922d. Letter dated March 18, 1922 to Dr. Carl V. Kent. CMNAC/1996-077, Series A—R.M. Anderson Collection, Box 67, Folder 18, Archives, CMN.

————. 1922e. Letter dated March 20, 1922 to Professor J.J. O'Neill. CMNAC/1996-077, Series A—R.M. Anderson Collection, Box 67, Folder 33, Archives, CMN.

————. 1922f. Letter dated March 25, 1922, to W.L. McKinlay. CMNAC/1996-077, Series A—R.M. Anderson Collection, Box 68, Folder 13, Archives, CMN.

————. 1922g. Letter dated April 11, 1922 to Professor J.J. O'Neill. CMNAC/1996-077, Series A—R.M. Anderson Collection, Box 67, Folder 33, Archives, CMN.

————. 1922h. Memorandum prepared for the press (published in part by the *Des Moines Sunday Register*). Typescript, 34 p. CMNAC/1996-077, Series A—R.M. Anderson Collection, Box 69, Folder 31, Archives, CMN.

————. 1923a. Letter dated January 18, 1923 to W.L. McKinlay, Glasgow. CMNAC/1996-077, Series A—R.M. Anderson Collection, Box 68, Folder 13, Archives, CMN.

————. 1923b. Letter dated April 24, 1923, to C. Camsell, Deputy Minister of Mines. CMNAC/1996-077, Series A—R.M. Anderson Collection, Box 53, Folder 22, Archives, CMN.

————. 1923c. Letter dated April 25, 1923, to C. Camsell, Deputy Minister of Mines. CMNAC/1996-077, Series A—R.M. Anderson Collection, Box 53, Folder 22, Archives, CMN.

————. 1923d. Letter dated May 11, 1923, to C. Camsell, Deputy Minister of Mines. CMNAC/1996-077, Series A—R.M. Anderson Collection, Box 53, Folder 22, Archives, CMN.

————. 1923e. Letter dated June 13, 1923, to E. de K. Leffingwell, Pasadena, California. CMNAC/1996-077, Series A—R.M. Anderson Collection, Box 63, File 37, Archives, CMN.

————. 1923f. Letter dated December 31, 1923 to Aarnout Castel, Seattle. CMNAC/1996-077, Series A—R.M. Anderson Collection, Box 65, Folder 31, Archives, CMN.

————. 1924a. Letter dated January 21, 1924 to Dr. Charles Camsell. CMNAC/1996-077, Series A—R.M. Anderson Collection, Box 65, Folder 22, Archives, CMN.

————. 1924b. Letter dated July 29, 1924 to F. McVeigh, Department of Marine and Fisheries. Box 53, Folder 17, Archives, CMN.

————. 1924c. Letter dated November 4, 1924 to G.J. Desbarats. CMNAC/1996-077, Series A—R.M. Anderson Collection, Box 68, Folder 36, Archives, CMN.

————. 1925. Letter dated May 13, 1925 to F. McVeigh, Dept. of Marine and Fisheries. CMNAC/1996-077, Series A—R.M. Anderson Collection, Box 68, Folder 16, Archives, CMN.

————. 1926. Letter dated March 4, 1926 to Roald Amundsen. MG30 B40 vol. 10, LAC.

————. 1928a. Letter dated July 14, 1928 to H.F. Osborn, New York. CMNAC/1996-077, Series A—R.M. Anderson Collection, Box 70, Folder 1, Archives, CMN.

————. 1928b. Letter dated December 13, 1928 to Charles S. Elton, Oxford. CMNAC/1996-077, Series A—R.M. Anderson Collection, Box 68, Folder 38, Archives, CMN.

————. 1944. Letter dated May 10, 1944 to Captain R.A. Bartlett, New York. CMNAC/1996-077, Series A—R.M. Anderson Collection, Box 71, Folder 29, Archives, CMN.

Anderson, Mrs. R.M. 1913. Letter dated November 7, 1913 to Dr. R.M. Anderson. MG30 B40, Vol. 1, File 15, LAC.

————. 1925. Letter dated February 4, 1925 to Mrs. J.T. Crawford. Crawford Papers in R.M. Anderson Collection. MG30 B40, LAC.

————. No date. Incomplete manuscript. MG30 B40, Folder 5, LAC.

Auditor General Report. 1920. Naval Service Department Expenditures. CMNAC/1996-077, Series A—R.M. Anderson Collection, Box 67, Folder 21, Archives, CMN.

Bartlett, R.A. 1922. Letter dated February 6, 1922 to Dr. R.M. Anderson. CMNAC/1996-077, Series A—R.M. Anderson Collection, Box 67, Folder 28, Archives, CMN.

Baur, W.J. 1918. Arctic diary. MG30 B8, LAC.

Beuchat, H. 1913. Letter dated July 14, 1913 to C.M. Barbeau. Barbeau Collection, B-Mc-2159, Canadian Centre for Folk Culture Studies, CMC.

Brett, G.P. 1913. Letter dated May 6, 1913. President, Macmillan Publishing Company, to R.M. Anderson. MG30 B40, Vol. 1, File 1-12, LAC.

Brock, R.W. 1913a. Letter dated February 4, 1913 to W.J. Roche, Minister of Mines. Gordon Papers Series 3, File 2117 (RLB) on microfilm FB674 reel 90, Robarts Library, University of Toronto.

————. 1913b. Letter dated May 28, 1913 to G.J. Desbarats. RG42 Vol. 475, File 84-2-1, vol. 1, LAC.

————. 1913c. Letter dated June 4, 1913 to V. Stefansson. (Copy in CMNAC/1996-077, Series A—R.M. Anderson Collection, Box 66, Folder 1, Archives, CMN.)

————. 1913d. Letter dated June 25, 1913 to V. Stefansson. in Nome. (Copy in CMNAC/1996-077, Series A—R.M. Anderson Collection, CMN.)

─────. 1914. Letter dated May 7, 1914 to V. Stefansson. (Copy in CMNAC/1996-077, Series A—R.M. Anderson Collection, Box. 65, Folder 34, Archives, CMN.)

─────. 1923. Letter dated April 24, 1923 to Dr. C. Camsell, Deputy Minister of Mines. RG45, File 4078C/57, LAC. (Copy in CMNAC/1996-077, Series A—R.M. Anderson Collection, Box 66, Folder 31, Archives, CMN.)

Bruce, Corporal W.C. 1917. Letter dated January 10, 1917 to Dr. R.M. Anderson. CMNAC/1996-077, Series A—R.M. Anderson Collection, Box 67, Folder 24, Archives, CMN.

Camsell, C. 1922. Memo dated January 15, 1922 to [L.L.] Bolton [Deputy Minister, Department of Mines, and Acting Director Victoria Memorial Museum]. (Copy in CMNAC/1996-077, Series A—R.M. Anderson Collection, Box 66, Folder 1, Archives, CMN.)

─────. 1923a. Letter dated April 16, 1923 to Dr. R.W. Brock. Copy in CMNAC/1996-077, Series A—R.M. Anderson Collection, Box. 66, Folder 31, Archives, CMN.

─────. 1923b. Letter dated April 26, 1923 to Professor E.E. Prince. CMNAC/1996-077, Series A—R.M. Anderson Collection, Box 66, Folder 31, Archives, CMN.

Canadian Arctic Expedition Committee 1913–1916. 1918. Report No. 14 of the Editorial Committee, February 15, 1918. CMNAC/1996-077, Series A—R.M. Anderson Collection, Box 53, Folder 23a, Archives, CMN.

Chafe, E.F. 1917. "The voyage of the Karluk and its tragic ending." Unpublished manuscript. Maritime Museum, Victoria, British Columbia.

Chipman, K.G. 1913. Letter dated July 18, 1913 to W.H. Boyd. MG30 B66, LAC.

─────. 1916. Arctic diary, 1913–1916. MG30 B66, LAC.

─────. 1920. Letter dated September 24, 1920 to Dr. R.M. Anderson. RG42, Vol. 471, File 84-2-7, LAC.

Cox, J.R. 1916. Field notes, Canadian Arctic Expedition, 1913–1916. MG30 B123, LAC.

Desbarats, G.J. 1913. Letter dated May 29, 1913 to V. Stefansson. (Copy with CAE instructions in MG30 B40, vol. 1, File 13, LAC.)

─────. 1914a. Letter dated April 30, 1914 to Dr. R.M. Anderson. CMNAC/1996-077, Series A—R.M. Anderson Collection, Box 68, Folder 36, Archives, CMN.

─────. 1914b. Letter dated April 30, 1914 to Dr. R.M. Anderson. RG42, Vol. 478, File 84-2-33, vol. I-2, LAC.

─────. 1918. Letter dated March 27, 1918 to Dr. R.M. Anderson. CMNAC/1996-077, Series A—R.M. Anderson Collection, Box 68, Folder 36, Archives, CMN.

─────. 1919. Letter dated February 26, 1919 to Fred Cook. CMNAC/1996-077, Series A—R.M. Anderson Collection, Box 53, Folder 23a, Archives, CMN.

Donohue, P. 1922. "Remarks on the Canadian Arctic Expedition, 1913–1918, by Peter Donohue, member of crew of *Polar Bear*, from 1915 to 1918." (Typed copy, 5 pages, in CMNAC/1996-077, Series A—R.M. Anderson Collection, Box 68, Folder 37, Archives, CMN.)

Douglas, R. 1921a. Letter dated June 16, 1921 to G.J. Desbarats, Deputy Minister of the Naval Service. RG42, Vol. 471, File 84-2-7, LAC.

─────. 1921b. Letter dated June 23, 1921 to G.J. Desbarats, Deputy Minister of the Naval Service. RG42, Vol. 471, File 84-2-7, LAC.

Foran, W.M. 1913. Letter dated May 3, 1913 to Dr. R.M. Anderson. MG30 B40, Vol. 1, File 13, LAC.

Grosvenor, G.H. 1912. Letter dated December 17, 1912 to Dr. R.M. Anderson. CMNAC/1996-077, Series A —R.M. Anderson Collection, Box 66, Folder 4, Archives, CMN.

─────. 1913. Letter dated February 11, 1913 to V. Stefansson. DCL.

Hadley, J. 1918. Arctic diary, 1917–1918, MG30 B2, LAC.

Jenness, D. 1914a. Letter dated June 29, 1914 to Professor G.W. von Zedlitz. Jenness Papers, Hocken Library, University of Otago, Dunedin, New Zealand. (Photocopy in Jenness Papers, MG30 B89, LAC.)

————. 1914b. Archaeological notes on Eskimo ruins at Barter Island on the Arctic Coast of Alaska, excavated by D. Jenness. Unpublished manuscript, no. 85, vol. 1, 106 pp. Archaeology Collection, Archives, CMC.

————. 1916. Arctic diary, 1913–1916 (3 vols.). MG30 B89, LAC.

————. 1922. Letter dated August 19, 1922 to Dr. R.M. Anderson. CMNAC/1996-077, Series A—R.M. Anderson Collection, Box 66, Folder 13, Archives, CMN.

Jenness, S.E. 1995. Central Arctic names and their origins. 190 pp. Private report. (Copy at Canadian Permanent Committee on Geographical Names, Geometrics Canada, 615 Booth Street, Ottawa.)

————. 1997a. Victoria Island and Stefansson Island: Names and their origins. 112 pp. Private report. (Copy at Canadian Permanent Committee on Geographical Names, Geometrics Canada, 615 Booth Street, Ottawa.)

————. 1997b. Banks Island names and their origins. 69 pp. Private report. (Copy at Canadian Permanent Committee on Geographical Names, Geometrics Canada, 615 Booth Street, Ottawa.)

————. (compiler). 2003. Photographs and films of the Canadian Arctic Expedition 1913–1918—One of Canada's little known historic treasures. An annotated catalogue with brief biographies. 228 pp. Private report. (Copies at Canadian Museum of Nature (Archives Section), Canadian Museum of Civilization (Archives Section), Photographic Section of Geological Survey of Canada, and Photographic Section of Library and Archives Canada.)

Johansen, F. 1926. Fishes of Arctic America. CMNAC 95-043. Incomplete manuscript. Archives, CMN.

Karluk Chronicles. 1913. A collection of daily typed newsletters prepared by Burt M. McConnell, Editor, and Dr. A. Forbes-Mackay, Associate Editor, between June 18 and July 7, 1913 on board the *Karluk* during its voyage from Esquimalt, B.C. to Nome, Alaska (20 pp. plus four supplementary pages). MG30 B40, Vol. 10, File 6, LAC.

La Nauze, Inspector C.D. 1917. Letter dated February 10, 1917 to Dr. R.M. Anderson. CMNAC/1996-077, Series A—R.M. Anderson Collection, Box 68, Folder 3, Archives, CMN.

LeRoy, O.E. 1913? Memo to Dr. R.W. Brock (undated but probably early May 1913). MG30 B40, R.M. Anderson Papers, vol. 20, LAC.

LeRoy, O.E. 1914? Memo to R.G. McConnell (undated but with another memo from LeRoy and W.H. Boyd dated October 29, 1914). MG30 B40, R.M. Anderson Papers, Vol. 20, LAC.

Mamen, B. 1914. Arctic diary, 1913–1914. MG30 B20, LAC. (Copy in CMNAC/1996-077, Series A—R.M. Anderson Collection, Box 68, Folder 10, Archives, CMN.)

Maurer, F.W. 1915. Remarks made by F.W. Maurer to Mrs. R.M. Anderson in Kent, Ohio (signed later for authenticity by Dr. R.M. Anderson). Typed memorandum in CMNAC/1996-077, Series A—R.M. Anderson Collection, Box 68, Folder 11, Archives, CMN.

Maurer, F.W., Wilkins, G.H., and Jenness, D. No date. Comments on a memorandum in CMNAC 1996/077, Series A—R.M. Anderson Collection, Box 68, Folder 11, CMN.

McConnell, B.M. 1914. Arctic diary, 1913–1914. MG30 B24, LAC.

McKinlay, W.L. 1914. Arctic diary, 1913–1914. MG30 B25, LAC.

————. 1915. Letter dated July 31, 1915 to G.J. Desbarats. RG42, Vol. 475, File 84-2-28, LAC.

————. 1922a. Letter dated February 8, 1922 to Mrs. R.M. Anderson. CMNAC/1996-077, Series A—R.M. Anderson Collection, Box 68, Folder 13, Archives, CMN.

————. 1922b. Letter dated March 30, 1922 to Dr. R.M. Anderson. CMNAC/1996-077, Series A—R.M. Anderson Collection, Box 68, Folder 13, Archives, CMN.

————. 1925. Letter dated May 13, 1925 to Mrs. R.M. Anderson. CMNAC/1996-077, Series A—R.M. Anderson Collection, Box 68, Folder 13, Archives, CMN.

—————. 1926. Letter dated May 27, 1926 to Mr. F. McVeigh, Department of Marine and Fisheries, enclosing his Arctic diary, Part 2, March 22, 1914 to June 25, 1914. RG40, Vol. 475, File 84-2-28, LAC.

Munro, John. 1914. Arctic diary, 1913–1914. With W.L. McKinlay Fonds, MG30 B25, LAC.

Noice, H. 1924. Letter dated December 24, 1924 to Joe [Bernard?]. (Copy in CMNAC/1996-077, Series A—R.M. Anderson Collection, Box 69, Folder 22, Archives. CMN.)

O'Neill, J.J. 1914. Letter dated July 4, 1914 to O.E. LeRoy. MG30 B66, Vol. 1, File 4A, LAC.

—————. 1916. Field notes, Canadian Arctic Expedition, 1913–1916. MG30 B171, LAC.

Pedersen, C.T. 1913. Telegram dated May 5, 1913 to Dr. R.M. Anderson, New York. MG30 B40, Vol. 1, File 1-12, LAC.

Perley, The Hon. G.H. 1913. Confidential letter dated March 21, 1913 to J.D. Hazen, Minister of Fisheries and Marine. RG42, Vol. 475, File 84-2-29, LAC.

Prince, E.E. 1923. Letter dated April 19, 1923 to Dr. C. Camsell. (Copy in CMNAC/1996-077, Series A—R.M. Anderson Collection, Box; 66, Folder 31, Archives, CMN.)

Roche, W.J. 1913a. Telegram dated February 8, 1913 to V. Stefansson, American Museum of Natural History, New York. RG42, Vol. 475, File 84-2-29, LAC.

—————. 1913b. Telegram dated February 10, 1913 to H.F. Osborn, President, American Museum of Natural History, Washington, D.C. (Mentioned in letter dated February 12, 1913 from Osborn to V. Stefansson.) (Copy in CMNAC/1996-077, Series A—R.W. Anderson Collection, Box 69, Folder 31, Archives, CMN.)

Seymour, W. 1917. Letter dated December 24, 1917 to Dr. R.M. Anderson. CMNAC/1966-077, Series A—R.M. Anderson Collection, Box 69, Folder 40, Archives, CMN. (An additional copy is in Box 68, Folder 36.)

Stefansson, V. 1913a. Letter dated February 28, 1913 to the Hon. George H. Perley, House of Commons. RG42, Vol. 475, File 84-2-29, LAC.

—————. 1913b. Letter dated March 1, 1913 to Dr. R.W. Brock, (Copy in CMNAC/1996-077, Series A—R.M. Anderson Collection, Box 61, Folder 13, Archives, CMN.)

—————. 1913c. Cable dated March 28, 1913 to Dr. R.M. Anderson, CMNAC/1996-077, Series A—R.M. Anderson Collection, Box 61, Folder 13, Archives, CMN.

—————. 1913d. Contract dated June 12, 1913 between V. Stefansson and Dr. R.M. Anderson. MG30 B40, Vol. 10, File 3, LAC.

—————. 1913e. Letter dated June 17, 1913 to G.J. Desbarats, Deputy Minister of the Naval Service. RG42, Vol. 476, File 84-2-29, vol. 2, LAC.

—————. 1913f. Letter dated July 26, 1913 to Chief Accountant, Department of the Naval Service. RG42, Vol. 476, File 84-2-29, vol. 2, LAC.

—————. 1913g. Letter dated October 18, 1913 to G.J. Desbarats, Deputy Minister of the Naval Service. RG42, Vol. 476, File 84-2-29, vol. 2, LAC.

—————. 1914a. Letter dated January 4, 1914 to G.J. Desbarats, Deputy Minister of the Naval Service. RG42, Vol. 476, File 84-2-29, vol.2, LAC.

—————. 1914b. Letter dated January 19, 1914 to Mrs. R.M. Anderson. MG30 B40, Vol. 2, Folder 1, LAC.

—————. 1915. Letter dated February 13, 1915 to G.J. Desbarats, Ottawa. RG42, Vol. 476, Folder 84-2-29, vol. 3, LAC.

—————. 1916. Letter dated February 13, 1916 to G.J. Desbarats, Ottawa. RG42, Vol. 476, Folder 84-2-25, vol. 3, LAC.

—————. 1918. Arctic diary, 1913–1918. MG30 B81, LAC.

—————. 1919a. Letter dated July 10, 1919 to G.J. Desbarats, Deputy Minister of the Naval Service. RG42, Vol. 476, Folder 84-2-29, vol. 3, LAC.

―――. 1919b. Letter dated December 10, 1919 to Dr. C. Camsell, Deputy Minister of Mines. (Copy in CMNAC/1996/077, Series A—R.M. Anderson Collection, Box 62, Folder 9, Archives, CMN.)

―――. 1919c. Letter dated December 10, 1919 to G.J. Desbarats, Deputy Minister of the Naval Service. RG42, Vol. 473, File 84-2-13, LAC.

―――. 1920. Letter dated September 7, 1920 to G.J. Desbarats, Deputy Minister of the Naval Service. RG42, Vol. 471, File 84-2-7, LAC.

―――. 1921. Letter dated June 1, 1921 to G.J. Desbarats, Deputy Minister of the Naval Service. RG42, Vol. 471, File 84-2-7, LAC.

―――. 1922. Letter dated May 29, 1922 to G. Grosvenor. Stefansson Collection, Correspondence 1928 file, Rauner Special Library, Dartmouth College, Hanover, N.H.

―――. 1923a. Letter dated April 18, 1923 to Dr. C. Camsell. (Copy in CMNAC/1996-077, Series A—R.M. Anderson Collection, Box 69 Folder 31, Archives, CMN.)

―――. 1923b. Letter dated May 5, 1923 to Dr. C. Camsell. (Copy in CMNAC/1996-077, Series A— R.M. Anderson Collection, Box 62, Folder 9, Archives, CMN.)

Storkerson, S. 1915. First sled trip surveying part of [the] north coast of Victoria Island, October 10 to December 4, 1915. MG30 B17, vol. 5, pp. 1–17, LAC.

―――. 1917. Second sled trip to survey the remaining unsurveyed N.E. coast line of Victoria Island, April 16, 1917 to July 31, 1917. MG30 B17, vol. 5, pp. 18–47, LAC.

Underwood, J.J. 1921. Letter (copy) dated December 20, 1921, to R.M. Anderson. CMNAC/1996-077, Series A—R.M. Anderson Collection, Box 70, Folder 46, Archives, CMN.

Wilkins, G.H. 1913. Letter dated May 3, 1913 to R.M. Anderson. R.M. Anderson fonds, MG30 B40, Vol. 10, LAC.

―――. 1916. Arctic diary, 1913–1916. V. Stefansson collection, Wilkins Papers, Box 1, Folders 2–12, Rauner Special Collections Library, Dartmouth College, Hanover, N.H.

INDEX

A

aerial photos reveal two islands, 177
Agnavigak, Ambrose, 273 fn 16, 282, 285, 299
Agnavigak and family accompany Chipman, 290
Aiyakuk, assists Jenness at Barter Island, 233
Aksiatak, family, 104, 232
Alaska, 220, 247, 273, 282, 299, 303
    gramophone, 49 fn 6
    Southern Party's schooner, 157
    lacks engineer, 157
    heads east from Herschel Island, 241
    aground at Baillie Islands, 245
    returns to Bernard Harbour, 273
    leaves Bernard Harbour for Nome, 299
    reaches Nome, 302
    sold, 302
Alaska and *Mary Sachs* to meet, 228
Algak Island, 289, 290
Alingnak, Iñupiaq man, 159
    and wife go to Melville Island, 167
    to go with Castel to *North Star*, 190
    and Guninana, at Cape Grassy, 186
Allan, Capt. Alexander, owner of *El Sueno*, 159
    agrees to take supplies east, 159
Allen, Gertrude, Stefansson's secretary, 28
    returns to U.S.A., 34
Allstrand, Mae Belle, weds Dr. Anderson, 5
Aluk family, 232
Alunak, daughter of Wikkiak, 283 fn 21
American Coastal Pilot (book), 65 fn 2
American Museum of Natural History, 5
    funds 1908 Stefansson Expedition, 5
    denies Stefansson funds, 6
    matches Nat.Geog.Soc., 7
    and Geol.Surv.Canada benefit, 341
Ammalurtuq Lake, 261
Amund Ringnes Island, 180
Amundsen Gulf, 287
    Capt. Lane knew ice conditions, 157
Amundsen, Roald, 3
"an adventure is a sign of incompetence," 165
Anadyr, 71 fn 9

Andersen, Karsten (Charlie), 212
    assists Storkerson, 161
    sights "Second Land," 178
    discovers petrified wood, 179
Anderson, Dr. R.M., zoologist, 219, 271, 310
    leads Southern Party, xiii
    ensured success of expedition, xix
    marries, 5
    on Stefansson's new expedition, 7
    regrets going on new expedition, 7
    assumes extra duties, 11
    offered job, Geol.Surv.Canada, 14
    attempts to resign from expedition, 22
    signs binding contract, 24
    tries to organize ship loads, 28
    dislikes Stefansson's plans, 106
    angered by Stefansson's interference, 106
    has different version of meeting on trail, 106
    fails to make copy of his letter, 111
    denies Stefansson's accusation, 111
    asks for list of supplies purchased, 113
    suggests Stefansson head Southern Party, 114
    responds to Stefansson's charges, 115
    instructions from Stefansson, 122
    to send *Mary Sachs* to Cape Kellett, 122
    to send *North Star* to Norway Island, 122
    his quandary at Herschel Island, 134
    told not to try to rescue Stefansson, 134
    told not to send Chipman for Stefansson, 134
    sends *Mary Sachs* to Banks Island, 135
    asks Wilkins to command *Mary Sachs*, 135, 239
    executive head, Southern Party, 219
    employed by National Museum of Canada, 220
    hunting up Hulahula River, 236
    receives conflicting instructions, 238
    lacks confidence in Capt. Bernard, 239
    asks Wilkins to look for Stefansson, 2439
    decides to take *North Star* east, 240

disputes Stefansson's name for camp site, 244

returns west with *Alaska* for coal, 245

heads east by sled for Bernard Harbour, 245

meets Chipman near Keat's Point, 246, 254

reaches Bernard Harbour, 246, 251

grieving the death of his son, 253

departs with Castel for Fort Norman, 254

heads west with Palaiyak, 263

encounters Chipman party, 264

teaches Johansen how to type, 264

sends instructions to Capt. Sweeney, 264

refuses to give *North Star* to Wilkins, 269

refuses Wilkins's compromises, 269

accepts Cox's compromise, 269

shoots barren ground bear, 278

unable to get mail to Fort Norman, 285

leads party up Croker River, 287

heads east to Cape Barrow, 291

encounters O'Neill at Cape Wollaston, 291

obtains Inuktitut names for islands, 291

seeks muskoxen specimens, 292, 298

meets Chipman at Cape Barrow, 292

meets Ilavinirk at Coppermine River, 292

returns to Bernard Harbour, 294

stores supplies for Northern Party, 302

leaves equipment at Barrow, 302

appointed editor of Expedition reports, 307, 310

indignant over Jenness controversy, 319

devotes much time to editing CAE reports, 321

correspondence over CAE reports, 325 fn 38

asks to join Stefansson in Arctic, 340

to study and collect birds and mammals, 340

assisted by Ilavinirk previously, 341

married, after first expedition to Arctic, 342

writes results of first Arctic expedition, 342

initiates letter-writing campaign, 343

attacks Stefansson's actions, 343

feels need to retaliate publicly, 346

needs to defend his companions, 346

needs to defend Geological Survey, 346

Anderson party, Dr., reaches Tree River, 280

reaches Bernard Harbour, 281

accomplishments of, during 1915, 281

Anderson, Mrs. R.M., returns to family, 33

greets Southern Party men, 303

unheralded contributor to CAE, 333

Anderson, Sandy, to establish camp, 56

gets near Herald Island, 58

Andreasen, Martin, 107, 112

Andreasen, Ole, 238

willing to go with Stefansson, 119

on ice party for money, 123

troubled by wolves at coast, 130

his trapping camp, 143

arrives from Cape Kellett, 144

first to walk on new land, 150

leaves expedition, 160

trading post at Shingle Point, 208

hired for 1918 ice trip, 208

Angutisiak (*see* Bolt, Ikey Angutisiak)

Anna Olga, icebound, 220

Annarihopopiak, Añayu, 273 fn 16

"Red Calico," 282

aranga, Chukchee driftwood house, 69

archaeological sites, Barter Island, 223 fn 13

destroyed in 1950s, 223

Arctic char, 274

Arctic Editorial Committee created, 310

Arctic Red River, 232

Arctic Sound, 278, 292

Arden, D'Arcy, former topographer, 290

accompanies Chipman, 290

guides Insp. La Nauze, 297

Arey, H.T. (Ned), American trader, 258

partner of "Scotty" McIntyre, 258

Arctic Biological Committee, 308

Argo, schooner, 258

Armstrong Point, 170

Polar Bear near, 160

base camp moves from, 189

Asetsaq, Jimmy, joins Expedition, 35

quits Expedition, 104

Atkoon, missionary's schooner, 159

blown ashore at Clifton Point, 282

Avingak, Uloksak, arrested, 297

taken to Edmonton, 297

Avingak, Uloksak and Sinnisiak,

tried for murder, 297

found not guilty of murder, 297

retried at Calgary, 297

found guilty of murder, 297

sentenced to death, 297

sentences reduced to life imprisonment, 297

imprisoned at Fort Resolution, 297

return to their people, 297

assist the police, 297

Avrunna, son of Higilak, 249

Axel Heiberg Island, 178

B

Baillie Islands, 200, 241, 242, 245, 272

open water around, 157

no news of Wilkins, 158

Banks Island, 113, 153, 162,170, 194

Banks Peninsula, 279

Banksland, Billy, 192 fn 6

Barker, Charles, 56

Baron Kleist, Emma Harbor, 71

"Barry Island," 280

Barry Islands, 289

Barter Island, 212

former Inuit meeting place, 117

only archaeological collection from, 233

Bartlett, Capt. Robert A., in his late 30s, 72

papers in Bowdoin College Library, xiv

world's best ice-master, 18

hired as captain of the *Karluk*, 18

no experience in Western Arctic, 18

questions condition of *Karluk*, 19

steers *Karluk* into ice, 37

safety of Northern Party, 43

lack of confidence in, 45

supported by some, 45

worries over *Karluk's* safety, 48

interest in classic literature, 48

tee-totaller, 50

orders "Abandon ship," 51

plays music while *Karluk* sinks, 52

favourite book, 55

offers dogs to Mackay party, 58

inventory at "Shipwreck Camp," 59

leaves for Wrangel Island, 59

prepares instructions for Munro, 62

starts journey to Siberia and Alaska, 65

first night at Skeleton Island, 65

second night at Rodger's Harbor, 65

reaches Blossom Point, 65

has American Coastal Pilot book, 65 fn 2

uses novel way to cross leads, 67

lands on Siberian coast, 68

encounters first Chukchee family, 68

reaches Koliuchin Island, 70

suffers from snow-blindness, 70

pays native for transport to East Cape, 70

abandons his worn sled, 70

abandoned by Chukchee native, 70

taken to house of trader Olsen, 70

given American money by Hadley, 70 fn 7

taken by Corrigan to Emma Town, 70

completes journey to Bering Strait, 71

decides to go to Anadyr, 71

visited by Baron Kleist, 71

gets tonsilitis and fever, 72

leaves dogs at Emma Town, 72

goes with Baron Kleist to Emma Harbor, 72

hears of *Herman* and Capt. Pedersen, 73

heads for Nome on *Herman*, 73

reaches Saint Michael, 73

sends wireless about *Karluk* to Ottawa, 74

receives medical treatment at St. Michael, 75

proceeds to Nome by motor boat, 75

leaves Nome to rescue *Karluk* survivors, 75

pays off Kataktovik at Point Hope, 75

meets McConnell at Point Barrow, 75

orders survivors to board *Bear*, 97

publishes early account of *Karluk*, 323

base camp on east side of Liddon Gulf, 186

Bathurst Inlet, 259, 269, 270, 278

Baur, W.J. "Levi," 197, 241

beacon, erected on Norway Island, 127

Bear, U.S. Revenue Cutter, 75, 91

Bear returns to Nome for coal, 75

Capt. Bartlett is passenger on, 94

takes Wrangel I. survivors on board, 97

looks for survivors at Herald Island, 97

Beaufort Sea, search for "Crocker Land," 9

Bell Island, 161

Belvedere, whaling ship, 5

steam whaler and freighter, 104

holds tons of expedition supplies, 105

icebound east of Martin Point, 117, 220

Berens Islands, 267

Bernard, Capt. Joseph (Joe), 242, 243, 253

provides map of good harbour, 242

Bernard, Capt. Peter, 107, 163, 241

retained as ship's captain, 33

on the spelling of his name, 33 fn 2
injured from sled accident, 120
in charge of Cape Kellett camp, 163
body never found, 175
with Olsen freights supplies, 229
uncle of Capt. Joe Bernard, 242
Bernard Harbour, 173, 244, 248, 249, 273, 294, 296, 298, 299
   Southern Party base camp, 243, 246
   name proposed for, 244 fn 30
   Stefansson rejects name for, 244
   revision of use of name, 245 fn 34
   men gather driftwood regularly at, 247
   nearby lake supplies fresh water, 247
   lighting facilities for, 247
   cooking arrangement at, 247
   Inuit name for camp at, 248
   petty thieving by Inuit at, 248
   sense of humour of Inuit at, 249
   Christmas at, 251
   average January temperature, 253
   average wind speed, 253
   strangers at, 282
Bernard Island, 127, 128, 143
Bernard River, 127
Bernier, Capt. Joseph, 187
   cache left by, 187
Beuchat, H., considered all on *Karluk* lost, 44
   unsuited for Arctic life, 45
   reported near death, 59
Beaufort Sea, 163, 167
Binder, O., left at Cape Kellett, 175
   with Crawford on *Challenge*, 196
   becomes engineer of *Challenge*, 198
   Noice, and Carroll buy *Challenge*, 200
bituminous shales for fuel, near camp, 187
Blackjack, Ada, later Wrangel I. survivor, 97 fn 17
Blond Eskimos, 4, 5, 261
"Blowhole," 229
Blue, Daniel, 245
   engineer on *Alaska*, 157
   dies of scurvy and pneumonia, 157, 282
Bluenose Lake, source of Croker River, 287, 288
   discovered in 1948, 291 fn 2
Bolt, Ikey Angutisiak, 245, 273, 287
   helps Stefansson's first ice trip, 120
   Point Hope Iñupiaq, 222

Angutisiak, 222
   slept in tent at Bernard Harbour, 246
   accompanies O'Neill, 288
   not used to building snowhouses, 289
   leaves Expedition, 299
   plans to marry Etna Klengenberg, 299
books, both parties well supplied with, 253
   those with Northern Party lost with *Karluk*, 253
   those with Southern Party, on *Alaska*, 253
Borden, Prime Minister Robert, 8
Boyd, W.H., unheralded contributor to CAE, 333
Bradford Point, 193
Brady, John, 56, 98
Breddy, G., 62
   Breddy walks to Cape Waring, 82
   shot and killed, 85
Breynat, Rt. Rev. Gabriel, 297
British Admiralty Chart, 128, 131, 147, 155, 178, 181, 183, 244
   shows Franklin's data, 279
   errors on, 147
British Mountains, 228
Brock, Dr. Reginald W., 8, 258 fn 6, 331, 332
   urges growth of Expedition, 8
   island named for, 173
   writes Stefansson not to interfere with Southern Party, 238, 242
   writes Dr. Anderson not to send Chipman after Stefansson, 238
   offers job to Dr. Anderson, 342
Brock River, 258
Brockie, Const. J., accompanies Stefansson, 209
   returns to Herschel Island, 209
Brooks, Charles, 223
Brower, Dora, 211
Brower, Charles, 211
Bruce, Cpl. W.V.,
   accompanies Dr. Anderson, 285
   leaves *Alaska* at Herschel Island, 299
Bush, Pres. George H.W., 274
Byam Martin Channel, 193

C

cache pilfered by Inuit, 259
caches left at Clifton Point and Clinton Point, 250

CAE scientists, reactions to diary requirements, 22
  request statement from Stefansson, 24
  weights, heights, and ages of, 27, 28
  discuss their grievance with Minister, 350
Cambridge Bay, DNA of elders, 4
Camden Bay, 38, 118, 221
Cameron, John, 310 fn 5
"Camp Hospital," 193
camp on sandspit, Banks Island, near river
    mouth, 128
Camsell, C., approves Jenness report, 318
  rebukes Stefansson, 319
Canadian Arctic Expedition Committee
    1913–1916, 307
Canadian Arctic Expedition 1913–1916, xiii
Canadian Arctic Expedition 1913–1918, xiii
Canadian Arctic Expedition of Vilhjalmur
    Stefansson, 9
CAE scientists, stay at James Hotel, 15
Canadian govt., will sponsor new expedition, 9
  requires other sponsors withdraw support, 9
  changes name of Stefansson's expedition, 9
  adds Geological Survey scientists, 10
  creates dual authority over expedition, 10
  creates dual duties for personnel, 10
  creates two expedition parties, 10
  protects senior administrators, 352
  makes limp statement, 352
canoe, 273, 281, 290
  given to Ikpukhuak, 298
canyons, on Firth River, 228
Cape Barrow, 270, 274, 275, 291
  safe arrival at, 280
Cape Bathurst, Dr. Anderson left at, 5
Cape Bexley, 251, 266, 288
Cape Cardwell, 265
Cape Dazhnev, 65 fn 1
Cape George Richards, 178, 186
Cape Grassy, 173, 190
  camp, 186
  better base than Cape Murray, 186
  fuel supply near, 186
Cape Halkett, four houses near, 102
Cape Isachsen, 177, 178, 186, 193
Cape Kellett, 180, 212
  men constructing house near, 132
  dog teams, only two good ones at, 143
Cape Krusenstern, 259, 266, 288

Cape Leopold M'Clintock, 147, 149
  latitude and longitude readings, 149
Cape Lyon, 242, 264
  North Star leaves cache, 242
Cape M'Clure, 196
Cape Murray (Brock Island), 173, 176, 184, 186
  base for fourth ice trip, 184
  no one camped there, 184
Cape Murray (Victoria Island), 294
Cape North (Mys Shmidta), 68 fn 4, 69, 71
Cape Parry, 242
Cape Prince Alfred, 124, 145, 165, 166, 167
Cape Ross, near entrance of Liddon Gulf, 169
  Storkerson establishes camp, 187
Cape Serdze (Mys Netan), 70
Cape Smyth, 35
Cape Waring, 94
  rescue of *Karluk* survivors from, 94
Cape Wollaston, 291
Cape Young, 250, 259
Caraieff, Mr., of Cape North, 69
  invites Bartlett for night, 69
  gives Bartlett letter of introduction, 69
Caraieff, Mr., of Emma Town, 70
caribou head is tastiest part, 127
caribou herds, changed migration route, 253
caribou meat, 139
Carpendale, C., Australian trader
    at Emma Town, 72
Castel, Aarnout, 121
  and M. Kilian discover large bay, 168
  finds two sleds at Mercy Bay, 175
  finds Thomsen's body, 175
  finds strangers at Cape Kellett, 175
  and Masik repair *Mary Sachs*, 175
  left report at Cape Grassy, 186
  left message on unknown island, 186
  visits Vasey Hamilton Island, 186
  heads for Liddon Gulf, 186
  crosses M'Clure Strait to Mercy Bay, 190
  looks for Thomsen, 190
  and Capt. Bernard to fix *Mary Sachs*, 190
  to close Cape Kellett and sail to Nome, 190
  and others, repair *Mary Sachs*, 197
  takes charge of *Challenge*, 199
  helps Wilkins repair *North Star*, 235
  sailor from *Belvedere*, 242
  works with Wilkins on *North Star*, 269

cat, secreted onto *Karluk*, 24
   named Nigeraurak, *Karluk's* mascot, 43
   missing as *Karluk* sinks, 51
   rescued by Maurer, 51
   why it was saved by crew, 55 fn 1
   taken back to U.S.A., 95 fn 12
CGS *Arctic*, 193
Chafe, E.F., moving supplies for Sandy
      Anderson, 58
   gets near Herald Island, 58
   returns to Icy Spit without sled, 79
   walks to Cape Waring, 82
   publishes early account of *Karluk*, 322
Challenge, taken to Minto Inlet, 174 fn 12
   appears at Cape Kellett, 198
   catches up to *Polar Bear*, 199
   wrecked, coast of Amundsen Gulf, 201
change of dates on CAE reports to 1913–1918,
      310
   Stefansson may have demanded the change,
      310
   Arctic Biological Committee rejects
      change, 309
   Desbarats overrules committee, 310
   McConnell won't intervene in conflict,
      310
Chantry Island, 247
   little driftwood east of, 247
Chapman Islands, thick volcanic flows on, 289
Charlie Brower, "The King of the Arctic," 102
Chipman, K.G., 30, 230, 231, 232, 247, 248, 249
   and Cox threaten to go south, 113
   told to search for Stefansson, 134
   geographer, Southern Party, 219
   employee of Geological Survey, 220
   charts Peel River to Fort McPherson, 232
   unwilling to search for Stefansson, 239
   and O'Neill undertake initial field work,
      250, 251
   leaves for Darnley Bay to start mapping, 256
   Klengenberg camp deserted, 256
   leaves O'Neill at Wicksuak's camp, 258
   finds trading camp of McIntyre and Arey,
      258
   gives Wicksuak note for supplies at *Alaska*,
      258
   names Hornaday River and Brock River, 258
   extends mapping to Bernard Harbour, 259

   will go home if without *North Star*, 269
   maps Bernard Harbour, 269
   finds O'Neill party, 276
   volunteers to stay alone at Kater Point, 278
   food supply very low, 280
   encounters Dr. Anderson at Tree River, 290
   obtains permission to leave Expedition, 290
   completes mapping to Bathurst Inlet, 290
   starts trek south, with Arden, 293
   accompanied by Uloksak Meyok family, 293
   crosses Great Bear Lake to Fort Norman, 293
   reaches Ottawa, 293
Chipman party, head for Pedersen's cabin, 231
   treks west to Black Mountain, 231
   uses Pedersen's whaleboat, 231
   charts east branch Mackenzie Delta, 231
   first to reach Herschel I. by boat, 236
   meets Dr. Anderson near Darnley Bay, 246,
      250
Christmas dinner held early for Stefansson, 140
Christmas at Collinson Point 1913, 222
Christmas at Bernard Harbour, 1914, 251,
Christmas at Bernard Harbour, 1915, 283
Chukchees, coughing of, 69
   tuberculosis suspected, 69
Chukchee food and drink, 69
Chukchi Sea, has only two islands, 49
Clarence Lagoon, 236
Clark, John, photographer on *Elvira*, 235
Clifton Point, 250
"coal" deposits, 184 fn 28,
   Amund Ringnes Island, 184 fn 28
   Lougheed Island, 184 fn 28
   Melville Island, 184 fn 28
"coal-mine" on Lougheed Island, 184
coal-oil is lost in transit, 144
coastal mapping, 228
   Alaska boundary to Mackenzie Delta, 228
Cochran, Capt. Claude, 75, 91, 97
   captain of U.S. Revenue Cutter *Bear*, 75
Collinson Point, 106, 107, 226, 228
   in Camden Bay, 221
   Southern Party seeks shelter at, 221
   elongated sandspit, 221
   vacant house at, 221
   controversial meeting at, 112
Collinson, Richard, captain of *Enterprise*, 161
Colville River, 104, 232

compromise suggested by Dr. Camsell, 349

concern for safety of wives, Cape Kellett, 145

controversial meeting, Victoria, 20

    Nome, 30

    Collinson Point, 112

controversy over *Karluk* getting into ice, 36

controversy over remains on Herald Island, 98

    fn 20

Cook, Capt., mapped Siberian coast, 69

copper-bearing rocks in Bathurst Inlet, 290

Copper Inuit gathered at Coppermine River,

    292

Copper Inuit tent camp, Berens Islands, 267

Coppermine River, 244, 254, 255, 269, 273,

    289, 286, 292

Coppermine River Inuit, visit Bernard

    Harbour, 282

    move to Liston and Sutton Islands, 282

    wives sewing fur garments for, 282

Coronation Gulf, 259, 266, 267, 268, 269, 270,

    288

Corrigan, good Siberian hunter, 70

Corwin, 91, 92, 95

    old U.S. Revenue Cutter, 65 fn 2

    mapped Wrangel Island, 65

cost of Expedition embarrasses government,

    323

cost of Expedition expands tenfold, 323

Cottle, Capt. S.F., 107, 112

Cox, J.R., field notes of, xvi

    geographer, Southern Party, 219

    Geol.Surv.Canada employee, 220

    maps coast east from Demarcation Point,

        228

    talks with O'Neill about Firth River, 228

    maps course of Firth River, 228

    maps from Firth River to Escape Reef, 229

    completes coastal mapping, 229

    repairs launch purchased by Stefansson, 229

    maps west branch of Mackenzie River, 232

    maps to mouth of Peel River, 232

    maps coast, Bernard Harbour to Rae River,

        259

    encounters Dr. Anderson party, 259

    caches supplies, 259

    returns to Bernard Harbour, 259

    renews coastal mapping, 259

    reaches mouth of Rae River, 259

    proceeds up frozen Rae River to source,

        259

    crosses height of land, 259

    proceeds downstream to Stapylton Bay, 259

    encounters many caribou, 259

    meets Chipman party at Cape Young, 259

    completes survey to Bernard Harbour, 259

    directs Wilkins to Bernard Harbour, 266

    his compromise over *North Star* accepted,

        269

    repairs damaged launch, 280

    accompanies O'Neill, 288

    checks readings, International Boundary,

        302

    joins Canadian Army, 308

    returns from war, 310

    resigns Geol.Surv. and goes to India, 344

Crawford, James, 145, 241, 198

    trapping cabin of, 106

    heavy drinker, 145

    brings schooner to Victoria Island, 189

    has news of Cape Kellett, 174 fn 12, 189

    sells *Challenge* to Stefansson, 198

crew of *Karluk*, 19

    three members depart in month, 19

Croker River, 287

    named by Sir John Richardson, 287

    emerges from steep-walled hills, 287

"Crocker Land," 6, 7, 9, 11 fn 26, 44, 115, 124,

    325

    non-existent, 180

Crocker Land Expedition, 181

    based at Etah, 181

Cross Island, 36, 211, 212, 220

"crowbills," 88

cutting route through pressure ridges, 60

D

Danish Sound, 183

Darnley Bay, 245, 250, 256, 258, 259, 292

Dealy Island, 153

    depot left by Capt. H. Kellett, 193

    large stone house at, 193

    monument at, 193

Deans Dundas Bay, 160, 161

Dease and Simpson monument, 280, 281

Deas Thompson Point, 250, 263

death of Breddy unsolved, 86

Demarcation Point,221, 228, 233
Department of the Naval Service, 10, 220, 331
Desbarats, G. J., 92
      Dep.Min. of the Naval Service, 10
      agrees to publication of reports, 307
      states furs are govt. property, 138 fn10, 236
      Stefansson receives instructions from
            Desbarats, 188 fn 37
      writes Stefansson not to interfere with
            work of Southern Party, 236
      writes Dr. Anderson to send someone to
            look for Stefansson, 238
      insists on change in titles of CAE reports,
            310
      appointed Dep. Min. National Defence, 311
      proposes Stefansson be CAE editor, 320
Detention Harbour, 292
Dezhnev (Emma Town), near East Cape, 70 fn 8
Diamond Jenness Peninsula, 266
diaries, instructions from Stefansson regarding
      them, 22
Dick, Inuk from Herschel Island, 228
did Stefansson write Hadley's story? 159 fn 3
disagreements of scientists vs. Desbarats, 309
disastrous attempt to recover supplies, 79
dissent and drunkenness on *Mary Sachs*, 241
DNA studies deny link with lost Norsemen,
      319
dolomite, a calcium-magnesium limestone, 259
      lacks fossils and copper in area, 259
Donahue, Peter, writes of *Polar Bear* activities, 205
      states *Polar Bear* easily refloated, 206
      carpenter on *Polar Bear*, 207
Donaldson, schooner, 98
      captained by Harold Noice, 98
Dolphin and Union Strait, 157, 248, 249, 253,
      259, 263, 270
      Inuit build snowhouses on, 249

E

East Cape, 65 fn 1, 69, 70, 71, 226 fn 6
Edison recording phonographs, 249
      one phonograph destroyed by water, 249
Edna, power launch for Chipman, 139, 231
Eight Bears Island, 152
Ekalulia Island, 289
Ekullialuk Island, 280
Ellef Ringnes Island, 177, 178, 181, 183

Elvira, icebound, 220, 235
Emerald Island, 152, 185
Emiu ("Split-the-Wind"), 212
      Iñupiaq man, 159
      accompanies Stefansson, 163
      takes message to Storkerson, 173
Encyclopedia Britannica, a set with each party,
      253
Endicott Mountains, 226
Enterprise, HMS, 161
Escape Reef, 232
Etah, Greenland, 181
Eunice, wife of Sweeney, leaves with baby, 302
Evinrude motor fails, 273
exchanges on conflict in *Science*, 349
expedition, Arctic, name change, xiii
Expedition accomplishments, 324–328
      those of Northern Party, 324–326
      those of Southern Party, 326–328
Expedition costs, 323
Expedition, finds copper deposits are low grade,
      327
      establishes culture of Copper Inuit, 327
      expanded knowledge of Arctic, 328
      received little publicity, 328

F

fates of two Inuit prisoners, 297
feelings of men on Meighen Island, 180
Findlay Island, 177, 178, 179, 181, 183
      discovered by Lieut. S. Osborn, 183
Finnie, R. S., reports fatherhood of Stefansson,
      119
fish net, to acquire food for dogs, 274
Fitzgerald, Annie, daughter of Unalina, 286
Fitzgerald, Sgt. F.J., 282
      died on "Lost Patrol", 282 fn 20
five govt. departments involved with
      Expedition, 10
five CAE reports added later, 310
First Land, later named Brock Island, 173
Firth River, mapping of, 225
Flaxman Island, 36, 215
      Leffingwell's cabin at, 38
Fletcher, Dennis, 284 fn 22
folklore, obtained from Guninana, 167
food cache left at Bloody Fall, 285
Fort Confidence, 296

Fort Macpherson, HBC's supply schooner, 159

Fort McPherson, 108, 110, 112, 208, 230, 232

Fort Norman, 285

Fort Resolution, 297

fossil forests on Axel Heiberg Island, 185
  on Ellesmere Island, 185

Franklin, Sir John, 7, 57 fn 2, 129, 152, 183, 244, 279

Franklin Bay, 242

French, Insp. F.H., leads patrol, 297

Frost, Mary Carpenter, xx fn 9, xxiv, 283 fn 21

Fry, Mr. W. Henry, 208, 209

Fry, Mrs. Henry, 208

fuel-oil supply exhausted at Icy Spit, 79

funding for CAE reports ceases 1926, 311

future of *Karluk* survivors looks hopeless, 89

G

games, *Karluk* men, New Year's Day, 50

Geological.Survey of Canada, 5
  funds part of Stefansson 1908 expedition, 5

Girling, Rev. Herbert, missionary, 159, 285, 292
  accompanies Jenness, 285
  wants to Christianize the Inuit, 286
  arrives at Bernard Harbour, 298
  establishes Bernard Harbour mission, 299

Gladiator, Fritz Wolki's schooner, 158
  sold to Stefansson, 158
  follows *Polar Bear*, 159

gold mines, near Cape North, Siberia, 69

Golightly, Edmund, (alias Archie King), 56

Gonzales, Capt. Henry, 165, 170
  captain of *Polar Bear*, 160
  was to take *Polar Bear* north, 170
  has news of Thomsen, 174
  arrives from *Polar Bear*, 189
  reaches Cape Kellett, 197
  has crew destroy *Mary Sachs*, 197

Gore Islands, 167

Goulbourn Peninsula, 290

government instructions, creation of, 10
  all men to leave Arctic in 1916, 285

gramophone on *Karluk*, 49 fn 6
  gift of Sir Richard McBride, 49 fn 6

granitic rocks lie east of Tree River, 276

Granville, Fred L., movie cameraman, 93
  took movies, rescue at Cape Waring, 94

Gray, Dr. D.R., xvi, xvii, xxiii, 321 fn 38
  contributes to CAE. publicity, 328

Great Bear Lake, 192, 254, 282, 290, 293, 296

Great Slave Lake, 279, 297

Greenland, lost Norse settlers, 3, 4
  disappearance disproved, 4

Grosvenor, G.H., Director, Nat.Geog.Soc., 7
  seeks Dr. Anderson for expedition, 7

Gumaer, G.Adelbert, 212 fn 2

Guninana, Iñupiaq woman, 159
  wife of Alingnak, 334
  source of much folklore, 167

H

Hadley, John, pilot for *Karluk*, 35
  diary of, 58 fn 6
  suspects Williamson killed Breddy, 85
  first mate, *Polar Bear*. 159, 160
  was ice pilot on *Fort Macpherson*, 159
  re. house construction, 160
  to write story of *Karluk*, 159
  wounded by polar-bear attack, 165
  writes note about *Mary Sachs*, 197
  has *Polar Bear* refloated, 215

Hall, E.S. Jr., 316

Haqungaq, third wife of Uloksak Meyok, 285

Harding, Hudson's Bay Company man, 209

"Harris Land", 124

Hassel Sound, 181

Hazen, Honourable J.D., 323 fn 50

HBC ship at mouth of Horton River, 250

Hecla and Griper Bay, 171, 173

Helen, daughter of Kuraluk and Keruk, 36, 59

Hepburn Island, 276, 288

Herald Island, 49, 56, 61, 62, 74, 77, 78, 98
  geographical details, 57 fn 5
  skeletons found in 1924 on, 59
  no *Karluk* survivors seen at, 96
  artifacts from, 98

Herman, 361
  owners charge for Bartlett, 73 fn 11
  leaves supplies at Cape Kellett, 196
  reaches Herschel Island, 240
  has news of *Karluk*, 240

Herona, son of Kullak, 154

Herschel Island, xix, xx, 11 fn 26, 18, 19, 31, 32, 34, 35, 38, 49, 104, 105, 107, 110, 125,

158, 159, 160, 168 fn 37, 202, 204, 207, 208, 215, 226, 236, 241, 246, 272, 302

Hewitt, Dr. Charles Gordon, 307, 315 fn 17

Higilak, shaman wife of Ikpukhuak, 249, 296, 330
    "adopts" Jenness, 261

Hoare, W.B.E., assistant to Rev. Girling, 159, 298

Hobson, Dr. George, xxiii, xxiv

Hoff, J.E., engineer from *Ruby*, 159
    hired by Stefansson for *Alaska*, 159, 282
    with Dr. Anderson to Tree River, 294
    and Siberia Mike take engines to base, 290

"hooch," pemmican boiled with flour or
    hardbread, 254
    variety created at Cape Kellett, 144

Hood River, 278, 326

hopes for rescue diminish, 87

Hopson, Alfred, 104
    "Brick," 104

Hornaday, Dr. William T., 258 fn 6

Hornaday River, 258

Horton River, 250

house at Armstrong Point, 161

Hunt Point, Wrangel Island, 65

hunting party, caught in storm, 43
    reaches Aksiatak's house, 102

Huntsman, Dr. A.G., 315
    feud with Johansen, 315

I

ice-drift party, to start from Cross Island, 211
    ends drift study in Harrison Bay, 215

Icy Spit, Wrangel Island, 61, 65

Igloopuk, 248

Iglorua Island, 289

Ikiuna, Iñupiaq woman, 159
    wife of Natkusiak, 186
    adopted daughter of Alingnak, 190

Ikpukhuak, Jenness' friend, 249, 333
    childless, 249
    "adopts" Jenness, 261
    his singing recorded, 284
    in dance costume, 296
    given Expedition canoe, 298
    in charge of Expedition supplies, 298

Ikugana (Anna), wife of Seymour, 160, 172

Ilavinirk, 211, 292, 293, 296, 304
    interpreter for police, 290
    Mackenzie Delta Inuk hunter, 341

Illun, 159, 160, 172
    member of *Polar Bear* crew, 190

Immaernirk, 293

International Boundary monument, 232

Intrepid, HMS, 148, 153

Investigator, HMS, 57 fn 5, 129, 131, 153, 161
    remains found, 153 fn 10

Inuit, build snowhouses near *North Star*, 256
    recall Capt. Klengenberg's ship, 161
    recall Capt. Mogg's ship, 161

Iñupiat women, to sew clothes, 159, 187
    for men at Cape Kellett, 159
    for men on *Polar Bear*, 159
    alter Capt. Gonzales' plans, 189 fn 1

Ipanna, assists Jenness at Barter Island, 233

Isachsen, Capt. Gunnar, 183
    saw King Christian land, 183

"Ireland's Eye," 149

Itayuk, baby son of Munnigorina, 288

J

Jenness, Diamond,
    ethnologist, Southern Party, xvi, 219
    greatest publication contribution to CAE,
        xvi
    first to reach Victoria, 20
    objects to diary requirements, 22
    considered resigning, 23
    may have suggested hunting trip, 38
    leaves *Karluk* with Stefansson, 38
    goes to Barrow, 102
    ague attacks, 103
    sent to fishing lake, 103
    returns to Barrow with "Brick," 104
    employed by Nat. Mus. of Canada, 220
    stays with Colville River family, 222
    studies artifacts Stefansson bought, 232
    goes east to Collinson Point, 233
    takes inventory at O'Connor's camp, 233
    excavates Inuit ruins at Barter Island, 233
    had single bunk at Bernard Harbour, 246
    does most of talking with Inuit, 248
    trades with Inuit at Bernard Harbour, 249
    trades for caribou and seal meat, 253
    and Palaiyak visit Inuit village, 255
    meets hunter Uloksak Meyok, 255
    sees no "blond Eskimos," 256
    trades with visiting Inuit, 256

Southern Party's authorized trader, 256
accused of murder, 256
attends trial and séance, 256
accompanies Cox as assistant, 259
to spend summer with Inuit family, 259
summer with Inuit on Victoria Island, 261
persuades Ikpukhuak to "adopt" him, 261
heads across Wollaston Peninsula, 261
camps at Lake Quunnguaq, 261
meets Inuit from Prince Albert Sound, 261
develops severe digestive disorder, 261
accumulates anthropological data, 261
amuses Inuit by bathing, 263
returns to Bernard Harbour, 263
has admiration for Copper Inuit, 263
brings caribou meat from Victoria Island,
     283
records Inuit songs, 284
records Ikpukhuak singing, 284
records Jennie Kannayuk singing, 284
visits Inuit at mouth of Bathurst Inlet, 286
learns that Uloksak has turned back, 286
obtains supplies from Tree River cache, 286
returns to Bernard Harbour, 286
records songs by Bathurst Inlet Inuit, 286
looks for Johansen, 295
joins Canadian Army and goes overseas, 308
returns from war in 1919, 310
publishes 6 CAE reports by 1926, 310
publishes 2 subsequent CAE reports, 311
rejects idea of "Blond Eskimos," 319
Jenness's *Life of the Copper Eskimos* was popular,
     311
Jennie Kannayuk (see Kannayuk, Jennie)
Jochimson, A.P., captain of *King and Winge,* 92
Johansen, Fritz, diary of, xvi
     publishes articles on Expedition, xvii
     tells Stefansson to go home, 113
     sounds for ocean depth, 120
     determines edge continental shelf, 121
     returns to mainland, 122
     marine biologist, Southern Party, 219
     hired by Dept. of the Naval Service, 220
     called "Prof," 223
     studies insects at Collinson Point, 233
     writes articles for Danish paper, 233
     finds bark beetles destroying trees, 254
     collects insects, plants, and fish, 264, 269

     dredges from *Alaska,* 264
     left in charge of Bernard Harbour, 273
     studies freshwater and marine life, 273
     prefers to work alone, 273
     explores south coast of Victoria Island, 294
     is accompanied by Adam Ovoiyuaq, 294
     discovers Silurian fossils, 295
     publishes early account of *Karluk,* 322
Jones, John J., dies of heart failure, 165, 335
Jorgensen, Capt, Christian Klinkenberg, 242,
     340
     Charles (Charlie) Klengenberg, 243 fn 26

K
Kamarkak, 302
Kannayuk (Jennie), 283 fn 21, 284, 284 fn 22,
     298
     daughter of Higilak, 249
Kanuyak Island, 279, 290
Karluk, 221,
     purchased for Arctic expedition, 12
     leaves for Nome, 25
     icebound, 38, 220
     loses its commander, 43
     may suffer fate of *Jeannette,* 44, 45
     prepared for winter, 46
     scheduled activities on, 46
     moving supplies from, 46
     risk of fire on, 46
     emergency house beside, 47
     maintaining morale on, 47
     reaches farthest north, 48
     Christmas 1913 on, 48, 49
     New Year's Eve activities, 50
     New Year's Day activities, 50
     listing to port, 50
     ice breaks through hull, 51
     water rushes into hold, 51
     ship sinks, 52
     depth measured where it sank, 52
     news of, 74, 135, 240,
Karluk Chronicles, news during voyage, 27 fn 28
Karluk survivors, kept cat alive, 55 fn 1
     safely on Wrangel Island, 61
     establish camp at Icy Spit, 61
     to go to Rodger's Harbor, 62
     divide into groups, 63
     have supplies for eighty days, 63

are rescued by *King and Winge*, 93, 94
    return to Nome on *Bear*, 97
Kataktovik, 56, 58,
    widower, 47
    accompanies Captain Bartlett, 65
    has fear of Siberians, 68
    waits for trip home, 72
Kater Point, 280, 289, 292
    Chipman alone at, 278
    men delayed at, 280
Kaudluak (Kaullu), widow with daughter, 267
    given in marriage to Natkusiak, 267 fn 12
    returned home with her brother, 268
Kay Point, 242
Keat's Point, 246
"Keenan Land," non-existent, 215
Kellett, Captain Henry, 57 fn 5
    captain of HMS *Intrepid*, 153
    left depot at Dealy Island, 193
Keruk, 43 fn 2, 50, 55, 60, 83, 86
    sewing clothing for men, 43
    carries daughter to Wrangel Island, 59
    called "Auntie" by McKinlay, 81
    invaluable to *Karluk* survivors, 330
Kilian, Herman, 161, 166, 171
    sailor on *Polar Bear*, 160
    on Storkerson's ice party, 212
Kilian, Martin, 162, 165, 166, 168, 169, 171,
    172, 173, 197
    sailor on *Polar Bear*, 160
    accompanies Stefansson, 162
    on Storkerson's ice party, 212
Kilian brothers, take Inuit to *Polar Bear* camp,
    175, 197
King, Archie, 88
    alias for E. Golightly, 56
King and Winge rescues men, Rodger's Harbor,
    94
    rescues survivors at Cape Waring, 95
    looks for survivors, Herald Island, 96
    meets *Bear*, 96
    transfers *Karluk* survivors, 97
    goes walrus hunting, 97
    owners bill Canadian govt., 97 fn 14
King Christian Island, 178, 183, 184, 185, 186
King Christian Land, 177, 183
King, William Lyon Mackenzie, 311, 334
Kittigazuit, 231

Klengenberg, Capt. Charles (Christian), 4, 161,
    231, 256, 299, 301
    camp of, 258
    allows son to join CAE, 258
    tells of blue-eyed Inuit, 340
Klengenberg, Etna, 256, 299
Klengenberg, Elvina (Uiniq, "Weena"), 330
Klengenberg family, 299
    taken to Baillie Islands, 299
    moves to Coppermine River, 299
    establishes trading post, 299
Klengenberg genealogy, 243 fn 28
Klengenberg, Gremnia Kemnik, 243
    wife of Capt. Klengenberg, 243
Klinkenberg Jorgensen, 242 fn 26
    becomes Charlie Klengenberg, 340 fn 8
Klengenberg, Patrick (Patsy), 258
    camping with sister Etna, 256
    hired by O'Neill, 258
    assists Chipman, 270, 327
    helps Dr. Anderson with specimens, 271
    accompanies Jenness, 285, 288, 295
    leaves Expedition, 299
Knight, E.L., 166, 189, 199, 212
    sailor on *Polar Bear*, 160
    leaves for *Polar Bear* camp, 165
    with Stefansson on last trip, 190, 192, 194
    shows symptoms of scurvy, 192
Knight Harbour, 195
Koliuchin Bay, 70, 72
Kohoktok, 288, 289
    brother of Mufpi, 285, 330
    husband of Munnigorina, 288, 330
    takes supplies to Tree River, 290
Koptana, younger of Uloksak's two wives, 266
Kullak, 154, 156
    spelling of name, 154 fn 11
    learns of Cape Kellett camp, 154
    gives slippers to Stefansson, 154
    meets Wilkins, 154 fn 12, 172
    angry man, 154 fn 12
    herds geese to Cape Kellett camp, 154
    loses wife at childbirth, 174 fn 12
    blames Stefansson, seeks revenge, 172
Kuraluk, 36, 56, 58, 59, 60, 62, 77, 78, 79, 80, 82,
    83, 84
    moves family to Skeleton Island, 81
    hunts birds, 85

builds kayak, 86
kills walrus, 86
kills bearded seals, 87
notices rescue ship, 94
Kutok, Iñupiaq woman, 159
second wife of Illun, 330

L

Lac Rouvière, priest's cabin at, 293
"Ladies' mutiny," 189 fn 1
Lady Franklin Point, 294
Lamont, Const. Alexander
has typhoid fever, 208
dies at Herschel Island, 208
hands Dr. Anderson money, 300
land sighted, from *Karluk*, 49
Land's End, 147
La Nauze, Insp. Charles Dearing (Denny),
290, 309
arrests two Copper Inuit men, 292
trades Expedition supplies for furs, 292
investigating missing priests, 297
leaves *Alaska* at Herschel Island, 299
takes Inuit prisoners with him, 299
Lane, Capt. L., finds remains of *Karluk* men,
59, 98 fn 16
hunting whales, 155
appears at Cape Kellett, 155
encounters Stefansson, 155
tells of rescue of *Karluk* survivors, 157
has no news of 8 missing men, 157
gives Stefansson first news of war, 157
hunted whales in Amundsen Gulf, 157
becomes trader, 158
lets Stefansson charter *Polar Bear*, 158
demands outrageous price, 158
friend of Stefansson, 160
is satisfied with payment, 160
large tree trunk, probable source of, 185
LeBourdais, D.M., at Herald Island, 98 fn 18
comments in book to D. Jenness, 350 fn 34
Lee, Hugh J., U.S. Marshall at Nome, 73, 74
Leffingwell, Ernest de Koven, 151, 211, 222,
247, 340
returns to Arctic, 33
pilots *Mary Sachs*, 35
cabin of, on Flaxman Island, 38
returns to U.S., 215

has Christmas with Southern Party, 223
gives Graphlex camera to Wilkins, 234
named Bernard Harbor in Alaska, 244
Leffingwell Crags, 151, 177, 225
Leffingwell Island, named by Storkerson, 325
now Stefansson Island, 325
Le Roux, Father Guillaume, 286
missing Oblate priest, 290
LeRoy, O.E., unheralded contributor to CAE,
333
a senior geologist, 10
killed in 1917 in war in France, 333 fn 69
letter of Dr. Anderson disappears, 111
Liddon Gulf, 168, 169, 170, 171, 173, 175, 178,
188, 189, 190, 192, 193, 196
base camp in, 186
Lindeberg, Jafet, Pres, Pioneer Mining
Company, 91
offers to send SS *Corwin* for survivors, 91
purchases *Polar Bear*, 216
Liston and Sutton Islands, 244, 249, 260, 282,
298
Copper Inuit village on ice near, 256
"living off the land," a concept of Stefansson,
326
Locker Point, 259, 286, 294
Long Strait, 65, 6, 97
Lopez, and wife Uttaktuak, 166, 168, 173, 188,
330
to Melville Island, 166
loss of life of CAE members, 335,
Lougheed Island, 184, 186, 193, 324, 325
Lougheed, Sir James Alexander, 184

M

Macallum, Dr. A.B., 307
McGhee. Dr. Robert, 151
identifies remains as Thule people, 151
Mackay, Dr. A.F., 27, 29, 31, 35, 38, 44, 47, 48, 51
leaves for Siberia, 57
on the use of dogs, 58 fn 6, 149 fn 3
refuses to return to "Shipwreck Camp," 59
Mackay party, departure from *Karluk*, 57
in trouble, 59
Mackenzie Delta, too shallow for ships, 232
Mackenzie River, 2, 10, 105, 108, 113, 114, 138,
232, 236, 244, 293, 294, 326, 340
Mackenzie River, Victoria Island, 294

MacMillan, D.B., rival expedition, 6
    seeks "Crocker Land," 6
    member of Peary's second expedition, 7
    beacon on Ellef Ringnes Island, 181
    expedition based at Etah, Greenland, 181
McBride, Sir Richard, Premier B.C., 24, 25,
    49 fn 6
M'Clintock, Capt. Leopold, 147, 150, 152
    left message in cairn, 148
    his journey one of Arctic's greatest, 149
M'Clure, Robert, captain of *Investigator*, 57 fn 5,
    131, 142, 147, 153, 167, 168, 170
    rescued by naval party, 57 fn 5, 152
    discovers Northwest Passage, 195
M'Clure Strait, 129, 141, 143, 145, 147, 152,
    169, 170, 171, 173, 174,175,190, 196
    distance across to Banks Island, 152
McConnell, B.M., diary comments, 23
    reaches Nome, 91
    asks permission to hire SS *Corwin*, 91
    invited to go on *King and Winge*, 92
    brings Jenness to Collinson Point, 113
    reaches first ice party with mail, 120
    wishes to join first ice trip, 120
    returns to mainland with Crawford, 122
    Stefansson's secretary, 222, 235, 235 fn 15
    stitches head wound on Capt. Bernard, 120
    one-year contract ends, 236
    relaxes at M. Andreasen's trading camp, 236
    goes to Barrow on *Anna Olga*, 236
    assists in rescue of *Karluk* survivors, 236
    publishes early account of *Karluk*, 322
McConnell, R.G., agrees to publish CAE
    reports, 307
    shows little interest in CAE reports, 309
    replaces Dr. R.W. Brock, 309 fn 3
McIntyre, Samuel "Scotty,"
    had cabin at Escape Reef, 229
    cabin used by Cox, 229
    trading camp in Darnley Bay, 258
    trapper, 287
McKinlay, W.L., only scientist survivor of *Karluk*,
    xiv, 219
    leaves for Wrangel Island, 59
    to apportion supplies, 63
    lacks authority over N. Party, 77
    asked to make inventory at Icy Spit, 63, 77
    bed-ridden with swollen feet, 78

    affected by snow-blindness, 80, 83
    walks alone to Rodger's Harbor, 80
    helps make grave for Malloch, 80
    loses sun goggles, 81
    returns exhausted to Icy Spit, 81
    moves Icy Spit gear to Cape Waring, 82
    gets his kit bag from Skeleton Island, 84
    falls off ladder, 85
    clothes in bad shape, 87
    and Chafe serve in First World War, 97
    took Hadley diary to Scotland, 153 fn 3
    publishes early accounts of *Karluk*, 322
McVeigh, Frank, replaces Desbarats, 311
Macoun, James M., 311
Malloch, finds *Karluk's* location, 10, 21, 43, 47
    sick at Skeleton Island, 78
    perishes at Rodger's Harbor, 80
mail, carried north by Capt. Bernard, 175
mail, for Wilkins left on *Polar Bear*, 167
Mamayauk, Mamie, wife of Ilavinirk, 211
Mamayauk, Violet, Iñupiaq woman, 159
    marries Capt. Gonzales, 165
Mamen, B., seeks route to Wrangel Island, 56
    returns to "Shipwreck Camp," 58
    moves supplies to Wrangel Island, 58
    dislocates knee-cap, 58
    leaves for Rodger's Harbor, 77
    Mamen's death, reported by Munro, 81
manak, device used to retrieve seals, 67, 147
Manning, Tom, 151
man-power vs. dog-power, 148
marine fossils, of Middle Jurassic age, 228
marine shells, indicating former sea level, 274
Markham Island, 186
Martin Point, 229
    a base established at, 109
    Stefansson heads for, 115, 117
    sandspit east of Barter Island, 117
    several tents, racks, caches at, 117
Mary Sachs, 33
    to be at Murray's disposition, 31
    retains former owner as captain, 33
    piloted by Leffingwell, 35
    wintering in Camden Bay, 102
    near Cape Kellett. 135
    ready to leave Arctic, 175
    destruction of, 196
    sails to Banks Island, 244

Masik, August, 197, 199, 212 fn 2
    Russian hunter from Nome, 175, 195
    left at Cape Kellett, 175, 195
    becomes captain of *Challenge*, 198
Mason, Willoughby, prospector, 108, 229
Maurer, returns to Wrangel Island, 97 fn 17
    publishes early account of *Karluk*, 322
Mecham, Lieutenant G. Frederick, 147, 149
meetings of Dr. Anderson, Chipman, Jenness, 322
Meighen, Arthur, 14 fn 36, 179
Meighen Island, haven for gulls, geese, 180
    Stefansson starts south from, 180
Melville Island, 129, 131, 148, 150, 152, 153, 158, 164, 166, 167, 168, 169, 170, 178, 185, 188 fn 37 189 fn 1, 193, 200
Memoganna, Roxy, hired to assist Cox, 229
    wife Monica and children, 229
Mercy Bay, 129, 152, 154 fn 10, 156, 165, 167, 168, 175, 189, 195, 198 fn 19, 266
Merritt, G.E., assistant to Rev. Girling, 159, 298
methods of MacMillan vs. Stefansson, 183
Meyok, Rosemarie, 284 fn 22
Meyok, Uloksak, 270, 333
    "the rich man," 255
    unusual possessions of, 255
    accuses Jenness of murder, 256
    holds seance, 256
    decides distant man murdered victim, 256
    accompanies Jenness on trip east, 285
    leads Bathurst Inlet Inuit west, 286
    shaman, at Coppermine River, 292
    leads Const. Wight to murder scene, 293
    has three wives, 293
    confusion over two Uloksaks, 298
Midway Islands, 205
Mines Dept., responsible for 4 CAE vols., 310
Minto Inlet, 154 fn 12, 161, 168, 171, 172, 189, 198 fn 19, 266
Mikkelsen, Ejnar, 33 fn 3, 108, 109, 211, 212, 325
Mikkelsen Islands, 325
Miles Islands 294
Milukattuk, wife of Avrunna, 249. 263
Minister refuses to seek public inquiry, 355
monument, built by Dease and Simpson, 281
Moose River, 230
Morris, T. Stanley, leaves for Siberia, 57, 58, 74, 98 fn 20

mosquitoes, 129, 184
    thick at Tree River, 274
    thick on Mackenzie Delta, 231
Mott, Hulin S., 112
"Mount Davy," examined by Chipman, 287
    composed of glacial gravel, 288
    more than 300 feet high, 288 fn 5
    is unique mega-kame, 288 fn 5
Mugpi, daughter of Kuraluk, 36, 59, 95
Mungalina, leaves Expedition, 299, 301, 329
Munnigorina, wife of Kohoktok, 288, 330
    has baby son Itayuk, 288
Munro, J., 19, 26 fn 24, 50, 52
    handy with sewing machine, 55
    leads men back to get supplies, 59
    in charge of Wrangel I. camp, 62
    unable to reach "Shipwreck Camp," 77
    unable to reach Herald Island, 62, 78
    returns without Chafe, 78
    searches unsuccessfully for Chafe, 79
    urges move to bay near Cape Waring, 82
    moves Williams to Cape Waring, 83
    moves to Rodger's Harbor, 84
    rescued, with Maurer and Templeman, 93
Mupfi, 269
    husband of Kilauluk, 273
    returns *umiak* to mouth of Tree River, 274
    dries fish for O'Neill party, 274
    takes most of O'Neill's dogs inland, 275
    takes supplies to Tree River, 285
    returns with scientists' specimens, 285
Mupfi's family, reaches coast from interior, 281
Murray, James, experienced oceanographer, 20
    wrote strong note to Stefansson, 30
    was to take charge of *Mary Sachs*, 35
    leaves for Siberia, 21, 57
    perishes, 219
murres, 88, 112 fn 14
muskoxen, have now re-populated Banks Island, 129
    killed by Storkerson's men, 188
    absence of, near Bernard Harbour, 292
muskoxen bones, bleached, piles of, 153
Mys Shmidta, 68 fn 4

N

Nahmens, Capt. Otto, cabin of, 223, 242
    leaves Expedition, 242

Naipaktuna, goes south with Stefansson, 209
National Geographic Society, 6, 9, 215, 342,
    344 fn 19
    funds Stefansson expedition, 7
Natkusiak, Billy, xx, 136, 141, 143, 146, 154,
    166, 175, 184, 235, 248, 266, 267, 272,
    275, 329
    shoots many caribou, 139
    traps foxes, 140
    freighting supplies on Banks I., 145
    refused to go far out on ice, 147
    alone at Cape Prince Alfred, 165
    explores centre of New Land, 174
    at Cape Grassy, 186
    marries Topsy Ikiuna, 186
    damages his sled, 185
    heads for Cape Murray 186
    accompanied by Alingnak family, 186
    to go with Castel to *North Star*, 190
    continues trapping foxes, 190
    moves to Baillie Islands, 191
    returns to Banks Island, 192 fn 6
    often called Billy Banksland, 192 fn 6
    with Wilkins to Banks Island, 241
    seeks wife in Coronation Gulf, 264
    serves as interpreter, 266
    trial marriage to Kaudluak fails, 268
    returns west on *North Star*, 272
    does not wish to return north, 276
    Alaskan Iñupiaq, 341
Naval Service Dept. responsible for 6 CAE
    vols., 310
navigation season of 1913 bad in Arctic, 35
Neriyok, pregnant wife of Kullak, 154
    dies giving birth, 154 fn 12
new land, discovered by Stefansson, 149
news about the *Karluk*, 135
newspaper interviews by Southern Party
    men, 346
New Year's Day (1914) activities, 50
New Year's Eve (1913), 50
nine ships in ice on Western Arctic, 220
no action (on grievance) would be taken, 351
no charges on deaths of Radford and Street,
    297
Noice, H., rescues Ada Blackjack, 98
    sailor on *Polar Bear*, 160
    assists Storkerson, 161

    accompanies Stefansson, 162
    leaves for *Polar Bear* camp, 165
    first to step on "Second Land," 179
    interprets Stefansson objective, 183
    with Stefansson on last exploration, 192
    shows symptoms of scurvy, 192
    discharged and paid by Stefansson, 200
    partners in buying *Challenge*, 200
    investigates archaeology of Pearce Point,
      201
    sells share of *Challenge*, 201
    two years with Inuit on Victoria Island, 201
    goes west with artifacts, 201,
    taken west by Pete Norberg, 201
    sells Inuit furs and artifacts, 201
    his artifacts sold to museums, 201
    takes charge of *Donaldson*, 201
    goes to Wrangel Island with supplies, 201
    rescues Ada Blackjack, 201
    has falling-out with Stefansson, 201
Nome, ice-blocked when *Herman* arrives, 73
Noqallaq, possessed religious objects, 286
Norberg, Pete, takes Noice to Herschel Island,
    201
    on *El Sueno*, 201
Norem, Andre, cook on *Mary Sachs*, 133
    suffered from bouts of despondency, 223
    disappears on Christmas Eve, 223
    found by Cox and Wilkins, 223
    shoots himself, 236
    buried at Collinson Point, 236
northeasterly trending islands are diabase, 276
Northern (Exploration) Party, 10
Northern Party, increased personnel, 160
    three scientists perish, 219
Norway Island, 122, 126, 128, 129, 134, 276,
    138, 162
North Star, 108, 112, 117, 122, 128, 157, 162
    purchased by Stefansson, 136
    too small for Stefansson search, 135, 137
    available for Southern Party, 136
    Wilkins sent to bring it north, 145
    leaves Cape Kellett for north, 160
    wintering north of Norway Island, 162
    sold to Billy Natkusiak, 190 fn 5
    destroyed in storm in 1928, 192
    leaves Herschel Island for east, 242
    proceeds to Tree River, 270

leaves supplies at Cape Barrow, 270

locates O'Neill party at Cape Barrow, 270

heads for Herschel Island, 272

no sign of Sandy Anderson party, 59

notes pencilled in book by Dr. Anderson, 345

notes refer to many inaccuracies, 345

Nuttaitok, visited Stefansson's camp, 161

O

O'Connor, Edward "Duffy," 108, 112, 113, 221, 226, 233, 235, 236

Old Crow, Yukon, 209

Olsen, American trader for O. Swenson, 70, 72

tells Baron Kleist about Bartlett, 72

Olsen, Louis, sailor, accompanies Stefansson, 105, 107, 222

helps move supplies, 110, 225, 229

with O'Neill on Firth River, 226

leaves Cox with supplies at Escape Reef, 229

with Chipman on Mackenzie Delta, 53, 109, 231,

O'Neill, Dr. J.J., field notes of, xvi

to map geology of Firth River, 110, 225

goes with sled to Herschel Island, 110, 225

meets Stefansson, 110, 226

gives Stefansson his chronometer, 111

gives Stefansson Dr. Anderson's reply, 111, 225

continues on to Herschel Island, 111, 230

to store supplies at Herschel Island, 225

geologist, Southern Party, 219

Geological Survey of Canada employee, 220

maps geology of Herschel Island, 226

starts up Firth River, 226

caught in blizzard, 226

maps to International Boundary, 227

unable to get launch *Edna* to work, 231

awaits supplies at Arctic Red River, 231

asks Klengenberg to let Patsy join CAE, 258

identifies rocks east to Bernard Harbour, 259

heads for Coronation Gulf with Cox, 269

has temporary camp at Tree River, 273

finds Pleistocene shells at 500 feet, 274

explores Tree River, 274

dries about 600 pounds of fish for dogs, 274

examines talc-chlorite schist deposits, 275

maps from Tree River to Cape Barrow, 275

leaves messages in stone cairns, 276

seeks to complete copper survey, 288

irritated by inconvenience of small baby, 288

starts for Bathurst Inlet, 288

meets returning Jenness party, 288

reaches cache at Cape Barrow, 290

encounters Victoria Island Inuit, 290

meets Dr. Anderson party, 290

resigns from Geological Survey of Canada, 344

goes to India, 344

"Operation Franklin" finds Stefansson's note, 179 fn 17

Osborn, F.F., Pres., Amer. Mus. Natural History, 6

Osborn, Lieut. Sherard, 183

discovered Findlay Island, 183

Otto Sverdrup Expedition, 337

ovibus, 129

Ovoiyuaq, Adam, 277 fn 16

accompanies Dr. Anderson, 290, 291

leaves CAE at Herschel Island, 300

dies at Herschel Island, 300 fn 18

P

Palaiyak, Silas, xxi, 165, 245

accompanies Wilkins, 172, 173

with Stefansson in 1908, 172 fn 10

shares tent with Ikey Bolt, 246

with Ikey Bolt takes mail to *Alaska*, 264

rejoins Northern Party, 273, 282

brother of Adam Uvoiyuaq, 282

arrives at Bernard Harbour, 296

leaves at Herschel Island, 300

Pálsson, Gisli, 4, 317 fn 24

Pammiungittok, old Minto Inlet man, 161

told Stefansson about *Enterprise,* 161

cannot recall visiting *Enterprise,* 161

had told of plundering *Investigator,* 161

Pannigabluk, xi, xii, 117, 172, 202, 211, 235, 330, 341

has son fathered by Stefansson, 117, 119

goes to Melville Island with Alex, 173

Pasley Cove, 266

Paterson Island, 177

Pauline Cove, 226

Payuraq, Jerry, 329
    joins Expedition, 35
    ends work with Expedition, 105
Pearce Point, 201, 242, 243, 245,246, 249, 256,
    299
Peary, Robert E., 6, 180, 181
    introduced name "Crocker Land," 6
    fails to reach North Pole, 6
    used special kind of sled, 97
Peddie Point, near base camp, 187
Pedersen, Capt. C.T., purchases *Karluk*, 12, 17,
    135, 240
    sails *Karluk* to Esquimalt, 12, 17
    selects best ship, 17
    selected to be master of *Karluk*, 18
    quits Expedition, 18
    visits Cape Kellett, 175 fn 12
Pedersen, Peder, 108, 109, 226, 228, 230, 231
Peel Point, 161, 194
pemmican, 29, 57, 59, 60, 62, 67, 74, 77, 81, 83,
    107, 181, 183, 246, 250, 261, 298
    trouble over content, 13
    composition of, 250
Pennock, Marjorie, fiancée of Chipman, 291
Phillips, G., acts for Dept. Naval Service, 12, 26
    advises sending scientists home, 31
Phillips, RNWMP Insp. J.W., 215, 229, 233
    agrees to let Stefansson leave, 209
Pikalu, Iñupiaq man, 159, 190,329, 330
    goes to Melville Island with wife, 173
    member of *Polar Bear* crew, 190 fn 1
Pipsuk, Iñupiaq man, 216, 329, 335
    drowns at Barter Island, 216
    buried east end of Barter Island, 216 fn 3
plaque, memorial to CAE dead, 334, 335, 336
Point Barrow, 36, 44, 117, 124, 205, 206
    Karluk within sight of, 44
Point Hope, 35, 39, 72, 75, 103, 222, 329
Polar Bear, icebound east of Martin Point, 117
    unexpected appearance of, 157
    arrival offers Stefansson new opportunities,
    157
    chartered by Stefansson, 158
    from Cape Kellett to Herschel Island, 158
    proceeds west to Herschel Island, 158
    bought by Stefansson, 158
    returns to Baillie Islands, 159
    second base camp at, 163

    not in Liddon Gulf, 189
    moves to Walker Bay, 189
    adds supplies at Herschel Island, 202
    driven ashore at Barter Island, 203
    left Arctic in summer, 216
    trip to Nome uneventful, 216
    repaired and sold at Saint Michael, 216
Polar Bear camp, 167, 180
    activities at, 171
polar bear, shot by Natkusiak, 169
polar-bear liver test, 169
polar oxen, 129
"Polynia Islands," 149
pressure ridges, 50, 52, 60, 62, 77, 92, 193
Prince, Dr. Edward E., 307, 313
    urges deletions from Jenness report, 318
Prince Albert Sound, 140, 266, 161, 261
Prince of Wales Strait, 141, 154, 160, 180,
    194, 265
Prince Patrick Island, 122, 145, 149, 162,
    211, 272
    Stefansson maps unmapped parts, 147
    no animal life seen, 147
problems from using Bernier's supplies, 187
publicity, important to Stefansson, 3, 157
Pusimmik, Iñupiaq woman, 159
    wife of Pikalu, 330

R

Radford, H.V., killed by Inuit, 297
radiocarbon dates, Barter Island, 316
Rae, John, 326
Rae River, 259, 269, 290, 292, 324
raftered ice, 66, 66 fn 3
Ramsay Island, Stefansson at, 162
reduced respect for Stefansson, 343
reindeer men, 69
removal of Williams' gangrenous toe, 79
repairs of launch *Edna* take ten days, 138
reports and personal mail, to Fort Norman, 254
request for government inquiry, 348, 349
Rev. Whittaker warns of "bad men," 113
Richard Island, 231
Richardson Islands, 294
Richardson, Sir John, 244, 256, 287, 326
    mapped Arctic coast in 1848, 244
Roberts, Helen H., 284 fn 22, 310 fn 5
Robillard Island, 272

Rodger's Harbor, 65, 75, 77
  desperate conditions at, 81
  Templeman alone at, 83
  flag at half mast, 93
  unfriendly reception at, 93
  rescue of *Karluk* men at, 93
Roman Catholic Oblate priests missing, 296
roses, interest McKinlay and Capt. Bartlett, 48
route of de Jong's icebound *Jeannette,* 207
route of Nansen's icebound *Fram,* 207
Rouvière, Father Jean-Baptiste, 286
  missing Oblate priest, 286
Rowley, Diana, xvi, xxiv
Royal Ontario Museum, 5, 201, 341
  funded Stefansson expedition, 5
Ruby, Hudson's Bay supply vessel, 157, 158, 245
  bringing supplies for expedition, 157
Russell Point, 170, 172, 194
Russian icebreakers ordered back to Anadyr, 75

S

Sapir, Dr. Edward, xxiii, 22, 318
Saryoak, accompanies Stefansson, 209
schist deposits provide rock for Inuit carving,
    275
scouting party from *Karluk,* 56
scraps of seal meat left to attract bears, 67
Second Stefansson-Anderson Expedition, 9
seal liver, unusually high in vitamin A, 170
seals, plentiful beyond continental shelf, 122
  retrieved by using a *manak,* 147
seamstresses, hired by Storkerson, 117
search for "Crocker Land" renewed, 177
"Second Land," 178, 179, 181
sedimentary rocks lie west of Tree River, 281
sentiments shared by W.L. McKinlay, 345 fn 21
seventeen people winter on Melville Island, 188
Seymour, Anna, accompanies Wilkins, 172
Seymour, W., 204, 205, 226
  husband of Ikugana (Anna), 160
  second mate on *Polar Bear,* 160
  master of *Polar Bear,* 202
  letter to Dr. Anderson, 204
sewing, *Karluk* men urged to commence, 43
Shackleton, E., 20, 31, 58, 250
Shaler Mountains, 168
Shannon, Anthony, on *Polar Bear,* 207
Shingle Point, 207, 208, 229

"Shipwreck Camp," 55, 56, 57, 58, 59, 60, 77, 95
Siberia Mike, 273, 301
  from East Cape, 226
  husband of "Sis," 226
  hired to assist O'Neill, 226
  with Dr. Anderson to Croker River, 287
  with Dr. Anderson to Tree River, 290
  leaves CAE at Herschel Island, 300
Simpson Bay, 261
Sinnisiak, from Victoria Island, 292
  arrested, 292
  taken to Edmonton, 297
  found guilty of murder, 297
"Sis," wife of Siberia Mike, 226, 242. 282, 330
  son Georgie Mike, 226, 242, 300
  death of, 330 fn 65
skeletons found on Herald Island in 1924, 59, 98
skis, usefulness of, 152
sleds, Peary and Nome types, 47
  built by Capt. Bernard, 196
  found west of Mercy Bay, 196
  construction of, 207
Smith, Donald (Lord Strathcona), 11
snow-blindness, affects Stefansson's men, 177
snowhouse, for tide-measuring, 140, 283
  built by Stefansson and photographed, 169
  for solar and stellar readings, 221
snowhouses, abandoned by Storkerson, 170
  built by Ikey and Kohoktok, 313
Southern (Scientific) Party, 10
  news at Barrow about, 102
  opposed to Stefansson taking supplies, 120
  has left the Arctic, 189
  one scientist perished, 219
  consisted of international scientists, 219
  main task of, 221
  completed assignments, 219
  Inuit names for scientific men, 248
  starts survey of Bathurst Inlet, 278
  field work completed, 292
Southern Party men sail for Seattle, 302
sovereignty, over Arctic Islands, 8
  concerns over, 336
Spy Island, hunting party marooned on, 40
  westernmost of Jones Islands, 101
Stapylton Bay, 243, 250, 259, 264, 266, 287, 292
start of Stefansson's second ice trip, 145
starving men, at Rodger's Harbor, 89

State University, now University of Iowa, 339 fn 2

Steckley, John, xxi, 4

Stefansson, Alex, 117, 119, 173, 202

Stefansson, in charge of Arctic Expedition, xiii

    leads Northern Party, xiii

    plans new Arctic expedition, 5

    his original plans for expedition, 6

    seeks funds from American Museum, 6

    changes Expedition objectives, 8

    must become naturalized British citizen, 9, 13

    seeks "Crocker Land," 9, 44

    serves without pay, 10

    seeks wireless stations, 11 fn 26

    goes to Europe, 11

    sells publication rights, 11

    citizenship of, 13

    swears allegiance to King, 14

    possible thyroid surgery, 14

    gives scientists their contracts, 15

    angers men at Victoria meeting, 20, 24

    gives unsatisfactory answers at Nome, 30

    attempts to replace Murray, 31

    fails to supply statement on authority, 31

    sends scientists to Teller, 31

    purchases Mary Sachs, 33

    claims he was asleep, 37

    decides to hunt caribou, 38

    separated from Karluk, 43

    did he desert Karluk?, 44

    re. Hadley's diary, 58 fn 6

    instructs men how to be warm, 101

    hunts caribou alone, 101

    wants Karluk news to reach media, 101

    leads hunting party to Barrow, 101

    arranges for Jenness to stay with Aksiatak, 102

    housed by Brower, 102

    dictates letters at Barrow, 102

    sends Wilkins to fishing lake, 103

    sends Jenness to fishing lake, 103

    reaches Teshepuk Lake, 104

    leaves "Brick" with Jenness, 104

    reaches Southern Party's base camp, 104

    learns Dr. Anderson took mail east, 104

    goes east to icebound Belvedere, 104

    hopes to catch Dr. Anderson and mail, 105

    meets Dr. Anderson at trapping cabin, 105

    won't take charge Southern Party, 106, 116

    Christmas with Capt. Cottle, 107

    has supplies moved from Belvedere, 107

    writes many cheques, 107

    buys Duffy O'Connor's supplies, 107

    buys Matt Andreasen's supplies, 107

    buys schooner North Star, 107

    reaches police depot at Herschel Island, 108

    goes to Fort McPherson, 108

    hires Storkerson for Northern Party, 108

    stays with P. Pedersen, W. Mason, 108

    buys Mason's launch, Edna, 108

    hires Mason to work with Chipman, 108

    buys launch from Captain Cottle, 108

    plans mapping of Mackenzie Delta, 108

    makes demands of Dr. Anderson, 108

    issues instructions for Dr. Anderson's men, 109

    tells Dr. Anderson to prepare ice-trip base, 109

    plans ice trip north from Martin Point, 109

    returns to Collinson Point, 110

    meets O'Neill west of Herschel Island, 110

    angered by Dr. Anderson's reply, 111

    claims letter stated men refused to help, 111

    accuses Dr. Anderson of taking letter, 111

    goes to Herschel Island with O'Neill, 111

    gets O'Neill to turn over chronometer, 111

    changes Dr. Anderson's orders, 112

    orders Capt. Bernard back to Collinson Pt., 112

    discusses exploration plans with others, 112

    angered his instructions not carried out, 112

    distributes mail from Fort McPherson, 112

    calls for meeting at Collinson Point, 112

    withholds sad letter for Dr. Anderson, 112 fn 14

    tells Southern Party of his plans, 112

    berates Dr. Anderson's actions, 112

    states exploration is main purpose of expdn., 112

    tells scientists about his new supplies, 112

    tells men about purchase of North Star, 112

    tells men of plans to use the Mary Sachs, 113

    ignores offer to take charge, 114

    determined to explore for land in Beaufort Sea, 115

heads for Martin Point, 115

Pannigaluk and Alex, 117

denies paternity rumours, 119

friend reveals Stefansson is father of Alex, 119

starts first ice trip from Martin Point, 120

sends first support party back, 120

takes injured Capt. Bernard to Martin Point, 120

has Wilkins photograph him seal-hunting, 120

sends Wilkins, Castel back to Martin Pt., 120

reaches edge of continental shelf, 120

writes new instructions to Dr. Anderson, 122

writes new instructions for Jenness, 122

sends instructions with McConnell, 122

remaining support party returns to shore, 122

three men, adrift on ice raft, 122

builds his first snowhouse, 123

starts travelling at night, 124

heads for Banks Island, 124

forced to go on half rations, 125

reaches latitude 74°N, 125

creates sledboat to cross water, 125

reaches Norway Island after 96 days, 126

assumes government thinks him dead, 126

kills 6 caribou, 127

carries caribou tongues back to camp, 127

places account of ice trip in beacon, 127

discovers first land but unconcerned, 128

wants a ship west of Banks Island, 128

wants a ship for Prince Patrick Island, 128

with Storkerson kills 42 bull caribou, 129

thinks caribou fat good to eat, 129

wonders why no muskoxen on Banks Island, 129

decides Victoria I. Inuit killed muskoxen, 129

dislikes word "musk-ox," 129

assumes his instructions will be carried out, 130

heads south to Cape Kellett, 131

sees boot imprint in mud, 132

sees ship's masts near Cape Kellett, 132

catches up on news, 133

learns of death of Andre Norem, 133

learns of fate of *Karluk*, 133

learns *North Star* is with Dr. Anderson, 134

wonders why Dr. Anderson disobeyed orders, 134

asks that Wilkins be put on govt. payroll, 137

discusses new plans with Wilkins, 137

new plans involve only exploration, 137

gives up study of "Blond Eskimos," 137

plans base camp near Norway Island, 138

asks Crawford to repair launch, *Edna*, 138

asks Wilkins to take Storkersons north, 138

with Natkusiak, kills 48 caribou, 139

decides to return to base camp, 139

leaves Wilkins to guard meat, 139

discovers stranded whale carcass 140

ferries whale meat to base camp 140

with Natkusiak, leaves for De Salis Bay, 140

returns without finding "Blond Eskimos," 141

tells Wilkins of plans for second ice trip, 141

asks Wilkins to lead freighting trips, 141

wants base camp at Cape Prince Alfred, 143

reaches Bernard River, 143

tells Wilkins of problems with his men, 145

asks Wilkins to get *North Star*, 145

starts over M'Clure Strait, 146

decides to return to shore, 146

accomplishments, second ice trip, 147

heads seawards to hunt seals, 147

places message in M'Clintock's cairn, 148

says Bartlett refused dogs to Dr. Mackay, 148 fn 3

discovers new land, 149

eager to get news of discovery to world, 150

takes solar readings on new land, 150

reaches summit of highest hill, 150

hoped to excavate an aboriginal village, 151

kills his first two muskoxen, 152

eager to reach Cape Kellett, 152

reaches Mercy Bay, 152

leaves sleds at Mercy Bay, 153

forgets alarm clock, 154

encounters Inuit, 154

reaches Cape Kellett camp, 154

completes second ice trip, 155

thought dead by Capt. Lane, 155

charters *Polar Bear*, 158

leaves note for Wilkins, at Baillie Is., 158

purchases *Polar Bear*, 158

agrees to supply Capt. Lane with ship, 158

purchases *Gladiator*, 158

plans new base camp on Melville Island, 158

plans to sail *Polar Bear* to Atlantic, 159

plans to cross Northwest Passage, 159

hires Hoff for *Alaska*, 159

hires Inuit men and women, 159

delayed at Baillie Islands, 159

finds note from Wilkins, 159

lends two men to missionary, 160

misses Wilkins at Cape Kellett, 160

gives *Gladiator* to Capt. Lane, 160

tries Prince of Wales Strait, 160

camp near Armstrong Point, 160

records Inuit stories, 161

leaves instructions for Storkerson, 161

hopes to establish new base camp, 162

demonstrates building a snowhouse, 162

leaves for Cape Kellett, 162

forms new exploration plans, 163

delayed as usual, 166

reaches *North Star* camp, 166

decides to abandon *North Star*, 166

sends supplies to Melville Island, 166

plans to remain in Arctic, 167

ends plans for third ice trip, 167

plans camp on Melville Island, 167

gets folklore from Guninana, 167

reaches Natkusiak's camp, 168

names Castel Bay, 168

goes to Mercy Bay, 168

finds letter at Mercy Bay, 168

names mountains for geologist, 168

will go to Melville Island, 168

turns back to *Polar Bear* camp, 170

goes north after Storkerson, 170, 171

injures ankle, 173

gets message from Storkerson, 173

makes Storkerson his right-hand man, 173

has men explore "New Land," 174

sends Castel south for 2 missing men, 175

denies benefits to widow, 175 fn 13

names part of "New Land," 177

names Borden Island, 177

names Mackenzie King Island, 177

gives instructions to Castel, 177

renews search for "Crocker Land," 177

takes depth readings, 178

troubled by men's food likes, 178

leaves note on "Second Land," 179

names Meighen Island, 180

finds shells on Meighen Island, 180

suggests land is rising, 180

corrects coastline of Amund Ringnes I., 180

finds note written by MacMillan, 181

replaces MacMillan's note, 181

his objective, "living off the land," 183

spends summer on "Third Land," 184

names it Lougheed Island, 184

returns to "Borden Island," 184

forced to change plans again, 185

heads south for Melville Island, 185

kills two muskoxen, 186

credits Castel with finding Lougheed I., 186

reaches camp near Cape Ross, 186

writes at Liddon Gulf camp, 187

gets ethnographic information, 188

instructed to return south, 188 fn 37

deliberately stayed north, 188 fn 37

final explorations, 189

sends supplies to Cape Grassy, 189

heads for Cape Grassy with Emiu, 190

sends Castel and men north, 190

sends 3 men south from Cape Mackay, 190

sends Castel and others south, 190

is prepared to sell *North Star*, 190

starts third ice trip, 192

soundings off "Borden Island," 192

forced to return to Cape Isachsen, 193

respite at "Camp Hospital," 193

with Emiu, kills herd of caribou, 193

heads for Lougheed Island, 193

goes around east side Melville Island, 193

discovers Capt. Bernier's cairn, 193

reaches Dealy Island, 193

crosses Viscount Melville Sound, 194

lands east of Peel Point, 194

crosses Prince of Wales Strait, 194

names Knight's Harbour, 195

finds cylinder with M'Clure note, 195

substitutes own note, 195

leaves sleds on Banks Island, 195 fn 14

sleds recovered years later, 195 fn 14

crosses Banks Island, 195

reaches Cape Kellett, 195

discovers treachery at Cape Kellett, 195

hears of death of Thomsen, 196

hears of death of Capt. Bernard, 196

buys *Challenge* from Crawford, 198

sails for Herschel Island, 198

accosts Capt. Gonzales re. *Mary Sachs*, 199

takes charge of *Polar Bear*, 199

discharges Capt. Gonzales, 200

puts Mamayauk ashore, 200

cancels Gonzales' pay for year, 200

discharges Noice, 200

heads for Herschel Island, 201

makes Seymour master of *Polar Bear*, 202

discharges Seymour at Herschel Island, 202

makes Hadley master of *Polar Bear*, 202

appoints Castel as first mate, 202

appoints Masik as second mate, 202

delays *Polar Bear* at Herschel Island, 202,
    203

heads *Polar Bear* for Nome, 202

aground at Barter Island, 203

lays charges on Dr. Anderson, 203 fn 33

ignores orders to leave Arctic, 203 fn 33

plans new ice trip, 205

plans to drift on ice mass, 205

contemplates westward ice drift, 205, 207

seeks dogs and men for new trip, 207

moves supplies to *Polar Bear*, 207

falls seriously ill, 208

taken to Herschel Island, 208

attended by missionary Henry Fry, 208

assisted by Abraham Stewart, 208 fn 2

had typhoid fever, 208

has relapse, 208

gets pneumonia, 208

moved to Leo Wittenberg's house, 209

calls Storkerson to Herschel Island, 209

asks Storkerson to take charge, 209, 211

develops pleurisy, 209

insists he go to Fort Yukon hospital, 209

hires men to take him to Fort Yukon, 209

men paid off at Old Crow, Yukon, 209

treated at hospital in Fort Yukon, 209

diagnosed with neurasthenia, 209 fn 4

convalesces at Fort Yukon, 209

returns south to Victoria, 209

lays charges against Capt. Gonzalez, 210

goes to Ottawa, then New York, 210

completes final Arctic expedition, 210

reaches Collinson Point, 222

criticisms by, irritate men, 222

tells of loss of Northern Party, 222

plans to raise new Northern Party, 222

spends Christmas at *Belvedere*, 222

proposes name "Fort Bacon," 244

created tribal divisions of central Inuit, 286

takes money from Dr. Anderson's mail, 300

publishes *The Friendly Arctic*, xiv, 312, 344

has many inaccuracies in his book, 345

his book is full of lies, 345

interferes with CAE reports, 316

demands deletion of parts of Jenness'
    report, 318

sees Camsell about Jenness' report, 319

seeks control of CAE reports, 320

delays map production, 320

reasons why he served without pay, 321
    fn 40

publication ban of, broken by CAE men,
    322

claims broken contract cost him money, 323

his views on loss of life, 324, 324 fn 53

discovers outer edge of Continental Shelf,
    325

confirms there was no "Crocker Land," 325

his flag planting was good for sovereignty,
    326

improvement of coastal mapping, 326

discovers last major Arctic islands, 328

claimed he was a Bachelor of Arts, 339 fn 4

graduate studies in religion at Harvard, 339

transfers to anthropology, 340

was asked to leave Harvard, 340 fn 6

not employed by Geological Survey, 340
    fn 6

Anglo-American Polar Expedition, 340

at Herschel Island in 1907, 340

meets Christian Klinkenberg Jorgensen,
    340

American Museum to sponsor him, 340

proposes one year to study Inuit, 340

works without salary, 340

first expedition, stayed in Arctic 4 years, 341
backed by Geol.Surv.Canada, 341
planned articles, books, and lectures, 341
assisted by Billy Natkusiak, 341
generally separated from Dr. Anderson, 341
accompanied by widow Pannigabluk, 341
had no moral sense, said Dr. Anderson, 341
gets partial backing of Nat.Geog.Soc, 342
persuades Dr. Anderson to go north, 342
goes to Europe, 342
leaves extra duties for Dr. Anderson, 342
criticizes government men, 343
retaliates via Can.govt. officials, 344
retaliates by charges in his book, 348
was "a liar and an imposter," 346
asks for similar inquiry, 350
admits charges were deliberate, 350
speaks to cabinet ministers, 350
loses respect of Geol.Surv. 353
Stefansson-Anderson feud, 339-353
both had Bachelor of Philosophy degrees,
339
both graduated from State University of
Iowa, 339
Stefansson's flag planting and Canadian sover-
eignty, 336
Stefansson's possessions left on *Karluk*, 44
Stefansson's hunting party reaches Barrow, 102
Stefansson's exploration main purpose, 106
Stefansson's ice raft drifts west, 124
Stefansson's sled damaged, 183
Stefansson's men at Dealy Island, 193
take items from depot, 194
Stefansson's plan to drift one year, 212
Stefansson's tribal divisions, 285
Akulliangmiut, 286
Kogluktomiut, 286
Noahognirmiut, 286
Puivlirmiut, 286
Stefansson-Anderson Arctic Expedition, 5
Stefansson-Dr. Anderson conflict over first
book, 313
Stefansson and Dr. Anderson, "each seek last
word," 314
Stewart, Abraham, accompanies Stefansson, 208,
208 fn 2
Stockport Islands, thick volcanic flows on, 289
stone tent rings, 153

Storkerson, Storker, xviii, 108, 109, 112, 117,
119, 123, 124, 127, 128, 130, 133, 138,
140, 145, 150, 153, 158, 162, 167, 170,
187, 197, 200, 212, 243, 246
builds camp at Martin Point, 119
willing to accompany Stefansson, 121
returns to Cape Kellett for coal-oil, 144
deathly ill from eating bear liver, 144
discovers new land, 149
leaves food depot at Peel Point, 161
mapping north coast of Victoria Island, 161
got two sleds from Mercy Bay, 165
was to meet Stefansson party, 165
goes to Melville Island, 165
sends note to Mercy Bay, 165
fails to complete mapping task, 168
discovers mountain range, 168
reaches *Polar Bear* camp, 172
to establish winter base in Liddon Gulf, 173
made Stefansson's right-hand man, 173
takes message to Capt. Gonzales, 174
starts south for Liddon Gulf, 174
makes base camp near Cape Ross, 186
heads for Cape Grassy, 190
instructed to move back to *Polar Bear*, 192
writes about *Mary Sachs* for Stefansson, 197
fails to complete Victoria I. mapping, 198
returns to Walker Bay, 198
arrives in time to catch *Polar Bear*, 198
moves supplies to *Polar Bear*, 207
given full responsibility for ice trip, 211
is responsible for ending Expedition, 211
asks for volunteers for a year, 211
to carry out scientific activities, 214
drift party kills 96 seals, 6 polar bears, 214
has asthma attack, 214
decides to return to land, 215
lands in Harrison Bay, 215
stays at Herschel Island until spring, 215
meets Stefansson at Banff, 215
ascertains "Keenan Land" does not exist, 215
Storkerson's main discoveries, 215
Storkerson, Elvina, 202, 211
camp of, 231
daughter Marina, 241
storm, separates Stefansson party from land, 121
Street, G.R., adventurer killed by Inuit, 297
St. Stephen's Episcopal Hospital, Fort Yukon, 209

Stuck, Archdeacon Hudson, 209
Sullivan, James "Cockney," 242, 245, 247, 250,
    256, 259, 264, 266, 268, 273, 285
    provides elaborate Christmas dinner, 284
sun's reappearance in mid-January, 254
supplies, moved from "Shipwreck Camp," 58
    on *Belvedere* were for Southern Party, 110
    at *North Star* for third ice trip, 167
    freighted to Cross Island, 211
    cached at Bernard Harbour, 298
Sutherland, Pat, evidence of Norse in Arctic,
    319 fn 34
Sverdrup, Otto, 183
Sweeney, Capt. Daniel, 241, 264, 273 fn 16
    sailor from *Belvedere,* 120
    returns with sled, dogs, Ikey, Taliak, 122
    replaces Capt. Nahmens, 242
    in charge with Blue of *Alaska,* 245
    married at Herschel Island, 282
Swenson, Olaf, of Seattle, 70, 91, 97
    sends *King and Winge* for survivors, 92

T

Taliak, helps Stefansson with first ice trip, 120
Taptuna, visited Stefansson's camp, 161
Taverner, P.A., rivalry with Dr. Anderson, 314
    publishes *Birds of Canada,* 314
Teddy Bear, 2435 fn 29, 244 fn 30, 245 fn 34
    icebound, 222
    at Baillie Islands, 245
Teddy Bear Island, 243 fn 29, 244 fn 30, 245
    fn 34
Teller, Alaska, 31, 33, 33 fn 1, 34, 35
Terror Island, correct on Admiralty Chart, 131
Teshekpuk Lake, Alaskan fishing lake, 102, 103
The Friendly Arctic, xiv, xviii, 8 fn 21, 37, 106 fn
    5, 111, 159, 171, 192, 199, 203 fn 33,
    204, 208 fn 4, 319, 324 fn 53, 325,
    349, 350
    not original title, 344 fn 19
    title suggested by G. Grosvenor, 344 fn 19
    re. Hadley's diary, 58 fn 6, 148 fn 3, 159 fn 3
    storm over, 344
    many inaccuracies, 345
theft of equipment, from Chipman, O'Neill, 30
"Third Land," 184
    Stefansson spends summer on, 184
    named Lougheed Island, 184

Thomsen, Charles (Karl), 33, 140, 145, 151, 154,
    166, 173, 196, 336
    leaves for *Polar Bear* camp, 165
    is to bring family north, 174
    is to gather Wilkins' specimens, 174
    failed to get Wilkins' specimens, 174
    frozen body discovered, 175, 196
Thomsen, Jennie, 211, 239, 241
    daughter Annie, 239
Thomsen's body, discovered by Castel, 196
Thomsen's family goes with two Kilians, 197
    to *Polar Bear* camp, 197
Thomsen River, 153
    named by Stefansson, 153
Thule people, 151
thunder, a rare event, 276
tide observations at Lougheed Island, 184
Titalik, daughter of Kullak, 154
Titanic, RMS, 3
topographic map, Collinson Point, 222
tragedy of Thomsen and Capt. P. Bernard, 174
treachery at Cape Kellett, 195
Tree River, 270, 271, 273, 275, 280, 294,
    famous for its Arctic char, 274
    geological fault at, 276
Tupik, too young to marry Natkusiak, 267

U

Ulipsinna, member of *Polar Bear* crew, 190, 325
Uloksak, confusion in identity of, 298
Uloksak Avingak, (see Avingak, Uloksak)
Uloksak Meyok, (see Meyok, Uloksak)
umiak, 35, 38, 62, 92, 269, 274, 275, 277, 278,
    279, 280, 281, 290
    at Tree River, 290
    left at Barrow, 300
Unalina, 273 fn 16
    wife of Ambrose Agnavigak, 273 fn 16
Unahak, wife of Wikkiak, 283 fn 21
    mother of Alunak, 283 fn 21
Underwood, J.J., meets Stefansson, 3
unfinished CAE report, by Johansen, 315
    by Jenness, 316
unheralded contributors to CAE story, 329
    Inuit, 329
    Deputy Ministers, 331
    Desbarats, G.J., 331–336
    Brock, Dr. Reginald W., 332

McConnell, Dr. R.G., 332
Camsell, Dr. Charles, 332
Boyd, W.H., 333
LeRoy, O.E., 333
Anderson, Mrs. Mae Belle, 333
Kuraluk, 329
Kataktovik, 329
Natkusiak,, Billy, 329
Alingnak, 329
Emiu, 329
Illun, 329
Palaiyak, Silas, 329
Bolt, Ikey, 329
Siberia, Mike, 329
Agnavigak, Ambrose, 329
Klengenberg, Patsy, 329
Guninana, 230
upland adjoining Croker River, 287
Uttaktuak, Iñupiaq woman, 159, 167
    Wife of Peter Lopez, 46, 169, 330

V

vegetation sparse on Croker R. upland, 288
    won't support wildlife, 288
Victoria, controversial meeting at, 20
Victoria Island, xxiii, 3, 8, 129, 140, 154, 161,
    192, 200, 202, 261, 262, 283, 294, 325,
    330
Victoria Memorial Museum, 5, 308
Victrola, offers music as *Karluk* sinks, 52

W

Walker Bay, 161, 189, 192, 198, 202
Walters, V., includes Johansen manuscript, 315
Weitzner, Bella, editorial assistant, 28
    returns to the U.S., 34
whale meat, in house owned by Kuraluk, 102
    for dog food, 140, 223
whale oil, *Karluk* smelled of it, 19
    formerly valuable, 158
    uses of, 158
    replaced by cheaper products, 158
whaling ships, 118, 119, 152, 155
    at East Cape, 69
whales, hunted in Amundsen Gulf, 158
    followed near-shore leads, 36
    hunted in Beaufort Sea, 158
    disappear from Arctic, 158

Whittaker, Rev. C.E., 113, 229
why no CAE volumes 1 or 2? 312
Wicksuak, 258
Wight, Const. J.E.F., 293, 304
    with Inspector La Nauze, 294
    backpacks overland with Ilavinirk, 297
Wikkiak, an Inuk, 283 fn 21
    husband of Unahak, 283 fn 21
    father of Alunak, 283 fn 21
    from southwest Victoria Island, 283 fn 21
wildflowers, 274
Wilkins, G.H., 162, 222, 223, 265
    leaves message on Spy Island, 102
    photographs Stefansson seal-hunting, 120
    unable to return to ice party, 120
    breaks contract of Gaumont Company, 135
    discovers Stefansson's note in beacon, 136
    sees Stefansson's men are healthy, 136
    would prefer to return to London, 137
    lacks hope for photographic fame, 137
    doubts feasibility of Stefansson's plans, 137
    suggests *Mary Sachs* to freight north, 138
    forced to unload launch and walk home,
        139
    joins Stefansson on hunting trip, 139
    asked to guard hunting camp, 139
    cleans 71 fox skins for Nat. Mus.
        Canada, 140
    undertakes first tidal studies Banks Island,
        140
    expects to be active with Stefansson, 141
    reaches Cape Prince Alfred, 143
    caches supplies near Cape Wrottesley, 143
    selects best site to cross M'Clure Strait, 143
    is to end Crawford's employment, 145
    to take *North Star* to Prince Patrick Island,
        145
    photographs Stefansson's snowhouse, 145
    photographs sledboat creation and use, 145
    starts south to Cape Kellett, 145
    reaches Baillie Island with *North Star*, 159
    starts north from Cape Kellett, 160, 272
    returns to Cape Kellett, 163
    takes load to *North Star* camp, 165
    guards *North Star* camp, 166
    upset over abandoning *North Star*, 166
    concerned about parent's health, 167
    wants to leave Arctic, 167

taking supplies to Melville Island, 167

moves to Mercy Bay, 168

wants to go to *Polar Bear* camp, 168

annoyed by Stefansson's new plans, 168

hopes to photograph muskoxen, 169

photographs building of snowhouse, 169

gets headache from eating polar-bear liver, 169

goes south to *Polar Bear* camp, 171

gets letter saying his father is dead, 171

photographs Inuit in Minto Inlet, 171

sees no evidence of "Blond Eskimos," 171

successful hunting trip, 171

starts with Palayiak for Bernard Harbour, 173

photographs Inuit at Bernard Harbour, 173

photographer and engineer, 219

answers to Dept. Naval Service, 220

shows moving pictures, 223

loses photo equipment on *Karluk*, 234

gets camera from Leffingwell, 234

gets moving picture camera from Clark, 234

goes to Barrow with Natkusiak, 235

instructed to repair *North Star*, 235

moves *North Star* to Herschel Island, 235

prepares to proceed to Banks Island, 235

hires Inuit seamstresses and sailors, 235

with Thomsen prepares body of Norem, 236

obvious choice to search for Stefansson, 240

gets letter from employer to return to England, 240

agrees to lead search party, 240

shows Cox how to run *North Star* engine, 240

sails *Mary Sachs* to Banks Island, 241

sleds from Banks Island to Bernard Harbour, 264

given letter with power of authority, 265

trades with Inuit at Minto Inlet, 266

encounters Cox near Cape Bexley, 266

finds Johansen, Castel at Bernard Harbour, 266

meets Uloksak Meyok, 266

films scenes of Copper Inuit, 266

decides to go to Coronation Gulf, 266

presents Stefansson's demands, 268

considered an adversary, 269

shunned by Southern Party men, 269

helps salvage *North Star*, 270

photographs Copper Inuit and birds, 270

takes *North Star* east to Cape Barrow, 272

heads west with *North Star*, 272

gets instructions at Baillie Islands, 272

discharges Crawford at Baillie Islands, 272

leaves mail at Baillie Islands, 272

picks up several Inuit, 272

proceeds to Cape Kellett, 275

waits for Stefansson, 272

starts north in *North Star*, 272

route blocked by sea ice, 272

Williams, H., falls through ice, 58

Williams Point, 266

Williamson makes return trip to Rodger's Harbor, 88

Wittenberg, Leo, 198

partner of Crawford, 189, 198

Winter Harbour, Melville Island, 187, 193

supplies left by Capt. Bernier, 187

Capt. Bernier's house west of, 194

Wolki, Fred, 212

Wolki, Fritz, sells *Gladiator* to Stefansson, 158

Wollaston Peninsula, 266

World War, Stefansson gets first news of, 157

Southern Party hears about it, 282

Wrangel Island, 49, 57 fn 5

geographical details, 56 fn 4

McKinlay copies map of, 57

arrival of *Karluk* survivors at, 61

six hundred miles from Nome, 92

Wynniatt, Lieutenant Robert, 161

Z

Zalibra, Charles, still-camera man, 93